D1348509

Plant Roots

Growth, Activity and Interaction with Soils

Peter J. Gregory

Director, Scottish Crop Research Institute, Invergowrie, Dundee
Visiting Professor of Soil Science, University of Reading

Blackwell
Publishing

Editorial Offices:
Blackwell Publishing Ltd, 9600 Garsington Road, Oxford OX4 2DQ, UK
Tel: +44 (0)1865 776868
Blackwell Publishing Professional, 2121 State Avenue, Ames, Iowa 50014-8300, USA
Tel: +1 515 292 0140
Blackwell Publishing Asia, 550 Swanston Street, Carlton, Victoria 3053, Australia
Tel: +61 (0)3 8359 1011

First published 2006 by Blackwell Publishing Ltd

ISBN-10: 1-4051-1906-3
ISBN-13: 978-1-4051-1906-1

Library of Congress Cataloging-in-Publication Data
Gregory, P. J.
Plant roots : their growth, activity, and interaction with soils / Peter J. Gregory.
p. cm.
Includes bibliographical references and index.
ISBN-13: 978-1-4051-1906-1 (hardback : alk. paper)
ISBN-10: 1-4051-1906-3 (hardback : alk. paper)
1. Roots (Botany) I. Title.

QK644.G74 2006
575.5'4--dc22
2005025239

A catalogue record for this title is available from the British Library

Set in Times 10/12.5 pt
by Sparks Computer Solutions Ltd, Oxford – www.sparks.co.uk
Printed and bound in India
by Replika Press Pvt Ltd, Kundli

The publisher's policy is to use permanent paper from mills that operate a sustainable forestry policy, and which has been manufactured from pulp processed using acid-free and elementary chlorine-free practices. Furthermore, the publisher ensures that the text paper and cover board used have met acceptable environmental accreditation standards.

For further information on Blackwell Publishing, visit our website:
www.blackwellpublishing.com

To Jane, Tom and George

'What greater stupidity can be imagined than that of calling jewels, silver and gold "precious", and earth and soil "base"? People who do this ought to remember that if there were as great a scarcity of soil as of jewels or precious metals, there would not be a prince who would not spend a bushel of diamonds and rubies and a cartload of gold just to have enough earth to plant a jasmine in a little pot, or to sow an orange seed and watch it sprout, grow, and produce its handsome leaves, its fragrant flowers and fine fruit.'

Dialogue on the Two Chief World Systems: Ptolemaic and Copernican, Galileo

Contents

Preface

Since about the age of ten, I have been fascinated by plants and their use for decoration as flowers, and for food. Much of my pocket money as a child came from the sale of plants and flowers and I quickly learned the practical benefits to be gained from controlling soil fertility in the garden and from good quality potting media in the glasshouse. It was this interest in plants, together with the misery of the famine in India during my teenage years, which led me to study soil science at the University of Reading, although my interest in plants was temporarily put on hold as much of my degree was essentially chemistry. For my PhD at Nottingham University, I was able to choose a topic that interested me, and after a false start on the kinetics of phosphate adsorption by soil minerals, I came across two papers in the library, one by Glyn Bowen and Albert Rovira, and the other by Howard Taylor and Betty Klepper, which enthused me with the possibility of combining my interests in plants and soils by studying roots and their interactions with soil. I quickly found that roots in soil were difficult to study, not least because they cannot be seen, but the satisfaction of patient discovery was considerable. The early encouragement in this endeavour by my supervisors David Crawford and Mike McGowan was essential, as was that of those who eventually became co-workers and colleagues, John Monteith, Paul Biscoe and Nick Gallagher.

Much of my professional career has been spent at the University of Reading where I was allowed the freedom by Alan Wild to continue and build my studies on root:soil interactions. Projects in the UK and overseas followed, and with a succession of PhD students and postdoctoral research workers I have been able to work on a wide range of crops and practical problems, all with a basis in the growth and activity of root systems. When I started my work, the emphasis was on how various soil properties affect the plant and its ability to take up water and nutrients, but recently the emphasis has changed, as it has come to be appreciated that plant roots also change the properties of soils and are not merely passive respondents.

The idea for this book first came in a conversation with my friend Rod Summerfield but for various reasons, including a career change in Australia, it is only now that I have had the determination to bring the project to a conclusion. In fact, I think it is a better book as a result because I believe that the recent development of techniques and the improved understanding of root:soil interactions make this a particularly exciting time to try and write such a book. I have tried to draw together information from diverse elements of the plant and soil literatures to illustrate how roots interact with soil, both to modify it and to obtain from it the resources required for the whole plant to grow. My emphasis has been on whole plants and root systems, although I have drawn on the growing body of literature at plant molecular and cellular levels as appropriate

A particular difficulty in the writing has been that roots of relatively few plant species have been studied and of these most are cereal crops such as maize and wheat. This means that the desire to generalize findings as one might in an introductory undergraduate textbook has had to be tempered with an appreciation of the paucity of information. I hope that I have been able to convey useful principles while at the same time indicating that plant species other than those studied might respond differently. A second area of caution is that many studies in the plant literature have been conducted on young, seedling roots in solutions or in non-soil media. Extrapolation of such findings to older plants, with roots of different anatomy, with fungal and bacterial associations, and with gradients of solutes and gases resulting from past activity, must be undertaken cautiously. Finally, there has been until recently a tendency to regard all roots on a plant as anatomically similar and functionally equivalent; this notion is beginning to be challenged as results indicating particular arrangements of cell types and functional specialisms appear. Measurements are few at present, but we may yet find that roots within a root system make particular contributions to the activities of the whole.

So, this is a personal view of the subject aimed at those who already have a background knowledge of soils and plants. Besides those I have already mentioned, I should like to thank Christopher Mott, Bernard Tinker, Dennis Greenland, Peter Cooper, Lester Simmonds, Ann Hamblin, Neil Turner and Derek Read for sustaining my enthusiasm in root studies at various points in my career, and to thank Michelle Watt, Glyn Bengough, Margaret McCully, John Passioura, Rana Munns, Sarah Ellis, Steve Refshauge, Mark Peoples, Ulrike Mathesius, Sally Smith, Ken Killham, Philippe Hinsinger, Richard Richards, Greg Rebetzke, Tim George, Manny Delhaize, Wolfgang Spielmeyer and John Kirkegaard for reading and suggesting improvements to various parts of the manuscript. I am very grateful to the University of Reading for giving me study leave to undertake this project, and to the Leverhulme Trust for a Study Abroad Fellowship that enabled me to spend a very productive period in Canberra, Australia. As ever, CSIRO Division of Plant Industry, Australia provided a challenging academic environment in which to work (my thanks to the Chiefs Jim Peacock and Jeremy Burdon) and I am indebted to Carol Murray and her staff, especially Michelle Hearn, at the Black Mountain Library for helping me locate reference materials. Finally, my thanks to my personal assistant, Tricia Allen, the staff of the ITS unit at the University of Reading and Ian Pitkethly at SCRI for help with the figures, and to Nigel Balmforth and the staff at Blackwell Publishing for seeing the manuscript through to publication.

Peter J. Gregory

Chapter 1

Plants, Roots and the Soil

This book focuses on vascular plants and their interactions with soils. It has long been appreciated that plants influence the properties of soils and that soil type can, in turn, influence the type of plant that grows. This knowledge of plant/soil interactions has been put to use by humans in their agriculture and horticulture. For example, Pliny The Elder quotes Cato as writing 'The danewort or the wild plum or the bramble, the small-bulb, trefoil, meadow grass, oak, wild pears and wild apple are indications of a soil fit for corn, as also is black or ash-coloured earth. All chalk land will scorch the crop unless it is extremely thin soil, and so will sand unless it is extremely fine; and the same soils answer much better for plantations on level ground than for those on a slope' (Rackham, 1950). Similarly, long before the nitrogen-fixing abilities of rhizobia were documented scientifically, Pliny The Elder noted that lupin 'has so little need for manure that it serves instead of manure of the best quality', and that 'the only kinds of soil it positively dislikes are chalky and muddy soils, and in these it comes to nothing' (Rackham, 1950).

This close association of soils and plants has led, too, to an ongoing debate as to the role of plants in soil formation. Joffe (1936) wrote that 'without plants, no soil can form' but others such as Jenny (1941, reprinted 1994) demonstrated that vegetation can act as both a dependent and an independent variable in relation to being a soil-forming factor. Ecologists find it useful to work with vegetation types and plant associations comprising many individual plant species; these associations are frequently linked to soil associations, and in this regard, at this scale, the vegetation is not an independent soil-forming factor. However, it is also appreciated that within a vegetation type, different plant species may have effects which lead to local variations in soil properties and where plants do act as a soil-forming factor. For example, in mixed temperate forests the pH of litter extracts of different species may range from 5.8 to 7.4, leading to different types of humus from the different species and hence different rates of mineral leaching. Similarly it is well documented that the planting of coniferous trees on several areas in Europe has increased rates of soil acidification in some areas, and resulted in podsol formation on soils that were previously earths (Hornung, 1985).

Although the focus of much plant and soil science has been on the return of leaves to the soil both as a stock of C in the soil and as a substrate for soil organisms, root returns to soil are larger than shoot returns in several regions. For example, early work by ecologists such as Weaver in the USA demonstrated that several grasses produced more organic matter below ground than above ground (Weaver *et al.*, 1935). This interest in carbon inputs to soils has been re-ignited with the current debate over sequestration of C by vegetation in an attempt to mitigate the greenhouse effect induced by rising CO_2 concentration of the atmosphere. For example, observations of deep-rooted grasses introduced into the grasslands of South America have demonstrated that they can sequester substantial amounts of carbon (100–500 Mt C a^{-1} at two sites in Colombia) deep in the soil (Fisher *et al.*, 1994). Roots and their associated flora and fauna are the link between the visible parts of plants and the soil, and are the organs through which many of the resources necessary for plant growth must pass. As part of the system that continually cycles nutrients between the plant and the soil, they are subject to both the environmental control of the plant and the assimilatory control of the plant as a whole.

This chapter examines the close connection between the root and shoot systems of vascular plants and what is known about the co-ordination of activities between the two systems. It also describes some of the main features of the interaction between roots and soils as a prelude to more detailed examination of changes to soil properties in the vicinity of roots in later chapters.

1.1 The evolution of roots

Roots and shoots are considered by most botanists to be entirely separate organs, although some developmental processes are shared, and some inter-conversion can occur (Groff and Kaplan, 1988). Raven and Edwards (2001) sought to define what constitutes a root of a vascular plant to distinguish it from a shoot, and concluded that the distinguishing features were 'the occurrence of a root cap, a more defined lineage of cells from the apical cell(s) to tissues in the more mature parts of the roots, the essentially universal occurrence of an endodermis, a protostele (i.e. a solid cylinder of xylem) sometimes with a pith, and endogenous origin of lateral roots from roots' (Table 1.1). These same features are shown diagrammatically in Fig. 1.1. Others (e.g. Gifford and Foster, 1987) have highlighted the uniqueness of roots because of their bidirectional meristem that produces both an apical root cap and subapical root tissues (see section 2.4.1).

The general structure and function of roots and shoots are so different that the two organs are often conveniently separated for the purposes of research. Functionally, roots absorb water and nutrients, and anchor the plant, while shoots photosynthesize and transpire, and are the site of sexual reproduction (Groff and Kaplan, 1988). Usually both root and shoot must occur together for a plant to function and grow, although there are some exceptions to this generalization. Roots and shoots gradually acquire their distinguishing features during the differentiation and growth of the embryo sporophyte, but are not usually recognizable until the apical meristems are differentiated.

The fossil record for the evolution of roots is less helpful than that for shoots, but recognizable root-like structures start to appear in Early Devonian times (410–395 million years ago). Fossilized remains of many early land plants are fragmentary, and delicate structures such as root caps may not have been preserved, so that evolutionary sequences are often difficult to date with certainty. Gensel *et al.* (2001) use the terms 'rootlike' and 'rooting structures' to describe fossil structures which resembled roots and were positioned such that they may have anchored the plant to a substrate; whether they also functioned as absorbers of water and nutrients is unknown. Raven and Edwards (2001) suggest that Lower

Table 1.1 The characteristics of early vascular plants (above and below ground) and those of shoots and roots of extant plants

Characteristic	Early vascular plants	Shoot of extant plants	Root of extant plants
Primary xylem	Protostele	Protostele in some pteridophytes; pith present in other vascular plants	Non-medullated protostele (except in some monocoyledons with central pith)
Root cap	Absent	Absent	Present
Endodermis in organs lacking secondary thickening	Absent (apparently)	Usually absent; present in many pteridophytes and some spermatophytes	Present in almost all cases and sometimes supplemented by an exodermis (an endodermis-like hypodermis)
Origin of branches	Superficial organ	Branch shoots are of superficial origin, while roots originating from shoots can be endogenous or exogenous	Branch roots arise endogenously, while shoots originating from roots can be endogenous or exogenous
Hairs	'Axis hairs'; mycorrhizas on below-ground parts	Varied 'shoot hairs' usually present	'Root hairs'; often supplemented by mycorrhizas

Adapted from Raven and Edwards, 2001.

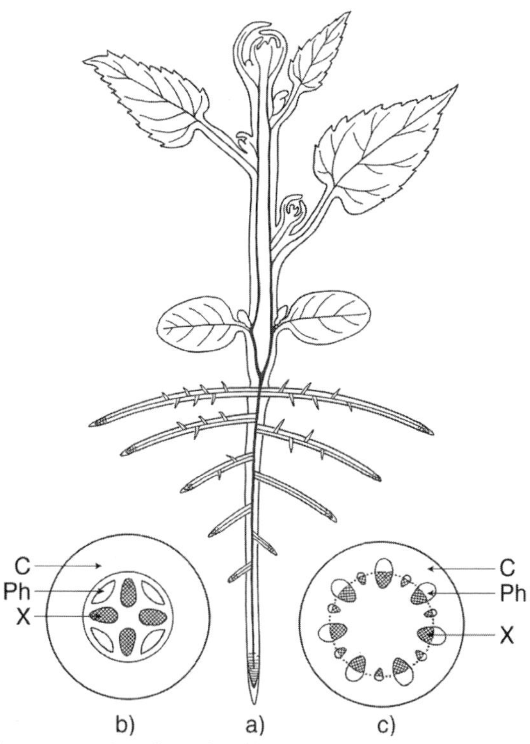

Fig. 1.1 Diagrammatic representation of a typical dicotyledon showing the characteristic properties of roots and shoots: (a) longitudinal view, (b) transverse section of a root, and (c) transverse section of a stem. The shoots bear leaves and daughter shoots that originate exogenously while lateral roots arise far from the root apex and are endogenous in origin. The arrangement of the cortex (C), phloem (Ph) and xylem (X) is shown. (Redrawn and reproduced with permission from Groff and Kaplan, *The Botanical Review*; New York Botanical Garden, 1988.)

Devonian sporophytes had below-ground parenchymatous structures which performed the functions of roots (anchorage, nutrient and water uptake), but they did not have root caps or an endodermis. Traces of dichotomous root-like structures 5–20 mm in diameter, 10–90 cm long, and penetrating into the substratum to nearly 1 m have been found in fossils of the late Early Devonian (375 million years ago) (Elick *et al.*, 1998), thus allowing the mining of nutrients from the rocks which supplied the increasing biomass of plants at this time. The distinguishing features of roots of vascular flowering plants (angiosperms) first appeared in several plant types such as lycopodia and some bryophytes in Mid Devonian times in a period of rapid plant diversification (Raven and Edwards, 2001). Brundrett (2002) suggests that as plants colonized the land they would have faced powerful selection pressure to increase the surface area of their absorptive surfaces in soil to parallel that occurring in their photosynthetic organs; interception of light and CO_2 would thereby be in balance with that of nutrients and water.

A possible evolutionary sequence of shoots and roots is shown in Fig. 1.2 (Brundrett, 2002) in which the evolution of roots emerged as a consequence of the differentiation of underground stems (rhizomes) into two specialized organs: (i) thicker perennial stems that form conduits to distribute water and nutrients, serve as stores and support above-ground structures; and (ii) thinner, longer structures to absorb water and nutrients. Root hairs may have evolved from the rhizoids of earlier plants to increase the volume of substrate available for exploitation, with mycorrhizal fungi also co-evolving with roots (Brundrett, 2002). The available evidence also suggests that while roots evolved first among the lycopsids, they also evolved on at least one other occasion during the evolution of vascular land plants. The suggestion that roots may have gradually evolved from shoots is supported by the observed

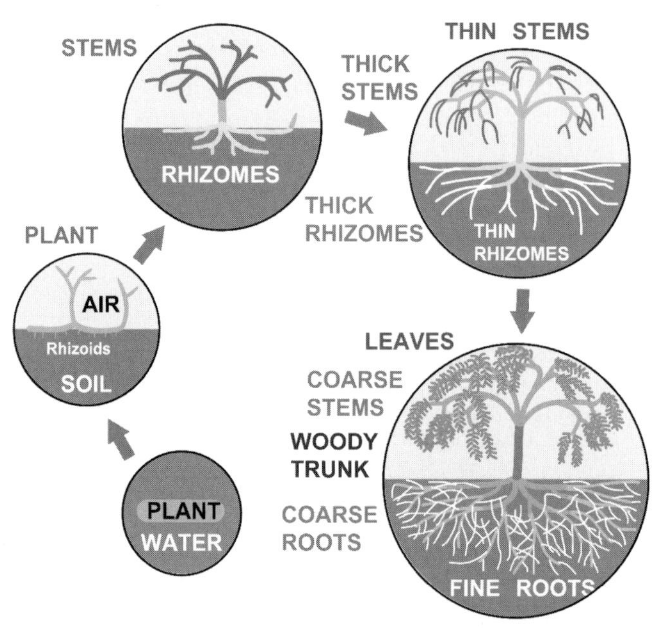

Fig. 1.2 A diagrammatic representation of the possible evolution of stems, rhizomes, leaves and roots from the thallus of an early bryophyte-like terrestrial plant, using a hypothetical final example with a woody trunk. (Reproduced with permission from Brundrett, *New Phytologist*; New Phytologist Trust, 2002.)

developmental and genetic similarities of shoot and root cell division and differentiation in *Arabidopsis*, although it is also possible that evolutionary convergence of the genetic mechanism occurred after evolution of the root (Dolan and Scheres, 1998).

1.2 Functional interdependence of roots and shoots

1.2.1 Balanced growth of roots and shoots

The different morphologies, anatomies, physiologies and functions of roots and shoots have frequently led to their being considered as two separate systems within the entire plant. Nevertheless, while each system grows and functions as a discrete site for the capture of specific resources (carbon dioxide, light, water and nutrients), the two systems are coupled together and their functions have to form an integrated system. Early explorations of this coupling led to theories based essentially on the size or weight of the two organs. Hellriegel in 1883 in a 'basic law of agriculture' (quoted by van Noordwijk and de Willigen, 1987) wrote that 'The total above-ground growth of plants is strongly dependent on the developmental stage of the root. Only when the root can fully develop will the above-ground plant reach its full potential'. From such writings came the notion that the size of both systems might be inter-related and the simpler notion that big shoots were associated with big root systems. Mayaki *et al.* (1976), for example, sought to determine a relation between rooting depth and plant height of soyabean as a means of estimating irrigation requirements, and shortly after dwarfing genes were introduced into cereal crops it was hypothesized that their root systems might be shallower as a result (e.g. Lupton *et al.*, 1974 for wheat). Such simple morphological equilibria were demonstrated to be non-existent.

In a set of experiments designed to investigate the equilibrium between root and shoot growth, Troughton (1960) and Brouwer (1963) observed that characteristic equilibria were attained depending on the conditions prevailing. Their experiments demonstrated the following:

(1) When root growth is limited by a factor to be absorbed by the root system, then root growth is relatively favoured; conversely, when the limiting factor has to be absorbed by the shoot, its growth is relatively favoured.
(2) Disturbance of the ratio of root:shoot brought about by either root removal or defoliation leads to changes in the pattern of growth so that the original ratio is rapidly restored (Fig 1.3).
(3) Transfer of plants from one environment to another causes changes in the pattern of assimilate distribution so that a new characteristic root:shoot ratio is established over a period.

The realization that disturbance led to plant activities that restored root:shoot balance, and that it was a combination of both growth and the activities of the root and shoot systems in capturing resources that determined the new equilibrium, led to the concept of a 'functional equilibrium' (Brouwer, 1963, 1983). According to this concept, the root and shoot respond not to the size of each other, but to the effectiveness with which the basic resources are obtained from the environment by the complementary organ. Photosynthate, then, is

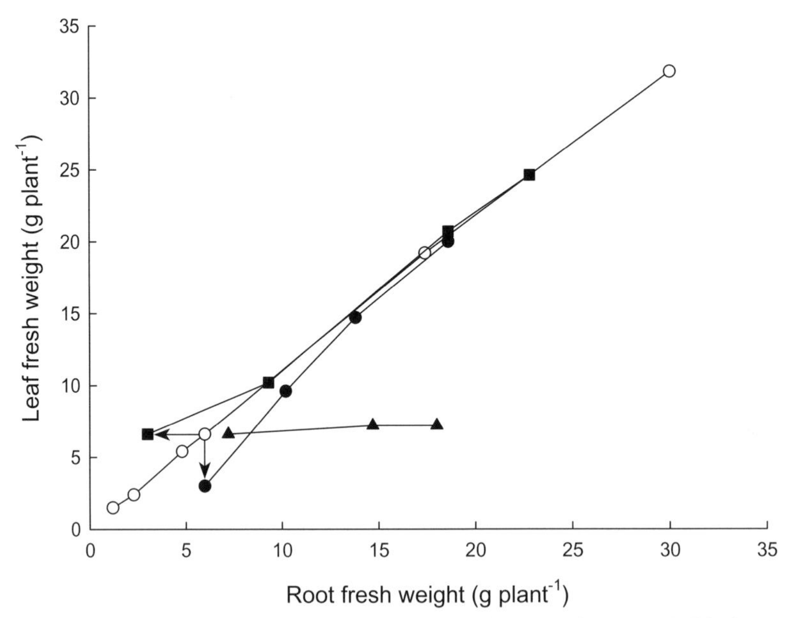

Fig. 1.3 Recovery to the original leaf:root ratio of common bean plants (F) after removal of the leaves (●) and the roots (■). No recovery of the ratio was found when the growing parts of the shoot were removed continuously (▲). (Redrawn from Brouwer, 1963.)

partitioned to roots and shoots in inverse proportion to the rates at which they capture resources. Davidson (1969) expressed this as:

$$\text{root mass} \times \text{specific root activity (absorption)} = \text{shoot mass} \times \text{specific shoot activity (photosynthesis)} \quad (1.1)$$

The consequence of this statement is that the root:shoot ratio of a plant will vary to compensate for changes in root and shoot activity induced by changes in the edaphic and atmospheric environments. It explains why small root systems may be sufficient for maximum plant growth when the supply of water and nutrients is optimal (e.g. in horticultural production systems) and why managing the soil to produce more roots may be counterproductive (van Noordwijk and de Willigen, 1987). Most investigation of this hypothesis has focused on nitrogen uptake and photosynthesis in young, vegetative plants (e.g. Thornley, 1972), with subsequent refinements to allow for dynamic responses to changes in environments. Johnson (1985) found that during balanced exponential growth, a relationship analogous to that proposed by Davidson (1969) applied, but that rates of uptake influenced partitioning through effects on substrate levels. The consequence was that there was no unique relationship between shoot and root activities (cf. Davidson's [1969] proposition), although over restricted ranges of root and shoot specific activities the linear relation implied by equation 1.1 held. A difficulty with exploring this concept further is that while shoot mass and photosynthesis can be measured relatively easily, the capture of below-ground resources and determination of the resource that is most limiting at a particular time poses many problems. Hunt *et al.* (1990) drew attention to some of the difficulties which include: (i) quantifying the fraction of a nutrient that is available at a particular time; (ii) allowing for the different nutrient contents of different soil layers; (iii) allowing for the differential rates of nutrient transport to different roots depending upon extant gradients of concentration; and (iv) allowing for the different depths and spatial distributions of roots when plants are grown in communities.

Farrar and Jones (2000) suggested that the functional equilibrium hypothesis is useful for describing how environmental factors such as light, water, N and P affect the relative growth of roots and shoots, but showed that it was inadequate for many other situations and that it lacks a physiological and mechanistic basis. Johnson (1985), too, commented on the lack of mechanistic understanding of resource and growth partitioning. Farrar and Jones (2000) proposed that acquisition of carbon by roots is determined by both the availability of, and need for, assimilate. This leads to the hypothesis that import of assimilates to roots is controlled by a range of variables in both root and shoot ('shared control' hypothesis) and that there is shared control of growth between leaves (the source of C) and roots (the sink for C). The mechanisms proposed to allow this control were phloem loading, and gene regulation by sugars and other resource compounds. However, while there is evidence for the coarse control of phloem loading in response to sink demand for carbohydrate, there is no evidence of fine control (Minchin *et al.*, 2002), and while there is some evidence of gene up- and down-regulation by sucrose in laboratory conditions, there is little evidence for gene regulation by resources in field-grown plants. In summary, while there is evidence to support the hypothesis that control of C flux to roots is shared between the many processes contributing to whole-plant C flux, no good mechanistic model of this phenomenon currently exists.

1.2.2 Communication between roots and shoots

Normal development of plants depends on the interaction of several external (e.g. light and gravity) and internal factors, with plant hormones being part of the internal factors that play a major role in regulating growth. Many hormones are produced in one tissue and transported to another where they influence the rate and nature of plant development and growth. For example, indole-3-acetic acid (IAA) is an auxin that is synthesized mainly in leaf primordia and young leaves, but plays a major part in the growth response of root tips to gravity (see section 5.2.1). Although IAA has been measured in root tips (typically in concentrations of about 150 $\mu g\ kg^{-1}$ fresh weight of root, or 5×10^{-7} M) most evidence suggests that it is not produced there but transported from the shoot via the vascular system (Torrey, 1976; Raven *et al.*, 1999). It has been shown that decapitated plants produce less auxin and also have decreased rates of root growth, and that application of auxin to the decapitated shoot tip restores root growth. Auxins also interact with shoot-produced gibberellins in regulating expansion of root cells (Dolan and Davies, 2004). Conversely, cytokinins (of which the most common is zeatin, 6-[4-hydroxy-3-methyl-2-transbutenylamino] purine) are synthesized primarily in root tips and transported to shoots via the xylem where they regulate cell division and, in more mature plants, the rate of leaf senescence.

In all, five groups of plant hormones (auxins, cytokinins, abscisic acid, gibberellins and ethylene) have long been recognized as regulating plant growth, but more recently other chemical signals have also been identified as important. These signals include the brassinolides (related to animal steroids) which stimulate cell division and elongation, salicylic acid which activates defence responses to plant pathogens, the jasmonates which act as plant growth regulators, and small peptide molecules such as systemin which activate chemical defences when wounding occurs (Raven *et al.*, 1999). A practical application of the role of hormones is the use of auxin to stimulate the initiation of roots from stem cuttings. This practice is widely used by horticulturists to ensure the vegetative propagation of many plants.

There is now strong experimental evidence that root signals to the shoot modulate growth responses of the shoot. This has been most thoroughly explored in relation to soil water deficits where abscisic acid (ABA) is believed to play a major role (see section 4.2.2), but other properties of the soil, especially its strength, also play a part. For example, Passioura and Gardner (1990) investigated the effects of soil drying on leaf expansion of wheat leaves, by measuring changes in soil water potential, soil strength and phosphorus availability of plants that were either pressurized to maintain high leaf turgor or unpressurized. Their results demonstrated no significant effects of the pressurization treatment on relative leaf elongation rate (RLER), and that the plants were sensing both the water status and the strength of the soil but not the availability of phosphorus. Figure 1.4 shows that RLER decreased as the soil dried even when plants were grown in loose soil (with penetrometer resistance <1 MPa), and that RLER of plants in drying, dense soil (penetrometer resistance 2–5.5 MPa) fell below that of the well-watered controls at a much higher soil water content (0.23 g g^{-1}, equivalent to about 100 kPa tension) than in loose soil (0.17 g g^{-1}, equivalent to about 270 kPa tension). These results suggest that the roots were sensing both the tension and strength of the soil and sent inhibitory signals to the shoots which reduced leaf expansion as either tension or strength increased. The precise nature of the signals is still unknown, although auxin and cytokinins have both been suggested to play a part (Davies and Zhang, 1991).

At the other end of the life cycle, leaf senescence is also affected by plant hormones, with cytokinins and ABA playing a direct role in the regulation of drought-induced leaf senescence (Yang *et al.*, 2003). Drought enhances ABA levels which increases carbon remobiliza-

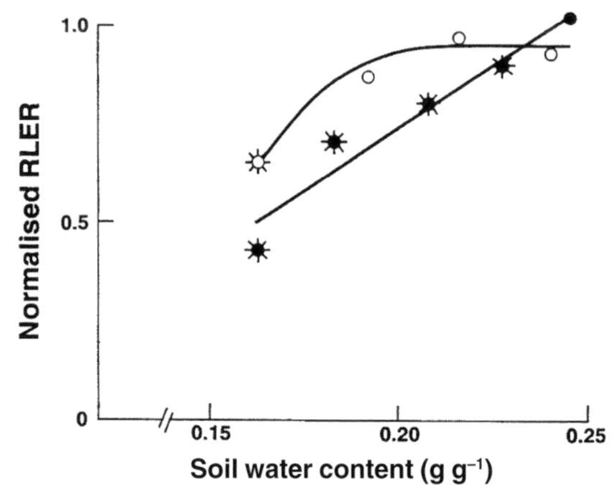

Fig. 1.4 Normalized relative leaf elongation rate as affected by soil water content and soil strength for plants grown in soil of low (F) or high (●) bulk density. The starred points denote that the elongation rates of wet and dry treatments differed significantly. (Reproduced with permission from Passioura and Gardner, *Australian Journal of Plant Physiology*; CSIRO Publishing, 1990.)

tion from senescing leaves to grains, and decreases cytokinin levels, which show a positive correlation with chlorophyll content and rates of photosynthesis. Cytokinins play a major role in the regulation of source-to-sink transitions with high levels promoting the activity of sugar transporters and cell division, and low levels inhibiting growth in older leaves. During drought, old leaves senesce, nutrients are translocated to young leaves, and the plant is able to withstand the drought until the good times return (Munné-Bosch and Alegre, 2004).

Long-distance signalling has also been shown to regulate the expression of shoot genes following changes in nutrient supply to roots independently of changes in nutrient delivery to the shoot. For example, Takei *et al.* (2002) showed that cytokinin metabolism and translocation were modulated by nitrogen availability in maize, suggesting that both nitrate and cytokinin were signals communicating nitrogen availability from root to shoot. Dodd (2005) suggests that the criteria for a root-to-shoot signal are that a compound must: (i) move acropetally in the plant via apoplastic (predominantly the xylem) or symplastic pathways; and (ii) influence physiological processes in a target organ (such as leaves or fruit) that is remote from the putative site of synthesis (the root). Most work has focused on compounds that are xylem-mobile (Table 1.2), but much remains to be explained about which signalling molecule is important in specific circumstances and how the observed plant response is brought about. The relatively slow progress in this area is perhaps unsurprising when it is appreciated that there are at least four kinds of signal by which stressed roots influence shoots (Jackson, 1993):

(1) Increase the output from roots of an existing signal compound or generate a new one (positive message).
(2) Decrease the output from roots of an existing signal compound (negative message).
(3) Reduce demand in the root for hormones or other compounds originating in the shoot leading to accumulation at source (accumulation message).
(4) Attraction of signalling molecules or assimilates away from the shoot such as occurs in the infection of roots by *Striga hermonthica* which increases root demand for assimilates (debit message).

In reality, two or more types of signalling are likely to coexist in a stressed plant and may interact (Jackson, 2002). For example, in flooded tomato plants, oxygen shortage at

Table 1.2 Xylem mobility of the classical plant hormones with plant species indicated for cytokinins

Hormone class	Xylem-mobile compounds
Abscisic acid	Abscisic acid (ABA), abscisic acid glucose ester (ABA-GE)
Auxin	Indole-acetic acid (IAA)
Cytokinins	Dihydrozeatin-9-glucoside (DHZ-9G), zeatin riboside (ZR), isopentenyladenine (iP) – sunflower
	ZR, zeatin (Z), iP-type cytokinins – pea
	ZR, dihydrozeatin riboside (DHZR), zeatin-O-glucoside (Z-OG), dihydrozeatin-O-glucoside (DHZ-OG), dihydrozeatin riboside-O-glucoside (DHZR-OG), nucleotides of Z, dihydrozeatin and DHZ-OG – common bean
Ethylene	Aminocyclopropane-1-carboxylic acid
Gibberellins	A large number of gibberellins

Adapted from Dodd, 2005.

the roots promotes cell expansion on the undersides of leaf petioles resulting in downward curvature of the leaf (epinasty). The principal signal responsible for this behaviour is a positive message caused by the transport of 1-aminocyclopropane-1-carboxylic acid (ACC) from the root to the leaves in the xylem. In oxygen-deficient roots, the oxidation of ACC synthesized in the root to ethylene is inhibited, but in the shoot where oxygen is abundant, the ACC is rapidly converted to ethylene by the enzyme ACC oxidase, whose activity is also increased in leaves soon after flooding. The signal resulting in increased ACC oxidase activity is unknown but the ethylene produced induces epinasty. Flooding, though, also decreases cytokinin and gibberellin concentrations in xylem sap (a possible negative message), and the large decrease in ABA transported to leaves may sensitize them to the action of ethylene. There is also some evidence that ABA may build up in leaves as an accumulation message, although this effect is probably too short-lived to have a major effect. Future progress in this area will depend on examination of a wider range of putative signalling compounds and of the interactions between them, and better measurement of the size and durability of signals in transit coupled with better measures of effectiveness at target sites (Jackson, 2002).

1.3 Roots and the soil

The growth of root systems in soils is affected by a wide range of soil properties but, in turn, the properties of soils are modified by roots. There is, then, a plethora of dynamic reactions occurring at the root surface whose consequences are felt at a range of temporal and spatial scales. Many classical approaches in soil science, especially in the use of soils for crop production, have served to minimize this dynamism by dealing with equilibrium measurements in homogenized soils (e.g. the use of chemical extractants on <2 mm sieved soil to approximate nutrient availability), but the situation is changing and techniques are increasingly being developed to explore the interactions of soils and roots. For example, Young and Crawford (2004) have drawn attention to the important role of microbes in the dynamic generation of soil structure and stressed the interactions of microbial and physical processes in soil and the self-organization that occurs in the soil–microbe system. Even this view of soils, though, is partial, ignoring as it does the overwhelming role of roots as a source of substrates for microbes and as agents for biological, chemical and physical changes in soils; roots are an essential component of soil biology.

Concerns for terrestrial biotic diversity are also giving rise to the need for greater understanding of soil:plant interactions, leading to an integrated biogeodiversity perspective in efforts to preserve landscapes. For example, agriculture and urbanization in the USA have resulted in a loss of soil diversity with about 4.5% of the nation's soils in danger of substantial loss, or complete extinction (Amundson *et al.*, 2003). In some instances rare or endangered plants are linked to rare or endangered soils so that arguments for soil and biodiversity preservation and planning are intimately linked.

1.3.1 The root–soil interface

The interface between the root and the soil is complex and frequently an ill-defined boundary. Products are released from roots into the soil which change its chemical and physical properties, and stimulate the growth of various microorganisms. Concurrently, the root tissues and associated root products also provide physical shelter for many microorganisms. This complex environment where root and soil meet is known as the rhizosphere. When Hiltner (1904) first coined this term, it was employed in the specific context of the interaction between various bacteria and legume roots in studies that he undertook on the value of green manures. It was quickly realized, though, that this was too limited and limiting a use of the word and it is now more widely used to describe the portion of the soil that forms the complex habitat of plant roots, the composition of which is altered by root activity. Roots and soil particles are frequently in intimate contact, with root hairs, mucilage and microbes forming a zone of multiple interactions between the plant and the soil (Fig. 1.5). Mucilage of both bacterial and plant origins is able to bind soil particles on drying, and to retain the particles on subsequent rewetting, although the binding by root mucilage seems to depend on 1,2 diols in component sugars whereas that by bacterial mucilage is likely to be protein-mediated (Watt *et al.*, 1993). Within the rhizosphere, some workers (e.g. Lynch, 1990) have sought to distinguish various regions such as the endorhizosphere (cell layers within the root colonized by microorganisms), the ectorhizosphere (the area surrounding the root containing root-associated microorganisms), and the rhizoplane (the root surface). However, the boundaries of these regions are themselves often diffuse and difficult to define.

Foster and Rovira (1976) were among the first to study the spatial relationships between microbial communities and the root by preparing ultra-thin sections of rhizospheres and examining them with transmission electron microscopy. In wheat, young roots were only sparsely colonized by microorganisms but by flowering, both the rhizosphere and the outer cortical cells and their cell walls showed considerable development of microorganisms. The bacteria present in the rhizosphere differed substantially from

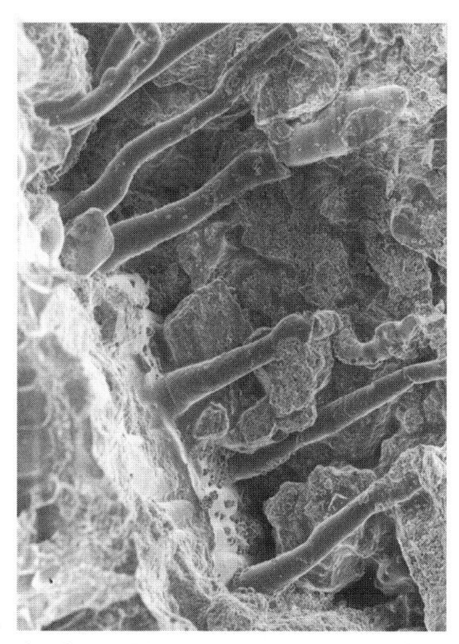

Fig.1.5 The root:soil interface of crabgrass (*Digitaria sanguinalis*) growing in the field. Clearly visible are root hairs in intimate contact with soil particles. (Reproduced with permission from McCully, *Physiologia Plantarum*; Blackwell Publishing, 1995.)

those in the bulk soil in several aspects. Not only were there more bacteria at the rhizoplane (120×10^9 ml^{-1} compared with 13×10^9 ml^{-1} at 15–20 μm from the root surface) but the number of types that could be recognized morphologically was also greater. Foster (1986) reported that of 11 morphologically distinct types that could be recognized, all occurred within 5 μm of the rhizoplane but only 3 types occurred at 10–20 μm from the root surface. There were differences in size too. Some 80% of the rhizosphere bacteria were >0.3 μm in diameter compared with only 37% in the bulk soil (Table 1.3). Furthermore, away from the rhizoplane, the bacteria tended to occur in isolated discrete colonies which, in the outer rhizosphere, were associated with organic debris. The largest colonies and the largest individuals were associated with cell wall remnants that still contained carbohydrate. Larger fungi and protozoa are observed less frequently in the rhizosphere although their total biomass may be as great as that of bacteria (Campbell and Greaves, 1990). Boundaries between rhizosphere and bulk soil are also less meaningful in relation to fungi, which may easily traverse the root, rhizosphere and, via pores, grow some distance into the bulk soil.

Root, soil and organisms interact to determine the rhizosphere environment. Because the amounts and types of microbial substrates are different to those in the bulk soil, there are different populations of bacteria, actinomycetes, fungi, protozoa, viruses and nematodes in the rhizosphere. Moreover, because a major function of roots is to acquire water and nutrients from the soil, the physicochemical properties of the rhizosphere are also different to the bulk soil. This leads to a wide range of potential habitats for organisms and equally to a wide range of microbially mediated processes occurring in the rhizosphere. As Lynch (1990) points out, the association between organisms and roots can be beneficial, harmful or neutral, but the outcome often depends on the precise conditions in the rhizosphere and bulk soil so that outcomes are frequently variable. This biologically active zone of soil means that root–root, root–microbe and root–faunal communications are likely to be continuous occurrences, although relatively little is known about these communication pathways. Walker *et al.* (2003) suggested that root exudates may act as messengers to convey a wide range of signals that initiate biological and physical interactions between roots and organisms including allelopathic root–root communication, and antibacterial compounds that interfere with bacterial quorum-sensors. Chapters 6 and 7 will explore these issues more thoroughly.

Table 1.3 Size classes of soil bacteria measured in transmission electron micrographs of ultra-thin sections of rhizospheres of several plant species compared to published values from bulk soil samples; over 900 bacteria in the rhizosphere were measured

Size range (μm)	Rhizosphere (%)	Bulk soil (%)
<0.3	20	63
0.31–0.5	49	31
0.51–0.9	25	6
0.91–2	5	0

From Foster, 1986.

1.3.2 Root-induced soil processes

As stated previously, vegetation is considered in the soils literature as being one of the six major factors (the others being parent material, relief, climate, time and human interventions) giving rise to different types of soil (Jenny, 1941 reprinted 1994). Many of these long-term effects are a consequence of the different properties of leaf and shoot components rather than roots *per se*, but over extended periods, roots have a major influence on the formation of soils. One effect of roots is to physically exploit cracks and fissures in rocks and, through repeated wetting and drying cycles and chemical modifications to the rhizosphere, to increase the soil volume. Roots growing in rock fissures are often morphologically adapted, with the outer cortex becoming flattened while the inner water-conducting tissues remain cylindrical (e.g. Zwieniecki and Newton, 1995). The smallest pore that can be entered is determined, then, by the size of the conducting tissues, which was about 100 μm in the woody shrubs studied by Zwieniecki and Newton. Many plant species obtain significant quantities of water from underlying rock formations by exploiting such fissures and drawing on reserves of stored water. For example, on the chalklands of southern England, which store substantial amounts of water, Wellings (1984) demonstrated that cereal crops were able to deplete water from the chalk/soil and chalk layers amounting to 71–80% of total profile depletion and 29–40% of seasonal crop water use. Similarly, Gregory (1989) estimated that upward flux of water to the root zone (i.e. from deeper than 0.9 m) contributed 8% of the shoot dry matter of winter cereals and 22% of that of spring cereals; over time, soil particles are washed into the cracks and the volume of soil material is increased.

Chemical weathering of minerals to form soil materials may also be enhanced by the presence of roots and their associated microflora. For example, root-induced vermiculitization of the mica phlogopite was measured under laboratory conditions by Hinsinger and Jaillard (1993) (see section 7.2.1 for details), and weathering of vermiculite has also been demonstrated by cultures of ectomycorrhizal fungi (Paris *et al.*, 1995). A range of processes may be involved including the release by roots of protons and organic anions, and the depletion of cations to concentrations low enough to destabilize crystal lattices. In the laboratory study of potassium release from phlogopite by ryegrass, the equilibrium concentration of the soil solution below which the mica became unstable was about 80 μmol K l^{-1}, but lower concentrations would be required if the dominant micas were the more commonly occurring dioctahedral soil minerals muscovite (equilibrium concentration 2–5 μmol l^{-1}) and illite (25 μmol l^{-1}) (Hinsinger and Jaillard, 1993). In contrast in rape, irreversible transformation of phlogopite to vermiculite was brought about by severe root-induced acidification of the rhizosphere, leading to acid dissolution of the phlogopite lattice (Hinsinger *et al.*, 1993). Plant roots, then, may be responsible for specific forms of weathering that are different to those occurring in the bulk of the soil and may explain why the clay mineralogy of the rhizosphere is sometimes reported to be different from that of the bulk soil (April and Keller, 1990).

Locally, in soils that are rich in calcium carbonate, root calcification can occur, leading to microstructures that contribute significantly to the genesis of some calcareous soils. In southern France, Jaillard *et al.* (1991) found that calcified roots were common on sites where the calcium carbonate content was >25%, the soils were dense or rich in fine particles, drainage was slow and soils were wet for a large part of the year. Calcified roots retained the structure of the original cells of the root cortex while the central conducting tissue was not preserved and was represented by an empty central channel. Each calcified cell had a central nucleus composed of an average of two calcite crystals and was coated with a thin calcareous layer that was poorly crystallized.

The development of water-stable aggregates is an important process in the genesis of soils because it strongly influences a range of soil characteristics including aeration, infiltration and erodability. Plant roots play a major role in this process. Their influence comes about indirectly through the release of carbon compounds which provide a substrate for microbes with all of their effects on structure (Young and Crawford, 2004), and directly through: (i) wetting and drying phenomena; (ii) the accumulation in some soils of inorganic chemicals at the root surface that act as cementing agents; (iii) the release of organic compounds that promote aggregation of particles; and (iv) the role of undecayed, senescent roots acting like steel rods in reinforced concrete. Tisdall and Oades (1982) showed that the water-stability of aggregates in many soils was dependent on organic materials, with roots and fungal hyphae (i.e. growing root systems) important in the stability of macroaggregates (>250 μm diameter). The numbers of stable macroaggregates decreased with organic matter content as roots and hyphae decomposed, and were related to management practices with increases under pasture and decreases under arable cropping. In contrast, the stability of microaggregates was determined by the content of persistent organo-mineral complexes and by more transient polysaccharides, leading to their relative insensitivity to changes in soil organic matter content caused by different management practices.

The role of different organic materials released by roots in promoting soil aggregation was investigated by Traoré *et al.* (2000) by mixing a luvisol with maize root mucilage, glucose, polygalacturonic acid and a 'model' soluble exudate comprising a mix of glucose, amino acids and organic acids, and incubating for 30 days at 25°C. Although the addition (2 mg C g^{-1} soil) was larger than the concentration of soluble exudates usually found in soils, there were substantial effects on soil structure. All additions increased the stability of water-stable aggregates from 7 days onwards, although the effect of glucose was small relative to the other amendments. The proportion of water-stable aggregates increased with time when mucilage and model exudates were added, but decreased in the polygalacturonic acid treatment. At 30 days, the proportions of stable aggregates were mucilage and model exudates (equal at about 0.7) > polygalacturonic acid (0.47) > glucose (0.36) > control (0.18); the associations between mucilage and model exudates and soil were very difficult to disrupt.

Cycles of soil wetting by rain and drying by soil roots also have a big effect on aggregation. Materechera *et al.* (1992) found that aggregation in two soils (a luvisol and a vertisol), initially dried and sieved to 0.5 mm, was influenced by soil type, plant species and wetting and drying cycles in a controlled experiment over a 5-month period. Denser and more stable aggregates were formed in the vertisol, but for both soil types wetting and drying cycles and higher root length increased the proportions of smaller aggregates and aggregate strength compared with unplanted soil. Root length was in the order ryegrass > wheat > pea, which was also the order of water-stable aggregates >0.25 mm diameter (Table 1.4). They concluded that the heterogeneity of water extraction by roots gave rise to

Table 1.4 Influence of plant species and water regime on the stability of aggregates

	Luvisol			Vertisol		
Plant species	Continuously wet	Wetting/ drying	Mean	Continuously wet	Wetting/ drying	Mean
Control	0.245	0.314	0.280	0.561	0.674	0.617
Pea	0.417	0.495	0.456	0.732	0.876	0.804
Wheat	0.468	0.556	0.512	0.793	0.853	0.823
Ryegrass	0.597	0.644	0.620	0.829	0.909	0.869
Mean	0.432	0.502		0.729	0.822	
LSD, $p = 0.05$						
Water regime	0.013			0.021		
Plant species	0.009			0.015		

The results are expressed as the proportion of the >0.25 mm fraction stable to the wet sieving treatment. From Materechera *et al.*, 1992.

tensile stresses which led to the production of small aggregates; compression also resulted from water extraction by roots leading to aggregates that were denser and of higher tensile strength than those in unplanted soils. Czarnes *et al.* (2000) examined the interaction of exudates and wetting and drying using two model bacterial exopolysaccharides (dextran and xanthan) and a root mucilage analogue (polygalacturonic acid) mixed with soil dried and sieved to 2 mm diameter. Xanthan and polygalacturonic acid increased the tensile strength of the soil over several wetting and drying cycles, suggesting that they increased the bond energy between particles. Polygalacturonic acid was the only material to affect water sorptivity and repellancy of the soil, resulting in slower wetting. Wetting and drying increased sorptivity and decreased repellancy except for the polygalacturonic acid-treated soils. Overall, then, polygalacturonic acid appeared to stabilize rhizosphere soil structure by simultaneously increasing the strength of bonds between particles and decreasing the wetting rate. Some caution is required in extrapolating these results to field conditions because polygalacturonic acid does not replicate exactly the behaviour of root mucilage (see results of Traoré *et al.*, 2000, above), and microbial degradation of polysaccharides released by roots and microbes may restrict their persistence in soils. Nevertheless, the interactive nature of exudates and of wetting and drying on the types and properties of structures produced matches qualitatively with field observations.

References

Amundson, R., Guo, Y. and Gong, P. (2003) Soil diversity and land use in the United States. *Ecosystems* **6**, 470–482.

April, R. and Keller, D. (1990) Mineralogy of the rhizosphere in forest soils of the eastern United States. *Biogeochemistry* **9**, 1–18.

Brouwer, R. (1963) Some aspects of the equilibrium between overground and underground plant parts. *Jaarboek Instituut voor Biologisch en Scheikundig Onderzoek van Landbouwgewassen, Wageningen* **1963**, 31–39.

Brouwer, R. (1983) Functional equilibrium: sense or nonsense? *Netherlands Journal of Agricultural Science* **31**, 335–348.

Brundrett, M.C. (2002) Coevolution of roots and mycorrhizas of land plants. *New Phytologist* **154**, 275–304.

Campbell, R. and Greaves, M.P. (1990) Anatomy and community structure of the rhizosphere. In: *The Rhizosphere* (ed. J.M. Lynch), pp. 11–34. John Wiley & Sons, Chichester.

Czarnes, S., Hallett, P.D., Bengough, A.G. and Young, I.M. (2000) Root- and microbial-derived mucilages affect soil structure and water transport. *European Journal of Soil Science* **51**, 435–443.

Davidson, R.L. (1969) Effect of root/leaf temperature differentials on root/shoot ratios in some pasture grasses and clover. *Annals of Botany* **33**, 561–569.

Davies, W.J. and Zhang, J. (1991) Root signals and the regulation of growth and development of plants in drying soil. *Annual Review of Plant Physiology and Plant Molecular Biology* **42**, 55–76.

Dodd, I.C. (2005) Root-to-shoot signalling: assessing the roles of 'up' in the up and down world of long-distance signalling *in planta*. *Plant and Soil* **274**, 251–270.

Dolan, L. and Davies, J. (2004). Cell expansion in roots. *Current Opinion in Plant Biology* **7**, 33–39.

Dolan, L. and Scheres, B. (1998) Root pattern: shooting in the dark? *Cell & Developmental Biology* **9**, 201–206.

Elick, J.M., Driese, S.G. and Mora, C.I. (1998) Very large plant and root traces from the Early to Middle Devonian: implications for early terrestrial ecosystems and atmospheric $p(CO_2)$. *Geology* **26**, 143–146.

Farrar, J.F. and Jones, D.L. (2000) The control of carbon acquisition by roots. *New Phytologist* **147**, 43–53.

Fisher, M.J., Rao, I.M., Ayarza, M.A., Lascano, C.E., Sanz, J I., Thomas, R.J. and Vera, R.R. (1994) Carbon storage by introduced deep-rooted grasses in the South American savannas. *Nature* **371**, 236–238.

Foster, R.C. (1986) The ultrastructure of the rhizoplane and rhizosphere. *Annual Review of Phytopathology* **24**, 211–234.

Foster, R.C. and Rovira, A.D. (1976) Ultrastructure of wheat rhizosphere. *New Phytologist* **76**, 343–352.

Gensel, P.G., Kotyk, M.E. and Basinger, J.F. (2001) Morphology of above- and below-ground structures in Early Devonian (pragian-Emsian) plants. In: *Plants Invade the Land,* (eds D. Edwards and P.G. Gensel), pp. 83–102. Columbia University Press, New York.

Gifford, E.M. and Foster, A.S. (1987) *Morphology and Evolution of Vascular Plants*, 3rd edn. W.H. Freeman and Co., New York.

Gregory, P.J. (1989) Depletion and movement of water beneath cereal crops grown on a shallow soil overlying chalk. *Journal of Soil Science* **40**, 513–523.

Groff, P.A. and Kaplan, D.R. (1988) The relation of root systems to shoot systems in vascular plants. *The Botanical Review* **54**, 387–422.

Hiltner, L. (1904) Über neuere Erfahrungen und Probleme auf dem Gebiete der Bodenbakteriologie unter besonderer Berücksichtigung der Gründüngung und Brache. *Arbeiten der Deutschen Landwirtschaftlichen Gesellschaft* **98**, 59–78.

Hinsinger, P. and Jaillard, B. (1993) Root-induced release of interlayer potassium and vermiculitization of phlogopite as related to potassium depletion in the rhizosphere of ryegrass. *Journal of Soil Science* **44**, 525–534.

Hinsinger, P., Elsass, F., Jaillard, B. and Robert, M. (1993) Root-induced irreversible transformation of a trioctahedral mica in the rhizosphere of rape. *Journal of Soil Science* **44**, 535–545.

Hornung, M. (1985) Acidification of soils by trees and forests. *Soil Use and Management* **1**, 24–28.

Hunt, R., Warren Wilson, J. and Hand, D.W. (1990) Integrated analysis of resource capture and utilization. *Annals of Botany* **65**, 643–648.

Jackson, M.B. (1993) Are plant hormones involved in root to shoot communication? *Advances in Botanical Research* **19**, 103–187.

Jackson, M.B. (2002) Long-distance signalling from roots to shoots assessed: the flooding story. *Journal of Experimental Botany* **53**, 175–181.

Jaillard, B., Guyon, A. and Maurin, A.F. (1991) Structure and composition of calcified roots, and their identification in calcareous soils. *Geoderma* **50**, 197–210.

Jenny, H. (1994) *Factors of Soil Formation: A System of Quantitative Pedology*. Dover Publications Inc., New York.

Joffe, J.S. (1936). *Pedology*. Rutgers University Press, New Brunswick, NJ.

Johnson, I.R. (1985) A model of the partitioning of growth between the shoots and roots of vegetative plants. *Annals of Botany* **55**, 421–431.

Lupton, F.G.H., Oliver, R.H., Ellis, F.B., Barnes, B.T., Howse, K.R., Welbank, P.J. and Taylor, P.J. (1974) Root and shoot growth of semi-dwarf and taller winter wheats. *Annals of Applied Biology* **77**, 129–144.

Lynch, J.M. (1990) Introduction: Some consequences of microbial rhizosphere competence for plant and soil. In: *The Rhizosphere,* (ed. J.M. Lynch), pp. 1–10. John Wiley & Sons, Chichester.

McCully, M.E. (1995) Water efflux from the surface of field-grown roots. Observations by cryo-scanning electron microscopy. *Physiologia Plantarum* **95**, 217–224.

Materechera, S.A., Dexter, A.R. and Alston, A.M. (1992) Formation of aggregates by plant roots in homogenised soils. *Plant and Soil* **142**, 69–79.

Mayaki, W.C., Teare, I.D. and Stone, L.R. (1976) Top and root growth of irrigated and nonirrigated soybeans. *Crop Science* **16**, 92–94.

Minchin, P.E.H., Thorpe, M.R., Farrar, J.F. and Koroleva, O.A. (2002) Source-sink coupling in young barley plants and control of phloem loading. *Journal of Experimental Botany* **53**, 1671–1676.

Munné-Bosch, S. and Alegre, L. (2004) Die and let live: leaf senescence contributes to plant survival under drought stress. *Functional Plant Biology* **31**, 203–216.

Paris, F., Bonnaud, P., Ranger, J., Robert, M. and Lapeyrie, F. (1995) Weathering of ammonium- or calcium-saturated 2:1 phyllosilicates by ectomycorrhizal fungi *in vitro*. *Soil Biology and Biochemistry* **27**, 1237–1244.

Passioura, J.B. and Gardner, A. (1990) Control of leaf expansion in wheat seedlings growing in drying soil. *Australian Journal of Plant Physiology* **17**, 149–157.

Rackham, H. (1950) *Pliny The Elder, Book XVIII*. Heinemann Ltd, London.

Raven, J.A. and Edwards, D. (2001) Roots: evolutionary origins and biogeochemical significance. *Journal of Experimental Botany* **52**, 381–401.

Raven, P.H., Evert, R.F. and Eichhorn, S.E. (1999) *Biology of Plants*, 6th edn. W.H. Freeman and Co., New York.

Takei, K., Takahashi, T., Sugiyama, T., Yamaya, T. and Sakakibara, H. (2002) Multiple routes communicating nitrogen availability from roots to shoots: a signal transduction pathway mediated by cytokinin. *Journal of Experimental Botany* **53**, 971–977.

Thornley, J.H.M. (1972) A balanced quantitative model for root:shoot ratios in vegetative plants. *Annals of Botany* **36**, 431–441.

Tisdall, J.M. and Oades, J.M. (1982) Organic matter and water-stable aggregates in soils. *Journal of Soil Science* **33**, 141–163.

Torrey, J.G. (1976) Root hormones and plant growth. *Annual Review of Plant Physiology* **27**, 435–459.

Traoré, O., Groleau-Renaud, V., Plantureux, S., Tubeileh, A. and Boeuf-Tremlay, V. (2000). Effect of root mucilage and modelled root exudates on soil structure. *European Journal of Soil Science* **51**, 575–581.

Troughton, A. (1960) Further studies on the relationship between shoot and root systems of grasses. *Journal of the British Grassland Society* **15**, 41–47.

van Noordwijk, M. and de Willigen, P. (1987) Agricultural concepts of roots: from morphogenetic to functional equilibrium between root and shoot growth. *Netherlands Journal of Agricultural Science* **35**, 487–496.

Walker, T.S., Bais, H.P., Grotewold, E. and Vivanco, J.M. (2003) Root exudation and rhizosphere biology. *Plant Physiology* **132**, 44–51.

Watt, M., McCully, M.E. and Jeffree, C.E. (1993) Plant and bacterial mucilages of the maize rhizosphere: comparison of their soil binding properties and histochemistry in a model system. *Plant and Soil* **151**, 151–165.

Weaver, J.E., Hougen, V.H. and Weldon, M.D. (1935) Relation of root distribution to organic matter in prairie soil. *Botanical Gazette* **96**, 389–420.

Wellings, S.R. (1984) Recharge of the Upper Chalk aquifer at a site in Hampshire, England. I. Water balance and unsaturated flow. *Journal of Hydrology* **69**, 259–273.

Yang, J.C., Zhang, J.H., Wang, Z.Q., Zhu, Q.S. and Liu, L.J. (2003) Involvement of abscisic acid and cytokinins in the senescence and remobilization of carbon reserves in wheat subjected to water stress during grain filling. *Plant, Cell and Environment* **26**, 1621–1631.

Young, I.M. and Crawford, J.W. (2004) Interactions and self-organization in the soil-microbe complex. *Science* **304**, 1634–1637.

Zwieniecki, M.A. and Newton, M. (1995) Roots growing in rock fissures: their morphological adaptation. *Plant and Soil* **172**, 181–187.

Chapter 2

Roots and the Architecture of Root Systems

Roots are complex structures that exist in diverse forms and exhibit a wide range of interactions with the media in which they live. They also exhibit a very wide range of associations with other living organisms with which they have co-evolved. Laboratory and field studies have revealed a great deal about this complexity, especially during the last 20 years or so when there have been several national programmes of research around the world focusing on below-ground processes. The purpose of this chapter is to describe the essential anatomical and morphological features of roots as a background to understanding the diverse forms of root systems and their functioning which follow in later chapters.

2.1 Nomenclature and types of root

Terrestrial plants produce roots of many types (e.g. aerial roots, storage roots, etc.), but in this book the focus is on roots that generally either originate from plant tissues located below ground or which function principally below ground or both. Many names are used to describe roots, some of which are confusing and inconsistently used. In part this is because appreciation of the full diversity of root types has emerged only slowly and the terms available differentiate only gross differences. For example, the word 'primary' is used variously in the literature to describe the first root to appear at germination, the first-formed branches of roots, and the largest root. The term 'adventitious root' is also commonly used elsewhere but is not used in this book because such roots are the norm in many plants (Groff and Kaplan, 1988). In this book, the nomenclature suggested by the International Society for Root Research (ISRR) has been adopted wherever possible, although because it has been common practice in the literature to use slightly different nomenclature for monocotyledonous and dicotyledonous plants, these names have also been employed if they were used by the original author of the work cited and are not confusing. For example, because the distinction between seminal (i.e. laid down in the seed) and other root axes is commonplace in the literature, this book adopts the terms seminal and nodal axes for graminaceous species. The ISRR nomenclature attempts to provide a uniform terminology based on the part of the plant from which the root grew rather than relying on knowledge of the tissue from which the root was initiated.

In most plants, the emergence of a root is the first sign of germination. The first root axis arises from cells laid down in the seed and for dicotyledonous plants is called the tap root; this term is also applied to any replacement root that may take over the role of this root if the original tap root is damaged. Subsequent root axes may arise from the mesocotyl or hypocotyl and are called basal roots, with roots arising from shoot tissues above ground called shoot-borne roots (Fig. 2.1a). In graminaceous plants such as maize, wheat and barley, the predominant nomenclature has been to refer to the first root and the other root axes arising from the scutellar node as the seminal axes, with axes arising from the mesocotyl called nodal axes because they arise from nodes at the bases of leaves; other terms such as crown, basal or adventitious have also been used for these axes in the literature (Fig. 2.1b). Although different names have been employed for the axes of monocotyledonous and dicotyledonous plants, these differences do not persist in naming the subsequent branches. Lateral roots (first order laterals) arise from the axes, and from these laterals other, second order, laterals arise. Hackett (1968) proposed an alternative nomenclature for laterals that is also widely used in the literature in which the first branches were termed primary, with subsequent branches being secondary and so on. The axis and its associated laterals are called a root, and all the roots of a plant together form the root system.

In gymnosperms and dicotyledons, the tap root and its associated laterals comprise the root system, and some workers, particularly in the ecological literature, have referred to such plants as tap-rooted species. In graminaceous plants, the multiple root axes and their laterals give the appearance of a more finely distributed or fibrous root system; this latter term is also used by some workers to distinguish types of root system. Some botanical textbooks state that the seminal roots of cereals live for only a short time; this despite many field measurements which show this statement is incorrect (Gregory *et al.*, 1978; McCully, 1999).

Some roots or root parts are specialized for a particular function (Plate 2.1). Such specialisms include the following.

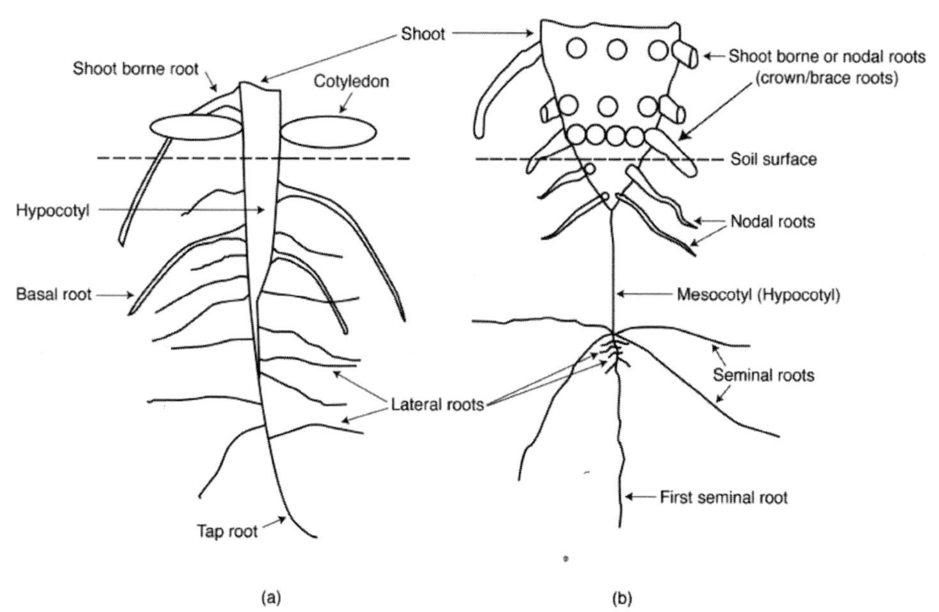

Fig. 2.1 Diagrammatic representation of generic (a) dicotyledonous, and (b) monocotyledonous plants with commonly used root nomenclature. (Redrawn from unpublished work of the International Society for Root Research.)

Storage roots. Parts of roots of plants such as carrot (*Daucus carota*), sugar beet (*Beta vulgaris*), sweet potato (*Ipomoea batatas*) and yams (*Dioscorea* spp.) are specifically adapted to store products photosynthesized in the shoot. The products are synthesized above ground and transported to the root in the phloem where they reside until needed to complete the life cycle. In biennial plants such as carrot and sugar beet, the storage organs are frequently harvested for human use before the life cycle is complete, but if allowed to mature, the stored materials are retranslocated to the shoot where they are used to produce flowers, fruits and seeds. The development of storage roots is similar to that of non-storage roots except that parenchyma cells predominate in the secondary xylem and phloem of the storage roots.

Aerial roots. Aerial or shoot-borne roots originate from a range of above-ground structures. In grasses such as maize, these roots act to prop or brace the stem but when they grow into the soil they may branch and also function in the absorption of water and nutrients (McCully, 1999). Many trees also produce prop roots – including the spectacular banyan tree (*Ficus macrophylla*) – which gradually invades new ground and 'takes over' surrounding trees. In plants such as ivy (*Hedera helix*), the aerial roots cling to objects like walls and provide support to the climbing stem. There are many adaptations in the aerial roots of epiphytes which allow the plants to live on, but not parasitize, other plants. In some genera (e.g. *Ansiella*, *Cyrtopodium* and *Grammatophyllum*) fine aerial roots grow upright to form a basket which collects humus which is then penetrated by other roots which utilize the nutrients. In epiphytic orchids, root tip cells contain chloroplasts as, in many cases, do the cortical cells. These perform photosynthesis and, in the case of leafless orchids of the genera *Taeniophyllum* and *Chiloschista*, are the only photosynthetic tissue of the plant (Goh and Kluge, 1989). A characteristic feature of the aerial roots of orchids is the outer layer of dead cells forming the velamen (Benzing *et al*., 1982). Many of the cells are water-absorbing while others are filled with air and facilitate the exchange of gases with the inner cortex. The physiological role of the velamen is not known with certainty. In some species it appears to aid the uptake of water and nutrients while in others it appears to be a structure for water conservation.

Air roots. In some trees that live in swamps, such as mangroves, parts of the roots develop extensions which grow upward into the air. These air roots or pneumatophores grow above the surface of the water and allow oxygen to be transported to the inner cortex of the root system, and CO_2 to escape from the root interior (Geissler *et al*., 2002). The primary structures allowing gas exchange through the pneumatophores are the lenticels, but several other structures, such as horizontal structures close to the apex, specific to particular species have also been identified (Hovenden and Allaway, 1994; Geissler *et al*., 2002).

Hair roots. These are produced by many heathland plants such as the Ericaceae and Epacridaceae and are the finest roots known (typically 20–70 µm diameter and <10 mm long). They are characterized by a reduction of vascular and cortical tissues, by the absence of root hairs, and by the presence, in what would be the root hair zone of other plants, of swollen epidermal cells occupied by mycorrhizal fungi (Read, 1996). The hairs develop as first order branches on normal root axes or as second or higher order branches on other hair roots (Allaway and Ashford, 1996). The hair roots form a dense fibrous root system and when excavated from soil, the roots have a coating of tightly bound soil particles (a rhizosheath – see section 2.4.2).

Proteoid or cluster roots. These specialized roots were first identified in members of the Proteaceae, but are now known to occur in other species from a diverse range of families (Purnell, 1960; Watt and Evans, 1999). Agriculturally important species include white lupin (*Lupinus albus*) and yellow lupin (*Lupinus cosentinii* Guss.). Cluster roots are bottle-brush-like clusters of hairy rootlets (each rootlet typically 5–10 mm long), and may appear as ellipsoidal-shaped clusters of roots (like small bunches of bananas) at intervals. There are many roots in each cluster, and the clusters may be separated by normally branched regions (Lamont, 2003). The formation of proteoid roots is typically associated with soils that are low in phosphate and/or iron, although lupins will form them in soils to which P fertilizers have been applied (Watt and Evans, 2003). Phosphate uptake in soils low in P is assisted by the exudation of a range of carboxylates with malate, malonate, lactate and citrate being common (Roelofs *et al.*, 2001) (see section 7.2.2 for more details).

Contractile roots. Contractile roots are widely distributed among monocotyledons and herbaceous perennial dicotyledons and serve to pull the shoot closer to the ground or, in bulbs, deeper into the soil. Contraction in many monocotyledonous species occurs when the inner cells of the cortex (contractile parenchyma) expand radially and contract longitudinally (Reyneke and van der Schijff, 1974). The consequence of the contraction in these cells is that the inner vascular tissues and the outer cortical cells become buckled longitudinally and the root appears wrinkled. This mechanism, though, is not universal in all species with contractile roots. Similarly, the factors which induce contractile root activity differ between species. For example, while light and temperature fluctuations appear important in inducing contractile behaviour in species such as *Nothoscordum inodorum*, *Narcissus tazetta* and *Sauromatum guttatum*, this is not the case in the ornamental day lily, *Hemerocallis fulva*, in which contraction appears to be a basic characteristic (Pütz, 2002).

Parasitic roots. In parasitic associations between higher plants, the connection between two plants is established via haustoria formation by the parasite (see section 6.3.3). For example, in *Striga* species, when the radicle makes contact with a host root, elongation ceases and the tip of the radicle swells to form a pre-haustorium. Sticky hairs develop on this structure, which results in parasite–host adhesion. After this, intrusive cells develop at the root tip, which penetrate the cortex and endodermis of the host root by secreting enzymes that cause separation of the host cells rather than effecting intra-cell penetration. Once in the stele, there is a rapid development of links between the parasite and the host xylem (Parker and Riches, 1993).

2.2 Root structure

The anatomy of roots is complex with very variable structures both between and within plant species. There are considerable differences among species (especially between angiosperms and gymnosperms), among habitats, and along the length of individual roots. Common examples of differences in structure include death of the epidermis and in some species the entire cortex, development of aerenchyma in the cortex, development of the endodermis and exodermis (with their Casparian bands, suberin lamellae, and thickened, modified walls), and the production of a periderm (Steudle and Peterson, 1998). It is important to note that most published work on root structure has been conducted with young plants grown in 'clean' environments. Whether such studies are useful in describing how a root growing in soil will look or understanding how a root system growing in soil functions are topics of lively scientific debate (e.g. McCully, 1995, 1999). This section draws mainly from literature on juvenile plants grown in solutions or sand.

2.2.1 Primary structure

In their primary stage of growth, roots show a clear separation between three types of tissue systems – the epidermis (dermal tissue system), the cortex (ground tissue system) and the vascular tissues (vascular tissue system). In most roots, the vascular tissues form a central cylinder, but in some monocotyledons they form a hollow cylinder around a central pith (Esau, 1977). These three tissue systems form a range of cells visible in transverse and longitudinal sections (Fig. 2.2).

Dermal tissue. In young roots, the epidermis is a specialized absorbing tissue containing root hairs which are themselves specialized projections from modified epidermal cells known as trichoblasts (Bibikova and Gilroy, 2003). Root hair formation is a complex process (see section 2.3.3) regulated by many genes and is also responsive to a variety of environmental stimuli. Root hairs markedly extend the absorbing surface of the root but they are often considered to be short-lived and confined to the zone of maturation. A thin cuticle may develop on the epidermis, and in some herbaceous species the cell walls thicken, suberin is deposited in them, and the epidermis remains intact for a long time as a protective tissue.

Ground tissue. In young plants, the cortex usually occupies the largest volume of most roots and consists mainly of highly vacuolated parenchyma cells with intercellular spaces between. The innermost cell layer differentiates as an endodermis and one or more layers at the periphery may differentiate as a hypodermis/exodermis.

Roots that undergo significant amounts of secondary growth (see next section), such as gymnosperms, often shed their cortex early in life, but in other species, the cortical cells develop secondary walls that become lignified. The intercellular spaces allow the movement of gases in the root and under particular conditions may develop into large lacunae (aerenchyma) in some species (e.g. rice, see section 5.4.3). The cortical cells have numerous interconnections both via the cell walls and via the plasmodesmata which link the protoplasm of each cell (Roberts and Oparka, 2003). Substances can move across the root, then, either via the cell walls (the apoplastic pathway) or via the cell contents (the symplastic pathway); these pathways are assumed to be important in the internal transport of water and nutrients (see section 4.2.3). In some species such as grasses, the epidermis may be shed together with all or part of the cortex as a normal part of the ageing process of roots or in response to adverse soil conditions (Wenzel and McCully, 1991). For example, when wheat roots were grown in dry soil, the upper portion of seminal axes had collapsed epidermal and cortical cells (Brady *et al.*, 1995). On re-wetting, dormant lateral roots grew rapidly to take up water and N but the seminal axes themselves did not appear capable of significant N uptake.

In contrast to the rest of the cortex, the endodermis lacks air spaces and the cell walls contain suberin in a band (the Casparian strip) that extends around the radial and transverse cell walls, which are perpendicular to the surface of the root. Three stages in the development of the endodermis can be discerned. First, radial and transverse endodermal cell walls are impregnated with lipophilic and aromatic compounds (Casparian strips) which restrict,

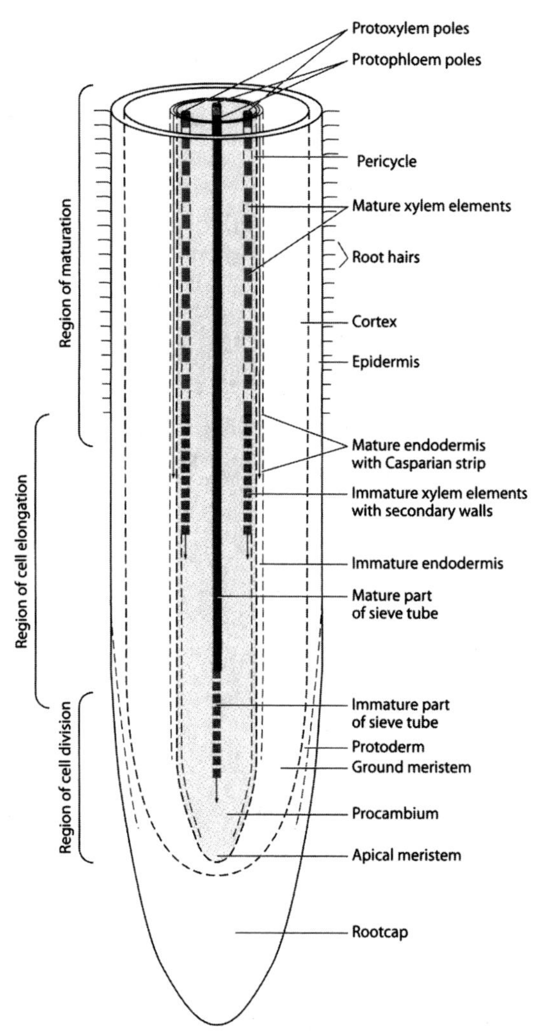

Fig. 2.2 Diagrammatic representation of the early stages of primary development of a root. The region of cell division extends for some distance behind the apical meristem and may overlap with the regions of cell elongation and cell differentiation/maturation. (Based on original work by Esau (1941); redrawn and reproduced with permission from Torrey, *American Journal of Botany*; Botanical Society of America Inc., 1953.)

but do not altogether stop, apoplastic movement of water and ions (see section 4.2.3). The second stage occurs especially in species in which the epidermis and cortex are shed and lateral roots emerge, and is characterized by the deposition of a thin, lipophilic suberin lamella to the inner surface of radial and tangential walls of endodermal cells. Finally, there may be considerably more deposition on the inner tangential and radial cell walls evident as U-shaped wall thickening (Schreiber *et al.*, 1999). These changes in the endodermis begin opposite the phloem strands and spread towards the protoxylem. Opposite the protoxylem, the cells may remain thin-walled with Casparian strips; these are called passage cells (Peterson and Enstone, 1996).

In many angiosperms, an exodermis differentiates from peripheral cortical cells. In a survey of >200 angiosperm species, 94% of all plants possessed a hypodermis and about 90% had a hypodermis with Casparian strips (Perumalla and Peterson, 1990; Peterson and Perumalla, 1990). In a small proportion of plants, the exodermis comprised more than one cell type with long, suberised cells and short cells in which deposition of suberin was delayed; these short cells act as passage cells for water and ions (Peterson, 1991). As cells in the epidermis mature, a hypodermis may differentiate in the outer cortical cells (Plate 2.2). Like the young endodermis, these cell walls are impregnated with lipophilic and aromatic compounds. Moreover, in response to environmental stresses such as drought, aeration and potentially toxic metals, Casparian strips and suberin lamellae form in the hypodermis just as occurs in the first stage of endodermis formation. A hypodermis with Casparian strips is called an exodermis (Perumalla and Peterson, 1986), and this forms a barrier of variable resistance to the flow of both water and nutrients across the root (Hose *et al.*, 2001). In summary, Casparian strips are a characteristic feature of primary endodermal cell walls, whereas they only form in hypodermal cell walls as a reaction to environmental factors (Schreiber *et al.*, 1999).

Vascular tissue. The vascular cylinder (stele) consists of vascular tissues (xylem and phloem) and one or more layers of non-vascular tissues, the pericycle, which surrounds the vascular tissues (Fig. 2.2).

The central vascular cylinder of most dicotyledonous roots consists of a core of primary xylem from which ridge-like projections of xylem extend towards the pericycle (Esau, 1977). Between the ridges are strands of primary phloem. If the xylem does not differentiate in the centre of the root (as happens in many monocotyledons), a pith of parenchyma or sclerenchyma (parenchyma cells with secondary walls) is present. The number of xylem ridges varies between species and among roots of the same species (see McCully, 1999, for a description of xylem development in maize) and this variation is captured by referring to roots as diarch, triarch, tetrarch, etc., depending on the number of ridges. The first xylem elements to lose their cell contents and mature (the protoxylem) are those next to the pericycle, while those closer to the centre are the typically wider metaxylem elements which mature later and commonly have secondary walls with bordered pits. As with the xylem, the phloem shows a centripetal order of differentiation with protophloem nearest the pericycle and metaphloem nearer the centre. Companion cells accompany the metaphloem but are less frequent in the protophloem, although in grasses each protophloem element is associated with two companion cells giving a consistent, symmetrical pattern in transverse sections. In contrast to the xylem, the phloem consists of living cells.

The pericycle is composed of parenchyma cells with primary walls but these may develop secondary walls as the plant ages. Lateral roots arise in the pericycle (see section 2.3.2), and in roots undergoing secondary growth, the pericycle contributes to the vascular cambium opposite the protoxylem and generally gives rise to the first cork cambium.

Maturation of the xylem so that it can conduct water may take some time and lignification is not a good indicator of maturity (McCully, 1995, 1999). For example, St Aubin *et al.* (1986) found that the large vessels of actively growing maize root did not mature and become open for conduction until at least 150 mm, and sometimes >400 mm behind the root tip. The narrower vessels started to mature about 40–90 mm from the tip and the very narrow protoxylem at about 10–20 mm (McCully, 1999). Similar results were summarized for other species including barley, banana, soyabean and wheat by McCully (1995), with a range of 110–1300 mm for the distance from the root tip at which late metaxylem vessels became open tubes. Immature, living xylem cells are highly vacuolated with a thin layer of cytoplasm. They do not conduct water but accumulate ions, especially potassium, in high concentration in the vacuole (McCully, 1994). The point of transition from closed to open (living to dead) large xylem vessels is important because it affects both water and nutrient uptake and transport in roots. Detached roots (used in many laboratory studies) tend to differentiate their tissues rapidly so that their behaviour may not represent that of more slowly maturing roots grown in field conditions.

2.2.2 Secondary structure

Secondary growth is characteristic of roots of gymnosperms and of most dicotyledons but is commonly absent from most monocotyledons. Secondary growth consists of (i) the formation of secondary vascular tissues dividing and expanding in the radial direction, and (ii) the formation of periderm, composed of cork tissue (Esau, 1977). Growth within the secondary vascular system is driven by the cambium, which consists of two morphologically distinct cell types: fusiform initials (greatly elongated in the axis of the root) and ray initials (which are cuboid) (Chaffey, 2002). Both cell types are thin-walled and highly vacuolated. The process starts with the initiation of vascular cambium by divisions of procambial cells that remain undifferentiated between the primary xylem and primary phloem. Thus, depending on the number of xylem and phloem groups present in the root, two or more regions of cambial activity are initiated. Soon the pericycle cells opposite the protoxylem elements also divide and become active as cambium so that cambium quickly surrounds the core of xylem (Plate 2.3). The vascular cambium opposite the phloem strands begins to produce secondary xylem toward the inside, so that the strands of primary phloem are displaced outwards. By the time that the cambium opposite the protoxylem is actively dividing, the cambium is circular and the primary phloem and xylem have been separated. By repeated divisions, secondary xylem and secondary phloem are added and files of parenchyma cells within these form rays. As the secondary xylem and phloem increase in width, so the primary phloem is crushed and disappears.

Periderm (analogous to bark) formation usually follows the initiation of secondary xylem and phloem formation to become the outer, protective covering of the root. Divisions of pericycle cells increase the number of layers of pericycle cells. In the outer cell layers, cork cambium is formed which produces a layer of cork on the outer surface and phelloderm toward its inner surface; these three tissues constitute the periderm. The remaining cells of the pericycle may form tissue that resembles a cortex. Lenticels may differentiate in the periderm to facilitate the passage of gases into and out of the root. With the formation of the periderm, the endodermis, cortex and epidermis are isolated from the rest of the root, die, and are sloughed off. A woody root remains.

While the preceding description applies to many roots, secondary growth may also result in roots with different appearance to that described above. For example, Plate 2.4 shows a young root of *Catalpa speciosa* in which appreciable secondary growth has occurred. The endodermis has expanded and is intact, and the cortex has outer cells which are differentiating to form a periderm.

2.3 Extension and branching

2.3.1 Extension

The extension growth of roots occurs in the apical regions of roots, and it is this extension of root axes and laterals into new regions of soil that expands the resource base and anchorage ability of the plant. Figure 2.3 shows that the zone of elongation is confined to the apical meristem where cell division occurs, and the region immediately behind this where predominantly longitudinal cell elongation occurs. Towards the tip of the apical meristem is a zone referred to as the 'quiescent centre' where cell division occurs rapidly during very early root growth but then becomes infrequent.

The cells in the root meristem are mainly cytoplasmic and have no clearly defined central vacuole (Barlow, 1987a). The patterns of cell division in this region are precisely regulated and determine the future characteristic form of the root (Fig. 2.3). Many of the cells divide in planes that are parallel to the main axis of the root and, in so doing, create files of cells which subsequently divide transversely to the axis of the root, thereby increasing the number of cells in each file (Barlow, 1987b). Groups of cells (packets) within a cell file can easily be seen and their ontogeny traced. For example, Barlow (1987b) followed the morphogenesis in the tap root of maize and by counting the number of cells in packets determined that the period between each round of cell division was fairly constant except in cortical and stellar cells around the quiescent centre where there was evidence for a steep gradient of rates of cell proliferation.

Cell division does not result in extension but rather provides the raw materials for subsequent cell expansion and so does not, itself, drive growth. In the elongating zone, outside the meristem, cells increase in length accompanied by a large increase in the size of the vacuole and an increase in the area of the lateral walls of the cell. Expansion of root cells requires the co-ordination of many processes including the control of ion (especially potassium) and water uptake into the vacuole, the production of new wall and membrane materials, and the increase in size of the cytoskeleton (Dolan and Davies, 2004). Root elongation occurs, then, as the sum of the individual cell expansions along a file of cells (i.e. in a single directional axis). During cell expansion, changes in cell wall properties enable the walls to be strong enough to cope with the internal pressure of the growing cell but flexible enough to allow growth (Pritchard, 1994). Cell walls can loosen rapidly during periods of accelerating growth and, conversely, tighten after exposure to stresses such as low temperature and high soil strength in ways that are still being explored. The cell wall consists of cellulose microfibrils, hemicellulose and pectin, together with various proteins. The microfibrils both provide a framework for the assembly of other wall components and influence the orientation of cell growth through their interaction with microtubules comprised of polymers of the tubulin protein (Barlow and Baluška, 2000). Cell expansion results from internal hydrostatic pressure (turgor) which expands the cell wall (Pritchard, 1994). However, the turgor pressure has no preferential direction so that the preferential longitudinal expansion which predominates in the zone of elongation is believed to be a consequence of differential depositions or modifications of cell wall materials mediated by microtubule-directed processes (Barlow and Baluška, 2000). Microtubules, together with actin microfilaments, form a cytoskeleton that confers structural order and stability to the

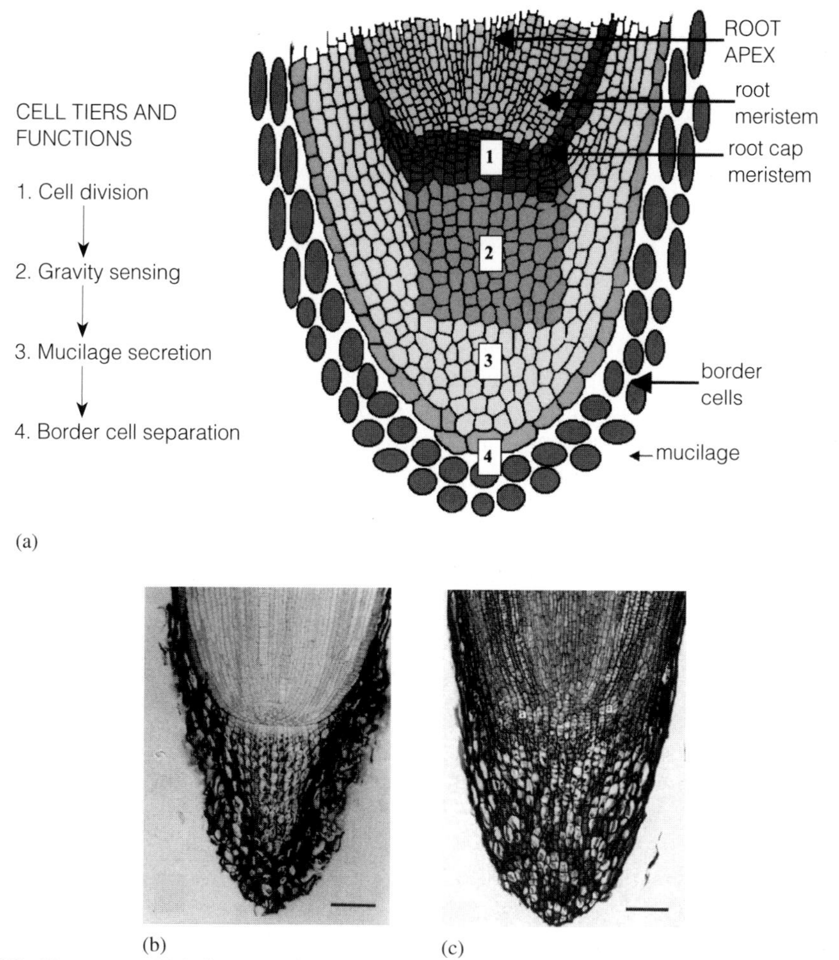

Fig. 2.3 The root cap. (a) A diagrammatic representation showing the root and root cap meristems. As cell division occurs in the root cap meristem so tiers of cells are displaced towards the periphery of the cap. As each tier is displaced, previous functions cease and new functions are initiated within the progressively differentiating cells. (Reproduced with permission from Hawes *et al.*, *Journal of Plant Growth Regulation*; Springer Science and Business Media, 2003.) (b) Micrograph of the root cap of *Zea mays* showing closed root cap meristem. (c) Micrograph of *Pisum sativum* showing an open root cap meristem. Scale bars: 100μm. (Figs b and c reproduced with permission from Barlow, *Journal of Plant Growth Regulation*; Springer Science and Business Media, 2003.)

cell interior, and also convey information to the peripheral regions of the cytoplasm where much cellular growth is controlled.

Van der Weele *et al.* (2003) used four species of flowering plant to demonstrate that there were two distinct regions of elongation at the root tip. In the apical, meristematic region, rates rose gradually with distance from the quiescent centre whereas in the zone of elongation rates increased rapidly with distance. Relative elongation rates in both zones were constant in each zone but changed in a step-wise manner from low in the meristem to values that were typically some three to five times greater in the zone of elongation. At the distal end of the elongation zone, relative elongation rate decreased to zero. These results imply that cell division and cell elongation parameters are regulated uniformly. Individual roots exhibit a determinate pattern of growth in which there is an initial period of accelerating elongation soon after emergence, followed by a period of steady growth, which is in turn followed by a decelerating phase leading to cessation of elongation (Chapman *et al.*, 2003). The organization of the apical meristem referred to above (Fig. 2.3) is not constant in all species throughout their life and Chapman *et al.* (2003) suggest that determinacy and organization of the apical meristem are linked. The cells of the apical meristem of roots may be capable of cycling for only a limited number of times, leading eventually to no new cells being produced and the cessation of axis elongation. As the root reaches its determinate length, so the apical meristem loses its organization. For example, the apical organization of five plant species with closed meristems changed to intermediate open during the deceleration phase (Chapman *et al.*, 2003). The frequency of plasmodesmatal connections between cells also decreases in this phase (suggesting reduced intercellular communication) and finally the cells become vacuolated and lose their meristematic identity.

2.3.2 Branching

Lateral roots (branches) originate in the pericycle, some distance behind the main root apices in partially or fully differentiated root tissues. Because they arise from deep within the root, they are described as arising endogenously and must traverse other living tissues before they emerge from the parent root (McCully, 1975) (Plate 2.5a and b). The detail of lateral root formation has only been studied in a few species, so much remains uncertain. In maize, the first indications of lateral formation are changes in the cytoplasm and cell walls of a few pericycle and stellar parenchyma cells close to a protoxylem pole, together with changes in endodermal cells tangential to the activated pericycle cells (McCully, 1975). Derivatives of both the parent pericycle and parent endodermis contribute to the tissues of the new meristem, although in many cases the derivatives of the endodermis are short-lived. In some plants the derivatives of the parent stellar parenchyma also contribute. In maize, the parent endodermis gives rise to the epidermis of the lateral and to the root cap. The new primordium grows through the root cortex, possibly using mechanical force and/or various enzymes to disrupt the cortical cells in its path. The lack of connectivity between the emerging lateral and the cortex of the parent may create a space into which microorganisms and pathogens can enter. Initially, the stellar tissues of the lateral and its parent are not connected but later they join as derivatives of the intervening parenchyma cells differentiate into xylem and phloem. Because laterals are initiated close to protoxylem, linear arrays of laterals appear along the root; this is especially obvious in many dicotyledons with small numbers of xylem poles.

Significant advances have been made in understanding the factors controlling lateral root initiation and emergence using the model plant *Arabidopsis thaliana*. In *Arabidopsis*, laterals are derived from a subset of pericycle cells adjacent to the two xylem poles known as founder cells. Genetic and physiological evidence suggests that auxin (particularly indole-3-acetic acid, IAA) is required to facilitate lateral root initiation and development (Casimiro *et al.*, 2003). For example, using the *stm 1* mutant, Casimiro *et al.* (2001) demonstrated that transport of IAA in the root to shoot direction (basipetal transport) was required during the initiation phase while leaf auxin transported to the root (acropetal transport) was required during the emergence phase. The linkage between lateral root development and auxin derived from the shoot apex (Reed *et al.*, 1998) may provide a means by which root and shoot response to environmental stimuli can be co-ordinated. Regulation of acropetal transport may be a mechanism by which environmental conditions perceived by the shoot can be communicated to control root development and growth. Other plant-produced chemicals, too, play a part in the development of lateral roots. Again in *Arabidopsis*, the plant hormone abscisic acid (ABA) inhibited lateral root development at the time at which the lateral root primordium emerged from the parent root, a response that was mediated by an auxin-independent pathway (De Smet *et al.*, 2003). Nutrients such as nitrate also have a large effect on lateral root development inducing proliferation in zones that are nitrate-rich (see section 5.5.1), and organisms such as mycorrhizal fungi can also modify branch numbers (e.g. Yano *et al.*, 1996).

Most studies of lateral root development and growth have been performed with either young plants and/or in growing media other than soil. The most comprehensive account of soil-grown roots is for maize by McCully and her co-workers, and summarized by McCully (1999). In contrast to laboratory-grown roots, the first order laterals of field-grown plants are short (mode $\leq$30 mm), with only about 2% exceeding 100 mm (Varney *et al.*, 1991; Pagès and Pellerin, 1994). Most roots reach their final length in <2.5 days, shortly before which the root cap is lost and tissues differentiate right to the tip with the surface cells at the apex often developing root hairs (Varney and McCully, 1991). These determinate roots persist for the life of the crop although they may become shorter if the root dies back from the end. The number of branches per unit length of axis in the upper part of the soil profile was also consistent (average 12 per 10 mm) and typical of the values found in other studies (7–12 per 10 mm) (McCully, 1999). Only about one-third of the first order laterals themselves branch, and the laterals produced are very short and sparse. Overall, the laterals constitute up to 30 times the length of the axial roots (Pagès and Pellerin, 1994). Varney *et al.* (1991) suggest that such short lengths may be a characteristic of maize rather than other cereals or grasses, yet a re-working by Gregory (1994) of data by Weaver *et al.* (1924) for winter wheat also demonstrated short lengths with an estimated mean length per root member (mainly first order laterals) of 10–23 mm between 10 and 70 days after planting. In maize, the branches have an epidermis, cortex and narrow stele. Much of the epidermis remains alive in moist soils, even in old roots, as does the cortex which contains the two specialized layers, hypodermis and endodermis, both with Casparian strips and suberized secondary walls (McCully, 1999). The diameter of xylem vessels ranges from <6 to about 60 μm, so that the axial conducting capacity of these branches varies by five orders of magnitude (Varney *et al.*, 1991).

2.3.3 Root hairs

Behind the zone of elongation is a zone of maturation (Fig. 2.2) in which root hairs are produced as specialized projections from modified epidermal cells. In many plant species (nearly all dicots, some monocots, and most ferns), all epidermal cells of the root seem capable of producing a hair, whereas in others there are cells that have the potential to become root hairs (trichoblasts), and others seem incapable of this development (atrichoblasts). The latter group of plants can also be divided into two. In the first group, root hairs form in the smaller cell produced in an asymmetrical cell division in the meristem, while in the second group (e.g. the Brassicaceae), root epidermal cells occur in files composed of either trichoblasts or atrichoblasts with the trichoblasts overlying the junction of two cortical cells (Gilroy and Jones, 2000; Bibikova and Gilroy, 2003) (Fig. 2.4). Studies with *Arabidopsis* on Petri dishes show that after epidermal cell fate has been specified in the meristem, the trichoblast elongates in the elongation zone and then growth is localized on a side wall as a root hair is initiated.

Initiation and subsequent growth of root hairs is under genetic, hormonal and environmental control with many regulators acting at several stages of development (see Bibikova and Gilroy, 2003). Initiation is evident as a bulge begins to form in the cell wall associated with microtubule rearrangements (Fig. 2.4b). The site of initiation is also precisely regulated so that in *Arabidopsis*, for example, root hairs always form at the end of the cell nearest the root apex (Gilroy and Jones, 2000). After a short transition period following initiation, the tip of the hair begins to grow. Deposition of new plasma membrane and cell wall material occurs at the elongating tip leading to a hair-like structure. Regulation of the elongation process appears closely linked to the gradient of cytoplasmic Ca^{2+} concentration within the cell which is much greater at the tip of the hair. The size of the Ca^{2+} gradient correlates well with the growth rate of individual root hairs and when hairs reach their final length, the gradient disappears (Gilroy and Jones, 2000). In this regard, root hairs demonstrate similar behaviour to other plant structures, as calcium concentration gradients are also important in the growth of other hair-like structures such as algal rhizoids and pollen tubes, and in the response to *Nod* factors during the formation of infection tubes in root hairs by *Rhizobium* (see section 6.2.1).

The development of root hairs is also greatly influenced by the surrounding environment. For example, when *Arabidopsis* was grown in a P-deficient soil, root surface area was increased sevenfold compared with plants grown under P-sufficient conditions, and root hairs constituted 91% of the total root surface area (Bates and Lynch, 1996). Availability of P also affects the rate of growth of root hairs, but there was no stimulation of root hair growth in *Arabidopsis* by deficiencies of K, B, Cu, Fe, Mg, Mn, S and Zn in the surrounding medium (Gilroy and Jones, 2000).

Root hairs can vary in length and frequency along a root but are typically 0.1–1.5 mm long, 5–20 μm in diameter, and vary from 2 per mm^2 on roots of some trees to 50–100 per mm of root length in some grasses and Proteaceae. Usually, the size of the root hair zone on roots is short because root hairs have a short life of a few days or weeks. For example, Fusseder (1987) found that the cytoplasmic structure of root hairs in maize grown in sand started to break down after only 2–3 days, although the walls can remain intact for some time after the hair has ceased to function. Nuclear staining with acridine orange suggested that the average life of the root hairs was 1–3 weeks. Lifespan will, though, be affected by several environmental factors including soil water status and nutrition.

Root hairs play an important role in root/soil contact through the formation of rhizosheaths (section 2.4.2) and in the acquisition of water and nutrients. High levels of H^+-ATPase activity have been demonstrated in root hairs together with their involvement in the uptake of calcium, potassium, nitrate, ammonium, manganese, zinc, chloride and phosphate (Gilroy and Jones, 2000). Several studies have shown the importance of root hairs in contributing to differences in P uptake between plant species and genotypes (e.g. Itoh and Barber, 1983; Gahoonia *et al.*, 1997) (see section 8.1.2), largely because long root

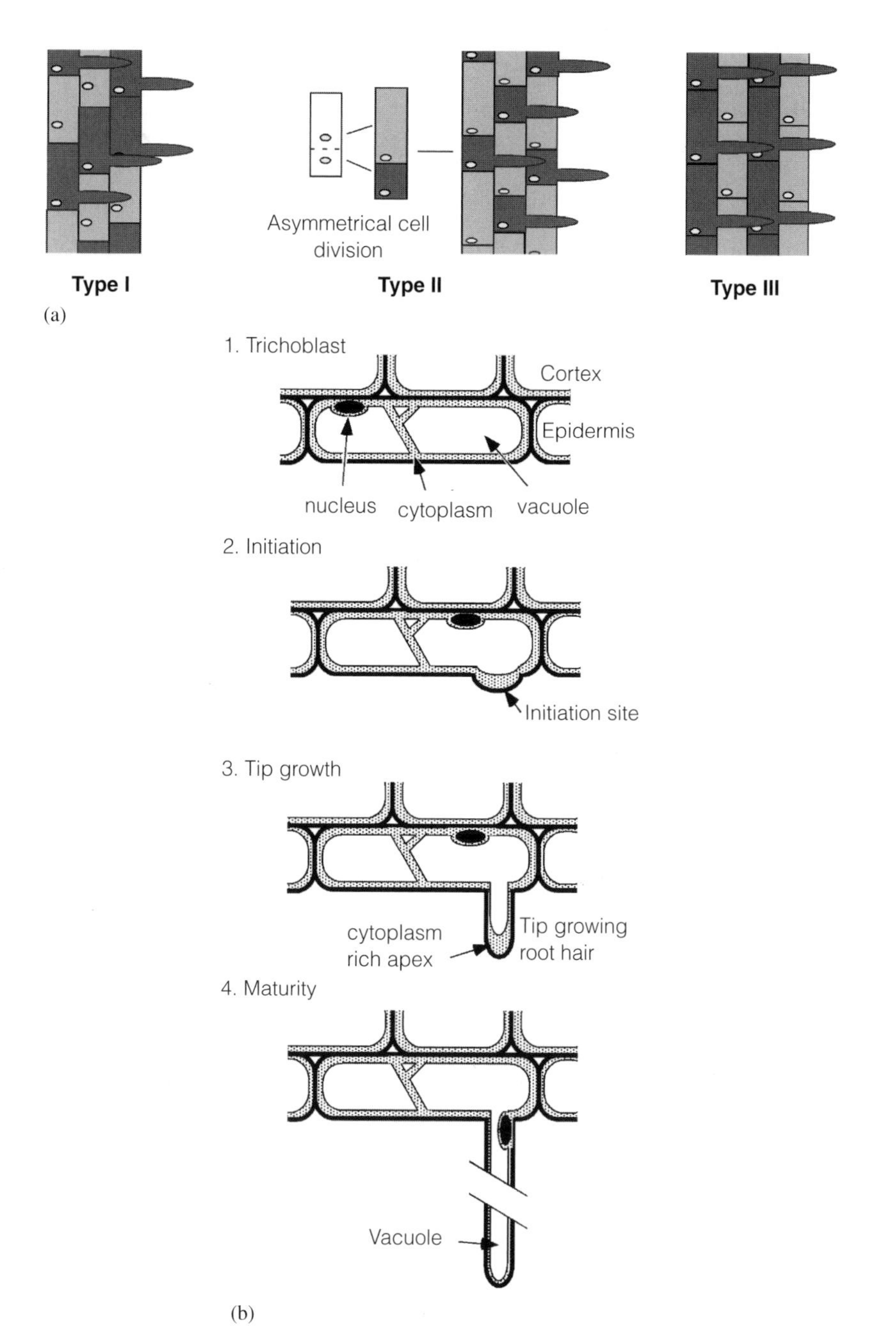

Fig. 2.4 Patterns of root hair development. (a) Three types of root epidermal differentiation occur in plant roots, trichoblast shown in dark grey and atrichoblast in light grey: type I, any cell can form a root hair; type II, the trichoblast is the product of asymmetrical cell division; type III, differentiation produces files of either trichoblasts or atrichoblasts. (Reproduced with permission from Bibikova and Gilroy, *Journal of Plant Growth Regulation*; Springer Science and Business Media, 2003.) (b) Cross-section of a trichoblast during root hair development showing nuclear movements and changes in the organization of the cytoplasm at each developmental phase. (Reproduced with permission from Gilroy and Jones, *Trends in Plant Science*; Elsevier, 2000.)

hairs of sufficient density can intercept P diffusing towards the root at some distance from actual root surface. Using a root hairless barley mutant (*bald root barley, brb*), Gahoonia and Nielsen (2003) demonstrated the important role that root hairs play in plant nutrition and plant survival when grown in soil. Wild-type barley with root hairs depleted twice as much P from the rhizosphere soil as *brb*, and in low-P soil *brb* did not survive after 30 days whereas the wild-type continued to grow. In high-P soil both the wild-type and *brb* maintained their growth, suggesting that root hairs are less essential under such conditions. Root hairs also appear to have effective internal mechanisms for the transport of P. Phosphate transporter genes of the *Pht1* family are expressed at high levels in the root hairs of barley plants grown in P-deficient conditions, consistent with a major role for these genes and root hairs in the uptake of inorganic P from soil solution (Plate 2.6) (Schünmann *et al.*, 2004). It is noteworthy that the phosphate transporter promoter was not present in other epidermal cells, suggesting that P uptake is primarily via root hairs rather than the whole of the root surface.

2.4 The root tip

2.4.1 The root cap and border cells

The root cap is an almost universal feature of growing plant roots with known exceptions among a few aquatic or marshland species, and parasitic plants whose tips develop as specialized penetrating haustoria (Barlow, 2003). Many species, including a number of annuals, normally have a cap but lose it as they become older and in tree roots that are surrounded by ectomycorrhiza, the roots become determinate. In older botanical books the cap is often portrayed as a mucilage-covered, bullet-shaped ram, but it is now appreciated that the cap is a much more complex and dynamic region. In addition to providing protection against mechanical damage to the root tip, the root cap is also involved in the simultaneous perception of several signals (e.g. pressure, moisture and gravity), and modifies or 'engineers' the properties of the surrounding soil. The root cap, then, responds to the soil environment to: (i) control the direction of movement; (ii) facilitate penetration into soil; and (iii) determine the microbial environment around the root (Hawes *et al.*, 2003).

While elongating roots possess a cap, the degree of continuity between the root meristem and the cap meristem differs (Fig 2.3b and c). In some species, the cells of the cap appear contiguous with those of the main part of the root (an open meristem, e.g. pea), while in others there appears to be no continuity and the cap develops separately from the root (a closed meristem, e.g. maize). A third type of meristem, intermediate open, has also been proposed with a relatively disorganized centre but with specific initials for the epidermis and root cap (Chapman *et al.*, 2003). There is a common form of development in root caps (Barlow, 1975), although the size of the cap may vary within a root system depending on the order of root and the environment. Caps of terrestrial plants are generally conical in shape and have a meristem at their proximal end from which new cells are derived. There is a non-meristematic, central columella zone containing cells which synthesize starch grains (statocytes) which participate in sensing gravity (Hawes *et al.*, 2003), surrounded by a lateral zone and an outer layer of mucilage-secreting cells (Fig. 2.3a). The cells of the root cap are generated in the cap meristem and differentiate progressively through a series of morphological changes that are correlated with specialized functions. As differentiation progresses, the starch grains in the columella cells degrade, and a polysaccharide mucilage is produced which is exuded to form a water-soluble coating around the cap (McCully and Boyer, 1997). As cells reach the outside of the cap, enzymes solubilize specific interconnections between cell walls to release single cells or small groups of cells with intact cell walls. These 'border cells' remain in the vicinity of the cap, often embedded in mucilage, and together fulfil many roles in the biology of the root and its interactions with the soil. Marked diversity occurs in the rate of border cell production, but it is a highly regulated process controlled by endogenous and environmental signals. Hawes *et al.* (2003) provide values ranging from about 100 in a 24-hour period for Solanaceae, to about 200 in many cereals, to about 4000 in many legumes, to 8000–11 000 in several Pinaceae.

Barlow (2003) concluded that the release of cap cells is under genetic control, and that the size of the root cap has probably evolved so that it is large enough to assist the passage of the root through the soil and to provide sufficient signals for the orientation of root growth (various tropisms), but not so large that detached border cells can smother the tip and deplete it of oxygen. The rate of cell detachment to form border cells may also be influenced by the rate of root elongation, thereby ensuring that the tip can advance continually through a coating of border cells and mucilage. Cell division in the cap meristem and cell separation at the cap periphery are regulated by two genes, *psugt1* and *rcpme1*, respectively. When gene expression is inhibited, then border cell development ceases (Hawes *et al.*, 2000).

The root cap has no boundary cuticle (cf. the epidermis), so that the outer cells are in direct contact with the soil. Barlow (1984) noted that it seemed that merely being at or near the surface of the cap was a sufficient condition to initiate enzyme induction, cell separation and mucilage production. The growth of underlying cells induces a stretch-activated response that results in mucilage production and cell separation, with the act of cell detachment resulting in new cells being uncovered which themselves are induced to detach, and so on. The surface of the cap is, then, like a perpetual open wound (Barlow, 2003). Typically, it takes about 3–10 days (6–7 days being common) under laboratory conditions for cells to flow from the cap meristem to the outer flanks of the cap (Barlow, 2003). Overall, the border cell/mucilage capsule links the external soil environment to cells in the cap and, thence to cells in the root meristem to form an intimate system of communication between the plant and the soil.

Our understanding of the processes and mechanisms involved in environmental perception and signal transmission by the root cap are only just beginning. Roots change their direction and rates of growth in response to a wide range of stimuli (see Chapters 5 and 6) including gravity, light, water, touch, nutrients, toxic metals and microorganisms. Roots are positively gravitropic and grow towards nutrients and water. The growth towards water (hydrotropism) modifies the response to gravity but the mechanisms of these responses and interactions with other stimuli are yet to be resolved (Takahashi, 1997). Conversely, roots grow away from toxic metals such as aluminium. For example, Hawes *et al.* (2000) show that the tap root of common bean (*Phaseolus vulgaris*) placed on agar containing no aluminium continued to grow downward, but when a toxic concentration of aluminium was added to the agar, the root grew straight into the air opposing the usual positive gravitropic response.

Border cells appear to play a major role in lubricating the passage of the root cap through the soil (Iijima *et al.*, 2000; see below for more details), and in the response of roots to the soil microflora. Hawes *et al.* (2003) document the biological effects of border cells on bacteria, fungi (pathogenic and non-pathogenic) and nematodes, and show that they produce and release an array of specific extracellular chemicals that can attract, repel and control growth and gene expression in some soil organisms. For example, Fig. 2.5 shows the ability of root border cells of pea to attract and immobilize the root knot nematode *Meloidogyne incognita* when placed on water agar. This immobilization is reversible within a few hours or days depending on conditions (Zhao *et al.*, 2000). No such attraction to root border cells occurred with common bean, and in alfalfa, a range of responses was obtained depending on cultivar but consistent with the known susceptibility of the cultivar to nematode damage; the resistant cultivar, Moapa 69, repelled nematodes.

In summary, to quote Hawes *et al.* (2003), 'The root cap is a multifunctional, molecular relay station that not only detects, integrates and transmits information about the environment to appropriate plant organs, but also functions to specifically modulate properties of the soil habitat in advance of the growing root. The cap maintains its own independent developmental patterns in response to the environment while simultaneously directing movement generated by the root meristem and region of elongation'.

2.4.2 Mucilage

The gel-like mucilage secreted by the cells of the root cap contributes to many interactions between the plant and the soil including root penetration, soil aggregate formation, microbial dynamics and nutrient cycling (McCully, 1999). As the root extends through the soil, so mucilage and associated root cap cells are left behind along the root–soil interface

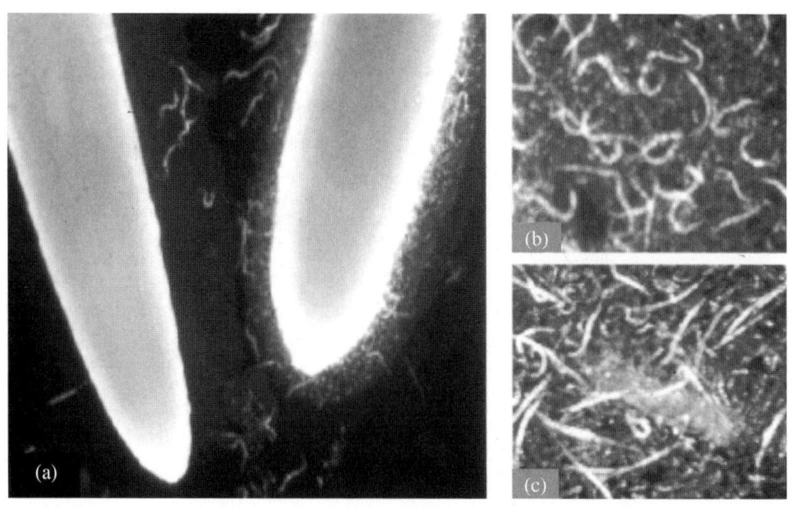

Fig. 2.5 Localized chemotactic attraction and induced quiescence of the nematode *Meloidogyne incognita* by pea roots. (a) When border cells were removed from the root tip prior to placing the root onto a plate containing actively motile second-stage juveniles, no accumulation occurred (left) compared with that after 5 minutes for a root with border cells (right). (b) When the root with border cells was removed, a high concentration of actively motile nematodes was found associated with clumps of detached border cells. (c) Within 30 minutes, most of the nematodes within the clumps of border cells appeared as rigid sticks and ceased movement. (Reproduced with permission from Hawes *et al.*, *Trends in Plant Science*; Elsevier, 2000.)

(Plate 2.7). Mucilage mixed with border cells penetrates between soil particles and into aggregates close to roots, and plays a major role in maintaining root–soil contact. In many grasses and some dicots, a coherent sheath (a rhizosheath) of soil permeated by mucilage and root hairs develops around the root (Fig. 2.6) and remains intact on root axes until the large xylem vessels mature and the epidermis disintegrates (Wenzel *et al.*, 1989). Watt *et al.* (1993) showed in a model system that the formation of the sheath requires a cycle of wetting, when the mucilage expands around soil particles, followed by drying, when contraction of the mucilage draws the particles together. They suggested that this wetting and drying cycle occurs by the release of small amounts of water from the root at night followed by water uptake during the day (Watt *et al.*, 1994; McCully, 1999). Further mucilage additions by mucilage-producing bacteria and wetting/drying cycles make the sheath more cohesive with time. Read *et al.* (1999) showed that the root cap cells were an integral part of the gel system. With cells present, the dynamic viscosity was 145 mPa s and maize mucilage behaved as a weak viscoelastic gel, whereas with the cells removed by filtration, the dynamic viscosity was lower (5–10 mPa s) and behaviour was that of a viscous liquid. This elasticity of the mucilage is an important property facilitating the drawing together of the root and soil particles.

Chemical analyses of mucilage have shown the presence of several polysaccharides and monosaccharides (Table 2.1), with fucose, arabinose, galactose and glucose being common after hydrolysis of the polysaccharides (Chaboud and Rougier, 1984; Read and Gregory, 1997). In contrast to earlier work (e.g. Bacic *et al.*, 1986), Read and Gregory (1997) found that glucose was the principal component of maize mucilage, and that the majority of the neutral sugars found in the crude mucilage were present as free monosac-

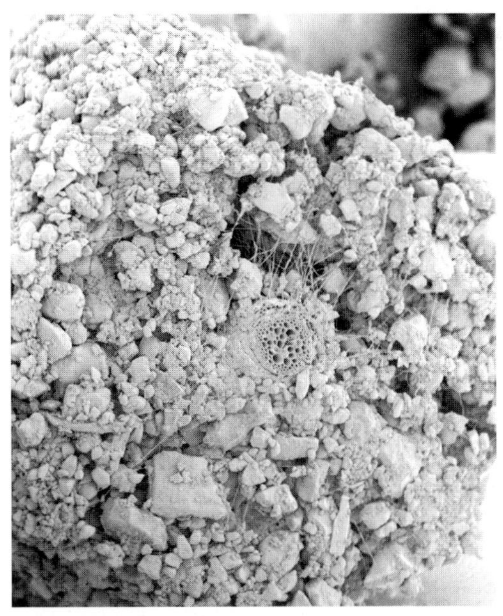

Fig. 2.6 Cryoscanning electron micrograph of the rhizosheath surrounding a buckwheat (*Fagopyrum esculentum*) root growing in the field. (I am grateful to Dr M. McCully for this previously unpublished figure.)

Table 2.1 Neutral monosaccharide analysis of maize (*Zea mays*) and lupin (*Lupinus angustifolius*) mucilages (concentration in mol %), listed in order of increasing gas chromatography retention time of the first peak for each sugar

	Maize		Lupin	
	Total soluble sugar	Polysaccharide fraction	Total soluble sugar	Polysaccharide fraction
Rhamnose	0	0.7	0	0.8
Fucose	4.0	10.9	36.9	34.7
Xylose	1.3	3.8	1.0	2.6
Arabinose	3.0	12.7	17.5	20.1
Galactose	3.9	8.9	15.5	23.2
Glucose	86.8	62.7	28.6	17.5
Mannose	1.0	0.3	0.5	1.1

From Read and Gregory, 1997.

charides and not bound within polysaccharides. However, high glucose contents were reported in unfractionated rice mucilage (Chaboud and Rougier, 1984). Such materials not only form a substrate for microorganisms close to the root, but the polysaccharide fraction with its large molecules which can twist and flex also contributes to the viscoelastic behaviour of mucilages described earlier.

While the mucilage *per se* has almost no capacity to store water in the rhizosphere (McCully and Boyer, 1997), the chemical and physical properties of mucilage influence the supply of water to the root. The water potential of root mucilage when fully hydrated is about –7 to –10 kPa (Guinel and McCully, 1986; Read *et al.*, 1999), and water is rapidly lost as the soil dries. Read and Gregory (1997) showed that the surface tension of both maize and lupin mucilages was reduced to about 48 mN m^{-1} at total solute concentrations >0.7 mg ml^{-1}, indicating the presence of powerful surfactants. Fatty acids and lipids (both surfactants) are common components of plant tissues and mucilage (e.g. Sukhija *et al.*, 1976) and subsequent analysis of maize, lupin and wheat mucilages showed that most of the plant-produced lipid present was phosphatidylcholine (Read *et al.*, 2003). Reduced surface tension will enhance the ability of the mucilage to wet the surrounding soil particles and may also change the moisture characteristic curve of the soil to reduce the water content at any particular tension by 10–50% depending on particle size distribution (Read *et al.*, 2003). Such alterations of physical properties suggest that if the root can maintain sufficient concentrations of surfactant close to the root, then it may be able to access water and nutrients from smaller soil pores than would otherwise be accessible to it.

Many functions have been ascribed to mucilage, but one of the most common is that it acts as a lubricant to ease the passage of the root through the soil. This role has been questioned (e.g. McCully, 1999), but it now seems likely that it is the presence of root cap cells within the mucilage that facilitates this role. The resistance to root penetration in a soil is the sum of the frictional resistance to root penetration plus the pressure required to form a cavity. Friction can be 80% of the penetration resistance experienced by roots as they move through soil, so reducing this resistance would be advantageous to plants in their exploration of soil resources. Bengough and McKenzie (1997) measured the frictional resistance experienced by metal rods, roots pushed into soil and growing roots, and found that the frictional resistance experienced by roots was a small, but not negligible, fraction of that experienced by metal probes. For example, in soil with a bulk density of 1.3 mg m^{-3}, the penetration force was about 330, 200 and 150 mN for a metal probe, pushed root, and growing root, respectively. They postulated that the friction was probably relieved by the detachment of root cap cells to form a low-friction lining to the cavity enlarged by the root. The border cells might act as a cushion between the soil particles and the root, to reduce local stresses and allow the maintenance of an intact mucilage layer over the root surface (Hawes *et al.*, 2003). For this concept to work in practice, there need to be sufficient border cells to cover a substantial proportion of the root cap surface. Iijima *et al.* (2000) found that the number of root cap cells of maize sloughed into sand increased 12-fold (from 60 to >700 per mm of root extension) as penetrometer resistance increased from 0.29 to 5.2 MPa. This increase in cell production was estimated to cover the whole of the root cap with detached cells in the compacted sand, compared with about 7% of the surface area of the cap in loose sand. In this case (see also Iijima *et al.*, 2003), sufficient border cells were released that this lubricating layer of sloughed cells and mucilage probably decreased the frictional resistance to root penetration; whether this occurs in species other than maize is not currently known.

2.5 Architecture of root systems

Root architecture, the spatial configuration of a root system in the soil, is used to describe distinct aspects of the shape of root systems. Lynch (1995) states that studies of root architecture do not usually include fine details such as root hairs, but are concerned typically with an entire root system of an individual plant. From the architecture both the topology (a description of how individual roots are connected through branching) and the distribution (the presence of roots in a spatial framework) can be derived, whereas neither topology nor distribution can be used to derive architecture. From the brief description of root branching given in section 2.3.2, it will be clear that root architecture is quite complex and varies between and within plant species. Drawings of excavated root systems of crops and other species show the differences in shape between monocotyledons and dicotyledons and allow some broad generalizations to be made about the depth of rooting and the relative distribution of roots (Kutschera, 1960) (Fig. 2.7). Nearly all such drawings show that, with the exception of the tap root which grows almost vertically throughout, most other root axes grow initially at some angle relative to the vertical but gradually become more vertically orientated. Gravitropic responses combined with responses to light, water and touch together, perhaps, with the predominance of vertical cracks in deeper soil layers, produce these patterns.

Root architecture's importance lies in the fact that many of the resources that plants need from soil are heterogeneously distributed and/or are subject to local depletion (Robinson, 1994). In such circumstances, in contrast to the shapes shown in Fig. 2.7, the development and growth of root systems may become highly asymmetric, and the spatial arrangement of the root system will substantially determine the ability of a plant to secure those resources (Lynch, 1995). Such ideas have been investigated in a series of experiments and models using common bean (Bonser *et al.*, 1996; Ge *et al.*, 2000). While root trajectories are essentially under genetic control, phosphorus deficiency was found to decrease the

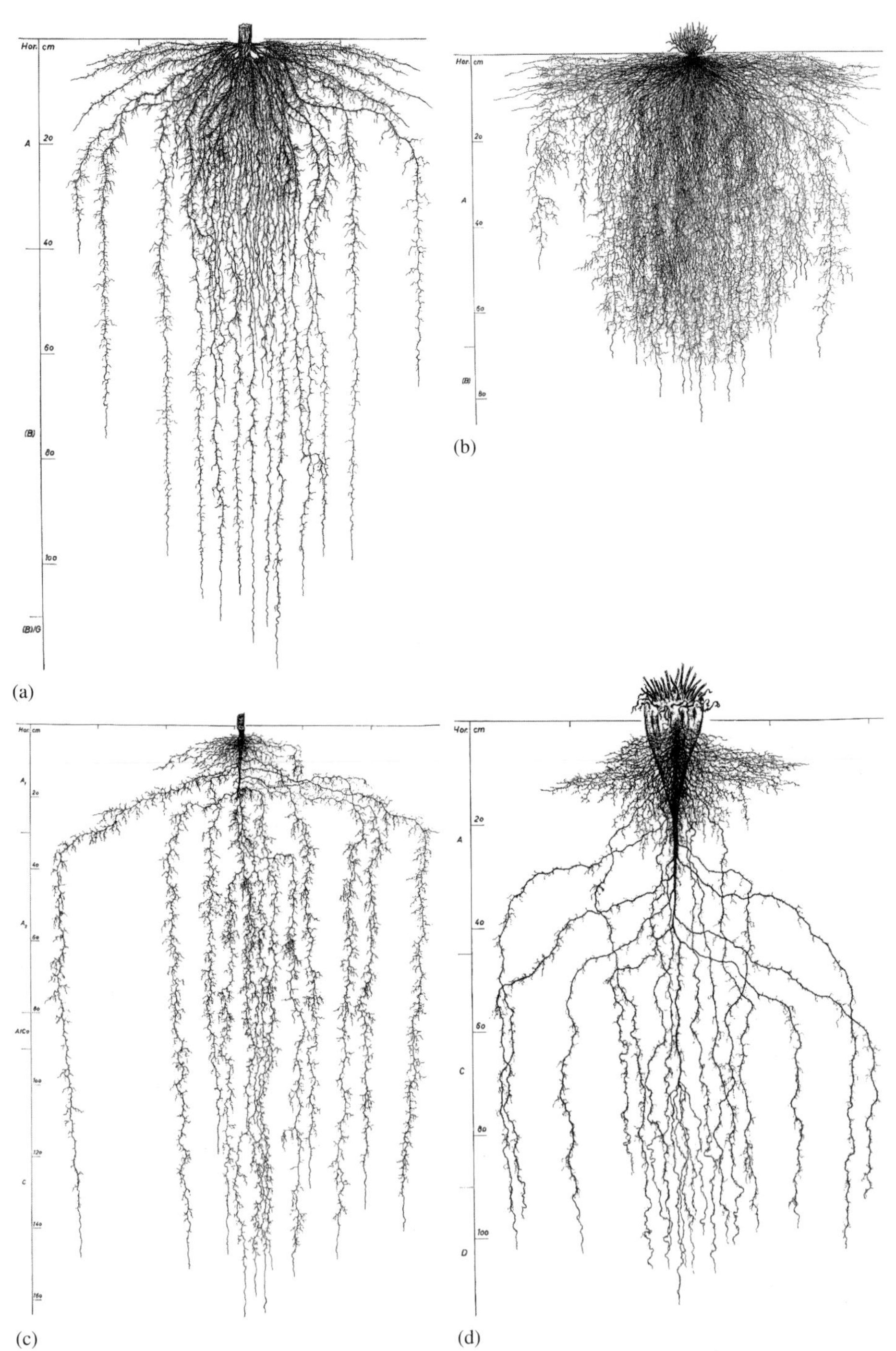

Fig. 2.7 Drawings of excavated root systems: (a) maize, *Zea mays*; (b) ryegrass, *Lolium multiflorum*; (c) oilseed rape, *Brassica napus*; and (d) sugar beet, *Beta vulgaris*. (Reproduced with permission from Kutschera, *Wurzelatlas*; DLG-Verlags-GmbH, 1960.)

gravitropic sensitivity of both the tap root and the basal roots, resulting in a shallower root system. It was hypothesized that the shallower root system was a positive adaptive response to low soil P availability by: first, concentrating roots in the surface soil layers where soil P availability was highest; and second, reducing spatial competition for P among roots of the same plant. This hypothesis was tested by modelling root growth and P acquisition by bean plants with nine contrasting root systems in which basal root angle was varied but not root length or degree of branching (Fig. 2.8). Shallower root systems acquired more P per unit carbon cost than deeper root systems and in soils with higher P availability in the surface layers, shallower root systems acquired more P than deeper root systems because of less inter-root competition as well as increased root exploration of the upper soil (Ge *et al.*, 2000). In practice, of course, the plant may have multiple resource constraints to contend with (e.g. heterogeneously distributed P and soil water) and will try to optimize its investment in roots. Ho *et al.* (2004) investigated this optimization with respect to beans grown under different combinations of water and P availability. They postulated that an optimizing plant would grow roots deeper into the profile until the marginal benefit exactly equalled the marginal cost and by modelling found that the basal root angle would be shallower for localized shallow P, and deeper for localized deep water than that obtained in the case of uniformly distributed water and P. When P was concentrated in the surface and water was located deep, the optimal basal root angle depended on the relative rates of change with depth in the values ascribed to the available resources. While useful in indicating general principles, it should be remembered that not all of the responses of roots to a heterogeneous environment (e.g. changes in branching frequency and root hair growth) are yet captured in such models; this is a substantial challenge.

The branching patterns (topology) of individual roots have implications not only for resource capture but also for the construction costs of roots (Fitter *et al.*, 1991). In topological analysis, the root can be considered as any other mathematical branching tree, with links that are either exterior (ending in meristems) or interior (i.e. internodes). The links have geometrical properties, including length, radius, angle and direction of growth, and are distributed in a defined pattern; as in most branching trees (e.g. the trachea in the lung), the diameter increases with increasing magnitude of the individual link. Fitter *et al.* (1991) em-

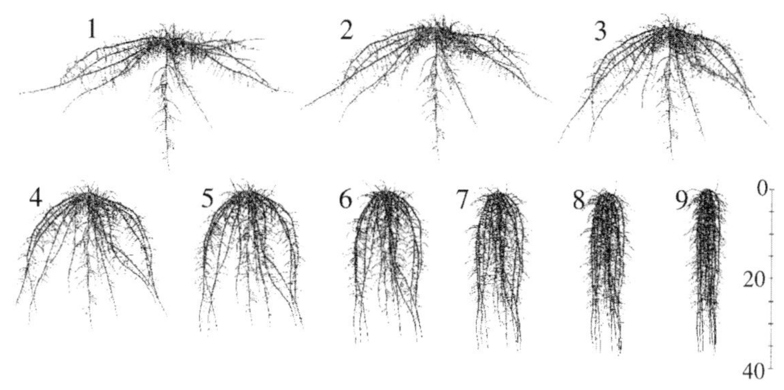

Fig. 2.8 Geometric simulation modelling of bean root systems that vary in basal root angle but are otherwise identical in length and branching. The variation illustrated is present among different genotypes of *Phaseolus vulgaris* and has been shown to be influenced by soil P availability. The scale on the right is from 0 to 40 cm. (Reproduced with permission from Ge *et al.*, *Plant and Soil*; Springer Science and Business Media, 2000.)

ployed a simulation model to demonstrate that a herringbone topology (where branching is on the main axis) with long interior and exterior links is associated with high exploration efficiency, although such a pattern is also characterized by large tissue volumes and hence high construction costs. Fitter *et al.* (1991) predicted that a herringbone topology would be favourable for slow-growing species in habitats where soil resources are scarce and that nutrient-rich soils or those with a nutrient-rich surface layer would encourage the formation of dichotomous topologies with their associated cheaper construction costs (Fig. 2.9). Such predictions were only partially borne out by the experimental results of Fitter and Stickland (1992) in which *Trifolium repens* L. became more herringbone-like as soil water content increased (contrary to prediction) but *Mercurialis perennis* L. responded as expected to both irrigation and N and P additions. Topological considerations alone, though, are unlikely to be the sole adaptive trait to particular soil environments. For example, Bouma *et al.* (2001) found that roots of Chenopodiaceae in a salt marsh changed from herringbone-like at low elevation to dichotomous at higher elevations but that the Gramineae showed no such relationship. Moreover, root diameter was not related to link magnitude thereby undermining the basis of the estimates of construction efficiency proposed by Fitter *et al.* (1991).

Not only does root topology and root system architecture respond to soil heterogeneity, but the form of the root system may, indeed, induce soil heterogeneity. In grassland and savanna systems, caespitose (i.e. tussock or bunch) and rhizomatous perennial grasses represent two distinct forms of grass. In rhizomatous grasses, nutrients can accumulate in the rhizomes but do not accumulate in the soil whereas in the caespitose grasses, both carbon and nitrogen accumulate in soils directly beneath plants resulting in fine-grained soil heterogeneity (Derner and Briske, 2001). The 'islands' of nutrients appear to accumulate beneath caespitose grasses even when they are small, suggesting that they are present throughout much of the plant's life. Plant-induced increases in nutrient concentrations do not form beneath the rhizatomous species and the large nutrient pool beneath such species in a semi-arid community was largely a consequence of niche separation for microsites characterized by deeper soils with higher amounts of water and nutrients.

In broad-scale agriculture where single crops are grown with inputs of fertilizers, there has been little consideration until recently of root architecture, but with the increasing emphasis on the more efficient use of water and nutrients in production systems, this is starting to change. For example, in soils where P availability is low, selecting genotypes with appropriate

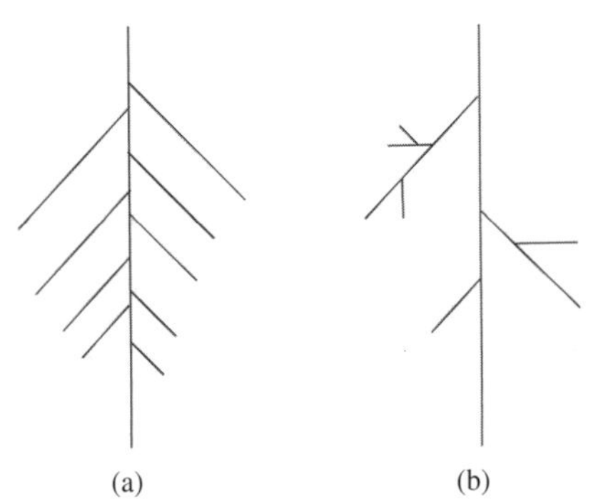

Fig. 2.9 Diagram showing the distinction between (a) herringbone, and (b) dichotomous branching patterns. (Modified and reproduced with permission from Fitter *et al.*, *New Phytologist*; New Phytologist Trust, 1991.)

architecture may increase soil exploration by roots and raise yields (Lynch and Beebe, 1995). Equally important in other areas is the ability of roots to capture nutrients such as nitrate that might otherwise leach from the soil profile into water courses. Dunbabin *et al.* (2003) have shown the role that root architecture may play in this regard and the importance of quickly producing a high density of roots in the topsoil on the sandy soils that they studied. In many parts of the world, though, mixed cropping is important either with crops grown together as intercrops or with different crops grown in sequence, as is the growing of trees and crops in agroforestry associations. In such systems, root architecture is important in determining both the spatial competition and spatial complementarity of root systems (van Noordwijk *et al.*, 1996). These ideas will be explored more in Chapter 9.

References

Allaway, W.G. and Ashford, A.E. (1996) Structure of hair roots in *Lysinema ciliatum* R. Br. and its implications for their water relations. *Annals of Botany* **77**, 383–388.

Bacic, A., Moody, S.F. and Clarke, A.E. (1986) Structural analysis of secreted root slime from maize (*Zea mays* L.). *Plant Physiology* **80,** 771–777.

Barlow, P.W. (1975) The root cap. In: *The Development and Function of Roots* (eds J.G. Torrey and D.T. Clarkson), pp. 21–54. Academic Press, London.

Barlow, P.W. (1984) Positional controls in root development. In: *Positional Controls in Plant Development* (eds P.W. Barlow and D.J. Carr), pp. 281–318. Cambridge University Press, Cambridge.

Barlow, P.W. (1987a) The cellular organization of roots and its response to the physical environment. In: *Root Development and Function* (eds P.J. Gregory, J.V. Lake and D.A. Rose), pp. 1–26. Cambridge University Press, Cambridge.

Barlow, P.W. (1987b) Cellular packets, cell division and morphogenesis in the primary root meristem of *Zea mays* L. *New Phytologist* **105**, 27–56.

Barlow, P.W. (2003) The root cap: cell dynamics, cell differentiation and cap function. *Journal of Plant Growth Regulation* **21**, 261–286.

Barlow, P.W. and Baluška, F. (2000) Cytoskeletal perspectives on root growth and morphogenesis. *Annual Review of Plant Physiology and Plant Molecular Biology* **51**, 289–322.

Bates, T.R. and Lynch, J.P. (1996) Stimulation of root hair elongation in *Arabidopsis thaliana* by low phosphorus availability. *Plant, Cell and Environment* **19**, 529–538.

Bengough, A.G. and McKenzie, B.M. (1997) Sloughing of root cap cells decreases the frictional resistance to maize (*Zea mays* L.) root growth. *Journal of Experimental Botany* **48**, 885–893.

Benzing, D.H., Ott, D.W. and Friedman, W.E. (1982) *Roots of Sobralia macrantha* (Orchidaceae) – structure and function of the velamen-exodermis complex. *American Journal of Botany* **69**, 608–614.

Bibikova, T. and Gilroy, S. (2003) Root hair development. *Journal of Plant Growth Regulation* **21**, 383–415.

Bonser, A.M., Lynch, J.P. and Snapp, S. (1996) Effect of phosphorus deficiency on growth angle of basal roots in *Phaseolus vulgaris*. *New Phytologist* **132**, 281–288.

Bouma, T.J., Nielsen, K.L., Van Hal, J. and Koutstaal, B. (2001) Root system topology and diameter distribution of species from habitats differing in inundation frequency. *Functional Ecology* **15**, 360–369.

Brady, D.J., Wenzel, C.L., Fillery, I.R.P. and Gregory, P.J. (1995) Root growth and nitrate uptake by wheat (*Triticum aestivum* L.) following wetting of dry surface soil. *Journal of Experimental Botany* **46**, 557–564.

Casimiro, I., Marchant, A., Bhalerao, R.P., Beeckman, T., Dhooge, S., Swarup, R., Graham, N., Inzé, D., Sandberg, G., Casero, P.J. and Bennett, M.J. (2001) Auxin transport promotes Arabidopsis lateral root initiation. *The Plant Cell* **13**, 843–852.

Casimiro, I., Beeckman, T., Graham, N., Bhalerao, R., Zhang, H., Casero, P., Sandberg, G. and Bennett, M.J. (2003) Dissecting *Arabidopsis* lateral root development. *Trends in Plant Science* **8**, 165–171.

Chaboud, A. and Rougier, M. (1984) Identification and localisation of sugar components of rice (*Oryza sativa* L.) root cap mucilage. *Journal of Plant Physiology* **116,** 323–330.

Chaffey, N. (2002) Secondary growth of roots: a cell biological perspective. In: *Plant Roots: The Hidden Half* (eds Y. Waisel, A. Eshel and U. Kafkafi), 3rd edn. pp. 93–111. Marcel Dekker Inc., New York.

Chapman, K., Groot, E.P., Nichol, S.A. and Rost, T.L. (2003) Primary root growth and the pattern of root apical meristem organization are coupled. *Journal of Plant Growth Regulation* **21**, 287–295.

De Smet, I., Signora, L., Beeckman, T., Inzé, D., Foyer, C.H. and Zhang, H. (2003) An abscisic acid-sensitive checkpoint in lateral root development of *Arabidopsis*. *The Plant Journal* **33**, 543–555.

Derner, J.D. and Briske, D.D. (2001) Below-ground carbon and nitrogen accumulation in perennial grasses: a comparison of caespitose and rhizomatous growth forms. *Plant and Soil* **237**, 117–127.

Dolan, L. and Davies, J. (2004) Cell expansion in roots. *Current Opinion in Plant Biology* **7**, 33–39.

Dunbabin, V., Diggle, A. and Rengel, Z. (2003) Is there an optimal root architecture for nitrate capture in leaching environments? *Plant, Cell and Environment* **26**, 835–844.

Esau, K. (1941) Phloem anatomy of tobacco affected with curly top and mosaic. *Hilgardia* **13**, 437–490.

Esau, K. (1977) *Anatomy of Seed Plants*, 2nd edn, pp. 215–256. John Wiley & Sons, New York.

Fitter, A.H. and Stickland, T.R. (1992) Architectural analysis of plant root systems. III. Studies on plants under field conditions. *New Phytologist* **121**, 243–248.

Fitter, A.H., Stickland, T.R., Harvey, M.L. and Wilson, G.W. (1991) Architectural analysis of plant root systems 1. Architectural correlates of exploitation efficiency. *New Phytologist* **118**, 375–382.

Fusseder, A. (1987) The longevity and activity of the primary root of maize. *Plant and Soil* **101**, 257–265.

Gahoonia, T.S. and Nielsen, N.E. (2003) Phosphorus (P) uptake and growth of a root hairless barley mutant (*bald root barley, brb*) and wild type in low- and high-P soils. *Plant, Cell and Environment* **26**, 1759–1766.

Gahoonia, T.S., Care, D. and Nielsen, N.E. (1997) Root hairs and phosphorus acquisition of wheat and barley cultivars. *Plant and Soil* **191**, 181–188.

Ge, Z., Rubio, G. and Lynch, J.P. (2000) The importance of root gravitropism for inter-root competition and phosphorus acquisition efficiency: results from a geometric simulation model. *Plant and Soil* **218**, 159–171.

Geissler, N., Schnetter, R. and Schnetter, M. (2002) The pneumathodes of *Laguncularia racemosa*: little known rootlets of surprising structure, and notes on a new fluorescent dye for lipophilic substances. *Plant Biology* **6**, 729–739.

Gilroy, S. and Jones, D.L. (2000) Through form to function: root hair development and nutrient uptake. *Trends in Plant Science* **5**, 56–60.

Goh, C.J. and Kluge, M. (1989) Gas exchange and water relations in epiphytic orchids. In: *Vascular Plants as Epiphytes: Evolution and Ecophysiology, Ecological Studies: Analysis and Synthesis*, Vol. 76 (ed. U. Lüttge), pp. 139–166. Springer-Verlag, Berlin.

Gregory, P.J. (1994) Root growth and activity. In: *Physiology and Determination of Crop Yield* (eds K.J. Boote, J.M. Bennett, T.R. Sinclair and G.M. Paulsen), pp. 65–93. ASA, CSSA, SSSA, Madison, USA.

Gregory, P.J., McGowan, M., Biscoe, P.V. and Hunter, B. (1978) Water relations of winter wheat.1. Growth of the root system. *Journal of Agricultural Science, Cambridge* **91**, 91–102.

Groff, P.A. and Kaplan, D.R. (1988) The relation of root systems to shoot systems in vascular plants. *The Botanical Review* **54**, 387–422.

Guinel, F.C. and McCully, M.E. (1986) Some water-related physical properties of maize root-cap mucilage. *Plant, Cell and Environment* **9**, 657–666.

Hackett, C. (1968) A study of the root system of barley. I. Effects of nutrition on two varieties. *New Phytologist* **67**, 287–299.

Hawes, M.C., Gunawardena, U., Miyasaka, S. and Zhao, X. (2000) The role of root border cells in plant defense. *Trends in Plant Science* **5**, 128–133.

Hawes, M.C., Bengough, A.G., Cassab, G. and Ponce, G. (2003) Root caps and rhizosphere. *Journal of Plant Growth Regulation* **21**, 352–367.

Ho, M.D., McCannon, B.C. and Lynch, J.P. (2004) Optimization modeling of plant root architecture for water and phosphorus acquisition. *Journal of Theoretical Biology* **226**, 331–340.

Hose, E., Clarkson, D.T., Steudle, E., Schreiber, L. and Hartung, W. (2001) The exodermis: a variable apoplastic barrier. *Journal of Experimental Botany* **52**, 2245–2264.

Hovenden, M.J. and Allaway, W.G. (1994) Horizontal structures on pneumatophores of *Avicennia marina* (Forsk.) Vierh. – a new site of oxygen conductance. *Annals of Botany* **73**, 377–383.

Iijima, M., Griffiths, B. and Bengough, A.G. (2000) Sloughing of cap cells and carbon exudation from maize seedling roots in compacted sand. *New Phytologist* **145**, 477–482.

Iijima, M., Barlow, P.W. and Bengough, A.G. (2003) Root cap structure and cell production rates of maize (*Zea mays*) roots in compacted sand. *New Phytologist* **160**, 127–134.

Itoh, S. and Barber, S.A. (1983) Phosphorus uptake by six plant species as related to root hairs. *Agronomy Journal* **75**, 457–461.

Kutschera, L. (1960) *Wurzelatlas mitteleuropäischer Ackerunkräuter und Kulturpflanzen*. DLG-Verlags-GMBH, Frankfurt, Germany.

Lamont, B.B. (2003) Structure, ecology and physiology of root clusters – a review. *Plant and Soil* **248**, 1–19.

Lynch, J.P. (1995) Root architecture and plant productivity. *Plant Physiology* **109**, 7–13.

Lynch, J.P. and Beebe, S.E. (1995) Adaptation of beans (*Phaseolus vulgaris* L.) to low phosphorus availability. *HortScience* **30**, 1165–1171.

McCully, M.E. (1975) The development of lateral roots. In: *The Development and Function of Roots* (eds J.G. Torrey and D.T. Clarkson), pp. 105–124. Academic Press, London.

McCully, M.E. (1994) Accumulation of high levels of potassium in the developing xylem elements in roots of soybean and some other dicotyledons. *Protoplasma* **183,** 116–125.

McCully, M.E. (1995) How do real roots work? Some new views of root structure. *Plant Physiology* **109**, 1–6.

McCully, M.E. (1999) Roots in soil: unearthing the complexities of roots and their rhizospheres. *Annual Review of Plant Physiology and Plant Molecular Biology* **50**, 695–718.

McCully, M.E. and Boyer, J.S. (1997) The expansion of maize root-cap mucilage during hydration. 3. Changes in water potential and water content. *Physiologia Plantarum* **99**, 169–177.

Pagès, L. and Pellerin, S. (1994) Evaluation of parameters describing the root system architecture of field grown maize plants (*Zea mays* L.). II. Density, length and branching of first-order lateral roots. *Plant and Soil* **164**, 169–176.

Parker, C. and Riches, C.R. (1993) *Parasitic Weeds of the World: Biology and Control,* pp. 28–33. CAB International, Wallingford, UK.

Perumalla, C.J. and Peterson, C.A. (1986) Deposition of Casparian bands and suberin lamellae in the exodermis and endodermis of young corn and onion roots. *Canadian Journal of Botany* **64**, 1873–1878.

Perumalla, C.J. and Peterson, C.A. (1990) A survey of angiosperm species to detect hypodermal Casparian bands. I. Roots with a uniseriate hypodermis and epidermis. *Botanical Journal of the Linnean Society* **103**, 93–112.

Peterson, C.A. (1991) Adaptations of root structure in relation to biotic and abiotic factors. *Canadian Journal of Botany* **70**, 661–675.

Peterson, C.A. and Enstone, D.E. (1996) Functions of passage cells in the endodermis and exodermis of roots. *Physiologia Plantarum* **97**, 592–598.

Peterson, C.A. and Perumalla, C.J. (1990) A survey of angiosperm species to detect hypodermal Casparian bands. II. Roots with a multiseriate hypodermis or epidermis. *Botanical Journal of the Linnean Society* **103**, 113–125.

Pritchard, J. (1994) The control of cell expansion in roots. *New Phytologist* **127**, 3–26.

Purnell, H.M. (1960) Studies of the family Proteaceae. I. Anatomy and morphology of the roots of some Victorian species. *Australian Journal of Botany* **8**, 38–50.

Pütz, N. (2002). Contractile roots. In: *Plant Roots: The Hidden Half* (eds Y. Waisel, A. Eshel and U. Kafkafi), 3rd edn, pp. 975–987. Marcel Dekker Inc., New York.

Read, D.B. and Gregory, P.J. (1997) Surface tension and viscosity of axenic maize and lupin mucilages. *New Phytologist* **137**, 623–628.

Read, D.B., Gregory, P.J. and Bell, A.E. (1999) Physical properties of axenic maize root mucilage. *Plant and Soil* **211**, 87–91.

Read, D.B., Bengough, A.G., Gregory, P.J., Crawford, J.W., Robinson, D., Scrimgeour, C.M., Young, I.M., Zhang, K. and Zhang, X (2003) Plant roots release phospholipid surfactants that modify the physical and chemical properties of soil. *New Phytologist* **157**, 315–326.

Read, D.J. (1996) The structure and function of the ericoid mycorrhizal root. *Annals of Botany* **77**, 365–374.

Reed, R.C., Brady, S.R. and Muday, G.K. (1998) Inhibition of auxin movement from the shoot into the root inhibits lateral root development in Arabidopsis. *Plant Physiology* **118**, 1369–1378.

Reyneke, W.F. and van der Schijff, H.P. (1974) The anatomy of contractile roots in *Eucomis* L'Hérit. *Annals of Botany* **38**, 977–982.

Roberts, A.G. and Oparka, K.J. (2003) Plasmodesmata and the control of symplastic transport. *Plant, Cell and Environment* **26**, 103–124.

Robinson, D. (1994) The responses of plants to non-uniform supplies of nutrients. *New Phytologist* **127**, 635–674.

Roelofs, R.F.R., Rengel, Z., Cawthray, G.R., Dixon, K.W. and Lambers, H. (2001) Exudation of carboxylates in Australian Proteaceae: chemical composition. *Plant, Cell and Environment* **24**, 891–903.

St Aubin, G., Canny, M.J. and McCully, M.E. (1986) Living vessel elements in the late metaxylem of sheathed maize roots. *Annals of Botany* **58**, 577–588.

Schreiber, L., Hartmann, K., Skrabs, M. and Zeier, J. (1999) Apoplastic barriers in roots: chemical composition of endodermal and hypodermal cell walls. *Journal of Experimental Botany* **50**, 1267–1280.

Schünmann, P.H.D., Richardson, A.E., Smith, F.W. and Delhaize, E. (2004) Characterization of promoter expression patterns derived from the Pht1 phosphate transporter genes of barley (*Hordeum vulgare* L.). *Journal of Experimental Botany* **55**, 855–865.

Steudle, E. and Peterson, C.A. (1998) How does water get through roots? *Journal of Experimental Botany* **49**, 775–788.

Sukhija, P.S., Sital, J.S., Raheja, R.K. and Bhatia, I.S. (1976) Polar lipids of spinach (*Spinacea oleracea*) roots. *Physiologia Plantarum* **38,** 221–223.

Takahashi, H. (1997) Hydrotropism: the current state of our knowledge. *Journal of Plant Research* **110**, 163–169.

Torrey, J.G. (1953) The effect of certain metabolic inhibitors on vascular tissue differentiation in isolated pea roots. *American Journal of Botany* **40**, 525–533.

van der Weele, C.M., Jiang, H.S., Palaniappan, K.K., Ivanov, V.B., Palaniappan, K. and Baskin, T.I. (2003) A new algorithm for computational image analysis of deformable motion at high spatial and temporal resolution applied to root growth. Roughly uniform elongation in the meristem and also, after an abrupt acceleration, in the elongation zone. *Plant Physiology* **132**, 1138–1148.

van Noordwijk, M., Lawson, G., Soumaré, A., Groot, J.J.R. and Hairiah, K. (1996) Root distribution of trees and crops: competition and/or complementarity. In: *Tree-Crop Interactions: A Physiological Approach* (eds C.K. Ong and P. Huxley), pp. 319–364. CAB International, Wallingford, UK.

Varney, G.T. and McCully, M.E. (1991) The branch roots of *Zea*. II. Developmental loss of the apical meristem in field-grown roots. *New Phytologist* **118**, 535–546.

Varney, G.T., Canny, M.J., Wang, X.L. and McCully, M.E. (1991) The branch roots of *Zea*. I. First order branches, their number, sizes and division into classes. *Annals of Botany* **67**, 357–364.

Watt, M. and Evans, J.R. (1999) Proteoid roots. Physiology and development. *Plant Physiology* **121**, 317–323.

Watt, M. and Evans, J.R. (2003) Phosphorus acquisition from soil by white lupin (*Lupinus albus* L.) and soybean (*Glycine max* L.), species with contrasting root development. *Plant and Soil* **248**, 271–283.

Watt, M., McCully, M.E. and Jeffree, C.E. (1993) Plant and bacterial mucilages of the maize rhizosphere: comparison of their soil binding properties and histochemistry in a model system. *Plant and Soil* **151**, 151–165.

Watt, M., McCully, M.E. and Canny, M.J. (1994) Formation and stabilization of rhizosheaths of *Zea mays* L. *Plant Physiology* **106**, 179–186.

Weaver, J.E., Kramer, J. and Reed, M. (1924) Development of root and shoot of winter wheat under field environment. *Ecology* **5,** 26–50.

Wenzel, C.M. and McCully, M.E. (1991) Early senescence of cortical cells in the roots of cereals. How good is the evidence? *American Journal of Botany* **78**, 1528–1541.

Wenzel, C.L., McCully, M.E. and Canny, M.J. (1989) Development of water conducting capacity in the root systems of young plants of corn and some other C4 grasses. *Plant Physiology* **89**, 1094–1101.

Yano, K., Yamauchi, A. and Kono, Y. (1996) Localized alteration in lateral root development in roots colonized by an arbuscular mycorrhizal fungus. *Mycorrhiza* **6**, 409–415.

Zhao, X., Schmitt, M. and Hawes, M.C. (2000) Species-dependent effects of border cell and root tip exudates on nematode behavior. *Phytopathology* **90**, 1239–1245.

Chapter 3

Development and Growth of Root Systems

This chapter concentrates on the growth of root systems under field conditions. The number of such studies is small compared with those on shoot systems because roots are obscured by soil and there are considerable labour and other costs involved in exposing and measuring them. No single technique has proved capable of providing all of the information about roots that is required, so a variety of methods has been used to collect measurements of particular facets of the system. For example, measurements of root mass at particular times are frequently obtained by washing roots from soil, whereas measurements of root longevity and turnover are commonly obtained by direct observation of roots growing against a transparent glass or plastic surface (a minirhizotron). The different methods employed sometimes make comparisons between studies problematic, but despite these various difficulties, sufficient studies now exist to make some generalizations possible and for the production of mathematical models of root system growth and behaviour.

3.1 Measurement of root systems

Soils are optically opaque so that continuous visual observation of growth is impossible, while disturbance of soil to expose roots substantially changes their environment which may, in turn, lead to modifications to growth and function. This conundrum is at the heart of the dilemma of selecting appropriate methods for assessing root growth and activity. The consequence is a variety of techniques each best suited to a particular purpose. A summary of methods for examining roots is given in books by Böhm (1979) and Smit *et al.* (2000a), but this section will focus on a few of the most commonly used techniques.

3.1.1 Washed soil cores

One of the commonest sets of measurements required by ecologists and agriculturists is that of the size of the root system, how it is distributed with depth, and how it changes with time. Because it is rarely possible to extract an intact root system from soil, typically either intact or disturbed samples of soil are collected and the roots are washed free of soil with water before the properties of the roots are measured. Because root systems vary spatially (see Fig. 2.7), a large number of replicate samples, or large individual samples, are necessary to obtain accurate estimates of root parameters. Agronomists and ecologists have found, by experience, the number and size of samples necessary to specify shoot mass with a certain precision and it is worthwhile reflecting on this experience when sampling roots. In agronomy, the growth analysis of a crop such as wheat typically requires the collection of a 1 m length of shoots replicated about five times (i.e. an area of about 1 m^2) in order to provide a mean value of shoot dry matter with a coefficient of variation of 10%; such a coefficient is normally sufficient to allow determination of statistically significant differences between treatments at the 5% level of probability. Soil samples for root measurements are typically only 5–10 cm in diameter (0.008–0.03 m^2), so it is not surprising that coefficients of variation are large if only a small number of small samples is taken. At the other extreme, taking a 1 m^2 sample of soil to a rooting depth of, say, 1 m depth is clearly impracticable both because of the mass involved (about 1.5 tonnes), and the destruction of the site. From experience, about 15–20 samples of 10 cm diameter are required from a structured soil to have a 90% chance of detecting significant differences at the 10% level of probability; significance at 5% often requires many more samples (60–90). These considerations mean that this technique is unlikely to detect small differences between treatments, and that it is unsuitable for estimating root turnover by frequent sampling schemes (van Noordwijk, 1993). Nevertheless, washed soil samples often give the best quantitative information about root mass and length so that the method is frequently used as the basis for calibrating other techniques (Oliveira *et al.*, 2000).

Because roots are not distributed uniformly with respect to distance from the plant, the position from which a sample is taken is important if results of different treatments are to be compared. This is especially important for trees where a few large roots may be sparsely distributed (Livesley *et al.*, 2000). In row crops a rectangular or square unit cell can be defined and ideally this shape of sample should be collected; in mixed or sparse plantings the difficulty of obtaining a representative sample is not as easily resolved. For practical reasons, circular cores rather than rectangular samples are normally collected but the exact location of these samples is important because it determines the bias introduced if results per core are extrapolated to provide results on a unit area basis. For example, van Noordwijk *et al.* (1985) showed that a sample comprising two 7 cm diameter cores, one taken centrally over the row and the other taken mid-way between rows of a winter wheat crop grown in rows 25 cm apart, could give large overestimates of up to 50% in root dry mass per m^2.

Separating the roots from the soil by washing brings its own difficulties. Various root washing devices are used ranging from as simple as a bucket, sieve and hosepipe, to as complex as closed systems using a combination of water pressure and compressed air (Smucker *et al.*, 1982; Pallant *et al.*, 1993). The main advantage of the enclosed systems is the standardization of the process thereby overcoming difficulties if different operators do the washing. Roots separate relatively easily from sandy soils but soil containing appreciable quantities of clay must often be pre-treated to disperse it. There is no agreed standard for the size of sieve mesh to be used when washing roots although it is evident that mesh size affects the recovery of roots. For example, Amato and Pardo (1994) found that wheat and faba bean roots washed on a 2 mm mesh sieve were an average of 55% of the mass, but only 10% of the length, of roots collected on a 0.2 mm mesh sieve. Similarly, Livesley *et al.* (1999) found that mesh size had large effects on the recovery of roots of *Grevillea robusta* and of maize (Fig. 3.1). Mesh ≥1 mm recovered about 80% of the total root biomass, but only 60% of root length, whereas mesh ≥0.5 mm recovered between 93 and 95% of grevillea and maize root biomass and between 73 and 98% of root length, depending on sample location. They concluded that sieves with mesh ≥0.5 mm were adequate for measurements of root mass but that mesh of 0.25 mm was required for accurate length measurements. No universal correction factor for the underestimation of mass or length could be applied when larger mesh was used because root recovery varied with plant species, soil depth, and distance from the plant (Fig. 3.1).

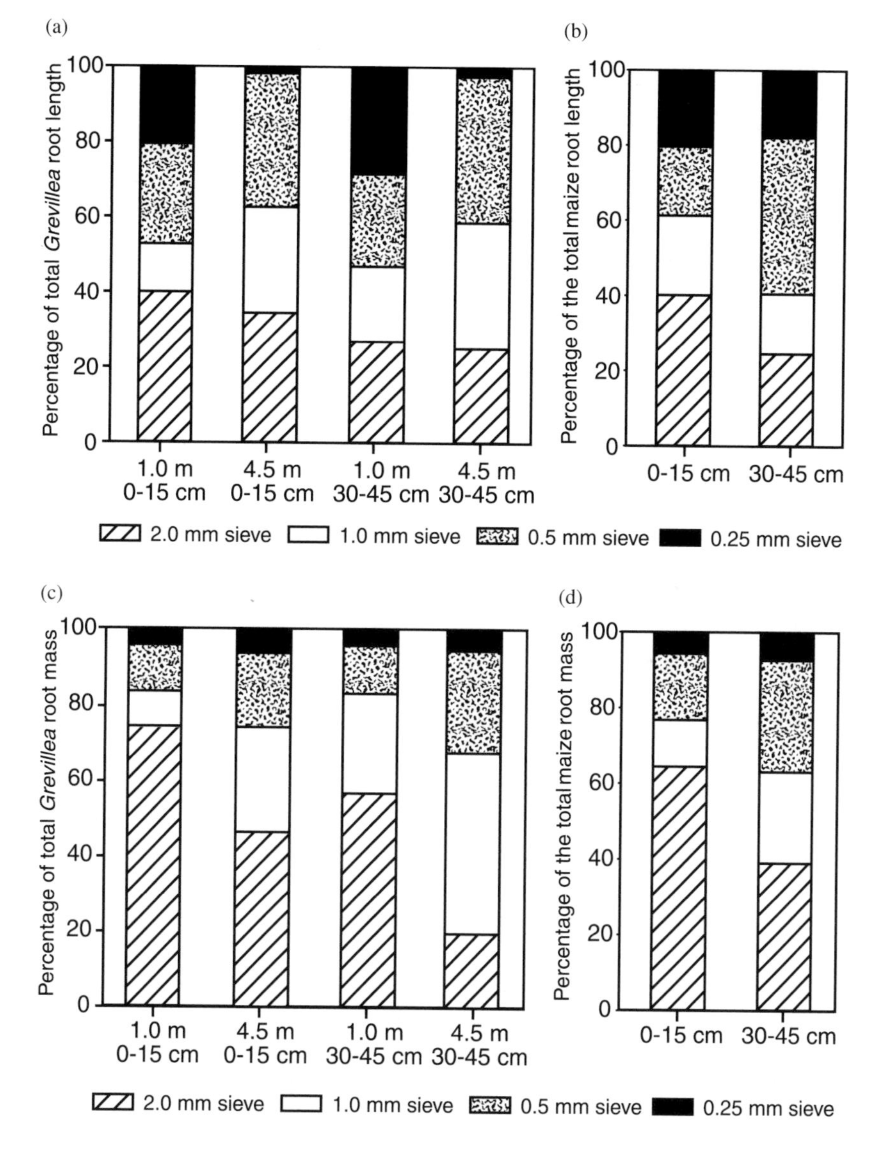

Fig. 3.1 Recovery of root length and root mass by sieves of different mesh size from cores of washed soil. The percentage of total root length recovered for (a) *Grevillea robusta* trees at depths of 0–15 cm and 30–45 cm at distances of 1.0 m and 4.5 m from the tree row, and for maize (b) at depths of 0–15 cm and 30–45 cm. Figures (c) and (d) show the corresponding recoveries of root mass. (Reproduced from Livesley *et al.*, *Plant and Soil*; Springer Science and Business Media, 1999.)

In addition to the failure to recover roots from the soil, root mass can also be lost during washing and storage. Root respiration may result in 5–10% loss of weight in the 24 hours after sampling unless samples are kept cool (4°C), and washing of wheat roots grown in solution culture resulted in 20–40% loss of dry weight depending on the method of storage and washing (van Noordwijk and Floris, 1979). Similarly, losses from sugar beet roots grown in nutrient solution were about 30%, with most of the loss from cell contents rather than cell wall material (Grzebisz *et al.*, 1989). Such studies suggest that there is a fraction of the dry matter that is lost readily (e.g. water-soluble sugars) while the remainder is stable. It is difficult to apply these results to soil-grown roots because such root tissues may be more or less resistant to leakage. Stronger roots might result in less leakage during washing, but the presence of microbes on the root surface might result in greater loss during storage. The practical response to these findings is to isolate roots from the soil as soon as possible after harvesting using the least water possible and to avoid storage in water. In reality it is almost impossible to avoid some period of root exposure to water since the outcome of washing is usually a mixture of live roots and other organic material whose separation is expedited by the presence of water.

Distinguishing live from dead roots can be difficult and, again, no universal standard is applied. The criteria are typically based on colour (separating white or pale brown roots from darker materials) and physical appearance (e.g. branched, able to bend, some elasticity). Some workers have attempted to use stains to separate live and dead roots but such attempts have been frustrated because old but living roots may stain only poorly while some types of dead organic materials stain readily. Physical separation techniques based on differences in density between live and dead materials have also had limited success because the differences in density are ill-defined. These problems mean that separation is usually a laborious job of hand labour in picking out debris by forceps from a sample that has been spread out in a shallow dish of water. These difficulties can be reduced in crop studies by siting them on land that has been fallowed in the previous season to reduce the quantity of dead roots, but for many studies this is impractical.

Root mass of live roots is normally determined after drying for 48 h at 80°C. Root length used to be measured manually either directly with a ruler or indirectly using a line-intercept technique (Newman, 1966; Tennant, 1975). For example, if roots are spread randomly on a sheet of graph paper with 1 cm grid lines and the number of intersects (N) with the grid lines is counted, then the length of roots (in cm) is equal to 0.786N. Such estimates are prone to human and other systematic errors (Bland and Mesarch, 1990). Recently, though, image analysis systems have become available for this measurement and also for simultaneous measurement of root diameter and, in some systems, angles of branching (Richner *et al.*, 2000). Scanned images can now be stored on compact discs for analysis with a variety of software types. In the edge discrimination method (Pan and Bolton, 1991), root length is calculated from the perimeter length of roots described by the number, orientation and length of the pixels describing the edge of the object. As with the earlier manual methods of length estimation, this technique assumes that roots are oriented randomly. Random placement of roots on a dish is – in practice – difficult to achieve with human operators, so Kaspar and Ewing (1997) developed an edge chord algorithm which traces chords along the object edge and is insensitive to object orientation. Their ROOTEDGE software has been widely used by root researchers. The other major difficulty of length estimation, root overlap, was tackled by Kimura *et al.* (1999) who developed the public domain NIH Image software to produce a thinned image (skeleton), and demonstrated good agreement between manual and image analysis measurements even when samples were overlapping.

3.1.2 Rhizotrons and minirhizotrons

Because of the destructive and time-consuming nature of washed soil methods, many researchers have attempted to use windows of glass or transparent plastics inserted into soils to observe roots. When installed carefully, such methods allow the dynamics of individual root growth and decay to be recorded, processes that are difficult to record in any other way. In a few instances, below-ground laboratories (rhizotrons) have been built (e.g. the facilities for measuring fruit tree roots at East Malling, UK [Rogers, 1939]; and that for measuring crop roots at Wageningen, The Netherlands [van de Geijn *et al.*, 1994]), and have provided long-term studies of a range of plants and treatments. For example, Smit and Groenwold (2005) used horizontal glass tubes to study differences between crops in the rates of downward movement of the rooting front and of root proliferation in a humic sandy soil. They found that the downward velocity of the root front was linearly related to thermal time (the summation over time of temperature above a specified base; see section 5.1.1) although there were substantial differences between species (Table 3.1). Roots of crops such as cereals and fodder radish grew downwards almost four times more rapidly than crops such as leeks and common velvet grass.

Rhizotrons, though, are expensive to construct and site-specific, so many researchers have preferred the relative cheapness and versatility offered by minirhizotrons (glass, transparent plastic, or flexible, pressurized tubes). Such systems, when combined with video cameras, allow characteristics such as growth rate, root orientation, branching, diameter, longevity, the presence of root hairs and infection with mycorrhizas or nematodes to be

Table 3.1 Estimated rates of downward movement of the rooting front (mm day^{-1}) for different crops measured in the Rhizolab at Wageningen, The Netherlands

Crop	Rate at 15°C	Rate at 20°C
Beetroot	24	31
Brussels sprouts	22	29
Common velvet grass	11	14
Faba bean	21	28
Fodder radish	38	51
Leek	10	14
Lucerne	19	25
Maize	15	20
Potato	15	20
Ryegrass	15	19
Spinach	29	38
Spring wheat	41	54
White lupin	24	31
Winter rye	26	35
Winter wheat	28	38

From Smit and Groenwold, 2005.

quantified (Taylor, 1987; Majdi and Nylund, 1996; Liedgens and Richner, 2001). A crucial factor in the use of the minirhizotron method is proper contact with the surrounding soil so that the glass/soil interface represents growth in the bulk soil rather than a zone of root proliferation or inhibition. In soils with hard layers, introduction of a tube may introduce a channel that promotes deeper rooting than would normally occur (Vos and Groenwold, 1987). If installed too tightly, smearing of soil against the tube prevents viewing of roots, while imperfect contact means that roots may proliferate preferentially in voids or track along the surface of the tube. Condensation on the surface of the tube and a tendency for roots to produce large numbers of long root hairs are frequent symptoms of poor tube/soil contact.

Observation of the roots with mirrors, cameras or video cameras has led to various attempts to estimate the rooting distribution in the surrounding soil. The basic issue is that of converting a measure in two dimensions to one in three dimensions. Root length on the minirhizotron surface (cm root cm^{-2} tube surface) can be estimated by counting the number of intersections on a grid inscribed on the tube (Tennant, 1975), and converted to a root length per unit volume of soil (L_v cm root cm^{-3} soil) by assuming a certain depth of view. Heeraman and Juma (1993) assumed a depth of view of 2 or 3 mm, although this appears an arbitrary choice. The number of roots at the minirhizotron surface (N_s) has also been converted to bulk soil root length using factors derived from theoretical models:

$$L_v = c.N_s \quad (3.1)$$

where c ranges from 1 to 3.8 depending upon assumptions in the model (Upchurch and Ritchie, 1983; Merrill and Upchurch, 1994). Bland and Dugas (1988) found that c = 2 provided the best estimate of L_v for cotton, and Smit *et al.* (1994) found the same result for a range of crops when minirhizotron counts were calibrated with washed soil cores. However, Bland and Dugas (1988) found no stable value of c for sorghum, and that c also varied with time for crops such as potatoes. Buckland *et al.* (1993) suggested that bias in estimating the number of roots at the minirhizotron surface could be minimized by counting only the first and last points of contact for each root. This method agreed well with washed cores for wild cherry roots, but not for roots of pasture species.

In short, there is no easy way of converting observations at the minirhizotron surface to bulk soil estimates of root length. Moreover, the distribution of root length estimated with minirhizotrons rarely replicates that obtained from washed soil cores. In general, less root length is found in the upper soil layers with minirhizotrons, while that in deeper layers is overestimated. For example, Heeraman and Juma (1993) found that L_v in the upper 10 cm was considerably underestimated by minirhizotrons in barley and faba bean, and considerably overestimated below 30 cm in faba bean (Fig. 3.2). Similar results of underestimation in surface layers and overestimation in deeper layers were found by Gregory (1979) for a wheat crop, and by Hansson and Andrén (1987) for crops of barley, lucerne and meadow fescue. Kage *et al.* (2000) also found underestimation in the surface layers beneath crops of cauliflower, although there was an acceptable correlation between minirhizotron and washed soil core results for the subsoil. Many reasons for underestimation in surface layers have been suggested including: (i) light effects; (ii) tracking encouraging deeper roots; (iii) surface rooting being more horizontal than vertical, resulting in

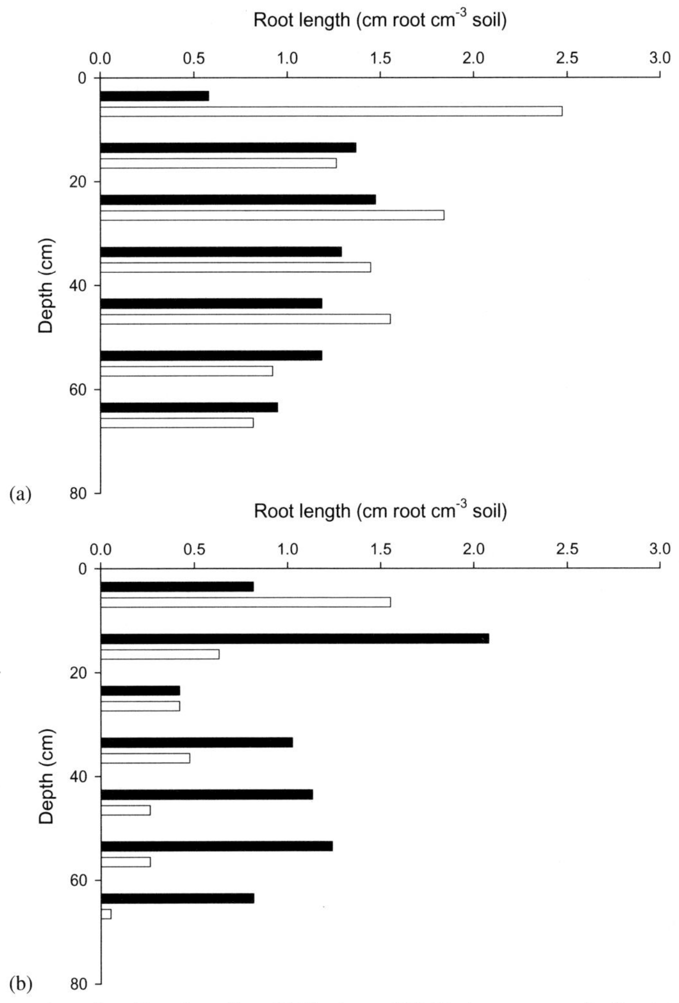

Fig. 3.2 Comparison of root length profiles of (a) barley and (b) faba bean measured with minirhizotron (filled) and washed soil core (open) techniques. (Redrawn and reproduced with permission from Heeraman and Juma, *Plant and Soil*; Springer Science and Business Media, 1993.)

different probabilities of roots intercepting the minirhizotron surface at different depths; (iv) temperature effects; and (v) disruption of soil structure and soil hydraulic properties. In summary, minirhizotrons are not as good as destructive methods such as washed soil cores for quantifying root length and mass especially in the upper soil (Heeraman and Juma, 1993; Smit *et al.*, 2000b).

Where minirhizotrons really come into their own, though, is in allowing repeated measurements at the same spot. This has allowed estimation of the periods of growth and death, and longevity of roots either as cohorts or in terms of cumulative curves of new and dead roots. Hendrick and Pregitzer (1992) introduced the idea of examining groups of roots (cohorts), defined as all individual roots appearing at about the same time, to give a population analysis of root longevity. For example, Hooker *et al.* (1995) observed poplar roots with a minirhizotron and found that within a cohort of non-mycorrhizal roots 10% survived only 14 days and a further 40% only 42 days. Other studies of root longevity using minirhizotrons are referred to in section 3.5.

3.1.3 Other techniques

Besides the measurement of root mass and length, root distribution with depth in the soil profile, and the longevity and turnover of roots, other features of root systems are also worthy of measurement. Radioisotopes (e.g. ^{32}P) injected into soil at specific depths and distances from plants have allowed rooting depth and lateral spread of roots to be measured without the need for soil sampling. The presence of isotope in the shoot means that roots are present and active at the site of injection (Bassett *et al.*, 1970). Alternatively, radioisotopes can be incorporated into the plant and autoradiography used to study the distribution of roots in soil (Fusseder, 1983) or the spatial relationships between the root systems of neighbouring plants (Litav and Harper, 1967). Such techniques are not used as widely as they were because of health and safety issues, and because non-uniform distribution of label can make interpretation of results difficult. For example, when grass (*Bouteloua gracilis*) and shrub (*Gutierrezia sarothrae*) plants were labelled with ^{14}C and ^{86}Rb, Milchunas *et al.* (1992) found that ^{14}C activity was concentrated near the soil surface, and ^{86}Rb activity was highly variable and randomly distributed so that neither technique produced the same estimate of root distribution as excavation on nail boards.

Digging a trench and observing the number of cut root ends (the trench-profile technique) can also provide a good measure of the spatial distribution of roots and is much quicker than washed soil cores (requiring about 20% of the time according to Vepraskas and Hoyt, 1988). The data can be displayed as dot maps, and careful washing of a thin, but known, layer of soil from the profile face may also allow estimation of root length *in situ* (Böhm, 1976). While destructive of a site, this technique can also be useful in observing interactions of root systems of different species (Gregory and Reddy, 1982), and for relating the position of roots to structural features in soils such as large pores, cracks and ped faces (Logsdon and Allmaras, 1991).

Non-invasive techniques such as X-ray micro-tomography for imaging roots growing in soils are under development, but limited to small plants growing in sieved soil in small containers (Heeraman *et al.*, 1997; Gregory *et al.*, 2003). Nevertheless, resolution of 50–100 μm is now readily achievable and developments in tube and detector design should allow imaging of larger samples and, in time, the analysis of field-grown roots in structured soils.

3.2 Root system development

Plants normally pass through a predictable sequence of stages between germination and maturity. While the rate of progress through this sequence and the duration of the various stages are affected by the environment, the order of the stages is largely genetically determined and only altered by extremes of environment. Several perennial plants have identifiable juvenile and adult stages and in many annual crop plants, there are distinctive developmental stages of the shoot (e.g. appearance of individual leaves, tillering, emergence of the reproductive structure) which have been codified to aid decisions about management of the crop (see Lancashire *et al.* [1991] for a range of temperate crops). Similarly, roots develop in an orderly manner, although much less is known about this and quantitative description is poor compared to schemes for shoots.

Root system development starts with the emergence of a root axis (the tap root) that later branches to form first order laterals, which in turn branch to form second order laterals and so on. Branching commences towards the base of an axis or lateral and proceeds towards the tip, whereas extension occurs only close to the tip. Topologically, the structure is that of a mathematical tree and never a network with loops (see section 2.5).

Several attempts have been made to relate root development to shoot development thereby avoiding the work involved in isolating root systems from soil. Klepper *et al.* (1984) developed a systematic identification scheme for root axes of wheat and then related their appearance and branching to the appearance of leaves and tillers. In wheat, seminal root axes appear first with subsequent axes appearing in pairs from nodes; more than one pair may appear from higher nodes located above ground. The scheme of Klepper *et al.* (1984) also specified the orientation of the root axes with respect to the midrib of the leaf present at the node. Figure 3.3 summarizes the integrated development of leaves and root axes for the main stem and tillers. For example, a main stem with 5 leaves could be expected to have a coleoptile tiller with 3 leaves, tiller 1 with 2.5 leaves, tiller 2 with 1.5 leaves and tiller 3 just visible. A pair of axes at node 3 would just be visible, all the seminal axes and a pair of coleoptile axes would have first order laterals, and the tap root and first pair of seminal axes would have young second order laterals. This scheme indicates the maximum number of root axes that will develop, although some genotypes of wheat may produce up to seven seminal axes. Such connectivity between shoot and root development allows the

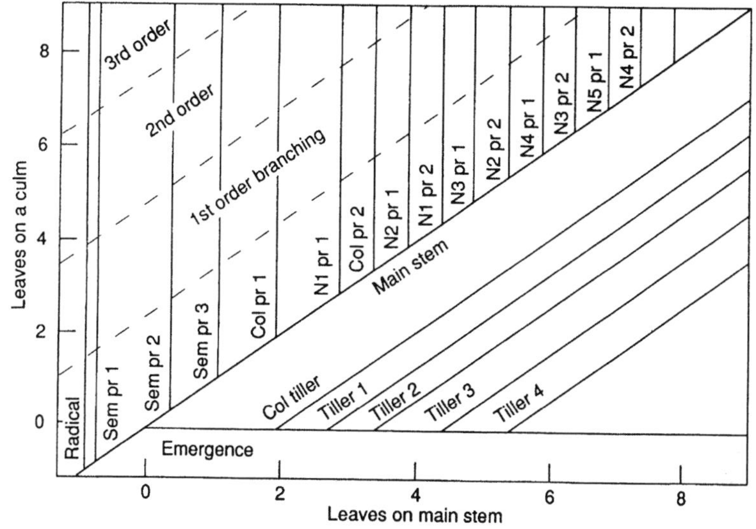

Fig. 3.3 Diagrammatic representation of the coordinated development of wheat leaves, root axes and tillers. The appearance time of tillers, root axes and root axis branching relative to appearance of leaves on the main stem is shown: seminal (sem), coleoptile (col), leaf node (n) and pair (pr). (Original research of Klepper *et al.*, 1984; reproduced with permission from Gregory, In: *Physiology and Determination of Crop Yield*; ASA, CSSA, SSSA, Madison, 1994.)

prediction of the number of root axes from shoot characters. Klepper *et al.* (1984) found that the number of nodal axes (y) on any culm was linearly related to the number of leaves (x) on the same culm by $y = 1.95x - 3.06$. Similarly, Gregory (1983) found a linear relation between the number of root axes and the number of leaves of pearl millet (*Pennisetum typhoides* S. & H.), a plant with a single culm, of $y = 1.42x - 2.26$.

In practice, genotype, crop management and the soil environment all affect the actual presence of tillers, axes and branches. For example, the coleoptile tiller is usually missing in wheat that is shallow-sown so that the number of root axes is correspondingly reduced. Similarly, drying of the upper soil has long been known to inhibit the appearance of nodal axes (Locke and Allen, 1924), and nutrient supply also has a big effect on the degree of branching (Forde and Lorenzo, 2001). Seed size can also have a large effect on the number of root axes that develop and on the rate of early growth. For example, Richards and Passioura (1981) found that the number of seminal axes of wheat cv. Kalyansona increased from 3.3 to 5.1 per plant as seed weight increased from 19 to 56 mg. Maximum xylem vessel diameter similarly increased from 62 to 75 μm, and the proportion of seminal axes with multiple metaxylem vessels from 0.15 to 0.54 over the same range of seed weight. Seed weight also affects the vigour of early shoot growth (Richards and Lukacs, 2002), and can be expected to have similar effects on early root growth.

3.3 Size and distribution of root systems

3.3.1 Mass and length

A worldwide database of measurements of root profiles showed that the average root mass ranged from about 0.2 kg m^{-2} for croplands to about 5 kg m^{-2} for forests and sclerophyllous shrubs and trees (Jackson *et al.*, 1996). Root mass in forest ecosystems ranged from 2 to 5 kg m^{-2} while that in croplands, deserts, tundra and grasslands was <1.5 kg m^{-2} (Table 3.2).

Table 3.2 Total root biomass, total fine root biomass (≤2 mm diameter) and live fine root area index for 11 terrestrial biomes

Biome	Total root biomass (kg m^{-2})	Total fine root biomass (kg m^{-2})	Live fine root area index
Boreal forest	2.9	0.60	4.6
Cropland	0.15		
Desert	1.2, 0.4	0.27	5.5
Sclerophyllous shrubs and trees	4.8	0.52	11.6
Temperate coniferous forest	4.4	0.82	11.0
Temperate deciduous forest	4.2	0.78	9.8
Temperate grassland	1.4	1.51	79.1
Tropical deciduous forest	4.1	0.57	6.3
Tropical evergreen forest	4.9	0.57	7.4
Tropical grassland/savanna	1.4	0.99	42.5
Tundra	1.2	0.96	5.2

The two values for desert root biomass are for cold and warm deserts. The apparently greater fine root biomass than total root biomass of the temperate grassland is a consequence of two different data sets being employed and the fact that most grassland roots are fine. (Data from Jackson *et al.*, 1996, 1997.)

In croplands and temperate grasslands, most of the roots are fine roots (≤2 mm diameter), but in many biomes coarse roots make up by far the majority of the root biomass. It is the fine roots that are of particular interest because they constitute the primary pathway for water and nutrient uptake. Jackson *et al.* (1997) used selected profiles from the database referred to previously together with information from about 100 other studies to estimate fine root biomass and surface area. Their estimates of fine root biomass ranged from 0.27 kg m^{-2} in deserts to 1.5 kg m^{-2} in temperate grasslands, with most values lying between 0.5 and 1.0 kg m^{-2} (Table 3.2). Fine root biomass was converted to live root biomass and thence to surface area using parameters such as root diameter and specific root length to give a root area index (m^2 m^{-2}) comparable with the concept of leaf area index used in shoot studies. Root area index of live fine roots was highest in grasslands (about 80 in temperate regions and 43 in tropical regions) and <12 for all other biomes (Table 3.2). These estimates suggest that root area index is at least comparable to leaf area index in all biomes, and greater than leaf area index in most biomes. In grasslands, root area index is about an order of magnitude greater than leaf area index.

Dry mass accumulation by root systems typically follows the sigmoidal pattern commonly observed with shoots, although the phases of growth in the root and shoot may not coincide exactly. Figure 3.4 shows changes in root mass with time for some annual crops. In annual crops, flowering (anthesis) appears to be a particularly important developmental stage after which assimilates are required to fill the growing grain leaving little for roots. In almost all studies with cereals, the mass of the root system has not been found to increase after flowering and in some studies a substantial decrease has been observed during grain-filling depending on soil conditions (for example, see Mengel and Barber, 1974). Most legumes are much less determinate than cereals, with flowering and grain-filling occurring over a more prolonged period so that the demand for assimilates by the grain increases gradually. In consequence the root mass of many legumes continues to increase during flowering and early grain-filling (e.g. Sivakumar *et al.*, 1977 for soyabean; Gregory and Eastham, 1996 for narrow-leaf lupin), although the degree of determinacy may influence this pattern (Mayaki *et al.*, 1976).

In grass swards, the pattern of root growth is complex and depends on the species and the grazing regime. Flowering of perennial ryegrass typically results in a decreased rate of initiation of new roots and increased senescence and decay of older roots, and reduces root growth relative to that of the shoot (Troughton, 1957). In controlled conditions, defoliation reduced the mass and length of the root system and root diameter of the grasses *Lolium perenne* and *Festuca ovina*, although the size of effects depended on N supply (Dawson *et al.*, 2003). Deinum (1985) postulated that a wide range of effects of grazing on root mass could be expected depending upon the frequency and severity of the defoliation with respect to tillering behaviour of the shoot. Severe defoliation usually reduces root mass because reserve carbohydrates are used to restore shoot growth, but if the defoliation regime is less intense and promotes tillering then associated new root axes may sustain the root mass. Similarly, cutting of the tap-rooted perennial forage legume crop, lucerne, produced an initial decrease in fine root mass followed by a recovery towards the end of the harvest cycle (Luo *et al.* 1995).

A sigmoidal pattern of growth is also evident for root length, although the patterns for mass and length may not always coincide. For example, Ford *et al.* (2002) measured live root

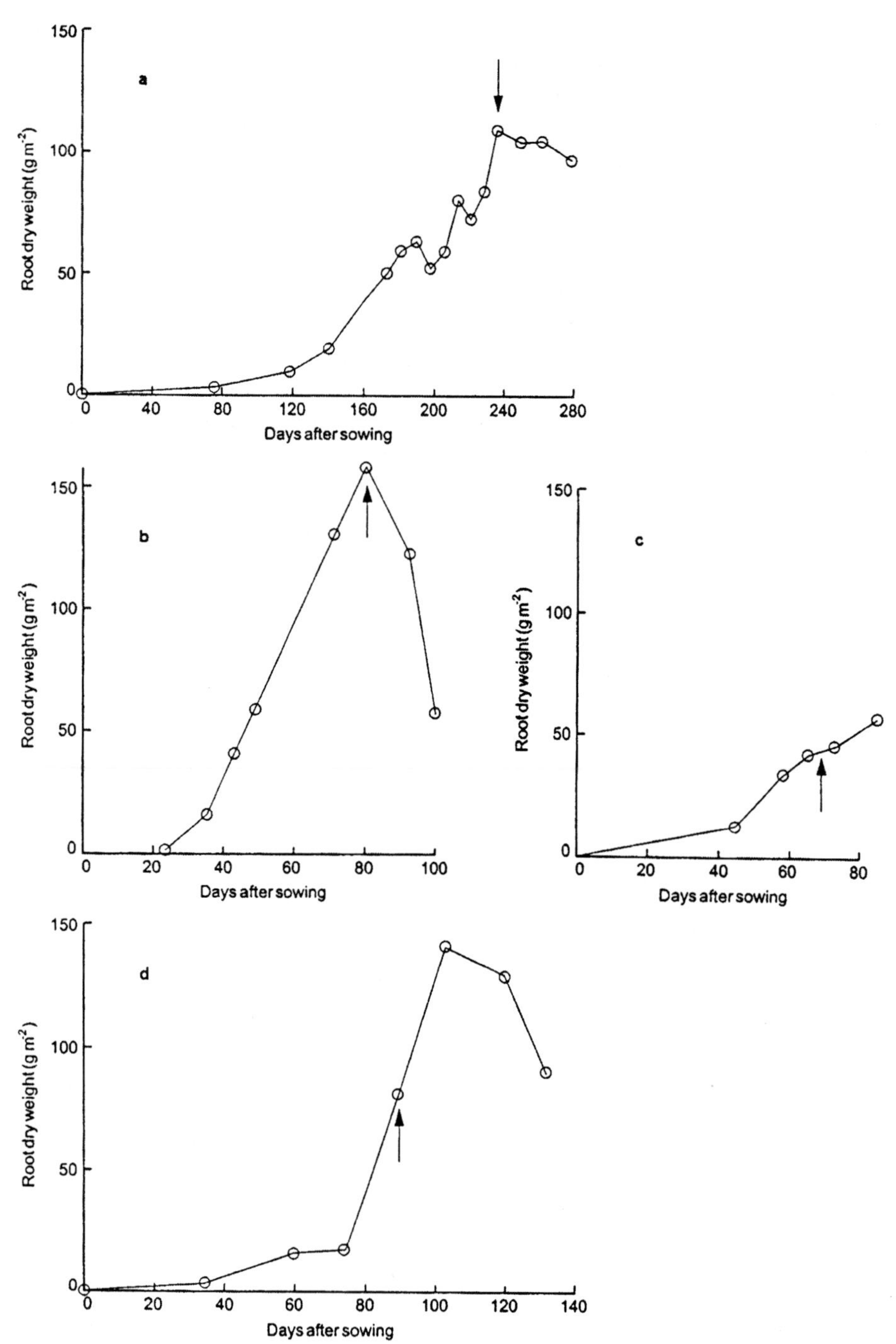

Fig. 3.4 Changes in root mass with time for crops of (a) winter wheat (Gregory *et al.*, 1978), (b) maize (Mengel and Barber, 1974), (c) soyabean (Sivakumar *et al.*, 1977), and (d) narrow-leaf lupin (Gregory and Eastham, 1996). The approximate time of the start of flowering is indicated with an arrow. (Reproduced with permission from Gregory, In: *Resource Capture by Crops*; Nottingham University Press, 1994a.)

mass and length of six cultivars of winter wheat on a sandy loam overlying sand at anthesis and 4 weeks later, and found that while root mass did not increase in this period, there was an increase in root length. The mean root length in the upper 0.3 m of the profile increased significantly from 3.70 to 5.34 cm cm^{-3} and that between 0.3 and 0.8 m increased from 0.96 to 1.27 cm cm^{-3} although this was not statistically significant. These results suggest that during grain-filling, proliferation of fine roots occurred while some of the thicker mature roots died. This, in turn, highlights a major deficiency in most estimates of root length and mass, namely that the technique most commonly used to obtain such data (soil coring followed by extraction of the roots by washing with water) gives a measure only of net growth (i.e. the balance between new roots produced and older roots that have died). There have been fewer studies of oilseed crop roots but, as with many cereal crops, root length of oilseed rape declined substantially in the 2 months after flowering (Barraclough, 1989).

Table 3.3 summarizes information from several experiments to show the maximum values of root mass and length recorded for crop species. The results were obtained from washed soil cores so, for the reason stated above, these values should be regarded as the maximum values of live roots recovered rather than true measures of total root mass production for the growing season. Despite this reservation, it is clear that cereals typically have heavier and longer root systems than legumes and other crops. Generally legumes have about one-half the root mass of cereals and about one-fifth the root length, although

Table 3.3 Maximum root dry mass and length of some common crops

Crop	Soil	Country	Dry mass (g m^{-2})	Length (km m^{-2})	Author
Barley	Typic Natrixeralf	Australia	98	5.3	Gregory *et al.* (1992)
Cauliflower	Loess loam	Germany		12.7	Kage *et al.* (2000)
Chickpea	Palexerollic Chromoxerert	Syria	45	4.5	Brown *et al.* (1989)
Faba bean	Chromoxerertic Rhodoxeralf	Syria	98	2.9	Manschadi *et al.* (1998)
Groundnut	Udic Argiustoll	USA	103	12.1	Ketring and Reid (1993)
Maize	Typic Argiaquoll	USA	160	15.1	Mengel and Barber (1974)
Pearl millet	Udic Rhodulsalf	India	63	4.2	Gregory and Squire (1979)
Pigeonpea	Udic Rhodulsalf	India	145	1.9	Devi *et al.* (1996)
Potato	Typic Haplorthod	USA	25.5	8.7	Opena and Porter (1999)
Potato	Polder soil	The Netherlands	77	7.1	Vos and Groenwold (1986)
Rape	Flinty loam	UK	163	28.3	Barraclough (1989)
Rice (lowland)	Maahas clay loam	The Philippines	86	20.7	Thangaraj *et al.* (1990)
Sorghum	Well-drained red earth	Australia	100	26.5	Myers (1980)
Soyabean	Typic Udorthent	USA	58	5.5	Sivakumar *et al.* (1977)
Subterranean clover	Entisol	Australia	109	17.5	Pearson and Jacobs (1985)
Sugar beet	Sandy clay loam	UK	76	9.7	Brown and Biscoe (1985)
Wheat	Vertic Argiudoll	Argentina	350	18.4	Savin *et al.* (1994)
Winter wheat	Sandy loam	UK	105	23.5	Gregory *et al.* (1978)

Table 3.4 Comparison of maximum root mass and length of different crop species at the same site

Soil	Site	Crop	Root mass ($g\ m^{-2}$)	Root length ($km\ m^{-2}$)	Author
Deep sand	Wongan Hills, Western Australia	Lupin		2.4	Hamblin and Tennant (1987)
		Pea		3.0	
		Wheat		12.3	
		Barley		14.0	
Deep sand	Merredin, Western Australia	Lupin		1.2	Hamblin and Tennant (1987)
		Pea		1.8	
		Wheat		7.7	
		Barley		5.3	
Vertisol	Jindiress, Syria	Chickpea	45	4.5	Brown *et al.* (1989)
		Barley	83	13.3	Brown *et al.* (1987)
Sand over clay	East Beverley, Western Australia (three seasons)	Lupin	166	1.33	Gregory and Eastham (1996)
			145	1.97	
			148	6.63	
		Wheat	107	4.84	
			95	3.45	
			86	10.42	

this is not always so (see Table 3.4 for an example of lupin root mass exceeding wheat root mass on the same site). Although specific root length (the mass per unit length) varies considerably and is situation-dependent, typical values are 10 m g^{-1} for trees, 50–200 m g^{-1} for dicotyledonous crops, 200 m g^{-1} for cereals and 400 m g^{-1} for fine-rooted grasses (van Noordwijk and Brouwer, 1991).

Conclusions from the results presented in Table 3.3 must be made cautiously because the studies involved differences in soils, climates and management practices as well as in crops. There have been few comparative studies of species grown under similar conditions and even fewer in which a species has been grown on the same site over several seasons. Table 3.4 summarizes results of some studies in which different crops were grown at the same location. As with Table 3.3 it confirms the general impression that the temperate cereals have longer root systems than legumes. For example, Hamblin and Hamblin (1985) found that the ranking of total root length was pasture legumes (subterranean clover and medics) > wheat > lupins and peas. Dry mass of cereal roots may, on occasion, be less than that of the legume (e.g. Gregory and Eastham, 1996) depending on the growing conditions and the effect that these have on both the size of the root system and the specific root length.

Inter-seasonal variation of root growth at a site has only been measured in a handful of studies. Welbank *et al.* (1974) compared the growth of spring barley roots over a 4-year period on a sandy silt loam soil in the UK. Their results showed little inter-seasonal difference in the maximum dry mass of roots recorded (95 g m^{-2} in the first year and 80 g m^{-2} subsequently) although there were differences in the rates of early growth from year to year. In contrast, Barraclough and Leigh (1984) working on two contrasting soil types in the UK found that winter wheat crops in one year produced considerably more root dry mass (154 g m^{-2}) than in the other (113 g m^{-2}), although there was no significant difference in root length (25.5 km m^{-2}). While inter-seasonal variation in climate is commonly low in the maritime, temperate UK, conditions elsewhere may vary substantially. Table 3.5 shows the year to year variation in root and shoot growth at anthesis of barley crops grown in the

Table 3.5 Seasonal growth of shoots and roots of barley crops (cv Beecher) grown on a Typic Calciorthid (clay loam) at Breda, northern Syria; the standard error of the mean is shown in brackets

Season	Rain (mm)	Shoot dry mass ($g\ m^{-2}$)	Root dry mass ($g\ m^{-2}$)	Root:total mass
1981/82	322	462 (29.6)	107 (10.6)	0.19
1982/83	272	342 (19.0)	58.5 (3.1)	0.15
1983/84	204	159 (11.7)	35.2 (4.9)	0.18
1984/85	264	393 (11.8)	57.4 (3.6)	0.13
1985/86	218	149 (54.1)	50.8 (1.8)	0.25
1986/87	228	399 (32.6)	61.4 (4.0)	0.13

From Gregory *et al.*, 1997.

Mediterranean climate of northern Syria. All crops were given moderate amounts of N and P fertilizers so that seasonal rainfall was the principal variable between years. Over the 6 years of the study, root and shoot masses differed between seasons by a factor of about three but root mass as a proportion of total plant mass differed less between seasons. The season with most rain had the heaviest root system and that with least rain, the lightest. The within-season pattern of rainfall had an influence on both growth and partitioning, so that there was no clear relation between total seasonal rainfall and the root:total mass ratio. A similar inter-seasonal comparison of the landrace barley, Arabic abiad, also undertaken at Breda in northern Syria, found that root mass differed between seasons (Gregory, 1994a). Root mass ranged from 42.5 to 90.6 $g\ m^{-2}$ and root length from 3.2 to 13.9 $km\ m^{-2}$ in the 5-year study. Root length and root mass at flowering were closely related so that the specific root length (mean 174 $m\ g^{-1}$) was similar despite the large seasonal differences in rainfall.

3.3.2 Depth of rooting

Genetic and environmental factors both influence the depth of rooting. The depth to which roots are able to grow has many implications for the hydrological balance and biogeochemical cycling of ecosystems. Canadell *et al.* (1996) summarized 290 observations of maximum rooting depth of 253 woody and herbaceous species from the major terrestrial biomes. They found that maximum rooting depth varied from 0.3 m for some tundra species to 68 m for *Boscia albitrunca* in the central Kalahari. Twenty-two species had roots that extended to 10 m or more but 194 species had roots that were at least 2 m deep. Table 3.6 shows the average maximum rooting depth for each biome, which ranged from 0.5 ± 0.1 m for tundra to 15.0 ± 5.4 m for tropical grassland/savanna. The results showed that deep root habits are quite common in woody and herbaceous species across most terrestrial biomes. Leaving aside the cropped areas, and grouping the species across biomes, the average maximum rooting depth was 7.0 ± 1.2 m for trees, 5.1 ± 0.8 m for shrubs, and 2.6 ± 0.1 m for herbaceous plants. These results emphasize the extent of root systems and that roots are not confined to the surface layer of soil.

The maximum depth of rooting on deep soils is genetically determined and differs between species grown under identical conditions. For example, Hamblin and Hamblin (1985) grew lupin (*Lupinus* spp.), pea and wheat on three deep sands (entisols) at differ-

Table 3.6 Average maximum depth of rooting and empirical distribution coefficient (β) for terrestrial biomes

Biome	Depth of rooting (m)	β
Tundra	0.5 ± 0.1	0.914
Boreal forest	2.0 ± 0.3	0.943
Cropland	2.1 ± 0.2	0.961
Temperate grassland	2.6 ± 0.2	0.943
Temperate deciduous forest	2.9 ± 0.2	0.966
Tropical deciduous forest	3.7 ± 0.5	0.961
Temperate coniferous forest	3.9 ± 0.4	0.976
Sclerophyllous shrubs and trees	5.2 ± 0.8	0.964
Tropical evergreen forest	7.3 ± 2.8	0.962
Desert	9.5 ± 2.4	0.975
Tropical grassland/savanna	15.0 ± 5.4	0.972

Data from Canadell *et al.* (1996) and Jackson *et al.* (1996).

ent sites in Western Australia and found that the maximum rooting depth was significantly different ($p<0.001$) between genotypes and species, but not between sites. Rooting depth averaged 1.9 m for lupins, 0.65 m for peas, and 1.13 m for wheats. Similarly, Greenwood *et al.* (1982) grew a range of vegetables on a sandy loam at Wellesbourne, UK and found that while onion and lettuce roots were confined to the upper 0.65 m, pea rooted to 0.75 m, broadbean to 0.85 m, and turnip, parsnip and cauliflower to >0.85 m. Minirhizotron studies of eight crops grown at different sites (soil predominantly a silt loam) over three seasons in North Dakota, USA showed considerable differences in rooting depth and total root length (Merrill *et al.*, 2002). Average maximum rooting depth was 1.6 m in safflower, 1.45 m in sunflower, 1.3 m in spring wheat, about 1.15 m in crambe (*Crambe abyssinica*) and canola (*Brassica rapa*), and 1.0 m in common bean, soyabean and pea.

In practice, environmental conditions may also play an important role in determining rooting depth either because of limited soil depth or hostile soil conditions. For example, on a deep vertisol at Jindiress, northern Syria, the depths of rooting and of water extraction of barley and chickpea crops was similar at 1.2 m (Gregory and Brown, 1989). In such regions of the Mediterranean, the depth of rooting is frequently determined by the depth of re-wetting by rainfall and varies with both site and season. Similarly, narrow-leaf lupin (*Lupinus angustifolius*) and wheat planted on a duplex (sand over clay) soil in Western Australia both rooted to 0.8 m because of physical impediments to growth in both the sand and clay layers, and because of the limited depth of wetting by rain (Dracup *et al.*, 1992).

Figure 3.5 gives some examples of the change of rooting depth with time for some annual crops. Downward rates of root extension are typically 10–40 mm day^{-1} during the phase of rapid downward extension depending on the crop and growing conditions. Gregory *et al.* (1978) found that the rate of downward growth of a winter wheat crop grown in the UK averaged 6 mm day^{-1} during the winter and 18 mm day^{-1} between early April and early June when temperatures were much warmer. In Kansas, the rooting depth of early summer-sown crops of sorghum and sunflower advanced at 25 and 41 mm day^{-1}, respectively (Stone *et al.*, 2001). Borg and Grimes (1986) analysed data from 48 crop species in 135 field studies. They found that in all cases, the increase in rooting depth with time fol-

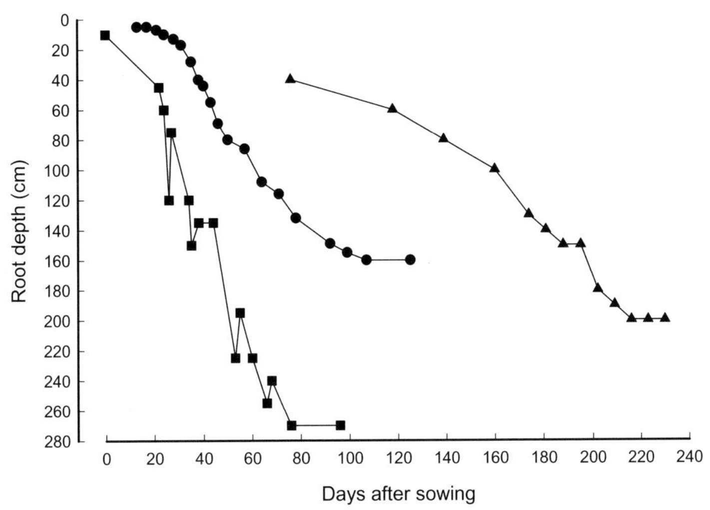

Fig. 3.5 Change in rooting depth with time for soyabean (●, Mayaki *et al.*, 1976), sunflower (■, data for two crops grown in consecutive seasons – Jaafar *et al.*, 1993) and winter wheat (▲, Gregory *et al.*, 1978).

lowed a sigmoidal pattern, although the final depth achieved, as well as the time required to achieve it, depended on the crop species and environmental conditions. They found that the actual rooting depth (RD) at a particular time could be estimated from:

$$RD = RD_{max} [0.5 + 0.5 \sin (3.03\, t_r - 1.47)] \quad (3.2)$$

where RD_{max} is the maximum rooting depth and t_r is the relative time elapsed between sowing and maturity. In practice, while the length of cropping cycle is usually easy to estimate, selecting RD_{max} requires some local knowledge.

3.3.3 Distribution of roots

Roots are not distributed evenly throughout the soil profile (see Fig. 2.7), and their distribution is important in determining the availability of water and nutrients to plants. To complement the global depth of rooting study referred to in section 3.3.2, Jackson *et al.* (1996) compiled a database of 250 root studies (mainly of root mass but some of root length) and fitted a simple asymptotic function to each to describe the distribution of roots with depth. The function defines the cumulative proportion of roots (Y) between the surface and depth (d, expressed in cm) as:

$$Y = 1 - \beta^{d} \quad (3.3)$$

where β is a dimensionless coefficient that is determined empirically. The higher the value of β, the greater the proportion of roots deeper in the soil profile. Table 3.6 shows the values of β obtained for the 11 terrestrial biomes examined. Tundra, boreal forest and temperate grasslands had the shallowest rooting profiles with 80–90% of root mass in the upper 0.3 m of the profile. The least pronounced profiles were in the deserts and temperate coniferous forests where only 50% of root mass was in the upper 0.3 m. Grouping the plant species across biomes also produced interesting differences. Grasses ($\beta = 0.952$) had 44% of their root mass in the upper 0.1 m of soil whereas shrubs ($\beta = 0.978$) had only 21% in that zone. Tropical and temperate trees ($\beta = 0.970$) had 26% of their root mass in the upper 0.1 m and 60% in the top 0.3 m. The global average distribution of root mass for all biomes and vegetation types was 30% in the upper 0.1 m, 50% in the upper 0.2 m, and 75% in the upper 0.4 m. Root mass in the upper 0.3 m of the soil profile was calculated from these rooting profiles and combined with a land-cover database to give a global map of the percentage root mass in the upper 0.3 m (Plate 3.1). The map shows that shallow root systems are predominant at high latitudes (associated with permafrost and waterlogging), and the locations of shallow-rooted grasslands and more deeply rooted woody ecosystems (including deserts, tropical savannas, and temperate coniferous forests).

Schenk and Jackson (2002) extended this analysis to a database of 475 rooting profiles from 209 locations and used a logistic curve to extrapolate the depth of rooting in those datasets in which roots clearly extended beyond the depth of sampling. The depth including 95% of all roots increased as latitude decreased from 80° to 30° but in the tropics there was no clear pattern of variation. In more than 90% of the rooting profiles, at least 50% of the mass (or length) was in the upper 0.3 m of soil (mean 0.18 m) and 95% was within the upper 2 m. Deep rooting depths were associated with water-limited environments. Using general linear models they found that annual potential evaporation and precipitation together accounted for the largest proportion of the variance in rooting depth globally (12% for the depth of 50% of roots and 16% for the depth containing 95% of roots). While differences in soil type had only a small effect on the depth containing 50% of roots (they accounted for only 5% of the global variance) they were a major contributor to variance in 95% rooting depth (17% of the global variance). Overall, the extrapolated depth containing 95% of all roots was deeper in sandy soils than in clay or loam soils for five of the six vegetation types in which such comparisons were possible (Table 3.7). The depth containing 95% of all roots was also deeper in ecosystems with shallow organic horizons compared with deeper organic horizons. The reasons for soil type appearing to be relatively unimportant in determining the depth containing 50% of the roots are probably because this depth is generally

Table 3.7 The extrapolated depths (m) containing 95% of root mass/length for six global vegetation types in soils of different texture

Vegetation type	Sand	Loam or clay
Boreal forest	0.80 (0.21, 0.20) **12**	0.51 (0.13, 0.12) **13**
Cool temperate forest	1.21 (0.20, 0.19) **24**	0.99 (0.10, 0.08) **42**
Semi-desert	1.64 (0.53, 0.49) **8**	0.95 (0.24, 0.22) **13**
Desert	1.66 (0.63, 0.55) **6**	0.90 (0.25, 0.23) **11**
Dry tropical savanna	1.52 (0.38, 0.36) **13**	1.29 (0.31, 0.29) **14**
Tropical evergreen forest	0.54 (0.13, 0.13) **13**	1.01 (0.10, 0.07) **46**

Values in brackets are the + and – 95% confidence intervals for the means and the numbers in bold are the numbers of samples. From Schenk and Jackson, 2002.

within the nutrient-rich topsoil and therefore comparatively unaffected by properties of the subsoil, coupled with the fact that sampling schemes are not usually fine enough to discern subtle differences in the upper 0.3 m of soil.

The amount of root length in layers within the soil profile is normally expressed in terms of a root length per unit volume of soil (L_v often with units of cm root cm^{-3} soil), sometimes referred to as a root length density. Typical values of L_v in the upper 0.1 m of soil are about 20 cm cm^{-3} in grasses, 5–10 cm cm^{-3} in temperate cereal crops, and 1–2 cm cm^{-3} in other crops. In the crop literature, it has frequently been found that roots are distributed in the soil such that their length and mass decrease exponentially with depth. Gerwitz and Page (1974) first proposed this model after reviewing literature on vegetable crops, cereals and grasses, and it has been widely adopted since (e.g. Robertson *et al.*, 1993; Zhuang *et al.*, 2001). Their relation ($Y = 1 - \exp[-\alpha d]$) is analogous to the analysis of Jackson *et al.* (1996) where $\exp(-\alpha) = \beta$, and means that a plot of Ln L_v with depth should give a straight line with gradient $\exp(-\alpha)$. Figure 3.6 shows that the distribution of roots of some crops (cauliflower and winter wheat) is well described by such a relation. In other crops (e.g. rape and sugar beet), while this relation can be found in the surface layers, there is a tendency for values of L_v in deeper soil layers to be almost constant. Whether this is strictly a property of the crop or a result of an interaction between the crop and soil properties remains to be established. Typically $\exp(-\alpha)$ changes rapidly during the early part of the growing season as the crop is establishing, but is fairly stable during the main phase of growth (see King *et al.*, 2003 for an example with winter wheat). This suggests that during the major period of growth, the relative extension of roots proceeds at a similar rate at all depths. Such relations mean that for some crops grown on deep,

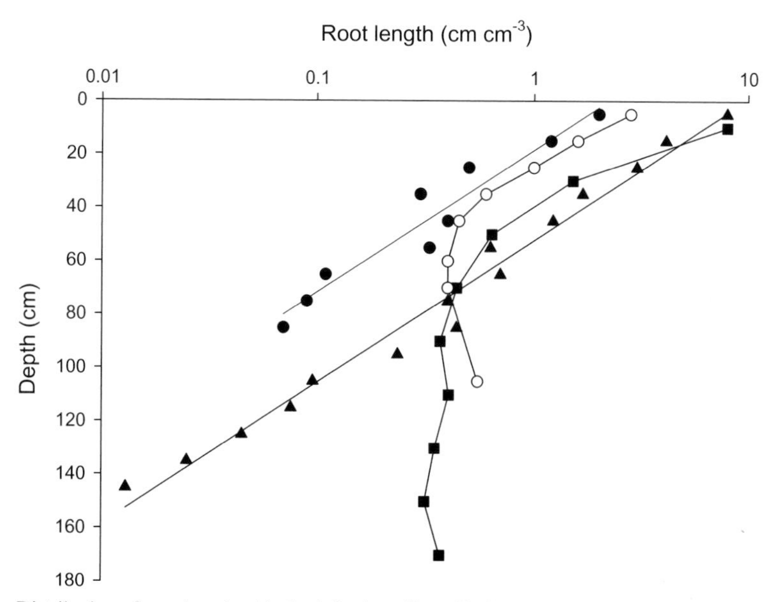

Fig. 3.6 Distribution of root length with depth in the soil profile for maturing crops of cauliflower (●, Kage *et al.*, 2000), oilseed rape (■, Barraclough, 1989), winter wheat (▲, Gregory *et al.*, 1978), and sugar beet (O, Brown and Biscoe, 1985). Linear regressions have been drawn for the distributions of cauliflower and winter wheat roots.

uniform soils, generalized functions of root distribution are possible (e.g. winter wheat – Zuo *et al.*, 2004), but for other plant species the paucity of data available does not yet allow robust generalization.

3.4 Root:shoot allocation of dry matter

The proportion of total plant dry matter allocated to roots differs substantially between different groups of plants and ecosystems. Jackson *et al.* (1996) found that mature crops had the lowest proportion of total dry matter in roots (averaging around 0.1) and that forest ecosystems also had only moderate proportions of total dry matter below ground (typically 0.2–0.35). Conversely plants in cold deserts, temperate grasslands and tundra frequently had more dry mass below ground than in the shoot so that root mass was typically 0.75–0.87 of total plant mass.

In annual plants, the allocation of dry matter to roots changes during their life cycle and with growing conditions. Typically, relatively more assimilates are channelled to roots during early growth but, as development proceeds, the growing reproductive structures come to dominate and the amount of assimilate translocated to roots decreases. This change in allocation has been observed in many crops (e.g. winter wheat – Gregory *et al.*, 1978; Barraclough, 1984; sugar beet – Brown and Biscoe, 1985; lupin – Gregory and Eastham, 1996) and is particularly pronounced in cereal crops as the stem elongates and the ear develops. Several studies have shown that the proportion of carbon translocated to roots decreases with time (Keith *et al.*, 1986; Gregory and Atwell, 1991; Swinnen *et al.*, 1995) as the ear grows and this is reflected in root mass. Barraclough (1984) found that the relation between root and shoot mass for several crops of winter wheat grown in the UK showed a distinct change in early spring at a shoot mass of 115 g m^{-2}. During the winter the relative rates of root and shoot growth were similar so that the slope of the allometric relationship (a plot of log root weight vs log shoot weight) was close to one, but from the time of N fertilizer application to anthesis it was about 0.33. The ratio of root:total plant mass decreased from about 0.3–0.5 over winter to about 0.10 at anthesis, and this has been found in many similar studies.

In general, a shortage of resources in the root environment causes a shift of assimilates in favour of the root system relative to the shoot and *vice versa* (Brouwer, 1983) (see section 1.2.1). This is clearly seen, for example, in the response of cereals to applications of N fertilizer where there was a sharp decline in root:total plant mass immediately following N fertilizer application (Barraclough, 1984). In drier conditions, too, fertilizer applications can have similar effects whereby shoot growth is increased substantially but root growth is less affected so that root:total mass ratio is decreased. For example, in barley crops in northern Syria, applications of P and N fertilizers decreased root:total mass at the start of stem elongation from about 0.35 without fertilizer to about 0.25 at two sites, although ratios were similar at anthesis (Brown *et al.*, 1987). At some locations there are multiple constraints to crop growth and interventions produce a range of root:total mass ratios. Hamblin *et al.* (1990) grew wheat in two seasons of different rainfall at a site in Western Australia subjected to a range of N fertilizer and tillage treatments. As conditions became harsher, root:total plant mass at anthesis increased from 0.1 to 0.6.

3.5 Root longevity and turnover

The root:total plant mass ratios described in section 3.4 are only a partial account of the investment made by plants in their root systems, because as well as assimilates for growth, assimilates are also required to maintain and renew the system. More details of carbon fluxes are given in section 7.1, but even considering just biomass, the investment in root production is considerable. Caldwell (1987) showed that more than half of the annual total biomass production of five ecosystems dominated by perennial species was below ground and that root:total plant mass ratio was not well correlated with root:total production. For example, in a deciduous forest dominated by *Liriodendron tulipifera* the root:total mass ratio was 0.1 and the root:total annual production ratio was 0.62, while the corresponding values for arctic tundra were 0.85 and 0.58. These different ratios reflect differences in root longevity, rates of root renewal and rates of metabolic activity.

The median lifespan of roots is highly variable, ranging from a few weeks in some plants (e.g. strawberry and annual crops like sorghum and groundnut) to many months (e.g. sugar maple) in others (Eissenstat and Yanai, 1997). Eissenstat and Yanai (1997) and Eissenstat *et al.* (2000) suggest that roots, like leaves, possess suites of inter-related traits that relate to their longevity. In fruit trees, apple roots, which are relatively small in diameter, and have low tissue density and little lignification of the exodermis, have substantially shorter longevity than the roots of citrus which have the opposite characteristics (Fig. 3.7). In grasses, thicker roots and high tissue density have also been associated with increased longevity. For example, a study of four grass species in The Netherlands found that the species from N-rich habitats (*L. perenne*) had significantly finer roots and shorter longevity than species from N-poor habitats. The range was 14 weeks for *L. perenne* roots to 58 weeks for *Nardus stricta* (van der Krift and Berendse, 2002). In nutrient-poor environments thicker roots with a longer lifespan may increase

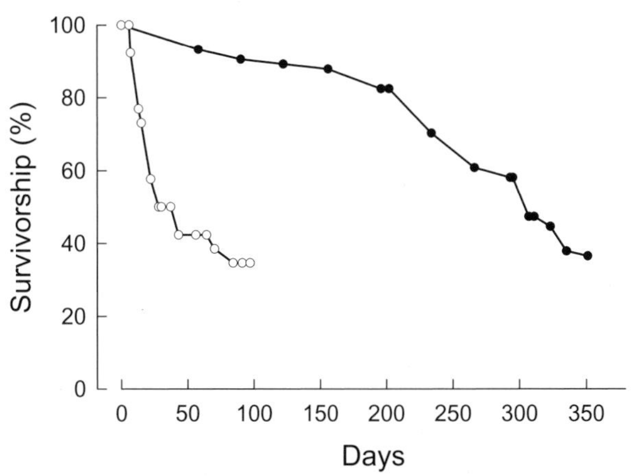

Fig. 3.7 Root survivorship of apple (○) and citrus (●) roots determined using minirhizotrons. Apple root longevity was determined on Red Chief Delicious on EMLA26 rootstock in Pennsylvania, USA for a cohort appearing in June, and citrus was for Red grapefruit on Sour orange rootstock in Florida, USA for a cohort appearing in April. Each curve comprises data from six trees with two minirhizotrons per tree. (Reproduced with permission from Eissenstat *et al.*, *New Phytologist*; New Phytologist Trust, 2000.)

the residence time of nutrients in the plant, and provide an important means of nutrient conservation (Eissenstat and Yanai, 1997).

Vogt and Bloomfield (1996) demonstrated that many factors contribute to the longevity of tree roots. It is affected by tree species, habit (deciduous or evergreen), and by structural and functional root classes in addition to environmental conditions. Perennial, coarse roots supporting a tree may be as old as the tree, whereas fine roots have high turnover rates. The limited data available suggest that fine root longevity of deciduous trees (<1 year) is shorter than that of evergreen species (<1 to 12 years). Longevity also varies within a species depending on growing conditions and the interaction with fungal symbionts and pathogens. Majdi and Persson (1995) found that nutrient applications to Norway spruce (*Picea abies* (L.)) significantly affected both net productivity of roots and the amount of root death. Application of ammonium sulphate increased the production of fine, white roots relative to the control but these had high mortality (60% compared with 30% in the control), whereas application of nitrogen-free fertilizer decreased production and mortality (8%). Tree roots without mycorrhizas often have a life limited to a few weeks whereas mycorrhizal roots may live for many years (Vogt and Bloomfield, 1996). Mycorrhizal colonization can decrease rates of root mortality through diverse effects such as improved nutrition, enhanced tolerance to drying soil, and reduced deleterious effects of pathogens and herbivory (Eissenstat *et al.*, 2000). Such benefits are by no means universal (for example, Hooker *et al.* [1995] found that arbuscular mycorrhizal infection reduced the longevity of poplar roots), and the interactions between roots, mycorrhizal fungi, non-mycorrhizal fungi and the soil biota are complex.

Gill and Jackson (2000) examined the influence of climate on rates of root turnover at a global scale using data from 190 studies. Root turnover was calculated as annual below-ground production divided by the maximum below-ground standing crop so that if all of the roots produced by a plant in a year were to die at the end of the growing season then the turnover would be 1.0 yr^{-1}. Although there was high variability in estimates of root turnover, there was a strong positive exponential relationship between turnover and mean annual temperature for shrublands, grasslands and all fine roots from all vegetation types combined, with a weaker relationship for forest fine roots. There was no significant relationship between turnover and any climate variable for coarse forest roots or for wetlands, and with temperature included in the regression model, there was no relationship with precipitation. Possible explanations for the association of greater turnover with higher temperatures include: first, the exponential increase of maintenance respiration with temperature; second, higher rates of nutrient mineralization with temperature; and third, increased pathogen and herbivore activity in warmer and freeze-free soils (Gill and Jackson, 2000). Turnover increased from boreal zones to the tropics, possibly reflecting the influence of seasonality and implying that if tropical systems are to maintain the same below-ground biomass as temperate or boreal systems then their below-ground productivity must be higher. In high latitudes, the average turnover for grasses, shrubs and fine tree roots combined was 13% of maximum root standing crop annually, rising to 40% in temperate zones and to 73% in tropical zones. The relative order of turnover between plant functional types was maintained within latitudinal zones with grasslands = fine tree roots = wetlands > shrublands > tree whole root system (Table 3.8). The turnover rate of tree roots decreased as root diameter increased. Turnover decreased from 1.2 yr^{-1}

Table 3.8 Rates of root turnover (yr^{-1}) for different vegetation types in different latitudinal zones

Vegetation type	High latitude	Temperate	Tropical
Trees: whole system	0.08	0.10	0.08
Trees: fine roots	0.40	0.64	0.77
Grasslands	0.27	0.52	0.90
Shrublands	0.13	0.26	0.63
Wetlands	0.46	0.58	0.69

From Gill and Jackson, 2000.

in roots 0–1 mm diameter, to 0.8 yr^{-1} in roots 0–2 mm diameter, to 0.52 yr^{-1} in roots 0–3 and 0–5 mm diameter, to 0.1 yr^{-1} in roots 0–10 mm in diameter. Assuming that fine roots (inadequately and variously defined in the forestry literature) turn over once per year, they would represent about one-third of global annual net primary productivity (Jackson *et al.*, 1997).

While such analyses are useful at a global scale, they are not necessarily useful predictors of inter-annual variability at individual sites. For example, Gill and Jackson (2000) found that variation in mean annual temperature did not explain the variation in inter-seasonal rates of turnover of shortgrass steppe at a long-term site in northeastern Colorado, USA but there was a strong positive relationship between turnover and the ratio of growing season precipitation:maximum mean monthly temperature. The ability of some species to respond to inter-annual variability may affect turnover at a single site and might be buffered by species-specific allocation patterns.

Root growth and root mortality occur throughout the year but the net balance is highly seasonal in perennial plants, with a burst of production in the spring and considerable mortality in the autumn. In temperate forests, the pattern corresponds approximately with duration of the canopy but the correspondence is inexact. It is not known whether the environmental cues that promote root and canopy growth in the spring, and root and canopy senescence in the autumn are related (Pregitzer *et al.*, 2000). Hendrick and Pregitzer (1993) used minirhizotrons to follow the fate of contemporaneous cohorts of fine roots (≤1.5 mm diameter) over two seasons in two temperate sugar maple (*Acer saccharum* Marsh.) forests. They found consistent differences in lifespan between the two forests with the warmer site having roots of shorter lifespan (as in the global study of Gill and Jackson, 2000) but growth and mortality were more continuous than leaf growth and senescence, which were highly determinate. A flush of root production in the spring occurred at all depths while the soils were still cool, suggesting that temperature alone may not be the sole factor inducing growth and that factors such as bud growth and the production of growth regulators may also play a role (Hendrick and Pregitzer, 1996). Seasonal patterns of root mortality differed from those of production such that it was distributed almost evenly through the year near the soil surface. Drying of surface layers during the summer did not result in increased root mortality.

Temporal patterns of root growth and mortality are also evident in perennial and annual crops. The general pattern of root system growth reported in section 3.3.1 represents a balance of new root growth and root senescence. Mean lifespan of annual crop roots has been reported in a range of studies including winter wheat (>125 days; Gibbs and Reid, 1992), groundnut (24–31 days; Krauss and Deacon, 1994) and sugar beet (60–130 days;

van Noordwijk *et al.*, 1994) but is affected by environmental conditions. For example, Huck *et al.* (1987) found that rates of appearance and death of soyabean roots were markedly affected by the availability of soil water as influenced by rain events and irrigation. The number of roots decreased after each rain event, increased during dry periods, and then declined again after another rainfall. Periodicity of fine root mass was also found in alfalfa by Luo *et al.* (1995) in association with shoot harvesting; mass declined after cutting then increased again towards the end of the period of shoot re-growth.

In pastures, the lifespan of roots may be relatively short. For example, 73% of white clover roots failed to survive for 21 days at a site at 440 m altitude in the Apennine Mountains of Italy compared with only 29% at a site at 120 m altitude near Aberdeen, UK (Watson *et al.*, 2000). Similarly, about 84% of Italian ryegrass roots failed to survive for more than 21 days in Italy compared with 38% in the UK. The reasons for these differences are difficult to unravel but may, in part, reflect differences in temperature at the two sites with lower temperatures enhancing longevity. In controlled conditions, Forbes *et al.* (1997) found that root longevity of the same cultivar of Italian ryegrass decreased as temperature increased. About 67% of roots survived for >35 days at 15°C compared with 42% at 21°C, and 16% at 27°C.

Much still remains to be learnt about the factors determining root longevity and mortality. The strategies of fine root production may differ between annual and perennial plants, and just as plants have different ways of deploying above-ground resources (e.g. deciduous vs evergreen) so they may differ below ground. Annual plants have a limited period in which they can grow and maintain all of the roots necessary to ensure reproduction but perennial plants have a much wider range of options open to them. In part the strategy may come down to a trade-off between the costs of maintaining a root against that of allowing it to die and then producing a new root. For example, to grow 1 g of root dry mass requires the respiration of 2 g CH_2O and maintenance respiration of 0.03 g C day^{-1}. In unfavourable conditions, then, maintaining 1 g of root for 66 days would cost as much as one cycle of root death followed by re-growth (van Noordwijk *et al.*, 1998).

3.6 Modelling of root systems

Many root simulation models have been developed for annual crop plants embracing a wide range of approaches. A basic difference in approach exists between those models that use architectural information to simulate the growth of individual roots within the root system, and those that do not. For ease of presentation, it is convenient to categorize the latter group of models under the headings of: (i) rooting depth models; (ii) root distribution models; (iii) root depth and distribution models that interact with the soil; and (iv) root depth and distribution models that interact with shoot growth and soil conditions. Rarely are the models employed merely to express the size and shape of the root system but rather to determine the surface available for the acquisition of water and nutrients from the soil. These aspects of root functioning will be dealt with in more detail in Chapter 4.

The simplest root models start from the assumption that the roots in each layer of soil take up the same amount of water and/or nutrients irrespective of the exact quantity of root present. Consequently they simulate only the advancing rooting front or the depth of the deepest root. The empirical observations and resulting model of Borg and Grimes (1986) were described in section 3.3.3, but several models of crop growth have incorporated a similar approach. For example, Chapman *et al.* (1993) calculated transpiration of sunflower as functions of the rate of downward progress of the rooting/water extraction front and the rate of water extraction from each later within the rooting front. As the rooting/extraction front reaches each layer, water is removed at a rate that decreases exponentially with time (see section 4.2.3). In effect, the rate at which water content decreases in the layer is the product of L_v and the hydraulic conductivity of the soil.

The utility of the above approach is often limited because roots may penetrate to great depths down cracks or worm holes without affecting resource acquisition appreciably and because of other research demonstrating the relations that exist between, for example, the amount of root and nutrient uptake (see section 4.3.4). Gerwitz and Page (1974) were the first to produce an empirical description of root distribution within the rooted zone and to show that root length decreased exponentially with soil depth for crops sown at normal densities. They also showed that it was possible to describe root growth and distribution in a manner analogous to the movement of solutes by diffusion. Page and Gerwitz (1974) observed that roots grow from zones in which they are present in high concentration to zones of lower concentration, and that individual roots might be regarded as growing in random directions as a result of repeated branching. The concentration of roots in the zone of higher concentration will, then, determine the rate of growth into the zone of lower concentration – a diffusion-like process. They showed that the consequence of the diffusion process was to produce a root system in homogeneous soil in which length decreased exponentially with depth. This approach has been refined more recently by de Willigen *et al.* (2002) to incorporate a first order sink term accounting for root decay, and different diffusion coefficients in the vertical and horizontal directions to allow for some soil heterogeneity. For a crop sown in a row perpendicular to the x direction with distance between rows of 2L and an infinite z downward direction:

$$\begin{matrix} 0 \leq x \leq L \\ 0 \leq z \leq \infty \end{matrix} \qquad \frac{\partial L_v}{\partial t} = \frac{D_x \partial^2 L_v}{\partial x^2} + \frac{D_z \partial^2 L_v}{\partial z^2} + kL_v \qquad (3.4)$$

where L_v is the root length per unit volume of soil, t is time, k is the decay constant, and D_x and D_z are the diffusion coefficients in the x and z direction, respectively. Figure 3.8 shows the simulated growth of a maize root system in time for a crop sown at a row spacing of 80 cm. The model uses an L_v of 0.1 cm cm^{-3} as the measure of root presence, and assumes the same value of diffusion coefficients in the vertical and horizontal directions. The isolines are elliptical at first but the horizontal gradient diminishes gradually to give almost straight lines. For the situation where dry matter enters through the complete surface (i.e. in a typical crop), a steady-state eventually occurs where L_v decreases exponentially with depth.

Many crop growth models simulate the exploration of the soil by the root system using two separate processes: first, downward growth of an abstract vertical axis; and second, proliferation of roots in specified soil layers throughout the depth achieved (e.g. Jones *et al.*, 1991). Models of the CERES family simulate root growth based on five relationships: (i) new root length is produced each day in proportion to the amount of shoot biomass produced that day; (ii) the root front descends at a rate determined by temperature; (iii) in the

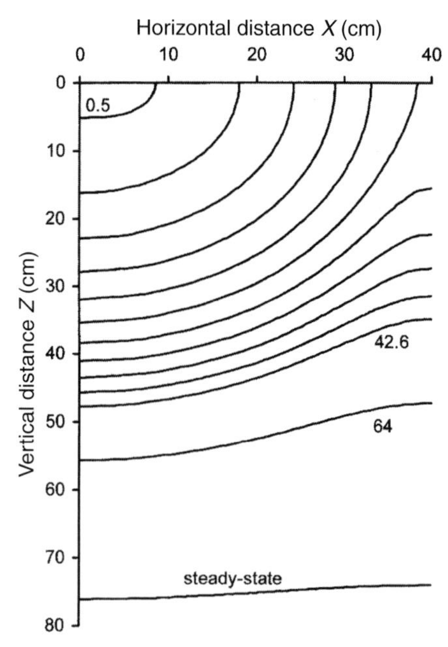

Fig. 3.8 Modelled isolines giving the position of a root length of 0.1 cm cm^{-3} for a maize crop at different times. The model assumes a constant input of root length and that the diffusion coefficients in the vertical and horizontal directions are the same. (Reproduced with permission from de Willigen *et al.*, *Plant and Soil*; Springer Science and Business Media, 2002.)

absence of soil limitations, the daily increase in root length is partitioned among soil layers in the root zone exponentially with depth; (iv) root proliferation in a soil layer is restricted by a multiplicative stress factor which acts as a proxy for soil strength and decreases linearly from 1.0 at 25% of extractable water to 0 when all extractable water is depleted; and (v) a fixed proportion of existing root length (say 0.5%) is lost due to senescence each day. Robertson *et al.* (1993) used this approach to simulate root growth of sorghum on drying soils in the sub-humid subtropics of Australia. Root distribution was well simulated but seasonal total root length was generally substantially underestimated for several reasons including the uncertainty of the conversion factor for daily shoot biomass increment to root length, and uncertainties in root length measurements used in the data sets. The model was limited to simulating root growth until early grain-filling because there is little quantitative information available about factors influencing root growth thereafter. A similar approach was used by Devi *et al.* (1996) to simulate the growth of a pigeonpea root system in central India, and by Manschadi *et al.* (1998) to simulate the growth of faba bean root systems in northern Syria. In pigeonpea, difficulties were encountered in getting simulated dry mass and length in layers to agree because the thick tap root contributes much to the mass in a layer but little to the overall length of the system.

The framework proposed by Jones *et al.* (1991) can also be modified to allow for the effects of other soil restrictions on growth by incorporating a wider ranging 'root stress factor' as part four of the framework outlined above. Stress factors for aluminium toxicity, calcium deficiency, aeration status and so on can be specified and then the minimum factor (i.e. most severe limitation) used in the simulation. Increasingly, too, models have become better at allowing for shoot/root interactions and the functional equilibrium that exists. For example, Asseng *et al.* (1997) included calculation of the daily available carbon for root growth according to a dynamic shoot:root ratio that changed with the developmental stage of the crop, and distributed the carbon supply over soil layers according to a 'top down principle'. This principle favours root growth in the upper soil layer, but under unfavourable soil conditions in that layer the carbon is given to the next deeper layer and so on. The result is an exponentially declining distribution of root length with depth in a model that is responsive to soil water, soil nitrogen, soil compaction, aeration and the history of root distribution. Simulating correctly the downward penetration, distribution of root length, total root length and root mass and the interaction with the shoot offers many challenges and much research still remains to be done (Savin *et al.*, 1994; Asseng *et al.*, 1997).

The preceding models make no attempt to describe the morphology and architecture of the root system. Early simulations of root architecture applied simple algebra to the developmental pattern of roots to give the numbers and lengths of root members with time in two dimensions. Rose's (1983) algebraic model took advantage of the fact that the extension and branching of young cereal roots grown in homogeneous media proceed at uniform rates so that each class of lateral could be characterized by a constant rate of extension and a constant branching density. These ideas, combined with those arising from the root development studies of Klepper *et al.* (1984) (see section 3.2), allowed Porter *et al.* (1986) to simulate the root distribution for crops of wheat based on cumulative thermal time. Because the model describes a system in which axes and branches are initiated at all potential sites (see section 3.2), it overestimated root length compared with that measured in field studies.

Diggle (1988) added to this approach by including geometric information about the axes and branches, and the characteristics governing the direction of root growth to produce

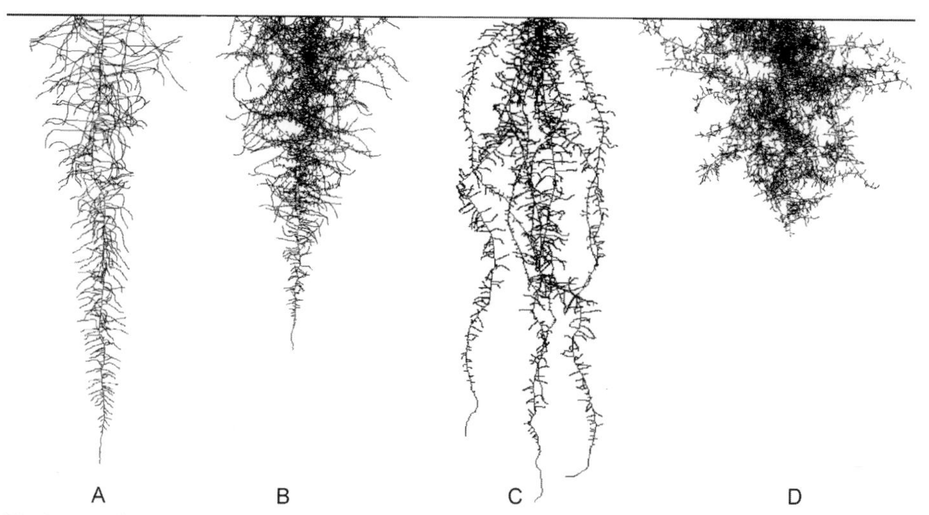

Fig. 3.9 Different root architectures generated by changing input functions to the ROOTMAP model: (A) a simple herringbone root system (a single axis and first order laterals only); (B) a tap-rooted architecture similar to *Lupinus pilosus* (a vertical axis with three orders of lateral branching); (C) a root system with multiple root axes (similar to the seminal axes of wheat) with first order laterals; and (D) a dichotomous root system in which there is no dominant axis and each branching order has the same probability of branching. (I am grateful to Dr V. Dunbabin for this previously unpublished figure.)

a three-dimensional simulation of growth. Essentially, the model (ROOTMAP) tracks the movement of root tips through the soil and the routes which these tips follow is the model's representation of the root system. The model does not allow for interaction with neighbouring plants and is therefore a description of an individual plant. Figure 3.9 shows examples of the different types of root architecture that can be generated by altering inputs in the model. Variations in the number of root axes, the geotropism index (degree of downward growth in response to gravity), deflection index (propensity to deviate from the current direction of growth), orders of root branching, and density of root branching are major determinants of the resulting root architecture. When coupled with information about the distribution of nitrate in the soil profile, the model was able to assess the optimum root architecture for minimizing nitrate leaching (Dunbabin *et al.*, 2003).

Tsegaye et al. (1995) compared the output of a modified ROOTMAP model with measurements of the number of pea roots striking the wall of a container under conditions where the mechanical resistance was maintained constant throughout the experiment, and where the soil was allowed to dry and mechanical resistance to increase with time. Under changing conditions, the predictions agreed well with the experimental observations but the predictions under constant conditions considerably overestimated the rate at which new contacts were made with the container walls. The difference between observation and prediction was probably a consequence of the dominance of second order branches during the later part of the experiment and the effect of the container wall on their development. In contrast, a similar comparison of a three-dimensional architectural model of the maize root system (Pagès *et al.*, 1989) with observed maps from the field gave good agreement overall (Pellerin and Pagès, 1996). Both the numbers of cross-sections of root axes and their spatial distribution were correctly predicted as was the number of cross-sections of lateral roots. Lateral roots were, though, more clustered around the axes on simulated maps than on observed maps.

Spatial heterogeneity of root processes and soil heterogeneity is increasingly being built into such architectural models. Lynch *et al.* (1997) allowed explicitly for the heterogeneity of physiological processes in the root system and adopted a kinematic approach to root axes that allowed the distinction between changes occurring as a function of growth from those occurring independently of growth such as changes in cell physiology as they mature. Although the soil was considered as a uniform medium, the approach allowed the visualization of root systems in one-, two- and three-dimensional space. Clausnitzer and Hopmans (1994) constructed a finite element model to simulate three-dimensional root architecture and the interaction between root growth and soil water movement that emphasized the effects of soil strength on root growth. Such models have been extended to include the influence of nutrient deficiency and toxicity of ions on root growth and allow the dynamic simulation of root growth and water and solute transport in three dimensions (Somma *et al.*, 1998). However, such models require quite large amounts of information from experiments to provide the models with the correct parameter values and functional relationships; at the present time computational capacity is often ahead of the experimental capacity to provide such information. Nevertheless, the power of such approaches is increasingly evident. For example, the Root Typ model of Pagès *et al.* (2004) takes into account, with the same level of detail, many processes involved in root architectural development, generalizes the concept of root type, and includes the effects of several soil properties on root development,

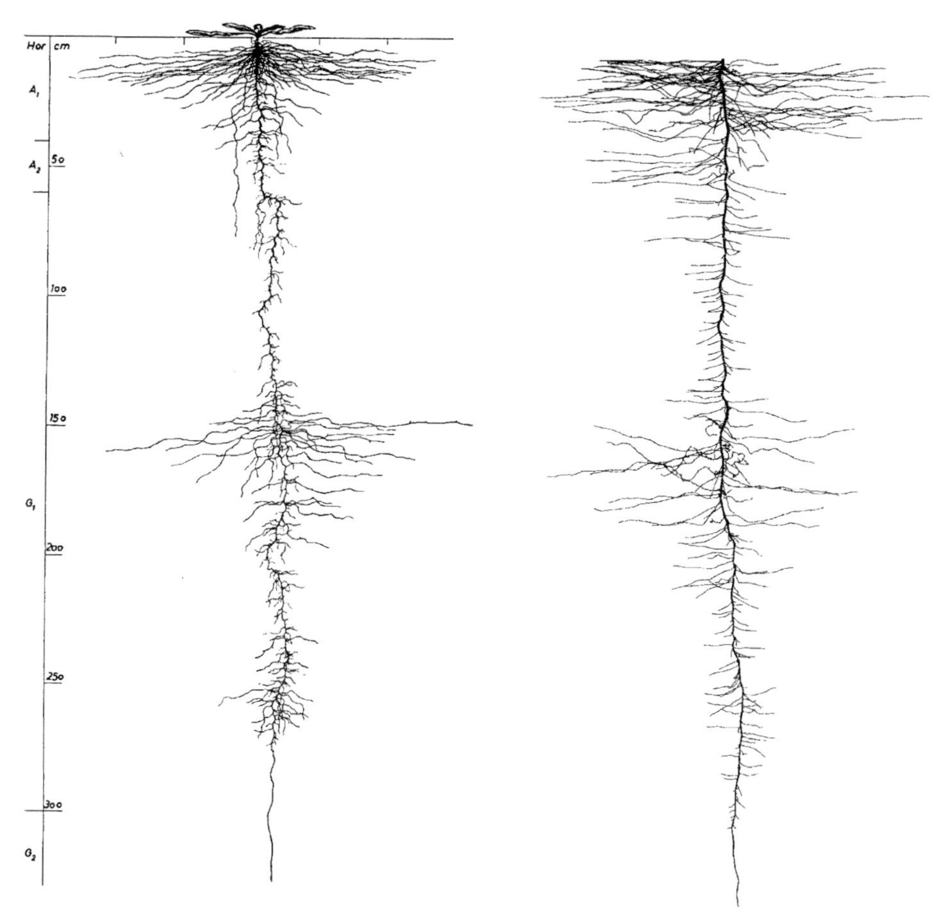

Fig. 3.10 Comparison of the observed root system of *Rumex crispus* drawn by Kutschera (1960) on the left, with a simulation by the Root Typ model of Pagès *et al.* (2004) on the right. (Reproduced with permission from Pagès *et al.*, *Plant and Soil*; Springer Science and Business Media, 2004.)

to provide visualizations of root systems grown under specific circumstances. The model includes details of root appearance, axial and radial growth, branching, formation of multiple root tips (reiteration), change of state (e.g. transition of a lateral root to become a tap root), and decay and abscission. Figure 3.10 compares measured and simulated images of curled dock (*Rumex crispus*) in which the simulation has allowed a varying soil parameter to operate at different depths such as may occur, for example, if dry soil conditions are experienced in the upper layers of the profile. In the drawing made by Kutschera (1960), it is clear that the plant has a deep root system with large differences in the length of lateral roots at different depths; these differences are reproduced well by the model.

References

Amato, M. and Pardo, A. (1994) Root length and biomass losses during sample preparation with different screen mesh sizes. *Plant and Soil* **161**, 299–303.

Asseng, S., Richter, C. and Wessolek, G. (1997) Modelling root growth of wheat as the linkage between crop and soil. *Plant and Soil* **190**, 267–277.

Barraclough, P.B. (1984) The growth and activity of winter wheat roots in the field: root growth of high-yielding crops in relation to shoot growth. *Journal of Agricultural Science, Cambridge* **103**, 439–442.

Barraclough, P.B. (1989) Root growth, macro-nutrient uptake dynamics and soil fertility requirements of a high-yielding winter oilseed rape crop. *Plant and Soil* **119**, 59–70.

Barraclough, P.B. and Leigh, R.A. (1984) The growth and activity of winter wheat roots in the field: the effect of sowing date and soil type on root growth of high-yielding crops. *Journal of Agricultural Science, Cambridge* **103**, 59–74.

Bassett, D.M., Stockton, J.R. and Dickens, W.L. (1970) Root growth of cotton as measured by P^{32} uptake. *Agronomy Journal* **62**, 200–203.

Bland, W.L. and Dugas, W.A. (1988) Root length density from minirhizotron observations. *Agronomy Journal* **80**, 271–275.

Bland, W.L. and Mesarch, M.A. (1990) Counting error in the line-intercept method of measuring root length. *Plant and Soil* **125**, 155–157.

Böhm, W. (1976) *In situ* estimation of root length at natural soil profiles. *Journal of Agricultural Science, Cambridge* **87**, 365–368.

Böhm, W. (1979) *Methods of Studying Root Systems. Ecological Studies: Analysis and Synthesis*, Vol. 33. Springer, Berlin.

Borg, H. and Grimes, D.W. (1986) Depth development of roots with time: an empirical description. *Transactions of the American Society of Agricultural Engineers* **29**, 194–197.

Brouwer, R. (1983) Functional equilibrium: sense or nonsense? *Netherlands Journal of Agricultural Science* **31**, 335–348.

Brown, K.F. and Biscoe, P.V. (1985) Fibrous root growth and water use of sugar beet. *Journal of Agricultural Science, Cambridge* **105**, 679–691.

Brown, S.C., Keatinge, J.D.H., Gregory, P.J. and Cooper, P.J.M. (1987) Effects of fertilizer, variety and location on barley production under rainfed conditions in northern Syria. 1. Root and shoot growth. *Field Crops Research* **16**, 53–66.

Brown, S.C., Gregory, P.J., Cooper, P.J.M. and Keating, J.D.H. (1989) Root and shoot growth and water use of chickpea (*Cicer arietinum*) grown in dryland conditions: effects of sowing date and genotype. *Journal of Agricultural Science, Cambridge* **113**, 41–49.

Buckland, S.T., Campbell, C.D., Mackie-Dawson, L.A., Horgan, G.W. and Duff, E.I. (1993) A method for counting roots observed in minirhizotrons and their theoretical conversion to root length density. *Plant and Soil* **153**, 1–9.

Caldwell, M.M. (1987) Competition between root systems in natural communities. In: *Root Development and Function* (eds P.J. Gregory, J.V. Lake and D.A. Rose), pp. 167–185. Cambridge University Press, Cambridge, UK.

Canadell, J., Jackson, R.B., Ehleringer, J.R., Mooney, H.A., Sala, O.E. and Schulze, E.D. (1996) Maximum rooting depth of vegetation types at the global scale. *Oecologia* **108**, 583–595.

Chapman, S.C., Hammer, G.L. and Meinke, H. (1993) A sunflower simulation model: I. Model development. *Agronomy Journal* **85**, 725–735.

Clausnitzer, V. and Hopmans, J.W. (1994) Simultaneous modeling of transient three-dimensional root growth and soil water flow. *Plant and Soil* **164**, 299–314.

Dawson, L.A., Thornton, B., Pratt, S.M. and.Paterson, E. (2003) Morphological and topological responses of roots to defoliation and nitrogen supply in *Lolium perenne* and *Festuca ovina*. *New Phytologist* **161**, 811–818.

de Willigen, P., Heinen, M., Mollier, A. and van Noordwijk, M. (2002) Two-dimensional growth of a root system modelled as a diffusion process. I. Analytical solutions. *Plant and Soil* **240**, 225–234.

Deinum, B. (1985) Root mass of grass swards in different grazing systems. *Netherlands Journal of Agricultural Science* **33**, 377–384.

Devi, G., Ito, O., Matsunaga, R., Tobita, S., Rao, T.P., Vidyalakshmi, N. and Lee, K.K. (1996) Simulating root system development of short-duration pigeonpea. *Experimental Agriculture* **32**, 67–78.

Diggle, A.J. (1988). ROOTMAP – a model in three-dimensional coordinates of the growth and structure of fibrous root systems. *Plant and Soil* **105**, 169–178.

Dracup, M., Belford, R.K. and Gregory, P.J. (1992) Constraints to root growth of wheat and lupin crops in duplex soils. *Australian Journal of Experimental Agriculture* **32**, 947–961.

Dunbabin, V., Diggle, A. and Rengel, Z. (2003) Is there an optimal root architecture for nitrate capture in leaching environments? *Plant, Cell and Environment* **26**, 835–844.

Eissenstat, D.M. and Yanai, R.D. (1997) The ecology of root lifespan. *Advances in Ecological Research* **27**, 1–60.

Eissenstat, D.M., Wells, C.E., Yanai, R.D. and Whitbeck, J.L. (2000) Building roots in a changing environment: implications for root longevity. *New Phytologist* **147**, 33–42.

Forbes, P.J., Black, K.E. and Hooker, J.E. (1997) Temperature-induced alteration to root longevity in *Lolium perenne*. *Plant and Soil* **190**, 87–90.

Ford, K.E., Gregory, P.J., Gooding, M.J. and Pepler, S. (2002) Root length density variation in modern winter wheat cultivars. In: *European Society for Agronomy VII ESA Congress,* 15–18 July 2002 (eds F.J. Villalobos and L. Testi), pp. 361–362. Cordoba, Spain.

Forde, B. and Lorenzo, H. (2001) The nutritional control of root development. *Plant and Soil* **232**, 51–68.

Fusseder, A. (1983) A method for measuring length, spatial distribution and distances of living roots *in situ*. *Plant and Soil* **73**, 441–445.

Gerwitz, A. and Page, E.R. (1974) An empirical mathematical model to describe plant root systems. *Journal of Applied Ecology* **11**, 773–782.

Gibbs, R.J. and Reid, J.B. (1992) Comparison between net and gross root production by winter wheat and by perennial ryegrass. *New Zealand Journal of Crop and Horticultural Science* **20**, 483–487.

Gill, R.A. and Jackson, R.B. (2000) Global patterns of root turnover for terrestrial ecosystems. *New Phytologist* **147**, 13–31.

Greenwood, D.J., Gerwitz, A., Stone, D.A. and Barnes, A. (1982) Root development of vegetable crops. *Plant and Soil* **68**, 75–96.

Gregory, P.J. (1979) A periscope method for observing root growth and distribution in field soil. *Journal of Experimental Botany* **30**, 204–214.

Gregory, P.J. (1983) Response to temperature in a stand of pearl millet (*Pennisetum typhoides* S. & H.). III. Root development. *Journal of Experimental Botany* **34**, 744–756.

Gregory, P.J. (1994a) Resource capture by root networks. In: *Resource Capture by Crops, Proceedings of the 52nd Easter School* (eds J.L. Monteith, R.K. Scott and M.H. Unsworth), pp. 77–97. Nottingham University Press, Nottingham, UK.

Gregory, P.J. (1994b) Root growth and activity. In: *Physiology and Determination of Crop Yield* (eds K.J. Boote, J.M. Bennett, T.R. Sinclair and G.M. Paulsen), pp. 65–93. ASA, CSSA, SSSA, Madison, USA.

Gregory, P.J. and Atwell, B.J. (1991) The fate of carbon in pulse-labelled crops of barley and wheat. *Plant and Soil* **136**, 205–213.

Gregory, P.J. and Brown, S.C. (1989) Root growth, water use and yield of crops in dry environments: what characteristics are desirable? *Aspects of Applied Biology* **22**, 235–243.

Gregory, P.J. and Eastham, J. (1996) Growth of shoots and roots, and interception of radiation by wheat and lupin crops on a shallow, duplex soil in response to time of sowing. *Australian Journal of Agricultural Research* **47**, 427–447.

Gregory, P.J. and Reddy, M.S. (1982) Root growth in an intercrop of pearl millet/groundnut. *Field Crops Research* **5**, 241–252.

Gregory, P.J. and Squire, G.R. (1979) Irrigation effects on roots and shoots of pearl millet (*Pennisetum typhoides*). *Experimental Agriculture* **15**, 161–168.

Gregory, P.J., McGowan, M., Biscoe, P.V. and Hunter, B. (1978) Water relations of winter wheat. 1. Growth of the root system. *Journal of Agricultural Science, Cambridge* **91**, 91–102.

Gregory, P.J., Tennant, D. and Belford, R.K. (1992) Root and shoot growth, and water and light use efficiency of barley and wheat crops grown on a shallow duplex soil in a mediterranean-type environment. *Australian Journal of Agricultural Research* **43**, 555–573.

Gregory, P.J., Palta, J.A. and Batts, G.R. (1997) Root systems and root:mass ratio – carbon allocation under current and projected atmospheric conditions in arable crops. *Plant and Soil* **187**, 221–228.

Gregory, P.J., Hutchison, D.J., Read, D.B., Jenneson, P.M., Gilboy, W.B. and Morton, E J. (2003) Non-invasive imaging of roots with high resolution X-ray micro-tomography. *Plant and Soil* **255**, 351–359.

Grzebisz, W., Floris, J. and van Noordwijk, M. (1989) Loss of dry matter and cell contents from fibrous roots of sugar beet due to sampling, storage and washing. *Plant and Soil* **113**, 53–57.

Hamblin, A.P. and Hamblin, J. (1985) Root characteristics of some temperate legume species and varieties on deep, free-draining Entisols. *Australian Journal of Agricultural Research* **36**, 63–72.

Hamblin, A.P. and Tennant, D. (1987) Root length density and water uptake in cereals and grain legumes: how well are they correlated? *Australian Journal of Agricultural Research* **38**, 513–527.

Hamblin, A., Tennant, D. and Perry, M.W. (1990) The cost of stress: dry matter partitioning changes with seasonal supply of water and nitrogen to dryland wheat. *Plant and Soil* **122**, 47–58.

Hansson, A.D. and Andrén, O. (1987) Root dynamics in barley, lucerne and meadow fescue investigated with a minirhizotron technique. *Plant and Soil* **103**, 33–38.

Heeraman, D.A. and Juma, N.G. (1993) A comparison of minirhizotron, core and monolith methods for quantifying barley (*Hordeum vulgare* L.) and fababean (*Vicia faba* L.) root distribution. *Plant and Soil* **148**, 29–41.

Heeraman, D.A., Hopmans, J.W. and Clausnitzer, V. (1997) Three dimensional imaging of plant roots *in situ* with X-ray computed tomography. *Plant and Soil* **189**, 167–179.

Hendrick, R.L. and Pregitzer, K.S. (1992) The demography of fine roots in a northern hardwood forest. *Ecology* **73**, 1094–1104.

Hendrick, R.L. and Pregitzer, K.S. (1993) Patterns of fine root mortality in two sugar maple forests. *Nature* **361**, 59–61.

Hendrick, R.L. and Pregitzer, K.S. (1996) Temporal and depth-related patterns of fine root dynamics in northern hardwood forests. *Journal of Ecology* **84**, 167–176.

Hooker, J.E., Black, K.E., Perry, R.L. and Atkinson, D. (1995) Arbuscular mycorrhizal fungi induced alteration to root longevity of poplar. *Plant and Soil* **172**, 327–329.

Huck, M.G., Hoogenboom, G. and Peterson, C.M. (1987) Soybean root senescence under drought stress. In: *Minirhizotron Observation Tubes: Methods and Applications for Measuring Rhizosphere Dynamics* (ed. H.M. Taylor), pp. 109–121. American Society of Agronomy Special Publication No. 50, Madison, USA.

Jaafar, M.N., Stone, L.R. and Goodrum, D.E. (1993) Rooting depth and dry matter development of sunflower. *Agronomy Journal* **85**, 281–286.

Jackson, R.B., Canadell, J., Ehleringer, J.R., Mooney, H.A., Sala, O.E. and Schulze, E.D. (1996) A global analysis of root distributions for terrestrial biomes. *Oecologia* **108**, 389–411.

Jackson, R.B., Mooney, H.A. and Schulze, E.-D. (1997) A global budget for fine root biomass, surface area, and nutrient contents. *Proceeding of the National Academy of Sciences, USA* **94**, 7362–7366.

Jones, C.A., Bland, W.L., Ritchie, J.T. and Williams, J.R. (1991) Simulation of root growth. In: *Modeling Plant and Soil Systems* (eds J. Hanks and J.T. Ritchie), pp. 91–123. American Society of Agronomy, Madison, USA.

Kage, H., Kochler, M. and Stützel, H. (2000) Root growth of cauliflower (*Brassica oleracea* L. *botrytis*) under unstressed conditions: measurement and modelling. *Plant and Soil* **223**, 131–145.

Kaspar, T.C. and Ewing, R.P. (1997) ROOTEDGE: Software for measuring root length from desktop scanner images. *Agronomy Journal* **89**, 932–940.

Keith, H., Oades, J.M. and Martin, J.K. (1986) Input of carbon to soil from wheat plants. *Soil Biology and Biochemistry* **18**, 445–449.

Ketring, D.L. and Reid, J.L. (1993) Growth of peanut roots under field conditions. *Agronomy Journal* **85**, 80–85.

Kimura, K., Kikuchi, S. and Yamasaki, S. (1999) Accurate root length measurement by image analysis. *Plant and Soil* **216**, 117–127.

King, J., Gay, A., Sylvester-Bradley, R., Bingham, I., Foulkes, J., Gregory, P.J. and Robinson, D. (2003) Modelling cereal root systems for water and nitrogen capture: towards an economic optimum. *Annals of Botany* **91**, 383–390.

Klepper, B., Belford, R.K. and Rickman, R.W. (1984) Root and shoot development in winter wheat. *Agronomy Journal* **76**, 117–122.

Krauss, U. and Deacon, J.W. (1994) Root turnover of groundnut (*Arachis hypogaea* L.) in soil tubes. *Plant and Soil* **166**, 259–270.

Kutschera, L. (1960) *Wurzelatlas mitteleuropäischer Ackerunkräuter und Kulturpflanzen*. DLG-Verlags-GMBH, Frankfurt, Germany.

Lancashire, P.D., Bleiholder, H., van den Boom, T., Langelüddeke, P., Stauss, R., Weber, E. and Witzenberger, A. (1991) A uniform decimal code for growth stages of crops and weeds. *Annals of Applied Biology* **119**, 561–601.

Liedgens, M. and Richner, W. (2001) Minirhizotron observations of the spatial distribution of the maize root system. *Agronomy Journal* **93**, 1097–1104.

Litav, M. and Harper, J.L. (1967) A method for studying spatial relationships between the root systems of two neighbouring plants. *Plant and Soil* **26**, 389–392.

Livesley, S.J., Stacey, C.L., Gregory, P.J. and Buresh, R.J. (1999) Sieve size effects on root length and biomass measurements of maize (*Zea mays*) and *Grevillea robusta*. *Plant and Soil* **207**, 183–193.

Livesley, S.J., Gregory, P.J. and Buresh, R.J. (2000) Competition in tree row agroforestry systems. 1. Distribution and dynamics of fine root length and biomass. *Plant and Soil* **227**, 149–161.

Locke, L.F. and Allen, J.A. (1924) Development of wheat plants from seminal roots. *Journal of the American Society of Agronomy* **16**, 261–268.

Logsdon, S.D. and Allmaras, R.R. (1991) Maize and soybean root clustering as indicated by root mapping. *Plant and Soil* **131**, 169–176.

Luo, Y., Meyerhoff, P.A. and Loomis, R.S. (1995) Seasonal patterns and vertical distributions of fine roots of alfalfa (*Medicago sativa* L.). *Field Crops Research* **40**, 119–127.

Lynch, J.P., Nielsen, K.L., Davis, R.D. and Jablokov, A.G. (1997) *SimRoot*: Modelling and visualization of root systems. *Plant and Soil* **188**, 139–151.

Majdi, H. and Nylund, J.E. (1996) Does liquid fertilization affect fine root dynamics and life span of mycorrhizal short roots? *Plant and Soil* **185**, 305–309.

Majdi, H. and Persson, H. (1995) A study on fine-root dynamics in response to nutrient applications in a Norway spruce stand using the minirhizotron technique. *Zeitschrift fur Pflanzenernahrung und Bodenkunde* **158**, 429–433.

Manschadi, A.M., Sauerborn, J., Stutzel, H., Gobel, W. and Saxena, M.C. (1998) Simulation of faba bean (*Vicia faba* L.) growth and development under Mediterranean conditions: model adaptation and evaluation. *European Journal of Agronomy* **9**, 273–293.

Mayaki, W.C., Teare, I.D. and Stone, L.R. (1976) Top and root growth of irrigated and nonirrigated soybeans. *Crop Science* **16**, 92–94.

Mengel, D.B. and Barber, S.A. (1974) Development and distribution of the corn root system under field conditions. *Agronomy Journal* **66**, 341–344.

Merrill, S.D. and Upchurch, D.R. (1994) Converting root numbers observed at minirhizotrons to equivalent root length density. *Soil Science Society of America Journal* **58**, 1061–1067.

Merrill, S.D., Tanaka, D.L. and Hanson, J.D. (2002) Root length growth of eight crop species in Haplustoll soils. *Soil Science Society of America Journal* **66**, 913–923.

Milchunas, D.G., Lee, C.A., Lauenroth, W.K. and Coffin, D.P. (1992) A comparison of ^{14}C, ^{86}Rb and total excavation for determination of root distributions of individual plants. *Plant and Soil* **144**, 125–132.

Myers, R.J.K. (1980) The root system of a grain sorghum crop. *Field Crops Research* **3**, 53–64.

Newman, E.I. (1966) A method of estimating the total root length of root in a sample. *Journal of Applied Ecology* **3**, 139–145.

Oliveira, M. do R.G., van Noordwijk, M., Gaze, S.R., Brouwer, G., Bona, S., Mosca, G. and Hairiah, K. (2000) Auger sampling, ingrowth cores and pinboard methods. In: *Root Methods: A Handbook* (eds A.L. Smit, A.G. Bengough, C. Engels, M. van Noordwijk, S. Pellerin and S.C.van de Geijn), pp. 175–210. Springer, Berlin.

Opena, G.B. and Porter, G.A. (1999) Soil management and supplemental irrigation effects on potato: II. Root growth. *Agronomy Journal* **91**, 426–431.

Page, E.R. and Gerwitz, A. (1974) Mathematical models, based on diffusion equations, to describe root systems of isolated plants, row crops, and swards. *Plant and Soil* **41**, 243–254.

Pagès, L., Jordan, M.O. and Picard, D. (1989) A simulation model of the three-dimensional architecture of the maize root system. *Plant and Soil* **119**, 147–154.

Pagès, L., Vercambre, G., Drouet, J.-L., Lecompte, F., Collet, C. and Le Bot, J. (2004) Root Typ: a generic model to depict and analyse the root system architecture. *Plant and Soil* **258**, 103–119.

Pallant, E., Holmgren, R.A., Schuler, G.E., McCracken, K.L. and Drbal, B. (1993) Using a fine root extraction device to quantify small diameter corn roots (≥0.025 mm) in field soils. *Plant and Soil* **153**, 273–279.

Pan, W.L. and Bolton, R.P. (1991) Root quantification by edge discrimination using a desktop scanner. *Agronomy Journal* **83**, 1047–1052.

Pearson, C.J. and Jacobs, B.C. (1985) Root distribution in space and time in *Trifolium subterraneum*. *Australian Journal of Agricultural Research* **36**, 601–614.

Pellerin, S. and Pagès, L. (1996) Evaluation in field conditions of a three-dimensional architectural model of the maize root system: comparison of simulated and observed horizontal root maps. *Plant and Soil* **178**, 101–112.

Porter, J.R., Klepper, B. and Belford, R.K. (1986) A model (WHTROOT) which synchronizes root growth and development with shoot development for winter wheat. *Plant and Soil* **92**, 133–145.

Pregitzer, K.S., King, J.S., Burton, A.J. and Brown, S.E. (2000) Responses of tree fine roots to temperature. *New Phytologist* **147**, 105–115.

Richards, R.A. and Lukacs, Z. (2002) Seedling vigour in wheat – sources of variation for genetic and agronomic improvement. *Australian Journal of Agricultural Research* **53**, 41–50.

Richards, R.A. and Passioura, J.B. (1981) Seminal root morphology and water use of wheat. I. Environmental effects. *Crop Science* **21**, 249–252.

Richner, W., Liedgens, M., Bürgi, H., Soldati, A. and Stamp, P. (2000) Root image analysis and interpretation. In: *Root Methods: A Handbook* (eds A.L. Smit, A.G. Bengough, C. Engels, M. van Noordwijk, S. Pellerin, and S.C.van de Geijn), pp. 175–210. Springer, Berlin.

Robertson, M.J., Fukai, S., Hammer, G.L. and Ludlow, M.M. (1993) Modelling root growth of grain sorghum using the CERES approach. *Field Crops Research* **33**, 113–130.

Rogers, W.S. (1939) Root studies VIII. Apple root growth in relation to rootstock, soil, seasonal and climatic factors. *Journal of Pomology and Horticultural Science* **17**, 99–138.

Rose, D.A. (1983) The description of the growth of root systems. *Plant and Soil* **75**, 405–415.

Savin, R., Hall, A.J. and Satorre, E.H. (1994) Testing the root growth subroutine of the CERES-Wheat model for two cultivars of different cycle length. *Field Crops Research* **38**, 125–133.

Schenk, H.J. and Jackson, R.B. (2002) The global biogeography of roots. *Ecological Monographs* **72**, 311–328.

Sivakumar, M.V.K., Taylor, H.M. and Shaw, R.H. (1977) Top and root relations of field-grown soybeans. *Agronomy Journal* **69**, 470–473.

Smit, A.L. and Groenwold, J. (2005) Root characteristics of selected field crops: data from the Wageningen Rhizolab (1990–2002). *Plant and Soil* **272**, 365–384.

Smit, A.L., Groenwold, J. and Vos, J. (1994) The Wageningen Rhizolab – a facility to study soil-root-shoot-atmosphere interactions in crops. II. Methods of root observations. *Plant and Soil* **161**, 289–298.

Smit, A.L., Bengough, A.G., Engels, C., van Noordwijk, M., Pellerin, S. and van de Geijn, S.C. (2000a) *Root Methods: A Handbook*. Springer, Berlin.

Smit, A.L., George, E. and Groenwold, J. (2000b) Root observations and measurements at (transparent) interfaces with soil. In: *Root Methods: A Handbook* (ed. A.L. Smit, A.G. Bengough, C. Engels, M. van Noordwijk, S. Pellerin, S.C. and van de Geijn), pp. 235–271. Springer, Berlin.

Smucker, A.J.M., McBurney, S.L. and Srivastava, A.K. (1982) Quantitative separation of roots from compacted soil profiles by the hydropneumatic elutriation system. *Agronomy Journal* **74**, 500–503.

Somma, F., Hopmans, J.W. and Clausnitzer, V. (1998) Transient three-dimensional modeling of soil water and solute transport with simultaneous root growth, root water and nutrient uptake. *Plant and Soil* **202**, 281–293.

Stone, L.R., Goodrum, D.E., Jaafar, M.N. and Khan, A.H. (2001) Rooting front and water depletion depths in grain sorghum and sunflower. *Agronomy Journal* **93**, 1105–1110.

Swinnen, J., van Veen, J.A. and Merckx, R. (1995) Carbon fluxes in the rhizosphere of winter wheat and spring barley with conventional vs integrated farming. *Soil Biology and Biochemistry* **27**, 811–820.

Taylor, H.M. (1987) *Minirhizotron Observation Tubes: Methods and Applications for Measuring Rhizosphere Dynamics*. ASA Special Publication 50. ASA, CSSA, and SSSA, Madison, WI, USA.

Tennant, D. (1975) A test of a modified line intersect method of estimating root length. *Journal of Ecology* **63**, 995–1001.

Thangaraj, M., O'Toole, J.C. and De Datta, S.K. (1990) Root response to water stress in rainfed lowland rice. *Experimental Agriculture* **26**, 287–296.

Troughton, A. (1957) The underground organs of herbage grasses. *Commonwealth Bureau of Pastures and Field Crops. Bulletin* No. 44, p. 163.

Tsegaye, T., Mullins, C.E. and Diggle, A.J. (1995) Modelling pea (*Pisum sativum*) root growth in drying soil. A comparison between observations and model predictions. *New Phytologist* **131**, 179–189.

Upchurch, D.R. and Ritchie, J.T. (1983) Root observations using a video recording system in minirhizotrons. *Agronomy Journal* **75**, 1009–1015.

van de Geijn, S.C., Vos, J., Groenwold, J., Goudriaan, J. and Leffelaar, P.A. (1994) The Wageningen Rhizolab – a facility to study soil-root-shoot-atmosphere interactions in crops. I. Description of main functions. *Plant and Soil* **161**, 275–287.

van der Krift, T.A.J. and Berendse, F. (2002) Root life spans of four grass species from habitats differing in nutrient availability. *Functional Ecology* **16**, 198–203.

van Noordwijk, M. (1993) Roots: length, biomass, production and mortality. Methods for root research. In: *Tropical Soil Biology and Fertility, A Handbook* (eds J.M. Anderson and J.S.I. Ingram), pp. 132–144. CAB International, Wallingford, UK.

van Noordwijk, M. and Brouwer, G. (1991) Review of quantitative root length data in agriculture. In: *Plant Roots and Their Environment* (eds H. Persson and B.L. McMichael), pp. 515–525. Elsevier, Amsterdam, The Netherlands.

van Noordwijk, M. and Floris, J. (1979) Loss of dry weight during washing and storage of root samples. *Plant and Soil* **53**, 239–243.

van Noordwijk, M., Floris, J. and de Jager, A. (1985) Sampling schemes for estimating root density distribution in cropped fields. *Netherlands Journal of Agricultural Science* **33**, 241–262.

van Noordwijk, M., Brouwer, G., Koning, H., Meijboom, F.W. and Grzebisz, W. (1994) Production and decay of structural root material of winter wheat and sugar beet in conventional and integrated cropping systems. *Agriculture, Ecosystems and Environment* **51**, 99–113.

van Noordwijk, M., Martikainen, P., Bottner, P., Cueva, E., Rouland, C. and Dhillion, S.S. (1998) Global change and root function. *Global Change Biology* **4**, 759–772.

Vepraskas, M.J. and Hoyt, G.D. (1988) Comparison of the trench-profile and core methods for evaluating root distributions in tillage studies. *Agronomy Journal* **80**, 166–172.

Vogt, K.A. and Bloomfield, J. (1996) Tree root turnover and senescence. In: *Plant Roots: The Hidden Half* (eds Y. Waisel, A. Eshel and U. Kafkafi), 2nd edn, pp. 287–306. Marcel Dekker Inc., New York.

Vos, J. and Groenwold, J. (1986) Root growth of potato crops on a marine-clay soil. *Plant and Soil* **94**, 17–33.

Vos, J. and Groenwold, J. (1987) The relation between root growth along observation tubes and in bulk soil. In: *Minirhizotron Observation Tubes: Methods and Applications for Measuring Rhizosphere Dynamics* (ed. H.M. Taylor), pp. 39–49. ASA Special Publication 50. ASA, CSSA, and SSSA, Madison, WI, USA.

Watson, C.A., Ross, J.M., Bagnaresi, U., Minotta, G.F., Roffi, F., Atkinson, D., Black, K.E. and Hooker, J.E. (2000) Environment-induced modifications to root longevity in *Lolium perenne* and *Trifolium repens*. *Annals of Botany* **85**, 397–401.

Welbank, P.J., Gibb, M.J., Taylor, P.J. and Williams, E.D. (1974) Root growth of cereal crops. In: *Report for Rothamsted Experimental Station for 1973, Part 2*, pp. 26–66. Rothamsted Experimental Station, Harpenden, UK.

Zhuang, J., Yu, G.R. and Nakayama, K. (2001) Scaling of root length density of maize in the field profile. *Plant and Soil* **235**, 135–142.

Zuo, Q., Jie, F., Zhang, R. and Meng, L. (2004) A generalized function of wheat's root length density distributions. *Vadose Zone Journal* **3**, 271–277.

Chapter 4

The Functioning Root System

The root system has to serve several functions simultaneously. It has to provide a stable platform for the shoot so that the photosynthetic organs can intercept sunlight, and also has to provide a network that can exploit the water and nutrient resources of the soil. As we shall see, the availability and movement in the soil of resources varies depending on the particular resource being considered, so that in contrast to the shoot which is essentially harvesting only two resources, light and carbon dioxide, the roots and root system have evolved to cope with a more challenging environment. The anatomy of roots described in Chapter 2 coupled with the patterns of growth described in Chapter 3 form the basis of the multi-functional system that is the subject of this chapter.

The successful functioning of root systems has ecological significance in terms of the competitive advantage of individual species in mixed communities but is also economically important in the plant-based industries of agriculture, horticulture and forestry. Much of what we know about root functioning has been learned from a small range of plants that are important in these industries. This, coupled with the fact that most physiological studies with roots have been conducted with seedling roots grown in solutions, means that while the general principles of how a root functions have become clearer, the exact activity of any particular element of a root within a system growing in soil is still largely unknown.

4.1 Root anchorage

Anchorage is firmly recognized as a major function of the root system, but our quantitative understanding of this function is slight compared to those of water and nutrient absorption. There are many reasons for this, not least the underlying complexity of the topic, but also included are three additional factors contributing to some common misconceptions highlighted by Ennos (2000). First, the early work of Pfeffer (partially described in section 5.3.1) examined the force required to pull roots out of soil and led to the misconception that roots are under load when in tension. As we shall see, pulling single roots from soil is not a good indicator of the performance of a root system. Second, the use in experiments of young root radicles coated with root hairs encouraged the view that root hairs were always important elements of anchorage. Finally, because single roots could withstand large pulling forces (10 kPa without root hairs and about five times greater with root hairs), the impression was created that sufficient anchorage could be gained by root systems as a by-product of their function as absorbers; this despite many observations of plants toppling over. It is now realized, though, that the need for anchorage has influenced the overall size and shape of root systems.

Many terms are used to indicate the consequences to plants of the failure of the anchorage function of roots including uprooting, windthrow and treefall (Schaetzl *et al.*, 1989). In this chapter, the terminology of Ennos (2000) will be used. Uprooting will be used to indicate the upward pulling of roots as, for example, occurs during the grazing of grass by cattle or the pulling up of weeds by a gardener. Overturning will be used to indicate the permanent failure of plants to remain upright as, for example, occurs with some trees during strong winds (windthrow) and with some crop plants often with a combination of strong winds and heavy rain (lodging).

4.1.1 Uprooting

The simplest form of anchorage failure to consider is that arising from an axial, uprooting, force. The failure behaviour of roots and soils is very different (Ennos, 1990). Unstrengthened roots in tension behave elastically with a typical Young's modulus of about 10^4 kPa, and stretch by about 10% before breaking at a stress of about 10^3 kPa. Soils, in contrast, have high initial cohesiveness, but fail at low stresses (typically 1–100 kPa for damp loams). They do not 'break' but continue to resist shear because of the friction and cohesion between particles. This difference in elasticity and shear strength between root and soil means that when an axial pulling force is first applied, the top of the root stretches and the root moves relative to the soil. Shear stress is concentrated initially around the top of the root, and the root/soil bond (or the soil around the root) fails first in this region before moving downwards as the uprooting force is increased. After failure at the top of the root, friction within the soil or between the root and the soil will continue to resist uprooting. The greater the axial force applied, the greater the area of soil that will fail, and the deeper in the soil will the root be stretched. The force required to cause displacement will, then, rise rapidly at first but then more slowly as uprooting proceeds (Fig. 4.1a). Catastrophic failure of the root occurs in one of two ways (Ennos, 1990): (i) if the failure front reaches the bottom of the root, the root will be pulled from the soil; or (ii) if the axial force applied exceeds the strength of the root then it will break rather than be pulled out. In practice, unstrengthened roots more than a few millimetres long usually break before they are pulled out of soil. For example, Ennos (1990) found that leek radicles could withstand a force of only 0.3 N and that they broke if longer than about 30 mm in soil with shear strength of 3.8×10^3 N m^{-2} (Fig. 4.1b).

The force required to pull a root from a soil (F) is dependent on the area of contact between the root and the soil, and the shear strength of the soil:

$$F = \pi D L \sigma \tag{4.1}$$

where D is the root diameter, L is the root length, and σ is the shear strength of the soil. The resistance to being pulled out is, then, greatest for long, thick roots in strong soil. Roots, though, may not be able to withstand this force (their breaking strength is proportional to their cross-sectional area) and may break before lower parts of the root are stretched. The

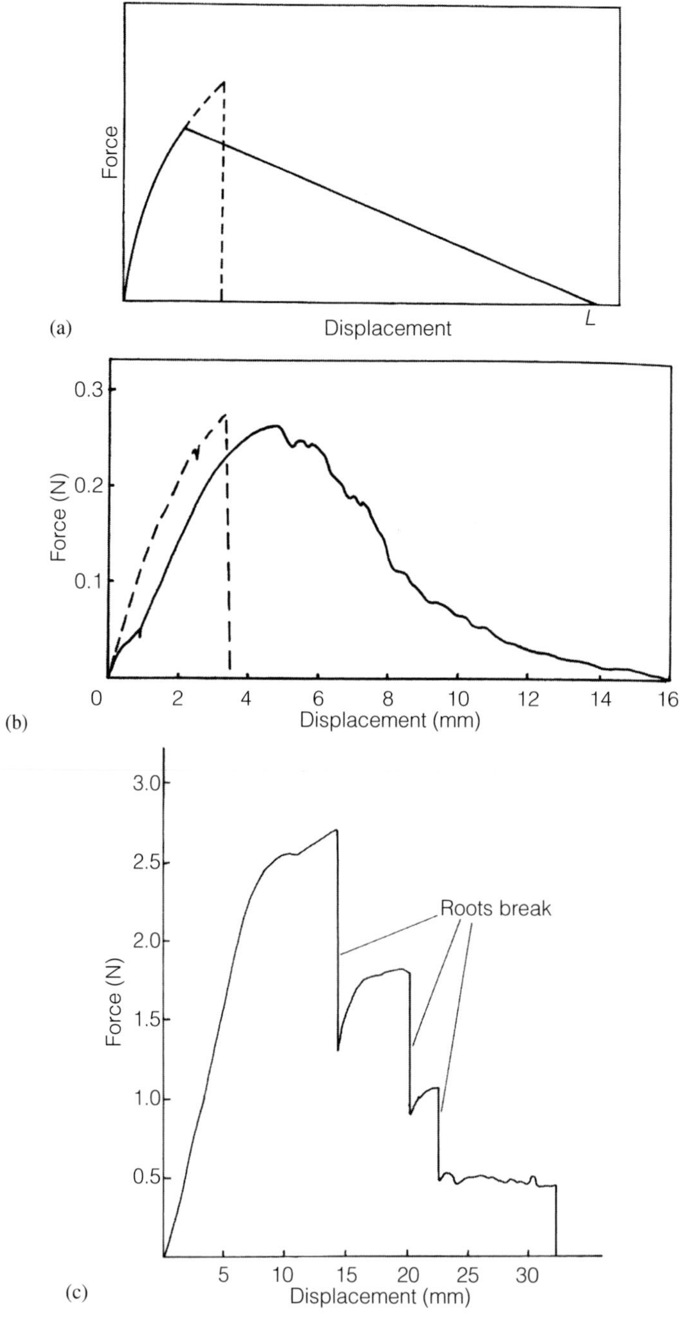

Fig. 4.1 Force/displacement curves for roots: (a) the predicted shape of the force/displacement curve for a long (dashed line) and short (solid line) root. The former breaks, while the latter produces force until it is pulled out of the ground (displacement = L); (b) force/displacement curves for a 41 mm long leek root which broke (dashed line) and for a 34 mm long leek root which pulled out (solid line); and (c) force/displacement curve for the uprooting of a 7-day-old wheat plant. (a and b reproduced with permission from Ennos, *Annals of Botany*; Oxford University Press, 1990; c reproduced with permission from Ennos, *Journal of Experimental Botany*; Oxford University Press, 1991.)

consequences of this are that for soil of a given strength, short roots will pull out while long roots will break, and that weaker soil will provide less force per unit length so the length at which the roots break will be longer.

Many plants, of course have more than one root but the principles of root uprooting are essentially similar. Figure 4.1c shows that the force/displacement curves for three unbranched seminal axes of 7-day-old wheat seedlings were similar in shape to those for the single root radicle of leek (Ennos, 1991a). However, the inner axes (axes 1, 2 and 3) with their predominantly vertical growth responded differently to the outer roots (axes 4 and 5) which grew at an angle of about 60° from the vertical. The inner roots broke at a displacement of about 5 mm, some 20–100 mm from their base, whereas the outer roots were pulled from the soil and only rarely broke. The measured extraction force of both the inner and outer roots was less than the force that should be resisted as calculated using equation 4.1 (Table 4.1), and the five roots of the plant's root system combined to resist almost as much force as the sum of the individual roots.

These uprooting tests lead to two important conclusions: (i) a plant cannot improve its anchorage just by increasing its root length or strengthening the bond between root and soil; and (ii) anchorage will be improved by strengthening the base of the root (e.g. by lignification and/or secondary thickening (Ennos, 2000). This strengthening of roots has been demonstrated in many plant species, but there is a cost to the plant in terms of carbon. Such costs can be minimized by:

(1) Strengthening only the basal areas – most roots are strengthened progressively towards their base (e.g. the strength of the three inner roots [axes 1, 2 and 3] of 21-day-old wheat plants increased linearly with distance from about 1 N at 140 mm from the base to about 3 N at the base) (Ennos, 1991a).
(2) Coating the top of the root with root hairs – root hairs are important in the establishment of seedlings (see Ennos [1989] for an account of sunflower radicles and root hairs) but in more mature roots they become less important because they die and new hairs are located close to the root tip remote from parts of the root that are likely to be subject to uprooting forces. However, the production of shoot-borne roots by grasses which then enter the soil and produce a profusion of root hairs and a strong rhizosheath is an example of a role for root hairs in the anchorage of an older plant.
(3) Root branching – the production of a fibrous root system rather than a single tap root is advantageous because many narrow roots have a larger surface area than a single root

Table 4.1 Comparison of the measured extraction force (in Newtons, N) to uproot roots of young wheat seedlings with the force that should be resisted (N) calculated using equation 5.1

	Inner three seminal axes		Outer two seminal axes	
Plant age (days)	Calculated	Measured	Calculated	Measured
7	1.06	0.84	0.50	0.29
21	2.08	1.39	0.82	0.63

From Ennos, 1991a.

of equivalent cross-sectional area and will transfer tension more readily to the soil. However, if there are too many roots, the soil will tend to fail as a 'root ball'.

4.1.2 Overturning

Most plants are rarely subjected to a vertical, uprooting force. Far more common is the application of a horizontal force by the wind (ultimately resulting in overturning), which is transmitted to the root system by the stem causing rotation in the soil. The anchorage systems of self-supporting plants must, then, be able to transmit rotational torque to the soil rather than transmit simple upward forces. This means that the fibrous root systems that are so good at preventing uprooting are much less good at preventing overturning because each root will simply bend at its base. Resistance to rotation requires at least one rigid element at the base of the stem to act as a lever; this can be provided by a tap root, or plate root systems, or by leaves growing at the base of the stem in a rosette, or by having several stems which grow horizontally along the ground before growing upwards (Ennos and Fitter, 1992). The importance of providing for anchorage can be seen in the comparative morphologies and investments in tap roots made by procumbent or climbing, rosette or multi-stemmed, and free-standing plants. Ennos and Fitter (1992) hypothesized that low, creeping and climbing plants would have fibrous root systems, as would rosette and multi-stalked plants. Single-stemmed, free-standing plants, though, would have a tap root or plate-root system with a greater proportion of their dry mass invested in their anchorage roots. This was, indeed, found to be the case in a field survey of 6 species of Brassicaceae, in a glasshouse study of 12 species of annual dicotyledons, and a field experiment of 17 different plant species. In all studies the lignified xylem of the tap root and the lignified stele of fibrous roots (the anchorage roots) were distinguished from the remaining non-lignified absorption roots. Results from the glasshouse study showed that the proportion of plant dry weight invested in anchorage roots increased from procumbent or climbing (1–4%) to rosette or multi-stemmed (3–7%) to free-standing (8–12%) plants, and that only free-standing plants invested significantly in a tap root (Fig. 4.2).

In addition to the tap-rooted systems of anchorage of free-standing plants, two other forms of root are important in resisting the horizontal forces that result in overturning. The shoot-borne upper roots (nodal/crown) of many grasses have already been referred to, and in many trees a thickened root base, or root plate, forms. Figure 4.3 summarizes the modes of failure due to horizontal forces found in three common types of root system. Each type will be considered in turn:

Tap root – a central tap root is a characteristic of most small dicotyledons. In many trees, the tap root dominates anchorage considerations when the tree is young, but its importance decreases as that of the near surface lateral roots increases (Coutts and Nicoll, 1991). There are two main components to the anchorage of such systems: (i) the resistance of the soil to compression on the leeward side; and (ii) the bending resistance of the tap root (Goodman *et al.*, 2001). The tap root acts like a foundation pile so that as the plant is overturned, the tap root is bent and rotates about a point at some distance below the soil surface. Engineering theory predicts that the maximum resistance (R_{max}) to lateral loading of a vertical pile is given by:

$$R_{max} = 4.5\,\sigma\,D\,L^2 \tag{4.2}$$

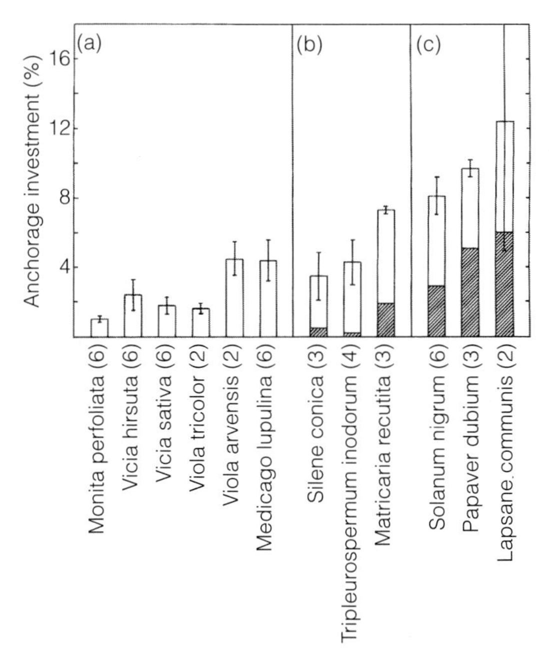

Fig. 4.2 Percentages of total dry mass invested in anchorage root (shaded area = tap root, open area = fibrous root) in 12 species of annuals of differing shoot form grown in a glasshouse: (a) procumbent or climbing; (b) rosette or multi-stemmed; (c) free-standing. Error bars show ± 1 SD of total anchorage investment, and the number of plants is given after each species name. (Reproduced with permission from Ennos and Fitter, *Functional Ecology*; Blackwell,1991.)

where D and L are the diameter and length of the rigid rod and σ is the shear strength of the soil. The result is that the shoot moves to leeward, and the plant part below the centre of movement moves to windward. Soil properties influence the exact mode of failure. If the soil is wet, then it will compress easily allowing rotation deep underground and the permanent leaning over of the plant, whereas if it is dry, the tap root or stem may fail (Ennos, 2000). In experiments with rape (*Brassica napus*) plants with pods and green seeds, Goodman *et al.* (2001) found that lateral roots originated below the centre of rotation of the root system (about 30 mm below the soil surface) and that the maximum anchorage moment was 2.9 ± 0.36 N m. On their sandy loam, tests at field capacity (shear strength about 46 kPa) showed that the soil resistance to compression accounted for about 60% of the anchorage moment and the tap root bending moment about 40%. In their soil, the anchorage moment and the stem bending moment at failure were similar, suggesting a similar resistance to root and stem lodging.

Plates – Trees commonly develop a plate root system (Coutts, 1983) consisting of large diameter roots which radiate almost horizontally from the base of the trunk before tapering and branching. These thick, horizontal roots may, in turn, develop 'sinker' roots which grow vertically downwards. In such systems, the resistance of the soil to downward movement of the roots is high because of their large area and the high resistance to compression of the soil. There are four main components to the anchorage of such systems: (i) the resistance of the soil; (ii) the resistance of the leeward hinge to bending; (iii) the resistance of the windward roots, especially the sinkers, to uprooting; and (iv) the mass of the root–soil plate (Coutts, 1986; Ennos, 2000). The leeward side of the root–soil plate acts as a cantilevered

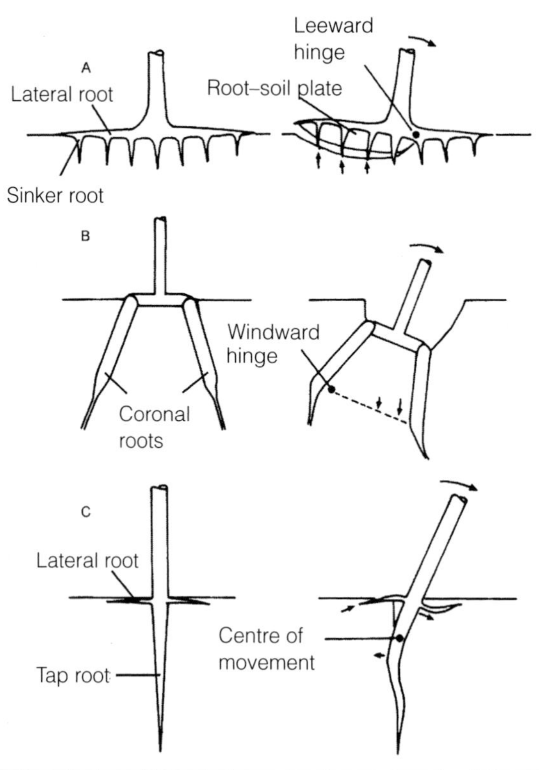

Fig. 4.3 Failure modes due to horizontal forces in three types of root systems. In widely spreading root systems with sinker roots (e.g. many trees and some herbaceous dicots), the system rotates up around a leeward hinge (A). In narrower systems (e.g. cereals and other monocots), rotation occurs about a windward hinge (B). Overturning of simple tap root systems occurs as the tap root is bent and rotates about a point directly beneath the stem at some distance below the soil surface (C). (Reproduced with permission from Ennos, *Advances in Botanical Research*; Elsevier, 2000.)

beam, and as force is applied on the windward side so upward movement of the root–soil plate on that side occurs accompanied by sequential breakage of roots and failure of soil. Eventually the tree overturns with a characteristically elliptical shaped root–soil plate attached; damage to leeside roots occurs nearer to the stem base than that to windward and largely at branching points (Coutts, 1983). In experiments with Sitka spruce on a peaty gley soil, Coutts (1986) found that the relative importance of the factors influencing anchorage changed with time. With a small horizontal force, the soil resistance was dominant, but with a large force components of anchorage were in the order roots > mass of plate > hinge resistance > soil resistance. The study highlighted the importance of laterally growing roots on the windward side in conferring stability to the tree.

From these considerations it is apparent that the stability of trees will be affected considerably by the depth of soil available for sinker roots to exploit, and by the architecture of the root system, so that predicting the anchorage failure of tree root systems is a complex matter. Shallow soils, and waterlogged soils, which in some tree species reduce the rootable soil depth (e.g. Lodgepole pine – *Pinus contorta*) (Coutts and Philipson, 1978), have received some attention in this regard because they are often marginal areas for crop production and forestry is the preferred land use. The stability of shallowly rooted trees can be strongly influenced by the symmetry of the root system and where woody lateral roots are poorly developed or absent, stability will be reduced (Coutts *et al.*, 1999). Trees, therefore, employ various modifications during their development to limit their liability to overturning. On shallow soils, the flexing of structural roots near the surface increases as height and movement in the wind also increase. To limit this movement, the size of the root–soil plate increases. For example, Nicoll and Ray (1996) and Ray and Nicoll (1998) found that for 46-year-old Sitka spruce on a peaty gley soil, there was an inverse relation between plate area and plate depth (Fig. 4.4). Thin plates, which developed over a shallow water-table, had a greater surface area than the thicker plates that developed where the water-table was deeper. They also found that anchorage was related to the rigidity of the plate, which affected the resistance to soil failure. Increasing rigidity had the effect of extending the leeward hinge away from the trunk and therefore increased the stability of the tree. On sloping sites, too, trees modify their architecture to enhance their anchorage. Chiatante *et al.* (2003) measured the root systems of five different woody plant species and found that their architecture changed from a symmetrical bell shape when grown on a plane to an asymmetrical bilateral fan shape when grown on a slope. The asymmetrical architecture on the slope was a consequence of preferential lateral root emergence and elongation in the up- and down-slope directions. Moreover, there were substantial modifications in the shape and organization of tissues in these roots compared with those growing on planar soil. These plant responses to mechanical stresses (thigmomorphogenetic responses) were similar to those obtained in other studies of plant responses to wind. For shoots, plant height is typically reduced and a bushier crown is produced with shoot branches mainly on the leeward side. For roots, a large number of changes may also occur. For example, Mickovski and Ennos (2003) found that when 4-year-old Scots pine (*Pinus sylvestris* L.) were subjected to regular unidirectional stem flexing for 6 months, there was an increase

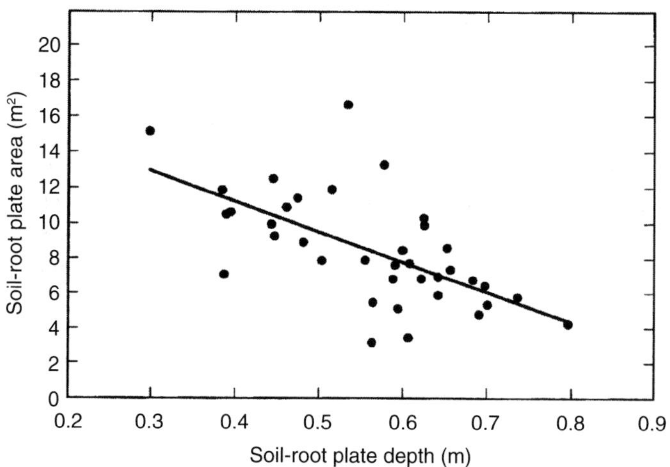

Fig. 4.4 The relation between soil–root plate area (normalized for tree size by dividing by stem mass) and soil–root plate thickness for Sitka spruce trees grown at Kershope Forest, Cumbria, UK. (Reproduced with permission from Nicoll and Ray, *Tree Physiology*; Heron Publishing, 1996.)

Table 4.2 Thigmomorphogenetic responses of flexed young *Pinus sylvestris* trees measured over a 6-month period

				Increase in		
Treatment	Increase in height (mm)	Root:shoot at harvest		Root cross-sectional area (mm^2)	Number of major laterals	Stem diameter (mm)
Control	31 ± 17	0.289 ± 0.047	par	5.09	0.11	0.16 ± 0.10
			per	2.22	0.22	0.17 ± 0.12
Flexed	13 ± 8	0.299 ± 0.035	par	14.74	0.88	0.66 ± 0.42
			per	3.65	0.50	0.48 ± 0.32

Values are means ± standard deviation either for the whole plant, or parallel (par) or perpendicular (per) to the direction of flexing. From Mickovski and Ennos, 2003.

in cross-sectional area and larger biomass allocation to the roots parallel to the plane of flexing. This, in turn, resulted in a larger number of major lateral roots with larger cross-sectional area in the plane of flexing (Table 4.2). In this study there were no significant differences in the proportion of biomass allocated to roots and shoots, or in the mechanical properties of the wood.

Prop roots – Many monocotyledons, including economically important crops such as maize and wheat, support their elongated stems with whorls of lignified roots which emerge from nodes on the stem and grow down into the soil like an inverted crown (hence they are sometimes referred to as crown roots; see Fig. 2.1). There are two components to anchorage in such systems: (i) the resistance of the soil to compression; and (ii) the buckling resistance of the windward roots (Crook and Ennos, 1993). In this anchorage system, which is substantially narrower than a root plate, when a horizontal force is applied to the stems they lean to leeward and deflection is mainly due to the rotation of the whole plant in the soil about a windward hinge. Engineering theory predicts that the root failure moment (M_R) will be given by:

$$M_R = 9/8\, \sigma \pi d^3 = 3.5\, \sigma d^3 \qquad (4.3)$$

where σ is the shear strength of the soil and d is the diameter of the root–soil cone. The result is that when a force is applied, a cone of roots and attached soil is levered into the ground on the leeward side. During small deflections, rotation is centred directly below the stem with the prop roots moving axially and bending at their base (Ennos, 1991b), but at greater stem deflections, the centre of rotation is located 20–30 mm below the soil surface at the windward edge of the root cone (Crook and Ennos, 1993).

Lodging resistance of cereals has been found by several researchers to vary between cultivars (Pinthus, 1973; Crook and Ennos, 1994; Berry *et al.*, 2003). Both stem and root lodging may occur depending on the circumstances (Baker *et al.*, 1998). In four cultivars of winter wheat, lodging was not related to the stiffness and strength of stems, which were adequate to withstand the prevalent forces, but to the failure of their root systems (Crook and Ennos, 1994). Resistance was associated with short and light stems, and with stronger, widely spread prop roots which produced larger soil cones during anchorage failure. The balance required, though, between root strength and size of cone was not resolved in this study; the cultivar Widgeon produced the largest cone and lodged most, while the cultivar Hereward had the smallest cone but did not lodge. Baker *et al.* (1998) calculated the risk of stem and root lodging of wheat from crop and soil parameters. They found that the theoretical value of 3.5 in equation 4.3 was 0.43 in their study (cf. the measured value of about 1 found by Crook and Ennos, 1993), and that crops were less prone to lodging on soils containing a large proportion of clay than crops grown on sandy soils. Their parametric analysis indicated that characteristics of the wheat crop could influence lodging as much, if not more, than the weather at the time of lodging, and demonstrated the strong correlation between different crop characters (e.g. root ball diameter and depth of the prop roots). Berry *et al.* (2003) assessed the lodging resistance of 15 winter wheat cultivars and found that there was only a weak correlation between the wind speeds resulting in stem and anchorage failure. Cultivar resistances to stem and root lodging were, then, different and improvement might be derived from breeding for both wide, deep root plates to reduce anchorage failure, and wide stems with high stem wall failure yield stress to reduce stem failure.

Many plants have root systems that are morphological combinations of the above forms and/or exhibit intermediate forms of behaviour. In particular, the exact location and angle of growth of roots close to the soil surface are often more important than the presence of a tap root or of prop roots in determining the mode of anchorage failure. For example, both sunflower (*Helianthus annuus*) with its large tap root and laterals that grow radially outwards and downwards, and Himalayan balsam (*Impatiens glandulifera*) with its prop root system, behave in failure like the plate systems of trees with a leeward hinge (Ennos *et al.*, 1993). In both plants, the wide spread of the lateral or prop roots restricts rotational or downward movement of the root system so that the system can only fail by being levered upwards.

Other plants produce a range of structures to aid anchorage. Several tree species of tropical rainforests, for example, produce root buttresses, triangular flanges joining the roots and lower trunk. Crook *et al.* (1997) undertook a comparative study of two buttressed (*Aglaia* and *Nephelium ramboutan* species) and one non-buttressed (*Mallotus wrayi*) tropical species in Malaysia. They found that whereas both leeward and windward laterals of buttressed trees contributed to anchorage strength, only the windward laterals of non-buttressed trees did so. The anchorage strength of buttressed trees (10.6 kNm) was almost twice that of unbuttressed trees (4.9 kNm), and the buttressed trees also possessed a well-developed tap root that supplied about 40% of the anchorage strength. Buttresses, then, contributed about 60% of the anchorage strength and had six times the anchorage moment of the thin laterals produced by *M. wrayi*.

4.2 Water uptake

Water is essential to the life of terrestrial plants and for the biota that live in the soil. It carries nutrients in the soil to the roots, is the solvent for, and medium of, most biochemical reactions within plants, and its loss from plants is an inevitable consequence of CO_2 exchange with the atmosphere. For most plants, the soil is the major source of their water, so that the acquisition of water from soils by roots has been, and continues to be, a major topic of soil/plant research. To understand fully the movement of water through soils to plant roots, uptake across the root, transport within the plant, and loss via evaporating surfaces to the atmosphere requires detailed knowledge of chemical, physical and biological concepts which are beyond the scope of this book (for more detail see Nobel, 1999; Tinker and Nye, 2000). Central, though, is the concept of chemical potential which is a measure of the energy state of a material or system, and thus of the ability to perform work. For a single component, if its potential at two points in a system are different, then the difference in potential gives a measure of the tendency of the component to move from the region of high potential to the region of low potential.

4.2.1 The concept of water potential

Various forms of energy can contribute to the total chemical potential of a material including its concentration (which may release energy on dilution), its compression (which may release energy on expansion), its position in an electrical field (which may release energy if the component is electrically charged and moves within the field), and its position in the gravitational field (which may release energy as it moves downwards – Tinker and Nye, 2000). In soil and plant studies, it is convenient to express these ideas not as a chemical potential but as a water potential, Ψ_w (water potential is the chemical potential divided by the molar volume of water), which has the advantage that the units in which the potential is expressed become those of energy per unit volume, that is, pressure. Water potential can, then, be expressed in a variety of almost equivalent units including energy per unit volume, energy per unit mass, and a pressure head: 1 MPa = 10^6 N m^{-2} = 10^3 J kg^{-1} = 10.2 m water head. Expression of Ψ_w in a soil or plant system requires definition of the height above a reference level and the pressure acting at the point with reference to the standard state. By definition, the potential of pure, free water is zero. Because water in soils and plants is normally neither pure nor free, its potential to do work is less than that of pure water, so its potential will normally have a negative value; it follows that a potential of –1 MPa is higher than a potential of –5 MPa. By convention (at least in the soils literature), the reference height usually used is that of the soil surface. The pressure acting at a point has to be carefully defined in relation to the various forms of energy outlined earlier that contribute to the total potential. For a solution:

$$\Psi_w = P - \Pi + \rho g h \qquad (4.4)$$

where P is the physical pressure, Π is the osmotic pressure (related to the concentration of solutes), ρ is the density of water, g is the acceleration due to gravity, and h is the height. Equation 4.4 requires some care when applied to soils and plants because of their particular properties. First, the pressure term, P, includes the pressures such as those that exist in the cells of plants, and those that occur in a saturated soil where water in lower soil layers is under a pressure head (a hydrostatic pressure). In soils that contain swelling clays, the pore water close to saturation starts to support the weight of the overlying soil, contributing to an overburden pressure. However, in an unsaturated soil (the condition that is commonly of interest in soil/plant studies), pores are only partially filled with water, and surface tension at the many air/water interfaces means that the water is under tension, or a negative pressure. This tension is called the matric potential, Ψ_m. So, in soils there cannot be a hydrostatic pressure and a matric potential at the same time.

Second, an osmotic pressure, Π, can only be generated if there is a semi-permeable membrane separating two solutions of different concentrations. Without a semi-permeable membrane, no physical pressure develops and there is simply a concentration gradient which drives diffusive fluxes of water and solute until the two solutions are uniformly mixed. In plants there are, of course, many semi-permeable membranes so that gradients of osmotic pressure are important in determining water flow. In soils, though, semi-permeable membranes to liquid water flow are rare and so osmotic pressure does not contribute to Ψ_w. For an unsaturated soil, then:

$$\Psi_w = \Psi_m + \Psi_g \tag{4.5}$$

where Ψ_g is the gravitational potential.

4.2.2 The soil–plant–atmosphere continuum

Central to the understanding of how water moves from soil to atmosphere via plants is the cohesion-tension theory first formulated over a century ago. Steudle (2001) summarized its features as follows:

(1) Water forms a continuous hydraulic system from soil, via plant, to the atmosphere. This system is analogous to an electrical system with several resistors arranged in series and in parallel (see below).
(2) Evaporation from leaves reduces their water potential causing water to move from the xylem to the evaporating surfaces; this, in turn, lowers the water potential of the xylem.
(3) Gradients of water potential within the plant result in water inflow from the soil into the roots and thence to the leaves.
(4) Water has high cohesion and can be subjected to tensions (negative pressures) up to several hundred MPa before the column will break. The pressure in xylem vessels is less than the equilibrium vapour pressure of free water at that temperature.
(5) Walls of vessels are the weakest part of the system and can contain air and/or water vapour. When a critical tension is reached in the xylem vessels, air can pass through pits in the walls resulting in cavitation (embolism).

The driving force, then, for the transfer of water from soil to plant to atmosphere is a gradient of water potential, Ψ_w, and most of this water flows in the xylem. Water can also flow in the phloem, often against this gradient in Ψ_w, and is driven by gradients in P. Typically, within the pores and cells of plants, Ψ_w is in the range 0 to –3 MPa, but in the atmosphere it is much lower, giving a very large gradient of potential between a leaf and the atmosphere. More importantly, however, the flux of water at this interface is affected by the change of phase, as water is converted from liquid to vapour. Energy is required to meet the latent heat of vaporization and it is the amount of energy available, together with the vapour pressure gradient between the leaf and the atmosphere and the rate at which the moistened air can be moved away, that govern the potential loss of water from a plant canopy covering the soil surface (Monteith and Unsworth, 1990). This external 'demand' for water distinguishes water uptake by plants from nutrient uptake in which it is the plant itself that determines demand and regulates uptake via membranes in root cells. For water, plants exercise their chief control at the site of water loss to the atmosphere via the stomata.

Despite the numerically small gradients of water potential within the soil–plant system, they are very important in inducing fluxes of water. That gradients of potential within the continuum drive a flux is similar in concept to an electrical resistance network, and the 'Ohm's law analogy' has been widely adopted as the basis for understanding and quantifying processes of water uptake and loss (Kramer, 1969; Campbell, 1985; Daamen and Simmonds, 1996). Each element of the pathway that water follows on its way from the soil to the atmosphere is characterized by a resistance that can be measured if the flux and potential gradient are known (Fig. 4.5). At its simplest, this analogue assumes a completely inelastic system without storage so that the inflow to a segment in the system equals the outflow, and the potential difference is proportional to the resistance of the segment:

$$q = (\Psi_s - \Psi_l)/(r_p + r_s) \tag{4.6}$$

where q is the rate of flow, Ψ is the water potential, r is the resistance to flow, and subscript s refers to soil, l to leaf, and p to plant. A clear limitation of equation 4.6 is that it ignores storage (although the capacitance effect can easily be added), which is likely to be particularly important in the stems of trees (Jarvis *et al.*, 1981). The processes of flow are also more complex than that suggested by this analogue because flow in soil is largely driven by a gradient of matric potential, that in the plant by a gradient in osmotic and matric potentials, and that to the atmosphere by a gradient of vapour pressure. The resistances, too, may not be constant but change with q especially in herbaceous plants (Jarvis *et al.*, 1981). However, the analogy has been useful in highlighting the stages in the continuum and in analysing the major resistances in the pathway between soil and atmosphere. Water flow in individual parts of the continuum will now be described.

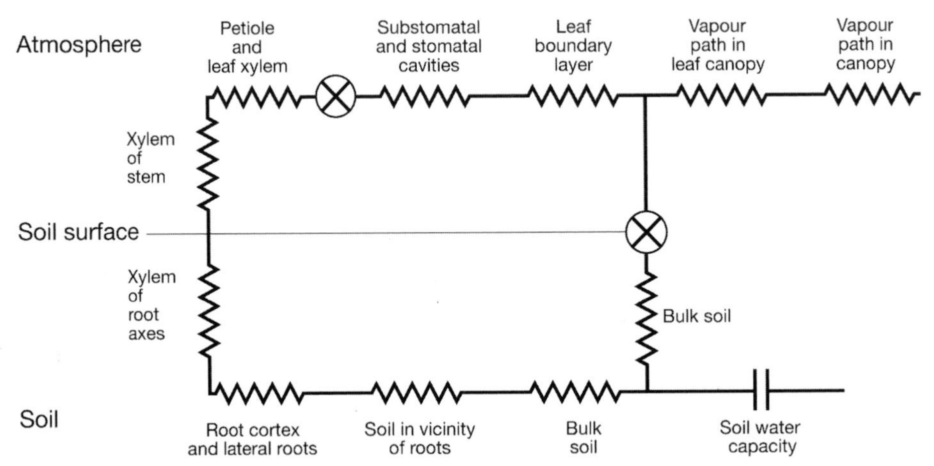

Fig. 4.5 The Ohm's law analogy of the soil–plant–atmosphere continuum showing the pathways of water transfer from bulk soil, either directly or via a plant, to the atmosphere. Resistances (wavy lines) and points of phase change from liquid to gas (cross in a circle) are shown.

Water movement in soil – as stated previously, soil water moves in response to a gradient of water potential, as described by Darcy's law:

$$q_w = -K\,(d\Psi_w/dx) \tag{4.7}$$

where q_w is the flux of water, K is the hydraulic conductivity (conductance is the reciprocal of resistance) and $(d\Psi_w/dx)$ is the gradient of water potential in the x plane. In saturated soil, K is constant so long as the soil structure remains stable, and the potential gradient is contributed by the hydrostatic pressure (i.e. the depth of the overlying water column). Typical values of K in saturated soils are 1 to 10 m d^{-1} in a sandy soil and 10^{-4} to 0.1 m d^{-1} in a clayey soil (Hillel, 1980).

In unsaturated soil, gradients of matric potential, Ψ_m, are the driving force for water movement in the x plane, and K decreases rapidly as water content (θ) decreases. Pores with larger radius empty first as the soil dries and this affects flow rate because the Poiseuille equation for flow in a tube shows that flow rate (f) is proportional to the fourth power of its radius (r). For a constant pressure gradient:

$$f = (\pi r^4/8\eta)\,dP/dx \tag{4.8}$$

where η is the viscosity and dP/dx is the pressure gradient. Typically, K decreases by five to seven orders of magnitude between Ψ_m values of 5 kPa and 1.5 MPa, so has a very low value in dry soils (see, for example, Gardner, 1960). A fundamental property of soils is the relation between Ψ_m and θ, known as the moisture characteristic curve. By taking the gradient of this characteristic, it is possible to express the flow in terms of the gradient of water content and the water 'diffusivity' (D_w):

$$q_w = -K\,(\partial\Psi_m/\partial x) = -K\,(\partial\Psi_m/\partial\theta)(\partial\theta/\partial x) = -D_w\,(\partial\theta/\partial x) \tag{4.9}$$

where $D_w = K\,(\partial\Psi_m/\partial\theta)$. The advantage of this formulation is that when combined with the continuity equation, the change of water content with time is in the same mathematical form as Fick's second law for which there are many analytical solutions:

$$\partial\theta/\partial t = \partial/\partial x\,[K\,\partial\Psi/\partial x] = \partial/\partial x\,[D_w\,\partial\theta/\partial x] \tag{4.10}$$

It has the practical advantage, too, that it is easier to measure soil water content than soil water potential. However, because D_w varies with θ, analogous to a concentration-dependent diffusion coefficient, analytical solutions may be difficult to obtain and numerical solutions may be necessary (Ross, 2003). In using the term diffusivity, it must not be forgotten, either, that it is a gradient of water potential that drives flow not a gradient of water content; water flow is a mass flow phenomenon under a pressure difference, not a diffusion process driven by random thermal motion. Flow is also complicated by the fact, usually ignored in most treatments, that hysteresis occurs in the moisture characteristic curve, so that D_w may vary even at the same θ.

The foregoing account strictly applies only to soils of uniform temperature. Temperature gradients are almost always present in soil profiles in the field, especially where radiant energy reaches the soil surface directly, and they are probably ignored too often as a cause of water movement.

When a temperature gradient exists, the driving forces for water transport are more complex than the gradients of matric and gravitational potential (equation 4.5) because as temperature increases, the vapour pressure is increased and the surface tension of water is decreased. Temperature, then, affects both the vapour pressure and the matric potential of the soil water, and the effect of a temperature gradient is to drive water from regions that are hottest to those that are coolest. Generally, the effect of temperature on liquid water flow in soils is much smaller than that on vapour flow, but the behaviour of these flow processes is complex with liquid water moving by mass flow and vapour by diffusion (Nassar and Horton, 1992). In a soil profile during the day, the temperature gradient results in flux of water vapour to the cooler, deeper soil where the vapour condenses; this, in turn, contributes to a counterflow of liquid water (with its dissolved salts) towards the warmer, surface soil. The accumulation of salts in the upper, warmer soil reduces still further the vapour pressure and thereby increases the vapour pressure gradient within the soil profile. At night, temperature gradients in the opposite direction may result in upward water movement. In laboratory studies, Nassar *et al.* (1992) showed that water moved from the hot to the cold end of a soil column with chloride moving in the opposite direction. In soil with low salt concentration (<0.1 molal, equivalent to 20% sea water) osmotic effects on water flow could be ignored, but at salt concentrations >0.5 molal, osmotic effects were pronounced.

Vapour flow of water directly to plant roots is unimportant because the relative humidity of soil even at Ψ_m of -1.5 MPa is 98.5%, but upward movement of water vapour at night in response to a temperature gradient may contribute to subsequent liquid water uptake the following day. In some desert succulents where an air gap forms the major resistance to water flow, then vapour transport may be important (Nobel and Cui, 1992).

Water movement from soil to root – Gardner (1960) was among the first to develop a quantitative framework for the flow of water to plant roots. His approach, which has subsequently become known as the single root or microscopic approach, assumed that steady-state flow inherent in equation 4.6 could be used to characterize flow towards a single cylindrical root in an infinite volume of soil. His analysis avoided the difficulties of defining the water demand of the plant and the root water potential by assuming that a root had a constant water potential at its surface. Unsaturated water flow from the soil towards a cylindrical root is given by:

$$(\partial\theta/\partial t) = (1/r)(\partial/\partial r)(r\,D_w\,(\partial\theta/\partial r)) \tag{4.11}$$

where r is the radial distance from the axis of the root. For boundary conditions given by a constant rate of water uptake per unit length of root (an inflow I_w), and an infinite soil volume, equation 4.11 solves as:

$$\Psi_s - \Psi_a = (I_w/4\pi K)(\ln(4D_w t/a^2) - 0.577) \tag{4.12}$$

where a is root radius and Ψ_a is the water potential at the root surface. This equation can be used to calculate whether a significant decrease in water potential could occur at the root surface, or, put another way, whether the soil could be an important resistance to water flow.

Gardner (1960) assumed in his calculations that K and D_w were constant and concluded that gradients of both water content and potential remain small until the water content approaches the wilting range. Passioura and Cowan (1968) recognized that roots normally have a limited volume of soil upon which they can draw so that Ψ_a will therefore decline faster than predicted by equation 4.12, and the mean values of θ and Ψ_s will also decrease steadily. If roots are assigned an 'equivalent cylinder' of soil and all water entering the root is assumed to originate in this cylinder, then steady-state and steady-rate approximations can be found (Passioura and Cowan, 1968). For a steady-state in which water is injected into the system at the periphery of the equivalent cylinder (radius b) then:

$$I_w = 2\pi r D_w (\partial\theta/\partial r) \quad (4.13)$$

and

$$\bar{\theta} - \theta_a \approx I_w/2\pi D_w (\ln b/1.65a) \quad (4.14)$$

where $\bar{\theta}$ is the mean water content of the cylinder. Tinker (1976) showed that assumptions about the values of I_w and D_w had important consequences for whether or not a significant soil resistance to flow would develop. The larger I_w, the greater the potential gradient in the soil and the lower the value of θ near the root surface. Gardner (1960) used a value of 10 ml m^{-1} d^{-1} in his calculations but this is a high value compared with those found for field-grown plants. For example, Gregory *et al.* (1978b) found that inflow to winter wheat roots averaged over the whole rooting depth ranged from 0.25 to 0.0066 ml m^{-1} d^{-1} and Taylor and Klepper (1978) similarly reported values <0.1 ml m^{-1} d^{-1} for cotton. Such results with root systems suggest that localized drying of soil around crop roots is unlikely but, as Reicosky and Ritchie (1976) and Hulugalle and Willatt (1983) pointed out, soils vary considerably in the relationship between conductivity and potential so that there could be considerable differences in resistance to flow between soils at a given Ψ_s. Their calculations showed that soil resistance would become appreciable when K decreased to 10^{-8} to 10^{-9} m d^{-1} for plants with root lengths of 1 cm cm^{-3} soil. However, not all of the root system may be equally active at taking up water because of differences in the extent of suberization, cortical degeneration, root/soil contact and root clumping, so that definite conclusions about the importance of the soil resistance to flow are currently impossible (Tinker and Nye, 2000). A good summary of the relative resistances of soil and plant in the soil–plant–atmosphere continuum is that of Hopmans and Bristow (2002): 'Under wet-soil conditions, the largest hydraulic resistance occurs in the leaf with water vapour diffusion in the surrounding air controlled by atmospheric conditions. Under these conditions, plant transpiration is at its potential rate, independent of the flow resistance of the plant, root, or soil. Transpiration is demand-controlled, rather than supply controlled. As the soil is depleted of water, its flow resistance increases, as controlled by the decreasing unsaturated soil hydraulic conductivity. Hence, … the soil resistance becomes the dominant factor controlling plant transpiration under dry soil conditions'.

Water movement in the root – To gain access to the root vascular tissues and thence to the rest of the plant, water has to flow radially across a series of concentric cell layers (see Fig. 2.2). While the flow of vapour from within the stomata to the atmosphere is usually the highest resistance in the soil–plant–atmosphere continuum, the highest resistance in the liquid part of the continuum is normally offered by this radial pathway. This means that at a given difference of water potential between the air and soil, the water potential of the shoot may be dominated by the difference in potential across the root (Steudle, 2000). It has been known for a long time that the hydraulic conductivity of roots is variable depending on such factors as water shortage, salinity and nutrient deficiency, and on the demand for water by the transpiring shoot (Weatherley, 1982; Clarkson *et al.*, 2000; Tinker and Nye, 2000). Generally, water flow to the transpiring shoot is adjusted (perhaps regulated) by the hydraulic conductivity of the roots, which can increase by up to 1000-fold as shoot demand increases.

Both water and nutrients can travel across the root via several pathways. In contrast to nutrient uptake (see section 4.3), active water transport in plants (i.e. flow linked to metabolic reactions) is negligible. This is because the permeability of membranes to water flow is several orders of magnitude larger than that of solutes (Steudle, 2000). The pathways of water flow have been the subject of considerable study but there is, as yet, only equivocal agreement on the importance of particular pathways (Passioura, 1988b). In part this is because of the complex anatomical structure of roots which varies between plant species and with age, and also because the anatomy of root tissues changes during stress thereby affecting flow. For example, suberization of the endodermis is more pronounced in trees than in herbaceous species, and accordingly tree roots are less permeable to water and solutes (Steudle and Heydt, 1997). Likewise, development of an exodermis in species such as maize, onion, sunflower, Rhodes grass and sorghum has been found to play a protective role against water loss from roots during soil drying (Taleisnik *et al.*, 1999). Additionally, there may also be changes in the permeability of membranes resulting from changes in the structure of water channel proteins, named aquaporins, which allow water to traverse membranes. Aquaporins are expressed in many membranes of plant cells including the plasma and vacuolar membranes, and have an ability to alter their permeability to water in a short time (a few hours to 2–3 days) in response to many stimuli (Fig. 4.6) (Javot and Maurel, 2002).

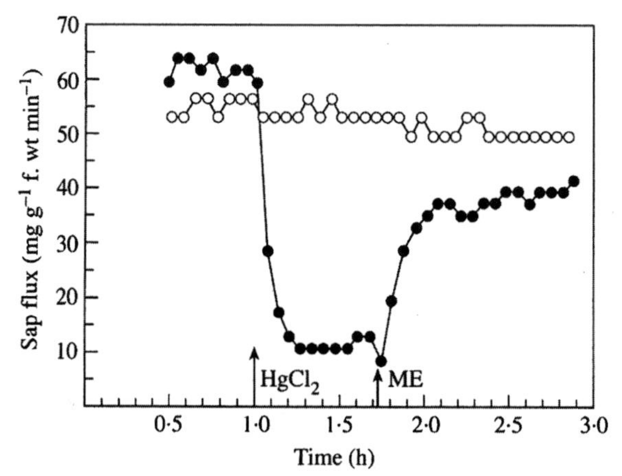

Fig. 4.6 Effects of mercury and β-mercaptoethanol (ME) on pressure-induced water transport in excised tomato roots: (○) control plants, (●) treated plants. The figure shows that $HgCl_2$ (a blocker of aquaporins) reduced sap flux by >70% and that this effect was partially reversed by application of ME. (Reproduced with permission from Javot and Maurel, *Annals of Botany*; Oxford University Press, 2002.)

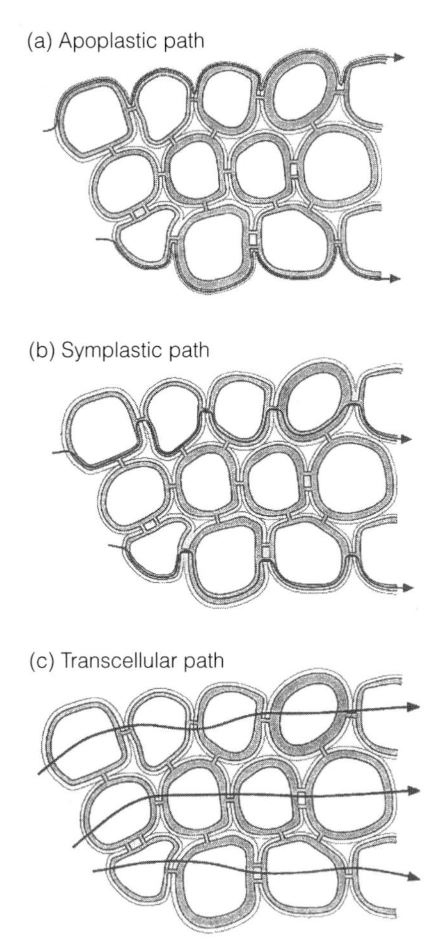

Fig. 4.7 Routes of water flow in plant tissue. The tissue is represented by four cell layers arranged in series. (a) is the apoplastic path (cell walls, grey) around protoplasts; (b) is the symplastic path mediated by plasmodesmata which bridge the cell walls between adjacent cells thereby maintaining cytoplasmic continuity; and (c) the transcellular path. In paths A and B no membranes are crossed, but in path C two plasma membranes have to be crossed per cell layer. Paths B and C (together forming the cell-to-cell path) cannot be distinguished experimentally but it is thought that water (but not solutes) uses path C because of its high membrane permeability. (Reproduced with permission from Steudle and Peterson, *Journal of Experimental Botany*; Oxford University Press, 1998.)

Three pathways exist for water and nutrient flow across living root tissues: (i) apoplastic flow, in which travel occurs through the extracellular matrix; (ii) symplastic flow, in which travel occurs intercellularly through the cytoplasm via the plasmodesmata that connect adjacent cells; and (iii) transcellular flow, in which flow occurs across membranes (Fig. 4.7). According to present understanding, the apoplast (pathway i) constitutes all components beyond the plasmalemma including the interfibrillar and intermicellar space of the cell wall, the xylem, and the water- and gas-filled intercellular space. The border of the apoplast is the outer surfaces of plants, namely the cuticle and rhizoplane (Sattelmacher, 2001; White, 2001). Pathways (ii) and (iii) are impossible to separate experimentally and are summarized as the 'cell-to-cell' component of transport (Steudle, 2000), with flow regulated by both the plasmodesmata and the aquaporins. Plasmodesmata are fluid, dynamic structures that can be modified structurally and functionally to separate the symplasm of specific cells and tissues into functional domains (Roberts and Oparka, 2003). The symplasm, then, is not just an open continuum, and plasmodesmal regulation can vary with developmental stage, tissue and plant species. Aquaporin-rich membranes may facilitate intense water flow across root tissues and may represent critical points where efficient, but spatially restricted, control of water flow can be exerted (Javot and Maurel, 2002). In some species (e.g. *Phaseolus coccineus* and barley) the transcellular component dominates cell-to-cell hydraulic conductivity but for other species the dominant pathway varies (Steudle, 2000). As roots develop, then, the relative contribution of the pathways to the total flow may change, especially as or when the endodermis and exodermis develop (Frensch *et al.*, 1996).

The development of suberin lamellae and Casparian bands in the apoplast of the endodermis, and in some plants the exodermis (see section 2.2.1), forms a substantial impediment to flow. In the endodermis, it is generally accepted that the Casparian band creates a major barrier to the flow of solutes so that they must pass into the symplasm and across at least two membranes before they reach the xylem (see section 4.3.3); the corollary is that the Casparian band also prevents the backflow of solutes from the stele. For water, though, the picture is less clear-cut. In older texts, the endodermis is often regarded as impermeable to water, but research in the last decade has challenged this view. First, there are now chemical analyses of the apoplastic barriers in the exo- and endodermis which show that the Casparian bands contain both aliphatic and aromatic suberin. The former is more hydrophobic and less permeable to water, while the latter is a lignin-like phenolic polymer esterified with hydroxycinnamic acids (Schreiber *et al.*, 1999). The lignin of Casparian bands is a relatively hydrophilic compound, suggesting that it would be impermeable to ions and large polar solutes but may allow some passage of water and small solutes (Schreiber *et al.* 1999). Hose *et al.* (2001) further showed that the apoplastic barrier in the exodermis exhibited a selectivity pattern because, depending on the stresses applied during growth, the barrier contained aliphatic and aromatic suberin and lignin in different amounts and proportions. They concluded that the apoplastic barriers in the exodermis presented a variable resistance to water and solute flow and that this allowed plants a means of regulating their uptake and loss. Second, there are experimental results that show that the endodermal barrier is incomplete. For example, Steudle *et al.* (1993) punctured the endodermis of young maize roots some 70–90 mm from the root tip with glass tubes 18–60 μm in diameter and found that while this substantially altered the reflection and permeability coefficients for NaCl (i.e. the endodermis was a barrier to NaCl flow), the hydraulic conductivity was unaltered. Moreover, the endodermis was discontinuous at the root tip and at sites of developing lateral roots. If the endodermis were impermeable to water, then the root would behave as an ideal osmometer and the hydraulic resistance of roots should remain invariant.

The composite structure of the root, the observation that root hydraulic conductivity is variable, and the two parallel pathways for water movement, namely the cell-to-cell and apoplastic paths, led Steudle *et al.* (1993) and Steudle (1994) to propose a 'composite transport model of the root' (Fig. 4.8). Depending on the structure of the root and the rate of transpiration, there will be both hydraulic water flow in response to hydraulic gradients, and osmotic water flow in response to osmotic gradients across semi-permeable membranes.

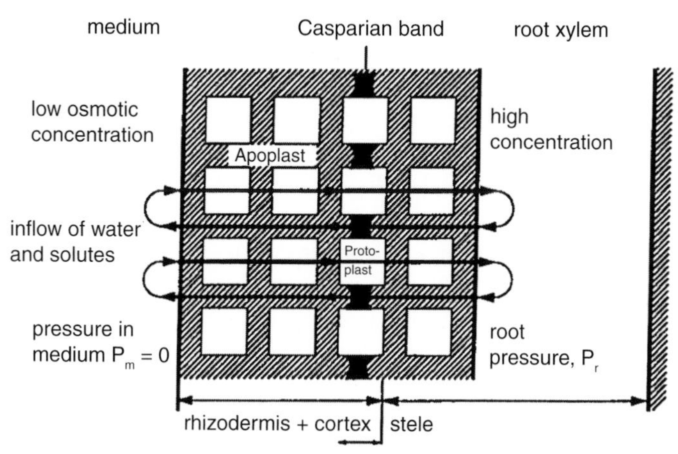

Fig. 4.8 The composite transport model of water transport in roots. In the apoplastic path there are no membranes and hence no osmotic water flow, but considerable hydraulic flow in the presence of hydrostatic pressure gradients. Across the cell-to-cell path both osmotic and hydraulic flow occur but this path has a relatively high hydraulic resistance because membranes have to be crossed. (Reproduced with permission from Steudle, *Journal of Experimental Botany*; Oxford University Press, 2000.)

Because there are no membranes in the apoplast, hydraulic flow dominates this path and the reflection coefficient (a coefficient describing the relative selectivity for water and solutes) is zero or close to zero. In contrast, in the cell-to-cell path there are many membranes with selective properties so that flow is in response to both hydraulic and osmotic gradients, and the reflection coefficient is close to unity. The dominant pathway of flow depends on whether there is a difference in either hydrostatic pressure (such as tension in the root xylem during transpiration) or osmotic pressure, and it is the switching between these different modes of transport that causes most of the variability in root hydraulic conductance. When the rate of transpiration is high, the cohesion/tension of water results in a strong hydrostatic pressure difference between the soil and the root xylem resulting in flow across the apoplastic and cell-to-cell pathways even in the exo- and endodermis. When transpiration is nil (e.g. at night), osmotic gradients within the cell-to-cell path will still allow water flow, although this path has a high hydraulic resistance. Such flow at night results in a positive root pressure and consequent guttation from leaf hydathodes. The switch from the hydraulic to the osmotic mode is purely physical depending on the forces acting in the system in response to the rate of transpiration. This model was derived from experiments in which excised roots (i.e. with no attachment to a shoot) were placed in nutrient solutions, so its applicability to live roots growing in soil is open to question and has yet to be thoroughly investigated.

Water movement from root to leaf – Once it has reached the xylem, water moves axially towards the shoot. The hydraulic resistance of the xylem cells can be estimated using Poiseuille's law (equation 4.8) which shows that the resistance offered to flow changes with the fourth power of the vessel radius. Generally, the resistance offered by differentiated (mature) xylem vessels (typically 1–10 mg^{-1} water m^{-1} root s MPa in cereals) is negligible compared with the radial resistance and that offered by plant membranes (Taylor and Klepper, 1978). For example, Steudle and Peterson (1998) calculated that a vessel with a diameter of 23 μm would need to be 24 km long to have a resistance comparable with that of the membranes in a single cell. However, estimates based on xylem radius alone may underestimate the resistance because the vessels are not uniform, smooth cylinders so that, for example, Yamauchi *et al.* (1995) found a 20-fold difference between measured and calculated axial resistance in cotton roots. Near the root tip of all plants, the xylem is undifferentiated, and even some distance from the tip it may be immature with considerable consequences for axial resistance and the flow of water (McCully and Canny, 1988; Steudle and Peterson, 1998). Axial resistance also becomes more significant in the flow of water within trees from deep roots, and may be so when crop roots extract water from a deep soil layer (e.g. Willatt and Taylor, 1978). Several fungal pathogens increase axial resistance substantially because they constrict the xylem through the presence of the fungal mycelium itself, through the release of polysaccharides and debris from xylem cells, and via wound reactions of tissues associated with the xylem.

Xylem conductance decreases in a predictable manner with decreasing water potential (Fig. 4.9) as air progressively enters the water-filled pore space (Sperry *et al.*, 2003). Air is eventually drawn through the narrow pits in the xylem resulting in cavitation and embolism in individual conduits (see earlier description of the cohesion-tension theory). Even in many well-watered crops, cavitation may be occurring on a diurnal basis. Sperry *et al.* (2003) estimate that xylem pressure would need to rise to within 0.1 MPa or so of atmospheric pressure before sap could be drawn into the xylem to dissolve an air bubble, and suggest that root pressure generated by gradients of osmotic potential plays an important role in the maintenance of hydraulic conductance. More work is required into the mechanisms of cavitation reversal because current theories do not fully explain the observations.

Water movement from root xylem to stomata – Water flow in the shoot is largely controlled by stomata which actively regulate the loss of water to the atmosphere in response to the water status of the shoot and signals, especially abscisic acid (ABA), from the root. The cuticle of leaves can be regarded as almost watertight, especially in young leaves, so the stomata are the main sites through which water vapour is lost to the atmosphere. Atmospheric factors determine the potential rate at which water can be lost to the atmosphere

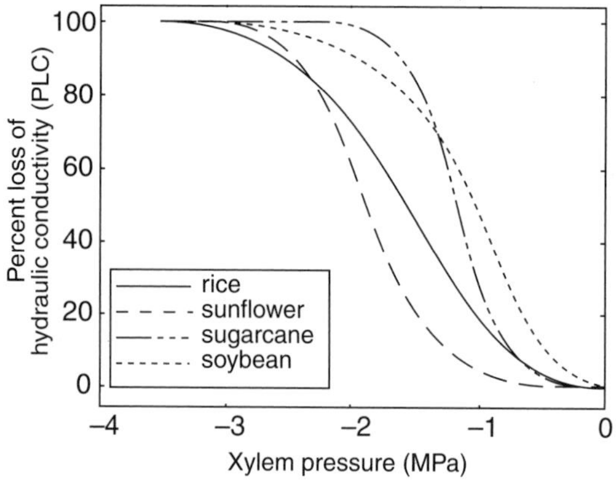

Fig. 4.9 Changes in the hydraulic conductivity of the xylem with xylem pressure for different crop species. (Reproduced with permission from Sperry *et al.*, *Agronomy Journal*; American Society of Agronomy, 2003.)

(for details see Monteith and Unsworth, 1990) but plants control the size of the stomatal opening so as to match the rate of loss to the supply available from the soil. Many environmental factors such as CO_2 concentration, irradiance, humidity and water potential of the leaf affect stomatal opening (Salisbury and Ross, 1992; Nobel, 1999), and older texts emphasized the importance of changes in leaf water potential that often accompany severe soil drying (e.g. Kramer, 1969). However, a common feature of the response in plants growing in drying soil is reduced growth, and sometimes stomatal conductance, before significant changes in leaf water potential occur (Passioura, 1988a; Zhang and Davies, 1989). In trying to explain such results, Passioura (1988a) concluded 'that the roots must have been sensing conditions in the soil, and transmitting controlling messages to the leaves, that overrode the effects of turgor'. Water flow through the soil–plant–atmosphere continuum, is not then simply a physical phenomenon.

There is evidence that ABA plays an important role as a chemical signal between the root system and shoot, modifying stomatal behaviour, water loss and growth, although this may not be the only hormone involved (Zhang and Davies, 1989; Wilkinson and Davies, 2002). Split root experiments with root systems grown in wet and drying soil have demonstrated reduced growth of the shoot and the generation of a chemical signal in the drying soil that could be reversed when the roots were removed from it (Blackman and Davies, 1985; Gowing *et al.*, 1990). Similarly in transgenic plants with different capacities to synthesize ABA, relative stomatal conductance was correlated with ABA concentration in the xylem although the results were also consistent with an additional root signal of soil drying modulating the stomatal response to this concentration (Borel *et al.*, 2001). ABA synthesis is increased in roots in response to soil drying and it is the movement of this ABA from root to shoot in the xylem that is commonly regarded as the main signalling pathway for water availability to the root system. However, ABA is also synthesized in leaves and in the soil, can move freely from soil to plant and *vice versa*, and moves rapidly through the plant in both the xylem and phloem. The long-distance transport of ABA in the soil–plant system is, then, more complex than the simple delivery of a chemical from root to leaf (Plate 4.1) (Sauter *et al.*, 2001). Much remains to be understood but a key development is an understanding of the partitioning of ABA in different plant compartments and tissues in response to pH. For example, the intensity of the ABA signal produced by soil drying can be enhanced by the simultaneous effects of drying on pH in various root compartments. Drying increases apoplast pH and acidifies the cytoplasm, which in turn prevents rhizosphere- and phloem-sourced ABA from entering the symplast. This induces more effective efflux of the root-synthesized ABA to the apoplast and thence to the xylem (Wilkinson and Davies, 2002). Similarly, in the leaf whether the ABA reaches the active sites in the guard cells of the stomata depends crucially upon the pH of the xylem sap and the consequent partitioning of the ABA between the apoplast and symplast of the mesophyll and epidermal cells (Sauter *et al.*, 2001; Wilkinson and Davies, 2002). It is also possible that ABA and nitrate can interact to influence stomata as a result of nitrate-induced changes in tissue pH and in N-containing molecules transported in the xylem sap (Wilkinson and Davies, 2002).

Root systems commonly experience soil zones of varying wetness, so how the plant integrates this information to regulate growth is of considerable interest. In horticulture, the adequate supply of water to fruits is a crucial part of the production process, particularly in crops such as tomato where suboptimal supplies can induce nutrient deficiencies that substantially reduce yield and fruit quality (e.g. blossom end rot caused by deficient calcium transport to the distal ends of fruits) (Tabatabaei *et al.*, 2004). Split root techniques and techniques of partial root drying are at an early stage of commercial development but offer prospects for making use of the chemical signalling system to reduce vegetative growth (e.g. in grapes), enhance fruit quality, and improve the efficiency with which irrigation water is used (Davies *et al.*, 2000, 2002).

Water movement from soil layer to soil layer via roots – While this section, so far, has focused on the transfer of water from soil to plant, it has been known for a long time that water can move from wet soil to dry soil via roots. The term 'hydraulic redistribution' is used to describe this phenomenon generally, but because dry soils frequently overlie wet soils, the term 'hydraulic lift' has also been used specifically for such situations (Burgess *et al.*, 2001; Ryel *et al.*, 2002). While roots generally behave as electrical 'rectifiers' that do not allow backwards flow of water, they are not perfect in this regard (see earlier description of the structure of the exo- and endodermis) and water movement within the soil profile by plant roots has been measured in many plant species in a range of habitats (Caldwell *et al.*, 1998; Jackson *et al.*, 2000). While the majority of field studies documenting hydraulic redistribution have been in semi-arid and arid environments, the phenomenon is not restricted to such regions (for an example in a moist coniferous forest, see Brooks *et al.*, 2002). Redistribution of water within the root zone of crop plants has also been known for many years (Baker and van Bavel, 1988) and satisfactorily modelled using the Ohm's law analogue described earlier in this section (Xu and Bland, 1993).

In most ecosystems, redistribution usually occurs late in the afternoon or at night when reduced transpiration allows xylem water potential to become greater than Ψ_s in drier soil layers, and results in distinct diel fluctuations in Ψ_s with low values during the day and higher values at night (Richards and Caldwell, 1987; Ludwig *et al.*, 2003). The importance of the process may depend on the weather in a particular year. For example, Ludwig *et al.* (2003) studied the water balance of *Acacia tortilis* trees in an East African savanna in a relatively wet year and a very dry year, and found that there was little evidence of hydraulic lift in the dry year (Ψ_s was below –5.0 MPa). In the wet year (Ψ_s typically –1 to –3 MPa), not only were diel variations in Ψ_s found up to 10 m from the tree, but stable oxygen isotope measurements of water extracted from xylem of grass growing beneath the trees confirmed the contribution that hydraulic lift can make to the growth of the understorey. Stable isotopes were employed, too, by Sekiya and Yano (2004) to investigate the potential of pigeonpea and sesbania in Zambia to lift water from groundwater and redistribute it to more shallowly rooted, intercropped maize. Both pigeonpea and sesbania produced a tap root that went to about 2.1 m depth but deuterium/hydrogen isotope ratios indicated that only pigeonpea redistributed water from the groundwater to maize.

While many studies demonstrate hydraulic redistribution in a range of ecosystyems, there have been relatively few attempts to quantify the amounts of water involved and their significance for the water balance. Jackson *et al.* (2000) estimated that hydraulic lift contributed 30–70 mm to the seasonal water use of a sugar maple forest (growing season 155 days) grown on a soil with a shallow water table. Ryel *et al.* (2002) employed a simulation model to evaluate the benefits of hydraulic redistribution to the transpiration of a stand of *Artemisia tridentata*, and found only a 3.5% enhancement of transpiration over a 100-day period, although the increase was up to 20% on some days. However, even if the contribution to the water balance is small, the benefits to the plant and other soil organisms from having a damp zone around roots in terms of nutrient diffusion and other processes of nutrient transformation in the rhizosphere are likely to be highly significant. Besides the ecological importance, genotypic differences in the ability to redistribute water have also been identified in crops (e.g. sorghum, Xu and Bland, 1993; maize, Wan *et al*., 2000). Wan *et al*. (2000) grew three maize hybrids in pots of soil and measured wetting of the upper soil by redistribution in the drought-tolerant hybrids but not in the drought-susceptible hybrid. The hydraulic lift allowed one of the drought-tolerant hybrids to achieve a peak transpiration rate 27–42% higher than the drought-susceptible hybrid on days with high evaporative demand.

4.2.3 Water uptake by plant root systems

There still remain a great number of uncertainties in our understanding of how water uptake by root systems occurs, and no single approach has yet found universal acceptance. This is perhaps not surprising given the complexity of both root structure and functionality, the differences in root system architecture between plant species, the interactions between roots and soils that modify root performance, and the different patterns of rainfall that occur relative to the growth patterns of different vegetation types. Nevertheless, some broad generalizations have been found useful in describing the activity of root systems in taking up water.

Measurements with annual crops have commonly shown that the most important feature of the root system relating to seasonal water use is the maximum rooting depth, because this determines the quantity of available soil water that is potentially accessible (Ehlers *et al*., 1991; Gardner, 1991). For example, Hamblin and Tennant (1987) compared water use by wheat, barley, pea and lupin crops grown on loamy sands and a sandy loam, and found that water loss from the soil profile over the growing season was better correlated with maximum rooting depth than with total root length. Inflow (uptake per unit root length) was highest in soil layers that were moist from recent rain or in which root length was expanding most rapidly. Because the maximum rate of uptake of the crop is determined by the atmosphere and legume crops typically have less length than cereal crops (see section 3.3.1), inflow in legumes was higher than in cereal crops although Ψ_s was similar in all crops (see Gregory and Brown [1989] for a similar result with barley and chickpea). In short, there is not a unique relation between the rate of water uptake from a given soil layer and the root length density in that layer. The relation changes as the soil dries and as the root system grows, ages and senesces.

Three other aspects of root system/soil water behaviour are also noteworthy. First, it is widely observed that in a moist soil profile, water uptake from the lower portion of the root system does not occur immediately and that significant quantities of water have to be removed from the surface layers before uptake from depth proceeds. This behaviour may not be immediately obvious in crops where the downward growth of the root system is closely associated with the downward movement of the drying front, but is obvious in perennial plants and in annual crops such as winter wheat where much of the root system is established before significant soil drying occurs. As an example, Table 4.3 shows the pat-

Table 4.3 Comparison of the relative use of water from different soil layers with the relative distribution of root length

Depth (cm)	Root length (% total)	Water uptake (%)	Root length (% total)	Water uptake (%)		
	8 April–13 May		13 May–29 July	13 May–17 June	17 June–8 July	8 July–29 July
0–30	71	100	66	71	39	86
30–60	18	0	21	16	17	6
60–100	9	0	11	13	25	2
>100	1	0	3	0	19	6

From Gregory *et al.*, 1978a and 1978b.

tern of water use beneath a crop of winter wheat grown in the UK; there was no clear correlation between the proportion of roots in a particular layer and the water extracted from that layer. The total length of the root system increased from 9.9 to 23.6 km m^{-2} between 8 April and 17 June but the proportion in each layer remained very similar throughout and especially from 13 May onwards. Initially water uptake was only from the upper 30 cm despite roots extending to >1 m depth. As the surface dried, uptake from deeper in the profile occurred. A large input of rain from 9 to 12 July resulted in uptake switching predominantly to the surface layer. Similarly the depth of maximum water inflow moved down the profile in a crop of winter wheat grown on stored soil water beneath a rain shelter (Weir and Barraclough, 1986). Second, root systems are rarely static and systems are characterized by the movement of roots towards new sources of water that are often deeper in the soil (Monteith, 1986). For an annual crop, the downward extension of the system is often at an almost constant velocity for a major part of the growing season (see section 3.3.2). Finally, measurements of changes in soil water content beneath crops growing on soil profiles that are uniformly wet, show that the shape of the extraction zone, once established, appears to remain about the same (Gardner, 1983). Figure 4.10 summarizes results from about 40

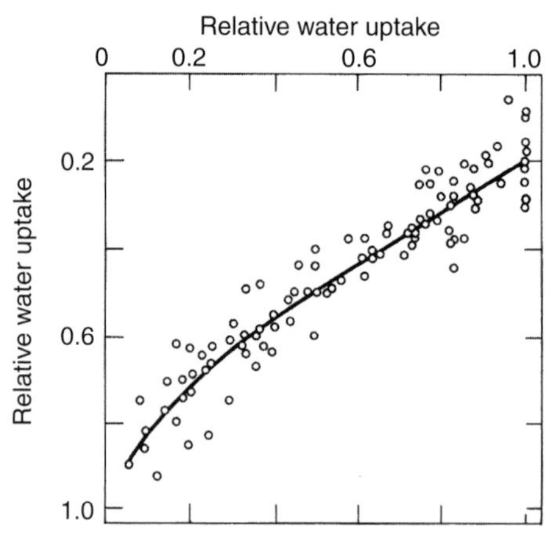

Fig. 4.10 Relative water uptake as a function of relative depth for nine different plant species grown on a range of soils. (Reproduced with permission from Gardner, *Irrigation Science*; Springer Science and Business Media, 1991.)

different experiments conducted in the USA from eight crop and one tree species, and soils ranging from sands to clays. Extraction from the upper 25 cm was set to 1.0 and uptake curves were normalized by plotting on a relative scale of 0 to 1.0 and scaling all the depths so that the curves all passed though the point 0.5, 0.5. The line drawn is part of a straight line chosen arbitrarily to decrease from 1.0 at depth 0.2 to 0 at depth 0.8. It indicates that 40% of the extraction was from the first 20% of the root depth, then 33%, 20% and 7% from successive layers. Extraction from the base of the profile is slightly underestimated by this 'rule of thumb', but the analysis reinforces the importance of rooting depth in determining total water uptake and provides a description of the root system as a 'sink' (see below).

There have been many different attempts to model water uptake by root systems, but two distinct lines of approach can be identified, each with their own assumptions and elements of empiricism. The first group are essentially a macroscopic version of the single root model outlined in section 4.2.2 (equation 4.12) involving exploitation of some form of relation between water extraction and the distribution of roots, together with soil hydraulic properties and gradients of water potential between soil and plant. However, the single root model proposed by Gardner (1960) has proven difficult to implement in practice so the second group avoid detailed consideration of roots and potential gradients and use the transpiration rate, and a plant root water extraction function (a 'sink' term) that is dependent on soil water status.

The single root model extended to the root system – The single root model can be extended to a root system by the simple expedient of apportioning to each root exclusive access to a volume of soil water. Complex descriptions can be employed (e.g. the Voronoi polygons used by Lafolie *et al.*, 1991), but if a root is ascribed to a cylinder of soil whose outer radius, b, is half the average distance between roots, then b is related to root length per unit soil volume (L_v) by (Tinker, 1976):

$$b = 1/(\pi L_v)^{0.5} \qquad (4.15)$$

Substituting this term into equation 4.14 shows that the rate of uptake from a given soil volume will be proportional to L_v, the diffusivity of soil water, and the difference in water content between the bulk soil and the root surface. While the bulk soil water content and the inflow will change with time, the water content at the root surface is essentially constant so that inflow will be proportional to the amount of water unextracted. If soil water diffusivity is constant (which it often is over a reasonable range of water content), then the mean water content of the soil volume (θ) will change with time (t) as follows (Passioura, 1983):

$$\theta = \theta^* \exp(-kL_v t) \qquad (4.16)$$

where k is a constant. Because $\theta = \theta^*$ at $t = 0$, and $\theta = 0$ when t is infinite, θ^* is the maximum amount of water that roots are capable of extracting from the surrounding soil. Monteith (1986) combined this approach with a mathematical function describing the descent of the root system to describe water extraction by crops grown on stored soil water. Fig. 4.11 shows results from a crop of soyabean grown on a deep silty loam to illustrate the applicability of equation 4.16 to the analysis of changes in soil water content during periods of continuous soil drying; a plot of the logarithm of soil water content with time gives a straight line once

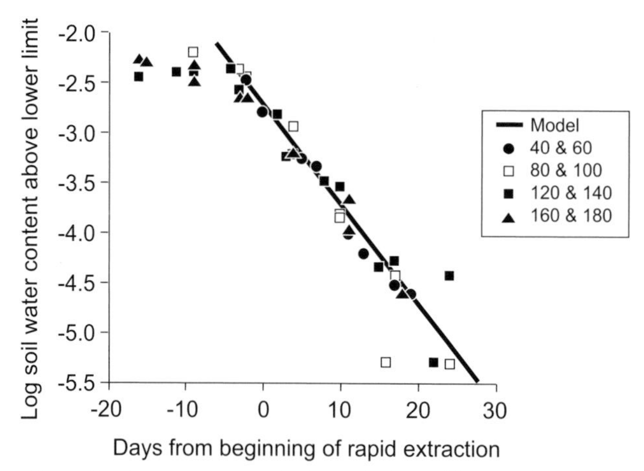

Fig. 4.11 Values of log soil water content above the lower limit for different soil layers with time for a crop of soyabean grown on a deep silt loam. (Reproduced with permission from Dardanelli *et al.*, *Field Crops Research*; Elsevier, 2004.)

water extraction from a layer commences. The quantity $1/kL_v$ can be regarded as a time constant and was used as an analytical framework by Robertson *et al.* (1993a, 1993b) to examine water extraction patterns of six sorghum crops under conditions of continuous soil drying. They found that while the mean rate of downward progress of the water extraction front was a conservative quantity (34 mm d^{-1}), the value of $1/kL_v$ varied considerably between crops depending upon L_v and atmospheric conditions such that it had no predictive value. Moreover, because some root proliferation occurred ahead of the extraction front and proliferation also occurred in layers during initial water extraction, uptake from some layers, especially in the middle of the soil profile, was better described by a sigmoidal curve.

All of these approaches assume that the root is a uniformly absorbing cylinder that is reasonably evenly distributed. Passioura (1983, 1988b) has outlined several reasons why such approaches may fail including the clumping of roots in soils with large structural units, the development of an interfacial resistance between the root and the soil, and the impermeability of portions of the root system. This has led many to conclude that, for practical purposes, it might be better to avoid measurement of roots altogether and focus on the outcome – i.e. the uptake of water (e.g. McIntyre *et al.*, 1995). The development of approaches using sinks is described below, but an approach similar to that described by Passioura (1983) has been developed by Dardanelli *et al.* (2004). Assuming that the roots in layers from which water is being extracted had reached an amount that allowed water extraction at a maximum rate, that the rate of atmospheric evaporative demand was greater than the rate of soil water supply, and that the soil water content at the lower limit of extractability (θ_{ll}) could be accurately measured or estimated, then the daily change in volumetric soil water content ($\partial\theta_d$) with time (t) is given by:

$$\partial\theta_d/\partial t = (\theta_{d-1} - \theta_{ll})\,C \qquad (4.17)$$

where C is an empirical constant. This equation describes an exponential decrease in soil water content over time (Fig 4.11) and Dardanelli *et al.* (2004) found that C was 0.096 for a range of crops and soil types. Values <0.096 were found for groundnut (0.064) and on soils with physical constraints and which swell and shrink. The utility of this approach requires further investigation.

Models with a sink term – Many studies of water extraction from soils have bypassed the requirement for a description of the root system and adopted a macroscopic approach in which extraction is represented by a sink term, S (Feddes *et al.*, 1976; Hopmans and Bristow, 2002). In the vertical, z, dimension the equation is:

$$\partial\theta/\partial t = \partial/\partial z\ [K\ \partial\Psi_w/\partial z] - S(z,t) \qquad (4.18)$$

where S is a function of depth and time, and when integrated over the rooting depth gives the transpiration by the plant. For non-stressed conditions, the maximum extraction for soil layer, i, is determined by the product of the non-stressed water extraction rate (T_{pot}) and the normalized active root distribution function (RF):

$$S_{max,i} = T_{pot,i}\ RF_i \qquad (4.19)$$

RF characterizes the depth distribution of potential root water uptake and is equal to 1.0 when integrated over the whole rooting depth. Various approaches have been used to define RF, including observed changes in soil water content (Feddes *et al.*, 1976), empirical approaches (Molz, 1981), a non-linear function of normalized root length distribution (Wu *et al.*, 1999) and explicit descriptions of root architecture (Wilderotter, 2003).

The approach adopted by Feddes *et al.* (1976) has been widely copied because it uses field data to inform modelling without the need to obtain measurements of roots and water potential. The sink term for each soil layer is calculated as the product of $S_{max,i}$ and a stress response function, α, which changes with soil matric potential (Fig. 4.12). The function describes the stress response of plants to wet and dry soils so that α has a value of 1 between field capacity (typically –5 to –30 kPa depending on the soil) and matric potentials ranging

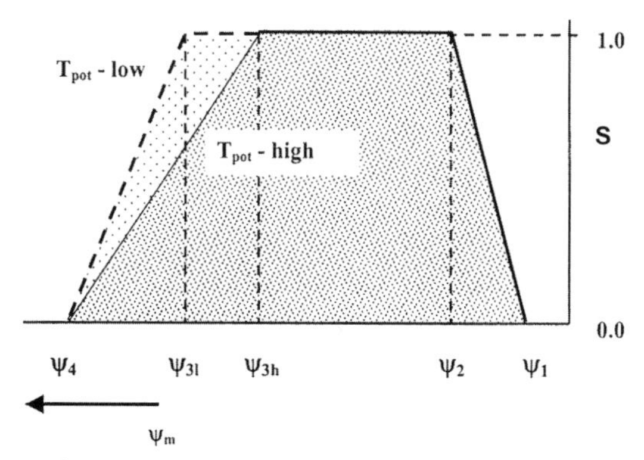

Fig. 4.12 The sink term for crop water uptake changes with soil water potential. At high values (near water-logged), the sink strength is 0 but increases rapidly as the soil drains to 1 at field capacity (Ψ_2) and remains at that level to a water potential of about –20 (Ψ_{3h}) to –100 kPa (Ψ_{3l}) depending on the crop and transpiration demand. As the soil becomes very dry, the sink strength falls to 0 again at about –1500 kPa (Ψ_4). (Reproduced with permission from Hopmans and Bristow, *Advances in Agronomy*; Elsevier, 2002.)

from –20 to –200 kPa, depending on crop and rate of potential evaporation, and then declines to a value of 0 at soil matric potentials of –1500 kPa (permanent wilting point). This approach to modelling root water uptake is capable of numerically simulating soil water profiles satisfactorily with measured evaporation as the only input (Luo *et al.*, 2003).

4.3 Nutrient uptake

There is a huge literature on the mineral nutrition of plants and the role of soils in supplying nutrients, especially to managed crops, grasslands and forests (e.g. Barber 1995; Marschner, 1995; Whitehead, 2000). The purpose of this section is to highlight the main processes operating at the root/soil interface and within the root that ensure a supply of nutrients to the plant shoot.

4.3.1 Nutrient requirements of plants and the availability of nutrients

Unlike water, there is no potential external demand for nutrients that can be readily calculated. Demand for nutrients is driven by the metabolic demands of the plant, and the plant exerts considerable, but not always perfect, control over the quantities of nutrients and other ions that are allowed to enter. In general, all higher plants require the same nutrients, although there are some minor variations. An element is essential to a plant if: (i) a deficiency makes it impossible for the plant to complete its life cycle; (ii) such deficiency is specific to a particular element and can be prevented or corrected by supplying this element; and (iii) the element is directly involved in the physiological or biochemical functions of the plant (Marschner, 1995). The elements considered as essential for higher plants are:

- Macronutrients: carbon, hydrogen, oxygen, nitrogen, phosphorus, potassium, calcium, magnesium and sulphur.
- Micronutrients: iron, manganese, copper, zinc, boron, molybdenum and chlorine.

Beneficial nutrients, which may assist the growth of some plants, and/or be essential for some plants include cobalt (required for some systems of biological nitrogen fixation), sodium (beneficial for sugar beet and essential for saltbush) and silicon (beneficial to many grasses and essential for *Equisetum arvense*). There have also been claims that nickel, tungsten and vanadium are essential, but the quantities involved are so small that it has been difficult to prove unequivocally.

With the exception of C and O, most of the other nutrients are obtained by plants from the soil and the basis of plant nutrition can be regarded as a series of chemical exchanges that drive nutrient, U, from the soil to the plant:

$$U_{\text{soil solid}} \leftrightarrow U_{\text{soil solution}} \leftrightarrow U_{\text{root surface}} \leftrightarrow U_{\text{root}} \leftrightarrow U_{\text{xylem}} \leftrightarrow U_{\text{shoot}}$$

This chain of processes is supplemented by two other processes that will receive particular attention in Chapters 6 and 7, namely: (i) the capacity to form symbiotic linkages with mycorrhizal fungi or N-fixing bacteria; and (ii) modification of the root environment to enhance nutrient supply (Aerts and Chapin, 2000). Nutrients are present in many different chemical forms in soils, and it is beyond the scope of this book to detail them (see standard texts such as Wild, 1988; Barber, 1995). Most nutrients, though, are taken up by plants from the soil solution so the processes of dissolution, desorption, exchange and mineralization are key initial steps in determining nutrient availability.

The concentration of nutrient ions in the soil solution is a factor in determining their rate of absorption by plant roots, and the composition of soil solutions has been widely analysed. On nutrient-poor soils, concentrations are very low and in some cases undetectable until recent analytical techniques were developed. In soils used for agricultural crops, concentrations of Mg, K and Na are usually between 10^{-4} and 10^{-3} M, those of Ca about 10^{-3} M, and of phosphate ions about 10^{-6} to 10^{-5} M. Nitrate concentration is generally higher than for other anions but both nitrate and bicarbonate concentrations are very dependent on microbial and root activity. Table 4.4 gives values for the cultivated layer from several soil types. In aridisols, concentrations of Ca, Mg and Na are higher than in other soils because plant uptake relative to the amounts in solution is less than for other ions, and there is insufficient water to leach them from the profile.

The composition of the soil solution at a particular location not only depends on the total quantities of nutrients present in the soil but is also highly dynamic. Soil water content is an especially significant factor because it affects many chemical equilibria between solid and liquid phases, and processes of organic matter mineralization. Generally nitrate and chloride and Na concentrations vary inversely with water content, K and bicarbonate concentrations are little affected by water content, and sulphate falls about midway between these patterns of behaviour (Wild, 1988). The increased activity of microorganisms after soil wetting typically increases the nitrate concentration in solution (Birch, 1958; Seneviratne and Wild, 1985), and in neutral or weakly acid soils Ca concentration increases to balance this. In some types of native vegetation (often on nutrient-poor soils), nutrients such as N may be conserved within the system by rapid recycling of the N contained in dead roots (Abbadie *et al.*, 1992). The residence time of mineral N in the soil solution arising from mineralization of dead roots is very short because the N is taken up rapidly by living roots.

4.3.2 Nutrient movement in soil solution

Roots are in direct contact with only a very small part of the nutrients in the soil solution, so that nutrients must move from the bulk soil to the root surface. This movement occurs through the

Table 4.4 Concentrations of ions (mM) in displaced soil solutions from different types of soil

Ion	Alfisols and Mollisols	Ultisol	Aridisols Normal	Aridisols Saline
Ca	1.5	1.65	3.3	37.0
Mg	2.5	0.50	1.94	34.0
K	0.15	0.22	0.70	0.40
Na	–	–	12.2	79.0
PO_4	0.0016	0.001	–	–
SO_4	–	0.27	4.93	47.0

From Barber, 1995.

processes of mass flow and diffusion. Mass flow (convection) occurs as a result of transpiration; dissolved ions are carried to the root surface in the hydraulic continuum formed by the soil–plant–atmosphere (section 4.2.2). As we shall see, though, the membranes of the root are highly selective about what enters the plant cells, and because of this selectivity concentration gradients of different strengths and directions are established around roots. Diffusion occurs when ions move along a concentration gradient established between the root surface and the bulk soil; ions diffuse towards the root if they are taken up faster than they are carried to the root surface by mass flow and away from the root if the converse pertains. A consequence of the differential mobility of ions in soil solution is that the zones of competition for different ions by a root system differ. For a mobile nutrient such as nitrate, the zone of competition between roots will be much larger than that for an immobile nutrient such as phosphate (Bray, 1954).

The supply of nutrients to a plant by mass flow can be estimated as the product of the quantity of water taken up and the concentration of ions in the soil solution. This calculation has been undertaken for a range of crops and trees (maize, Barber *et al.*, 1963; leek, Brewster and Tinker, 1970; winter wheat, Gregory *et al.*, 1979; beech trees, Prenzel, 1979) and the results show that, in general, mass flow will transport more than sufficient S, Ca, Na and Mg to the root surface (although not Ca on the acid soil, pH 4, used by Prenzel, 1979), significant but insufficient quantities of K and N, and insufficient P. The concentration of micronutrients in solution is highly dependent on pH and there are few experimental results, so generalizations are impossible. Calculations of mass flow are often based on values for the whole growing season and do not reflect the dynamic changes in nutrient concentrations either within the plant or the soil solution; they should, then, be treated cautiously.

When a nutrient such as potassium or phosphate is absorbed at a rate greater than that of water, its concentration at the root surface falls (Fig. 4.13), although less than it would if transpiration were not occurring. As a consequence, the soil solids release these ions into solution, which tends to buffer their concentration. Despite this, the concentration will fall and this establishes a concentration gradient between the root surface and the soil which causes these ions to diffuse towards the root. A zone of depletion develops which can be viewed in an autoradiograph (Fig. 4.14). Conversely, when mass flow supplies nutrients

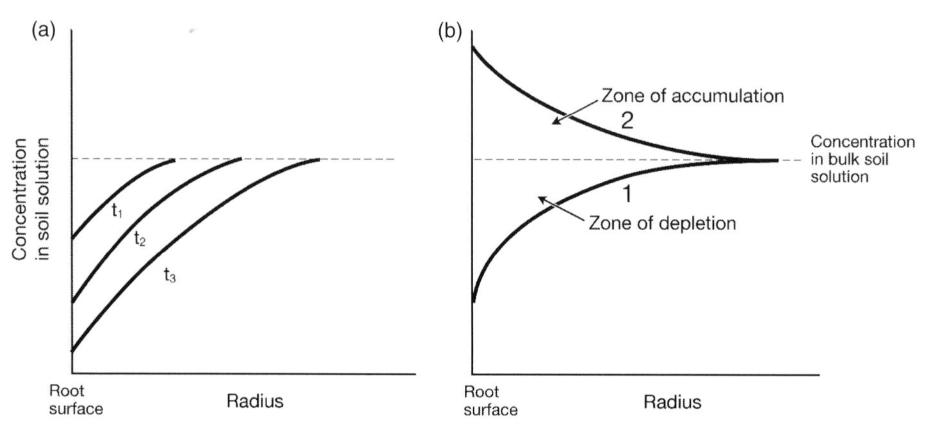

Fig. 4.13 Concentration of solute near a root surface: (a) diffusion alone with increasing time t_1, t_2 and t_3; and (b) with water absorbed relatively slower than solute a zone of depletion results (1) while when water is absorbed relatively faster than solute a zone of accumulation results (2).

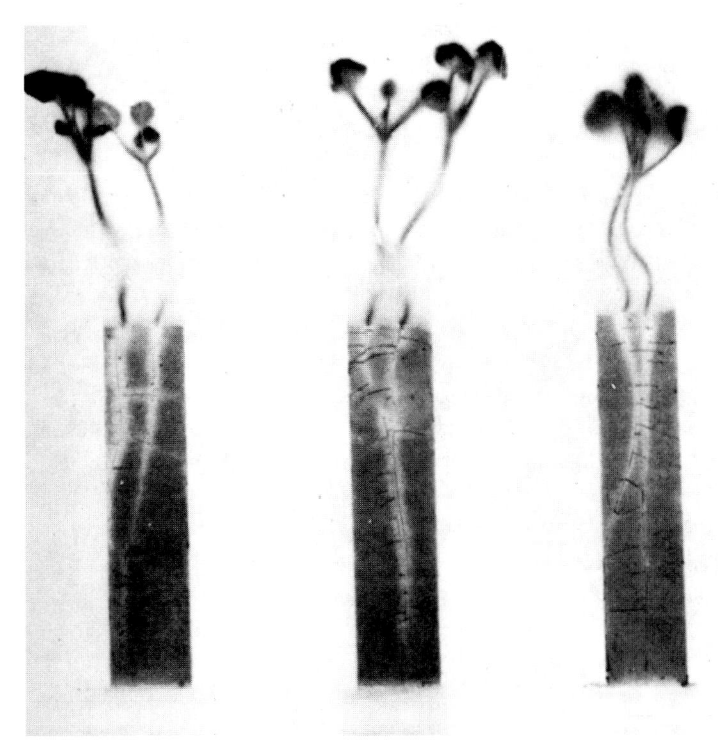

Fig. 4.14 Autoradiographs of roots of rape in a soil labelled with ^{33}P after 7 days of plant growth. The labelled soil (black) shows a zone of depletion around the roots (white), and accumulation within the roots (black). (Reproduced with permission from Bhat and Nye, *Plant and Soil*; Springer Science and Business Media, 1973.)

such as Ca at a rate faster than they are taken up by the plant, the nutrient accumulates at the root surface and diffuses away. When a concentration gradient is established, ions move by thermal motion (Brownian movement) from points of high concentration to points of low concentration until such time as equilibrium is re-established. The distance that the concentration gradient extends from the root surface depends on the rate of diffusion and the length of time for which the process persists (Fig. 4.13).

The application of diffusion theory to the supply of plant nutrients to roots is now well established and thoroughly documented (Barber, 1995; Tinker and Nye, 2000). Diffusion within solids is ignored in such approaches because it is small relative to that in solution. The amount of ion diffusing across unit area in unit time (F) with a concentration gradient dC/dx is given by Fick's first law of diffusion:

$$F = -D\,(dC/dx) \tag{4.20}$$

where D is the diffusion coefficient which can be calculated from the self-diffusion coefficient of the ion in free solution (D_o). In soil the diffusion coefficient (D_s) is $< D_o$ because: (i) solution does not occupy the whole soil volume; (ii) the diffusion path is more tortuous than in solution with diffusion confined to water films adhering to the solids; and (iii) many ions are adsorbed by surfaces and their replacement when depleted from the soil solution depends on the ease of exchange with the solid surface (characterized by the buffer power, b*). The diffusion coefficient in soil, D_s, can be written:

$$D_s = D_o \, \theta \, f \, dC_l/dC \qquad (4.21)$$

where θ is the volumetric water content, f is the impedance factor, C_l is the concentration of ions in the soil solution, and C is the concentration of ions per unit volume of soil. The term dC_l/dC is the reciprocal of the buffer power, b*, for that ion (i.e. 1/b*). The more strongly the ion is adsorbed by the soil, the smaller is 1/b*, and the smaller is D_s. For a non-adsorbed ion such as nitrate, b* = 1 but for a strongly adsorbed ion such as phosphate b* may be as high as 1000; values for ions such as ammonium and potassium will fall between these values. Typically, for a non-adsorbed ion, D_s at field capacity (10 kPa) will be about 0.2 D_o, while at permanent wilting point (1.5 MPa), it will be about 300 times smaller than at field capacity (Rowell *et al.*, 1967). In moist soil, D_s is of the order of 10^{-6} cm^2 s^{-1} for non-adsorbed ions such as nitrate, 10^{-7} cm^2 s^{-1} for cations such as calcium, magnesium and potassium, and 10^{-9} to 10^{-10} cm^2 s^{-1} for strongly adsorbed ions such as phosphate (Table 4.5). Since the average linear distance of diffusive movement with time t, will be a distance of $(2Dt)^{0.5}$ from its starting point, an ion such as nitrate (D_s 10^{-6} cm^2 s^{-1}) will move 4 mm in one day, while potassium (D_s 10^{-7} cm^2 s^{-1}) will move 1.3 mm, but phosphate (D_s 10^{-10} cm^2 s^{-1}) only 0.04 mm. These distances give some indication of the quantities of nutrients that are potentially available to plants by diffusion and of the difficulty of acquiring P relative to other nutrients. In contrast, a molecule of gas will move about 800 mm in a day in air and about 8 mm in solution (Tinker and Nye, 2000).

Although mass flow and diffusion occur simultaneously and interact, little is lost by assuming that they are additive processes provided that, as is usually the case, water taken up by the root is drawn from a zone beyond that from which nutrients have diffused. In such cases little error is introduced by assuming that mass flow acts in addition to diffusion (Nye and Marriott, 1969). Assuming a cylindrical root surface:

Table 4.5 Typical diffusion coefficients of ions in soils

Ion	Soil	Volumetric water content (%)	D_s (cm^2 s^{-1})	Authors
Cl^-	Sandy clay loam	40	9×10^{-6}	Rowell *et al.*, 1967
Cl^-	Sandy clay loam	20	2.4×10^{-6}	Rowell *et al.*, 1967
Na^+	Sandy clay loam	40	2.2×10^{-6}	Rowell *et al.*, 1967
Na^+	Sandy clay loam	20	0.5×10^{-6}	Rowell *et al.*, 1967
$H_2PO_4^-$	Sandy clay loam	40	3.3×10^{-9}	Rowell *et al.*, 1967
$H_2PO_4^-$	Sandy clay loam	20	0.3×10^{-9}	Rowell *et al.*, 1967
NO_3^-	Sand	5–20	$1.2\text{–}5.0 \times 10^{-6}$	Clarke and Barley, 1968
NO_3^-	Loam	10–26	$0.5\text{–}4.3 \times 10^{-6}$	Clarke and Barley, 1968
NH_4^+	Sand	5–20	$1.6\text{–}3.2 \times 10^{-7}$	Clarke and Barley, 1968
NH_4^+	Loam	10–26	$0.4\text{–}2.3 \times 10^{-7}$	Clarke and Barley, 1968
K^+	Sandy loam	40	2.3×10^{-7}	Vaidyanathan and Nye, 1968
Ca^{2+}	Sandy loam	40	3.3×10^{-7}	Vaidyanathan and Nye, 1966
Na^+	Sandy loam	40	1.0×10^{-5}	Vaidyanathan and Nye, 1966
$H_2PO_4^-$	Sandy loam	40	1.0×10^{-10}	Vaidyanathan and Nye, 1966

$$(\partial C_l/\partial t) = (1/r)(\partial/\partial r)(r\, D_s\, (\partial C_l/\partial r) + v\, a\, C_l/b^*) \qquad (4.22)$$

where v is the water flux to the root surface. This equation can be solved numerically for several conditions and has been widely used to calculate nutrient uptake under specified conditions (see Barber, 1995, and Tinker and Nye, 2000 for examples).

4.3.3 Nutrient uptake and movement across the root

Many of the considerations pertaining to water transport across the root to the xylem, such as the role of the cell walls, membranes, exodermis and endodermis, also apply to the uptake and movement of nutrients, although there are some important differences. A small proportion of the nutrients measured in plants are absorbed passively into 'free spaces' within cell walls as a consequence of cation exchange with the negatively charged cell walls; anions are excluded from this space. Values of cation exchange capacity are higher in dicot species than in monocot species but are not constant and are highly responsive to environmental conditions (Sattelmacher, 2001). Values ranging from 53 to 236 $mmol^+ kg^{-1}$ dry matter have been reported for monocotyledons, and 202 to 532 $mmol^+ kg^{-1}$ dry matter for dicotyledons (Crooke, 1964). Most uptake, though, is active and ion-specific against an electrical or concentration gradient. Uptake into the plant almost always involves the passage across a membrane at some point in the transfer between soil and xylem, and this requires the expenditure of energy provided via the process of respiration (Marschner, 1995). For example, Lambers *et al.* (1996) estimated that the proportion of respiration allocated to ion uptake was between 25% and 50% of the total for a group of 24 herbaceous species. Uptake of nitrate is among the most costly in terms of energy expenditure, and its uptake across the plasma membrane is energy-dependent over almost the whole range of nitrate concentrations found in soils (Forde and Clarkson, 1999).

The transfer from apoplast to symplast (cytoplasm) across the plasmalemma is a particularly important part of the pathway for nutrient transport to the xylem, while transfer from symplast to cell vacuole across the tonoplast is important for the storage of ions in cells. Uptake across all membranes is driven by specific energy-driven carriers or through ion channels embedded in the membranes together with the aquaporins described in section 4.2.2 used to transfer water. Although specific membranes may differ, they have many features in common (Salisbury and Ross, 1992; Marschner, 1995). Membranes are typically 7.5–10 nm thick, and consist largely of proteins (one-half to two-thirds of the membrane's dry weight), lipids (mainly phospholipids, glycolipids and sterols) and calcium. Each lipid has a 3-carbon glycerol backbone (this part is hydrophilic) to which two long-chain fatty acids (usually 16–18 C atoms long) are esterified (this part is hydrophobic). The hydrophilic parts of the lipids dissolve in water at one or other membrane surface, but the hydrophobic parts are forced towards the interior of the membrane forming a bilayer (Fig. 4.15). Sterols have a small hydrophilic head with a larger hydrophobic tail made of 6-C cycles which stabilizes the hydrophobic interior and prevents it from becoming too fluid as temperature rises. The lipids are fluid and move within their side of the bilayer. They rarely change sides but the sterols do. The proteins, collectively known as transport proteins, can be conveniently grouped under three headings: (i) catalytic proteins (enzymes) that use energy to pump protons (H^+)

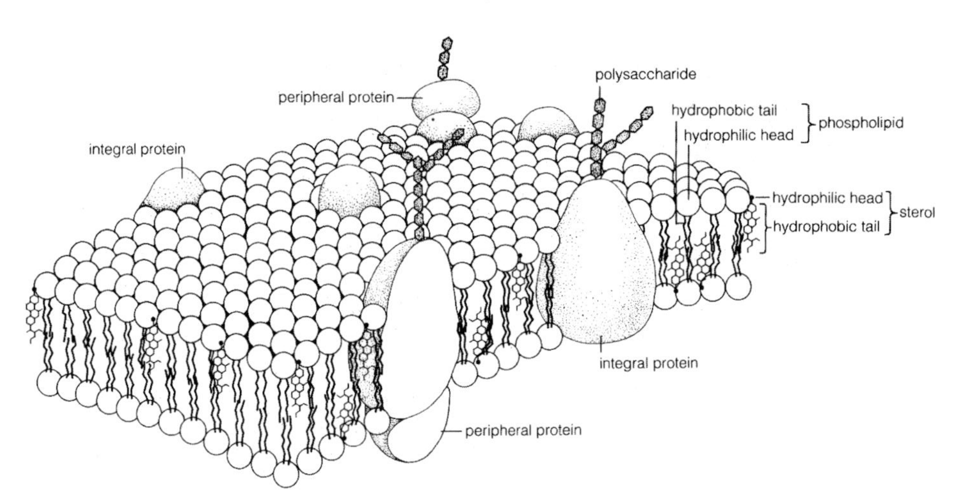

Fig. 4.15 The fluid mosaic model of membrane structure. Proteins are either continuous across the membrane or lie on either side. (Reproduced with permission from Salisbury and Ross, *Plant Physiology*; Wadsworth Publishing Company, 1992.)

across membranes; (ii) solute channels with gated holes between the protein molecules that can allow only one ion or several ions to pass across simultaneously; and (iii) carrier proteins which combine with specific ions on one side of the membrane and then rotate to release it on the other side. The role of calcium in membranes and cells is complex (White and Broadley, 2003), but it probably bonds the hydrophilic portions of the phospholipids to each other and to negatively charged parts of proteins in the membrane. Figure 4.15 shows that the proteins are embedded in the lipid bilayer. Some, intrinsic proteins, pass through the bilayer while others, extrinsic proteins (some with polysaccharides attached called glycoproteins), are bound to one or other surface.

Not only is there a chemical concentration gradient between the cell cytoplasm and the solution outside, but cytoplasm is also negatively charged (typically 100–150 mV), resulting in the attraction of cations and the repellance of anions. Several sources of energy associated with the phosphate molecule are used to transport ions across membranes against this electrochemical gradient (see Salisbury and Ross, 1992, for more details). One source of energy is that stored in adenosine tri-phosphate (ATP) which is liberated when ATP is hydrolysed to form adenosine di-phosphate (ADP) and inorganic phosphate. This reaction is catalysed by ATP phosphohydrolase (ATPase), which is one of the transport proteins mentioned earlier, and results in the transport of a proton from the cytoplasm to the cell wall. There are at least three kinds of H^+ transporting ATPases known to exist and several other forms of ATPases. The H^+ATPase is energized by catalysing a short-lived transfer of the terminal P from ATP located in the cytoplasm releasing ADP (Fig. 4.16). During the creation of the energy-rich ATPase, it changes shape to create a hole, combines with a proton and opens the hole to allow movement of the proton across the membrane. Immediately, water is used to hydrolyse P from the ATPase, and the proton moves through the membrane to the apoplast. The proton can, in turn, act as a pump making use of the pH and electropotential gradients across the membrane either with an accompanying anion via a symport or in exchange for a counterbalancing cation via an antiport (Fig. 4.16). In addition to this ATPase pump, the tonoplast has a pyrophosphatase pump facilitating proton transfer to the vacuole.

Given the many nutrients required by plants combined with the variety of plant life, it is not surprising that much remains to be found out about this very basic process of nutrient uptake into plants. Recent biochemical, electrochemical and molecular biology approaches have added substantially to our knowledge and it is clear that several transport proteins are used for each nutrient ion. As an example of the complexity that is involved, Fig. 4.17a and b provides a schematic representation of the transporters identified for calcium and chloride (White and Broadley, 2001, 2003). Transporters identified for nitrate and ammonium uptake are listed by Forde and Clarkson (1999) and for potassium uptake by White and Broadley (2000). These different transport proteins are not necessarily all active at the same time, being rapidly down-regulated or up-regulated in response to available levels of exogenous nutrients. Results of experiments in which the rate of uptake of ions is measured as the external concentration increases commonly produce a set of curves as shown in Fig. 4.18. Each curve can be analysed in terms of Michaelis-Menten kinetics (see Tinker and Nye, 2000), but the utility of this approach is now questionable. One interpretation of these curves is that each section is associated with the operation of different transport proteins. For example, kinetic data show that there are at least three uptake systems for nitrate uptake: (i) a constitutive low-affinity transport system important in the millimolar range; (ii) an inducible high-affinity system; and (iii) a constitutive high-affinity system important in the micromolar range (Forde and Clarkson, 1999). At the genetic level, each transporter is encoded by multiple members of the corresponding gene families (Glass *et al.*, 2001).

The routes used for water flow across root tissues to the xylem (section 4.2.2 and Fig. 4.7) are also employed by nutrients except that the transcellular pathway is thought to be negligible for solutes (Steudle and Peterson, 1998). As with water, the Casparian band of endodermal cells used to be considered a barrier to nutrient movement towards the xylem, with nutrients forced to enter the symplasm of the endodermis via channels in the

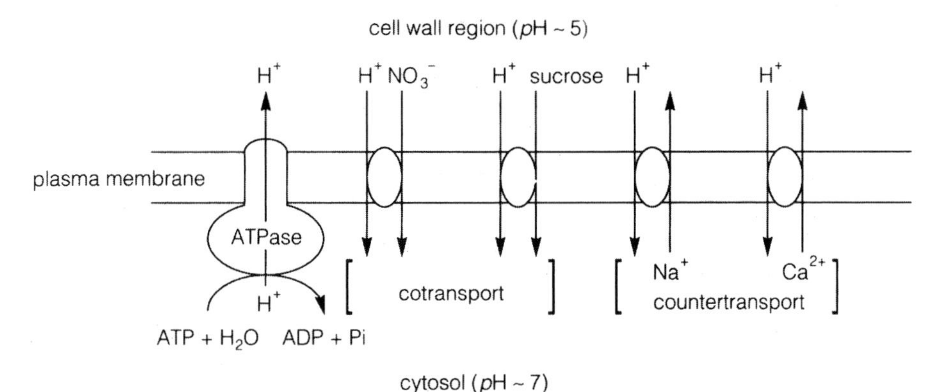

Fig. 4.16 Co-transport and counter-transport of solutes across the plasma membrane is driven by the energy in ATP. Proton-translocating ATPase (shown on the left) moves H^+ out of the cytosol. The protons can return to the cytosol in co-transport via carriers that simultaneously transport anions (e.g. nitrate) or sugars (e.g. sucrose). Counter-transport (antiport) also allows return of protons, but with exchange for an outgoing cation (e.g. Na^+ or Ca^{2+}). (Reproduced with permission from Salisbury and Ross, *Plant Physiology*; Wadsworth Publishing Company, 1992.)

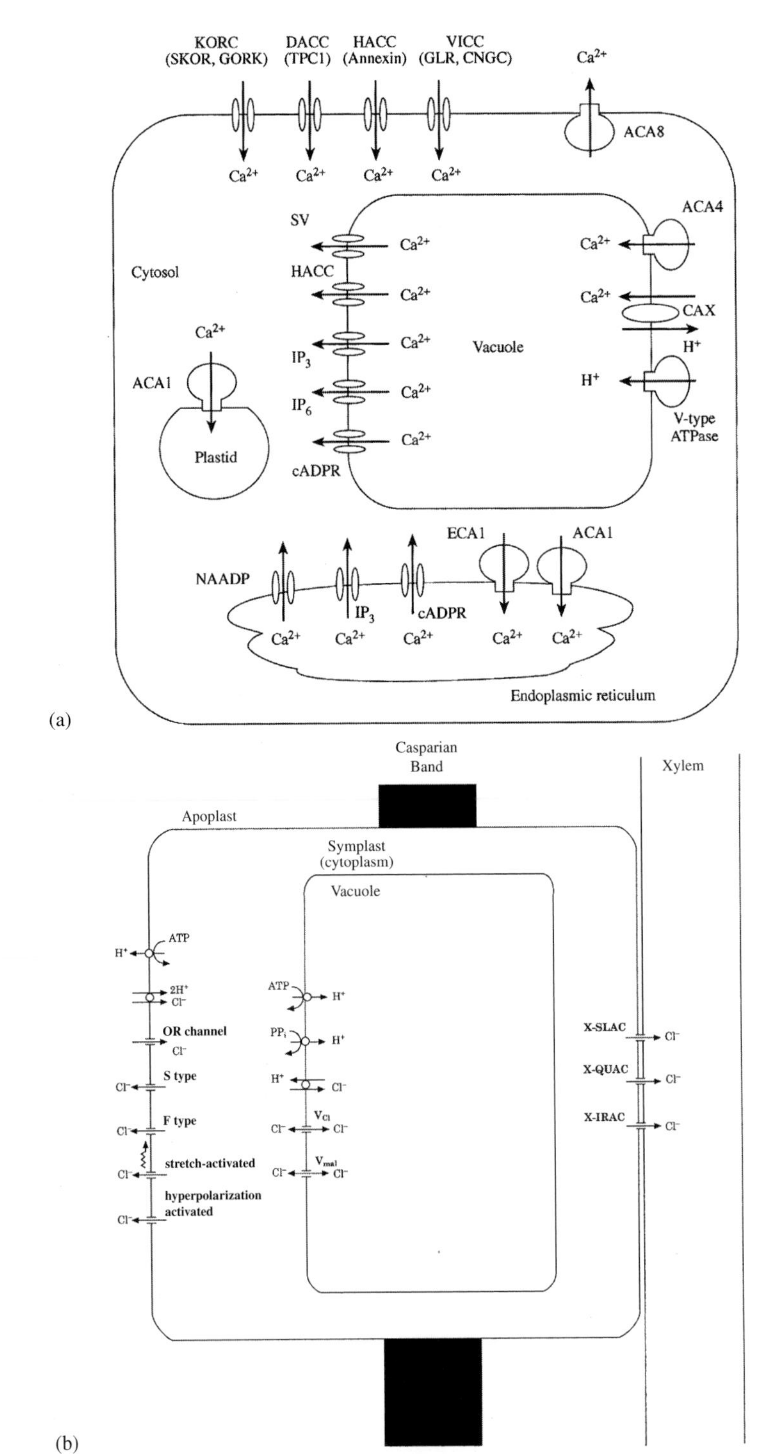

Fig. 4.17 Diagrammatic representations of the transporters involved in the transport of (a) calcium, and (b) chloride in plant roots. For details see original publications. (Reproduced with permission from White and Broadley, *Annals of Botany*; Oxford University Press, 2001 and 2003.)

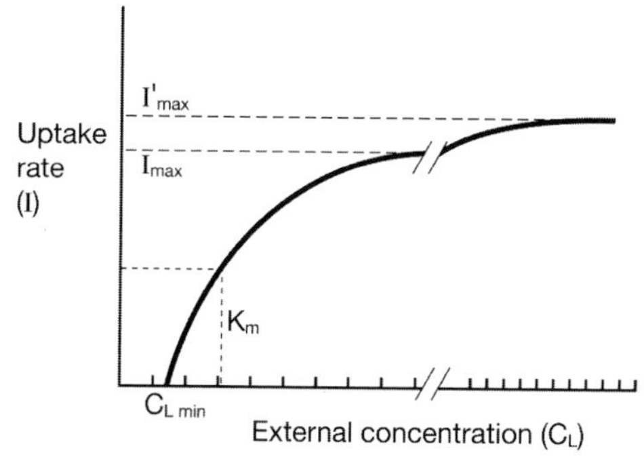

Fig. 4.18 A typical Michaelis–Menten relationship for ion uptake by a plant root. There is a minimum concentration below which uptake does not occur (C_{Lmin}); above this concentration the rate of uptake increases at low and medium concentrations to an asymptote (I_{max}) representing saturation of the high-affinity uptake mechanism. At high concentrations, the rate of uptake may again increase to an ill-defined maximum (I'_{max}) due to low-affinity processes. For the high-affinity process, the concentration equivalent to 0.5 (I_{max}) is the Michaelis constant (K_m).

plasmalemma to circumvent it (e.g. Marschner, 1995). However, there is some evidence to doubt this view (Steudle and Peterson, 1998; Schreiber *et al*., 1999). For example, White (2001) marshals observations on Ca^{2+} fluxes that suggest: (i) there is doubt as to whether the plasmalemma of endodermal cells could contain sufficient Ca^{2+}-ATPase to catalyse the observed flux; and (ii) the lack of selectivity or interactions between Ca^{2+}, Ba^{2+} and Sr^{2+} for transport to the shoot, together with the concentrations found in shoots suggest that these ions are unlikely to be transported across a plasmalemma (which would impart selectivity, competition and saturation phenomena) before delivery to the xylem. He concludes that substantial apoplastic bypass flow is indicated for transport of Ca to the shoot.

The relative importance of the apoplastic and symplastic pathways, though, depends on the nutrient under consideration. Studies using radioactive tracers showed that radial movement of phosphorus and potassium, and transport to the shoot could occur almost equally from root sections close to the root tip as from sections 400 mm from the tip. However, for calcium, movement to the shoot was restricted to the younger parts of the root system that were without suberin deposits in the endodermis (Russell and Clarkson, 1976). The measurements suggested that phosphorus, potassium, chloride, ammonium and nitrate moved predominantly by the symplastic pathway while nutrients such as calcium and iron moved in the apoplastic pathway. In practice, both pathways appear to operate for most nutrients but in varying amounts depending on the plant species, function of the nutrient, and environmental conditions. White (2001) and White and Broadley (2003) conclude, for example, that the functional separation of apoplastic Ca^{2+} fluxes (for transfer to the shoot) from symplastic Ca^{2+} fluxes (for cell signalling) would have the advantage of allowing the root to fulfil the shoot demand for Ca without compromising the intracellular signalling function of the Ca concentration in the cytoplasm. Similarly exposure to high concentrations of ions may alter the predominant pathway (from symplastic to apoplastic), as suggested for the hyperaccumulation of Zn when *Thlaspi caerulescens* is grown with high concentrations of Zn in the rooting medium (White *et al*., 2002).

The role of the exodermis in limiting the transport of nutrients is also a topic of ongoing research (Hose *et al*., 2001). As with water, as the exodermis develops, solute flow is reduced but not stopped. Hose *et al.* (2001) state that 'Broad generalizations about the barrier properties of the exodermis are unwise in the face of rather small amounts of experimental evidence'. They point out, too, that much of what is known about nutrient uptake has been learnt from the seminal axes of young cereal plants. The structure, physiology and function of lateral roots (and nodal axes in cereals) are largely unstudied territory, as are the transport processes operating in unexcised roots grown in soil.

Having traversed the root via the symplasm to the stellar tissues, nutrients must be downloaded into the xylem apoplast. This appears to be predominantly a thermodynamic passive process by ion channels (Wegner and Raschke, 1994) with the driving force generated by H^+ATPase in the xylem parenchyma cells, though much remains to be learned. Both inward and outward rectifying channels have been identified. For example, Köhler and Raschke (2000) identified three types of anion conductance in the plasmalemma of barley xylem parenchyma leading to xylem loading, and demonstrated both the role of cytoplasmic Ca^{2+} concentration in the function of two of them, and that the size of anion currents was sufficient to confirm xylem loading as a passive process. This downloading process to the xylem is separate from the uptake process into the cortex (Marschner, 1995; Sattelmacher, 2001) leading to a xylem sap that is highly variable in composition.

While this section has focused on transfer of solutes into the plant, as with water, efflux can also occur. This is particularly noticeable with some ions. NO_3^- efflux, for example, can play a significant role in determining net uptake especially at high external concentration of NO_3^- efflux, under conditions of stress, and when NH_4^+ is available. In some circumstances, as much as 80% of the NO_3^- influx can be simultaneously lost as efflux (Forde and Clarkson, 1999).

A large proportion of the nutrients within the plant are redistributed via the phloem from the shoot to the root so that the concentration of nutrients in the xylem does not necessarily reflect root uptake. This circulatory mechanism ensures charge balance through the plant and ensures that root tissues have adequate nutrition during periods of low uptake activity, as well as ensuring adequate cell turgor for root elongation (Sattelmacher, 2001). Depending on the plant species and the nutritional environment, up to 60% of the N (Cooper and Clarkson, 1989), 30% of the S (Larsson *et al.*, 1991), and 80% of the potassium (Jeschke and Pate, 1991) found in the xylem can be allocated to the cycled fraction. The exchange between xylem and phloem can be considerable, especially for nitrogenous compounds including amino acids and the combination of processes may lead to strong concentration gradients within the plant (higher at the base and lower in apical regions) that are correlated with changes in xylem pH (Sattelmacher, 2001). The xylem/phloem recirculatory system may form part of the signalling system controlling nutrient uptake by roots, and the partitioning of nutrients between the cytoplasm and vacuole.

4.3.4 Nutrient uptake by root systems

Much of the research on nutrient uptake by root systems has been undertaken with crop plants grown under conditions of monoculture and often in pots rather than fields. A wide variety of models is now available, each of which places particular emphasis on different parts of the uptake process (for reviews see Rengel, 1993; Silberbush, 1996; Tinker and Nye, 2000). The main elements of the processes affecting soil supply of nutrients to plants appear to be reasonably well understood, but a major difficulty with most current approaches lies in the inadequate description of plant 'demand' for particular nutrients. This is a complex topic, and the relationships between nutrient concentration in particular plant organs and other processes involved in growth and partitioning are not well understood. Typically, an overall mean nutrient concentration for a plant is specified because this is at least measurable (e.g. Greenwood, 2001), but this average value is the resultant of at least four separate processes (Tinker and Nye, 2000): (i) inherent differences between species in the nutrient concentration of the cytoplasm; (ii) changes in the nutrient composition of the vacuole – this concentration is more variable than that in the cytoplasm; (iii) differences in nutrient requirements of different types of tissue – structural tissues have lower concentrations than living tissues; and (iv) changes in the proportions of the different organs and tissues with time. These considerations mean that the relationship between concentration of a particular nutrient in a plant and the rates of growth and continued uptake of that nutrient are, in practice, complex. To make progress, several simplifying assumptions are often made which may or may not be correct in particular circumstances. A general equation that describes the underlying relationship between growth, nutrient concentration and uptake is as follows (Nye and Tinker, 1969):

$$dU/dt = d(WX)/dt = W\ dX/dt + X\ dW/dt = \bar{I}L \quad (4.23)$$

where W is the dry mass of the plant, X is the mean nutrient concentration, L is total root length, and $\bar{I}$ is the mean uptake per unit root length (mean inflow). This equation is based upon the conservation of nutrient mass. If the period to which it is applied is a few days or more, and there are no sudden changes in the availability of the nutrient U or of growth rate (i.e. it does not apply to circumstances where external concentrations are changed rapidly), then equation 4.23 can be further simplified. Usually W dX/dt will vary little compared with X dW/dt because X varies by only a factor of 2–5 over a growing season whereas W may vary by a factor of 1000 or more. This means that for a period such as a week or so, dX/dt = 1, so equation 4.23 simplifies to:

$$dU/dt = X/W.\ dW/dt = XR_w \approx \bar{I}.\ L/W \quad (4.24)$$

where R_w is the relative growth rate. Equation 4.24 shows that the mean inflow is proportional to the relative growth rate of the plant (not the absolute growth rate) and that for a plant in steady exponential growth with constant root:shoot partitioning, there will be a linear relation between $\bar{I}$ and X. Over periods of a month or so, these relations have been found in practice so, for example, Raper *et al.* (1977) found that uptake of P and K for a 25-day period was related to R_w in tobacco grown in solution culture, but because the concentrations of N, Ca and Mg decreased during growth, their relative accumulation rates were less than R_w (Table 4.6).

In practice both X and R_w decrease with time; X because as the plant grows it produces more tissues to support the structure of the plant and these have lower nutrient concentrations than, for example, leaves, and R_w because self-shading of leaves occurs. In the foregoing account, it is assumed that the plant demand for nutrients is met via the rate

Table 4.6 Relative growth rate (R_w) and relative accumulation rate (RAR) of nutrients for young plants of tobacco

	Temperature			Nutrient supply		
Variable	Cool	Warm	Hot	Low	Medium	High
R_w	0.122	0.129	0.129	0.124	0.127	0.129
RAR P	0.129	0.127	0.124	0.129	0.129	0.124
RAR K	0.127	0.122	0.122	0.124	0.124	0.124
RAR N	0.133	0.115	0.113	0.113	0.115	0.111
RAR Ca	0.113	0.117	0.115	0.117	0.120	0.106
RAR Mg	0.115	0.120	0.117	0.117	0.122	0.115

For P and K the two rates are similar, indicating that concentrations are constant throughout the experiment irrespective of temperature or external nutrient supply, while for N, Ca and Mg relative accumulation rates were less than relative growth rate, indicating that concentration decreased with time. From Raper *et al.*, 1977.

of uptake, $\bar{I}L$, that maintains its optimal rate of growth and ensures that X is equal to the concentration that just allows this growth rate. A further step in analysing demand can be made by relating root uptake to the concentration of ions in the external solution at the root surface (C_{la}) so that:

$$dU/dt = 2\pi \, \alpha a \, L \, C_{la} \tag{4.25}$$

where αa is the root demand coefficient (Nye and Tinker, 1969; Tinker and Nye, 2000). Values of αa have typically been determined in solution culture with values of about 10^{-10} $m^2 \, s^{-1}$; in soil, values are usually 10–100 times smaller than those found in solution culture (Brewster and Tinker, 1972; Tinker and Nye, 2000). While the root demand coefficient is empirically based, it has several uses: (i) it expresses plant demand at the root surface; (ii) it allows approximate solutions to equations that define concentrations at root surfaces when αa is assumed to be constant; and (iii) αa is often constant over a range of solution concentrations, especially in solution culture, so is useful in a range of nutrient uptake models.

Claassen and Barber (1976) were among the first to put together the components of a mechanistic model of nutrient uptake by root systems that was later elaborated by Barber and Cushman (1981) and Barber (1995) to allow competition between roots by defining a zero-transfer boundary at a distance midway from the next root surface. In essence the model has three components. First, nutrients are supplied to the root surface by mass flow and diffusion and the concentration at the root surface is calculated based on equation 4.22. Second, the inflow is assumed to be dependent on the concentration at the root surface and a Michaelis–Menten function is adopted to define the relation. Finally, uptake is calculated per unit of root length and summed over the root system (the model has an algorithm that allows total root length to increase either exponentially or linearly with time). The model has 11 parameters, all of which can be measured (Table 4.7). The Michaelis–Menten parameters (1–3) are measured in solution culture, the plant root parameters (4–7) are measured over time in the soil investigated, the water influx can be measured by weighing pots or from calculation of the soil water balance, and the soil parameters (9–11) can be measured in the laboratory. The model has been tested under both pot and field conditions. Barber (1995,

Table 4.7 Soil and plant parameters used in the Barber Cushman model of plant nutrient uptake showing some typical values used for K uptake by soyabean (cv. Williams 79) on Raub silt loam topsoil

Symbol	Parameter	Initial value
I_{max}	Maximal influx rate	7.05×10^{-4} µmol m^{-2} s^{-1}
C_{min}	Minimal concentration at which influx = 0	1.4 µM
Km	Concentration at which influx = 0.5 I_{max}	10.3 µM
L_0	Initial root length	0.25 m
K	Root growth rate	3×10^{-4} m s^{-1}
r_0	Root radius	1.5×10^{-4} m
r_l	Half distance between roots	2×10^{-3} m
v_0	Water influx to roots	5×10^{-9} m s^{-1}
D_s	Diffusion coefficient in bulk soil	3.47×10^{-12} $m^2 s^{-1}$
b*	Soil buffer power	24.0
C_{li}	Initial concentration in soil solution	280 µM

From Silberbush and Barber, 1983.

p. 119) lists 10 pot experiments with several different plant species on a range of soil types in which uptake of potassium (four experiments) and phosphorus (six experiments) were measured and modelled. Generally, the agreement was good with correlation coefficients ranging from 0.89 to 0.99 and the slope of the regression ranging from 0.79 to 1.19.

The Barber and Cushman model has been modified by several groups to allow for processes and features not included initially. For example, Itoh and Barber (1983a, 1983b) added a root hair submodel to improve to predictions of P uptake by six plant species with different root hair characteristics. Root hairs are usually clustered within a reasonably well-defined cylinder so that, to a first approximation, they can be regarded as extending outwards from the root surface the effective surface to which nutrients must diffuse. Li and Barber (1991) investigated the effects of rhizosphere pH change in three legume species by growing maize, alfalfa, Austrian winter pea (*Lathyrus hirsutus*) and faba bean in pots of silt loam amended with lime to give a range of pH values for periods of 12–28 days in a glasshouse. Using the unchanged Barber and Cushman model, good agreement was obtained between measured and calculated K uptake for all species (Fig. 4.19a), but P uptake was only well predicted in maize and underestimated in all of the legumes with the difference becoming larger as the initial soil pH increased (Fig. 4.19b). Because soil pH, as expected, had little effect on K uptake, the observed difference between K and P was most likely due to changes in rhizosphere pH. Li and Barber (1991) calculated the pH change in the rhizosphere required to get the predicted and observed P uptake to agree and estimated that the legumes reduced pH by 0.39 to 0.77 units and increased P availability by 21 to 242%.

In field-grown plants where conditions fluctuate and the period of growth is usually longer than that used in pot experiments, model outcomes have been more variable. Potassium uptake has been modelled with acceptable accuracy for some crops in some soils, but P uptake has almost always proved more challenging. For example, Silberbush and Barber (1984) grew five cultivars of soyabean (*Glycine max* L. Merr.) for 3 years on different field sites with two soils (one with high P concentration in solution and the other with low P) for about 80 days after germination until the plants had reached early to mid-reproductive

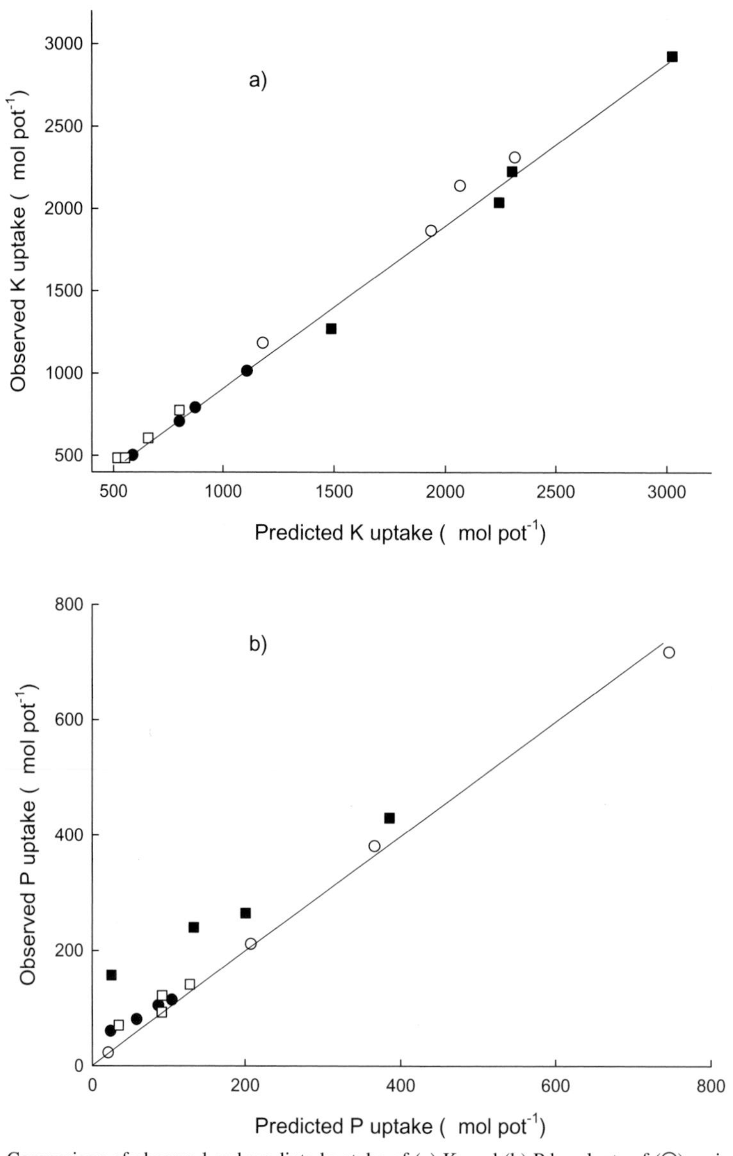

Fig. 4.19 Comparison of observed and predicted uptake of (a) K, and (b) P by plants of (○) maize, (●) faba bean, (□) Austrian winter pea, and (■) lucerne grown in pots. (Reproduced with permission from Li and Barber, *Communications in Soil Science and Plant Analysis*; CRC Press, 1991.)

stages of maturity. Potassium and P uptake was predicted separately for up to three soil layers because the supply parameters changed with depth. Predicted and observed K uptake agreed well on both soils with only 2 of the possible 10 comparisons showing significant differences between observed and predicted uptake (Table 4.8). It is noteworthy that a substantial proportion of the K (range 0.12–0.37) came from the subsoil (>0.3 m) of the Chalmers silt loam, whereas negligible amounts came from such depth on the Raub silt loam. For the Chalmers soil (52 μM P in solution), measured P uptake was significantly

Table 4.8 Predicted and observed K uptake (mmol m^{-2}) of five soyabean cultivars grown in the field on Chalmers silt loam in 1980 and on Raub silt loam in 1978

Soil	Soyabean genotype	Predicted uptake				Observed uptake
		0–15 cm	15–30 cm	30–76 cm	Total	
Chalmers	Amsoy 71	175	55	110	340	325
	Beeson 80	200	60	140	400	375
	Century	195	60	145	400	360
	Elf	220	75	145	440	450
	Williams	295	70	50	415	405
Raub	Amsoy 71	300	100	0	400	365
	Beeson 80	220	120	0	340	375
	Elf	280	130	10	420	430
	Wells	185	110	10	305	335
	Woodworth	250	110	0	360	430

The LSD ($p = 0.05$) between predicted and observed total uptake was 32 mmol m^{-2} for Chalmers silt loam and 54 mmol m^{-2} for Raub silt loam. From Silberbush and Barber, 1984.

higher ($p = 0.05$) than that predicted in four of the five cultivars and significantly lower than predicted in one, although Silberbush and Barber (1984) claimed that the model 'accurately predicted' P uptake on this soil. On the Raub soil (7.8 μM in solution), the model performed poorly predicting only 30–35% of the observed P uptake. The reasons for the under prediction of P uptake were not known with certainty. The effect of root hairs was included in the model but effects of mycorrhizas and pH changes in the rhizosphere were not. Similar difficulties with modelling P uptake were also obtained with maize grown at a range of P fertilizer applications (Macariola-See *et al.*, 2003). During vegetative growth (to stage V12–12 leaves) the model predicted 86–90% of observed P uptake in treatments with high P in solution (21–77 μM) but only 23–44% with low P in solution (1–16 μM). Between growth stage V12 and silking (R2) the model predicted only a small uptake of P, whereas measured uptake doubled in this period. Root hair growth, mycorrhizal infection, rhizosphere acidification, and root architecture were all suggested as reasons for model underperformance.

Generally, this model has been most successful in fast-growing plants with concentrations in solution well above deficiency levels or those likely to substantially reduce plant growth. Even with a relatively straightforward nutrient such as K, though, several problems remain to be resolved, especially because most of the parameters are not allowed to vary with time. First, correctly specifying the soil solution concentration over prolonged periods is difficult because soil solution will be replenished by both exchangeable and slowly exchangeable K, requiring detailed knowledge of mineralogy and K release dynamics to estimate the changing buffer power. For example, Brouder and Cassman (1994) used the Barber and Cushman model to investigate K uptake by cotton grown on vermiculitic soil of high K-fixation capacity, on which K deficiency late in the season is common. In pots fertilized with combinations of K and N and harvested at 25 days, initial model output produced both substantial over- and under-estimates of measured uptake and a correlation coefficient of 0.24. The model prediction was greatly increased when the buffer capacity was specified in terms of a Langmuir adsorption isotherm rather than the relationship between exchangeable solid-phase and solution-phase K pools used in the standard model. Second, despite sensitivity analysis of the model suggesting its relative insensitivity to the Michaelis–Menten parameters (Silberbush and Barber, 1983), other workers have shown that using these terms to, in effect, specify plant demand is problematic (Seward *et al.*, 1990; Brouder and Cassman, 1994). Seward *et al.* (1990) found that both the Barber and Cushman and Claassen *et al.* (1986) models overestimated K uptake, mainly because measured field influxes were almost an order of magnitude less than the value of I_{max} used in the models. Such findings lead to a fundamental question as to the circumstances in which either the plant or the soil determines the rate of uptake. Barraclough (1986) concluded that for crops of winter wheat growing on moist soils in the UK, nutrient transport in the soil was unlikely to limit uptake (i.e. the rate of plant growth drove uptake).

An alternative approach was adopted by de Willigen and van Noordwijk (1994a, 1994b) in which uptake by roots was assumed to be determined by plant demand as long as transport in the soil maintained the concentration at the root surface above zero. Thereafter, the root will behave as a zero sink so that soil transport processes then determine the rate of uptake. The root, then, either takes up nutrient at the required rate so long as the soil can transport nutrient to the root surface sufficiently rapidly, or it takes up nutrient at the rate at which it arrives at the surface. This approach is similar to that used to define rates of water uptake (Monteith, 1986), and has the advantage that Michaelis–Menten parameters are avoided. It requires, though, that nutrient demand is properly understood. While this may be sufficiently well approximated for several agricultural and horticultural crops (Greenwood, 2001; Greenwood *et al.*, 2001a, 2001b), as indicated earlier, our ability to state this quantitatively for most plants and trees is very limited.

References

Abbadie, L., Mariotti, A. and Menaut, J.-C. (1992) Independence of savanna grasses from soil organic matter for their nitrogen supply. *Ecology* **73**, 608–613.

Aerts, R. and Chapin III, F.S. (2000) The mineral nutrition of wild plants revisited: a re-evaluation of processes and patterns. *Advances in Ecological Research* **30**, 1–67.

Baker, C.J., Berry, P.M., Spink, J.H., Sylvester-Bradley, R., Griffin, J.M., Scott, R.K. and Clare, R.W. (1998) A method for the assessment of the risk of wheat lodging. *Journal of Theoretical Biology* **194**, 587–603.

Baker, J.M. and van Bavel, C.H.M. (1988) Water transfer through cotton plants connecting soil regions of differing water potential. *Agronomy Journal* **80**, 993–997.

Barber, S.A. (1995) *Soil Nutrient Bioavailability: A Mechanistic Approach*. John Wiley & Sons, New York, USA.

Barber, S.A. and Cushman, J.H. (1981) Nutrient uptake model for agronomic crops. In: *Modeling Waste Water Renovation-Land Treatment,* (ed. I.K. Iskandar), pp. 382–409. Wiley-Interscience, New York.

Barber, S.A., Walker, J.M. and Vasey, E.H. (1963) Mechanisms for the movement of plant nutrients from the soil and fertiliser to the plant root. *Journal of Agriculture, Food and Chemistry* **11**, 204–207.

Barraclough, P.B. (1986) The growth and activity of winter wheat roots in the field: nutrient inflows of high-yielding crops. *Journal of Agricultural Science, Cambridge* **106**, 53–59.

Berry, P.M., Spink, J.H., Gay, A.P. and Craigon, J. (2003) A comparison of root and stem lodging risks among winter wheat cultivars. *Journal of Agricultural Science* **141**, 191–202.

Bhat, K.K.S. and Nye, P.H. (1973) Diffusion of phosphate to plant roots in soil. I. Quantitative autoradiography of the depletion zone. *Plant and Soil* **38**, 161–175.

Birch, H.F. (1958) The effect of soil drying on humus decomposition and nitrogen availability. *Plant and Soil* **10**, 9–31.

Blackman, P.G. and Davies, W.J. (1985) Root-to-shoot communication in maize plants of the effects of soil drying. *Journal of Experimental Botany* **36**, 39–48.

Borel, C., Frey, A., Marion-Poll, A., Tardieu, F. and Simonneau, T. (2001) Does engineering abscisic acid biosynthesis in *Nicotiana plumbaginifolia* modify stomatal response to drought? *Plant, Cell and Environment* **24**, 477–489.

Bray, R.H. (1954) A nutrient mobility concept of soil-plant relationships. *Soil Science* **78**, 9–22.

Brewster, J.L. and Tinker, P.B. (1970) Nutrient cation flow in soil around plant roots. *Soil Science Society of America Proceedings* **34**, 421–426.

Brewster, J.L. and Tinker, P.B. (1972) Nutrient flow rates into roots. *Soils and Fertilizers* **35**, 355–359.

Brooks, J.R., Meinzer, F.C., Coulombe, R. and Gregg, J. (2002) Hydraulic redistribution of soil water during summer drought in two contrasting Pacific Northwest coniferous forests. *Tree Physiology* **22**, 1107–1117.

Brouder, S.M. and Cassman, K.G. (1994) Evaluation of a mechanistic model of potassium uptake by cotton in vermiculitic soil. *Soil Science Society of America Journal* **58**, 1174–1183.

Burgess, S.S.O., Adams, M.A., Turner, N.C., White, D.A. and Ong, C.K. (2001) Tree roots: conduits for deep recharge of soil water. *Oecologia* **126**, 158–165.

Caldwell, M.M., Dawson, T.E. and Richards, J.H. (1998) Hydraulic lift: consequences of water efflux from the roots of plants. *Oecologia* **113**, 151–161.

Campbell, G.S. (1985) *Soil Physics with BASIC: Transport Models for Soil-Plant Systems*. Elsevier, Amsterdam.

Chiatante, D., Scippa, S. G., Di Iorio, A. and Sarnataro, M. (2003) The influence of steep slopes on root system development. *Journal of Plant Growth Regulation* **21**, 247–260.

Claassen, N. and Barber, S.A. (1976) Simulation model for nutrient uptake from soil by a growing plant root system. *Agronomy Journal* **68**, 961–964.

Claassen, N., Syring, K.M. and Jungk, A. (1986) Verification of a mathematical model by simulating potassium uptake from soil. *Plant and Soil* **95**, 209–220.

Clarke, A.L. and Barley, K.P. (1968) The uptake of nitrogen from soils in relation to solute diffusion. *Australian Journal of Soil Research* **6**, 75–92.

Clarkson, D.T., Carvajal, M., Henzler, T., Waterhouse, R.N., Smyth, A.J., Cooke, D.T. and Steudle, E. (2000) Root hydraulic conductance: diurnal aquaporin expression and the effects of nutrient stress. *Journal of Experimental Botany* **51**, 61–70.

Cooper, H.D. and Clarkson, D.T. (1989) Cycling of amino-nitrogen and other nutrients between shoots and roots in cereals – a possible mechanism integrating shoot and root in the regulation of nutrient uptake. *Journal of Experimental Botany* **40**, 753–762.

Coutts, M.P. (1983) Root architecture and tree stability. *Plant and Soil* **71**, 171–188.

Coutts, M.P. (1986) Components of tree stability in Sitka Spruce on peaty gley soil. *Forestry* **59**, 173–197.

Coutts, M.P. and Nicoll, B.C. (1991) Orientation of the lateral roots of trees I. Upward growth of surface roots and deflection near the soil surface. *New Phytologist* **119**, 227–234.

Coutts, M.P. and Philipson, J.J. (1978) Tolerance of tree roots to waterlogging. I. Survival of Sitka spruce and Lodgepole pine. *New Phytologist* **80**, 63–69.

Coutts, M.P., Nielsen, C.C.N. and Nicoll, B.C. (1999) The development of symmetry, rigidity and anchorage in the structural root system of conifers. *Plant and Soil* **217**, 1–15.

Crook, M.J. and Ennos, A.R. (1993) The mechanics of root lodging in winter wheat, *Triticum aestivum* L. *Journal of Experimental Botany* **44**, 1219–1224.

Crook, M.J. and Ennos, A.R. (1994) Stem and root characteristics associated with lodging resistance in four wheat cultivars. *Journal of Agricultural Science, Cambridge* **123**, 167–174.

Crook, M.J., Ennos, A.R. and Banks, J.R. (1997) The function of buttress roots: a comparative study of the anchorage systems of buttressed (*Aglaia* and *Nephelium ramboutan* species) and non-buttressed (*Mallotus wrayi*) tropical trees. *Journal of Experimental Botany* **48**, 1703–1716.

Crooke, W.M. (1964) The measurement of the cation-exchange capacity of plant roots. *Plant and Soil* **21**, 43–49.

Daamen, C.C. and Simmonds, L.P. (1996) Measurement of evaporation from bare soil and its estimation using surface resistance. *Water Resources Research* **32**, 1393–1402.

Dardanelli, J.L., Ritchie, J.T., Calmon, M., Andriani, J.M. and Collino, D.J. (2004) An empirical model for root water uptake. *Field Crops Research*. **87**, 59–71.

Davies, W.J., Bacon, M.A., Thompson, D.S., Sobeih, W. and Rodriguez, L.G. (2000) Regulation of leaf and fruit growth in plants growing in drying soil: exploitation of the plants' chemical signalling system and hydraulic architecture to increase the efficiency of water use in agriculture. *Journal of Experimental Botany* **51**, 1617–1626.

Davies, W.J., Wilkinson, S. and Loveys, B. (2002) Stomatal control by chemical signalling and the exploitation of this mechanism to increase water use efficiency in agriculture. *New Phytologist* **153**, 449–460

de Willigen, P. and van Noordwijk, M. (1994a) Mass flow and diffusion of nutrients to a root with constant or zero-sink uptake. I. Constant uptake. *Soil Science* **157**, 162–170.

de Willigen, P. and van Noordwijk, M. (1994b) Mass flow and diffusion of nutrients to a root with constant or zero-sink uptake. II. Zero-sink uptake. *Soil Science* **157**, 171–175.

Ehlers, W., Hamblin, A.P., Tennant, D. and van der Ploeg, R.R. (1991) Root system parameters determining water uptake of field crops. *Irrigation Science* **12**, 115–124.

Ennos, A.R. (1989) The mechanics of anchorage in seedlings of sunflower, *Helianthus annuus* L. *New Phytologist* **113**, 185–192.

Ennos, A.R. (1990) The anchorage of leek seedlings: the effect of root length and soil strength. *Annals of Botany* **65**, 409–416.

Ennos, A.R. (1991a) The mechanics of anchorage in wheat *Triticum aestivum* L. I. The anchorage of wheat seedlings. *Journal of Experimental Botany* **42**, 1601–1606.

Ennos, A.R. (1991b) The mechanics of anchorage in wheat *Triticum aestivum* L. II. Anchorage of mature wheat against lodging. *Journal of Experimental Botany* **42**, 1607–1613.

Ennos, A.R. (2000) The mechanics of root anchorage. *Advances in Botanical Research* **33**, 133–157.

Ennos, A.R. and Fitter, A.H. (1992) Comparative functional morphology of the anchorage systems of annual dicots. *Functional Ecology* **6**, 71–78.

Ennos, A.R., Crook, M.J. and Grimshaw, C. (1993. A comparative study of the anchorage systems of Himalayan Balsam *Impatiens glandulifera* and mature sunflower *Helianthus annuus*. *Journal of Experimental Botany* **44**, 133–146.

Feddes, R.A., Kowalik, P., Kolińska-Malinka, K. and Zaradny, H. (1976) Simulation of field water uptake by plants using a soil water dependent root extraction function. *Journal of Hydrology* **31**, 13–26.

Forde, B.G. and Clarkson, D.T. (1999) Nitrate and ammonium nutrition of plants: physiological and molecular perspectives. *Advances in Botanical Research* **30**, 1–90.

Frensch, J., Hsiao, T. and Steudle, E. (1996) Water and solute transport along developing maize roots. *Planta* **198**, 348–355.

Gardner, W.R. (1960) Dynamic aspects of water availability to plants. *Soil Science* **89**, 63–73.

Gardner, W.R. (1983) Soil properties and efficient water use: an overview. In: *Limitations to Efficient Water Use in Crop Production* (eds H.M. Taylor, W.R. Jordan and T.R. Sinclair), pp. 45–64. ASA, CSSA, SSSA, Madison, WI, USA.

Gardner, W.R. (1991) Modeling water uptake by roots. *Irrigation Science* **12**, 109–114.

Glass, A.D.M., Brito, D.T., Kaiser, B.N., Kronzucker, H.J., Kumar, A., Okamoto, M., Rawat, S.R., Siddiqi, M.Y., Silim, S.M., Vidmar, J.J. and Zhuo, D. (2001) Nitrogen transport in plants, with an emphasis on the regulation of fluxes to match plant demand. *Journal of Plant Nutrition and Soil Science* **164**, 199–207.

Goodman, A.M., Crook, M.J. and Ennos, A.R. (2001) Anchorage mechanics of the tap root system of winter-sown oilseed rape (*Brassica napus* L.). *Annals of Botany* **87**, 397–404.

Gowing, D.J.G., Davies, W.J. and Jones, H.G. (1990) A positive root-sourced signal as an indicator of soil drying in apple, *Malus x domestica* Borkh. *Journal of Experimental Botany* **41**, 1535–1540.

Greenwood, D.J. (2001) Modeling N-response of field vegetable crops grown under diverse conditions with N-ABLE: a review. *Journal of Plant Nutrition* **24**, 1799–1815.

Greenwood, D.J., Karpinets, T.V. and Stone, D.A. (2001a) Dynamic model for the effects of soil P and fertilizer P on crop growth, P uptake and soil P in arable cropping: model description. *Annals of Botany* **88**, 279–291.

Greenwood, D.J., Stone, D.A. and Karpinets, T.V. (2001b) Dynamic model for the effects of soil P and fertilizer P on crop growth, P uptake and soil P in arable cropping: experimental test of the model for field vegetables. *Annals of Botany* **88**, 293–306.

Gregory, P.J. and Brown, S.C. (1989) Root growth, water use and yield of crops in dry environments: what characteristics are desirable? *Aspects of Applied Biology* **22**, 235–243.

Gregory, P.J., McGowan, M., Biscoe, P.V. and Hunter, B. (1978a) Water relations of winter wheat.1. Growth of the root system. *Journal of Agricultural Science, Cambridge* **91**, 91–102.

Gregory, P.J., McGowan, M. and Biscoe, P.V. (1978b) Water relations of winter wheat. 2. Soil water relations. *Journal of Agricultural Science, Cambridge* **91**, 103–116.

Gregory, P.J., Crawford, D.V. and McGowan, M. (1979) Nutrient relations of winter wheat. 2. Movement of nutrients to the root and their uptake. *Journal of Agricultural Science, Cambridge* **93**, 495–504.

Hamblin, A.P. and Tennant, D. (1987) Root length density and water uptake in cereals and grain legumes: how well are they correlated? *Australian Journal of Agricultural Research* **38**, 513–527.

Hillel, D. (1980) *Fundamentals of Soil Physics*. Academic Press, New York.

Hopmans, J.W. and Bristow, K.L. (2002) Current capabilities and future needs of root water and nutrient uptake modeling. *Advances in Agronomy* **77**, 103–183.

Hose, E., Clarkson, D.T., Steudle, E., Schreiber, L. and Hartung, W. (2001) The exodermis: a variable apoplastic barrier. *Journal of Experimental Botany* **52**, 2245–2264.

Hulugalle, N.R. and Willatt, S.T. (1983) The role of soil resistance in determining water uptake by plant root systems. *Australian Journal of Soil Research* **21**, 571–574.

Itoh, S. and Barber, S.A. (1983a) Phosphorus uptake by six plant species as related to root hairs. *Agronomy Journal* **75**, 457–461.

Itoh, S. and Barber, S.A. (1983b) A numerical solution of whole plant nutrient uptake for soil-root systems with root hairs. *Plant and Soil* **70**, 403–413.

Jackson, R.B., Sperry, J.S. and Dawson, T.E. (2000) Root water uptake and transport: using physiological processes in global predictions. *Trends in Plant Science* **5**, 482–488.

Jarvis, P.G., Edwards, W.R. and Talbot, H. (1981) Models of plant and crop water use. In: *Mathematics and Plant Physiology* (eds D.A. Rose and D.A. Charles-Edwards), pp. 151–194. Academic Press, London.

Javot, H. and Maurel, C. (2002) The role of aquaporins in root water uptake. *Annals of Botany* **90**, 301–313.

Jeschke, W.D. and Pate, J.S. (1991) Cation and chloride partitioning through xylem and phloem within the whole plant of *Ricinus communis* L. under conditions of salt stress. *Journal of Experimental Botany* **42**, 1091–1103.

Köhler, B. and Raschke, K. (2000) The delivery of salts to the xylem. Three types of anion conductance in the plasmalemma of the xylem parenchyma of roots of barley. *Plant Physiology* **122**, 243–254.

Kramer, P.J. (1969) *Plant and Soil Water Relationships: A Modern Synthesis*. McGraw-Hill, New York.

Lafolie, F., Bruckler, L. and Tardieu, F. (1991) Modeling root water potential and soil-root water transport: I. Model presentation. *Soil Science Society of America Journal* **55**, 1203–1212.

Lambers, H., Scheurwater, I. and Atkin, O.K. (1996) Respiratory patterns in roots in relation to their function. In: *Plant Roots: The Hidden Half*, (eds Y. Waisel, A. Eshel and U. Kafkafi), 2nd edn, pp. 323–362. Marcel Dekker, New York.

Larsson, C.M., Larsson, M., Purves, J.V. and Clarkson, D.T. (1991) Translocation and cycling through roots of recently absorbed nitrogen and sulphur in wheat (*Triticum aestivum*) during vegetative and generative growth. *Physiologia Plantarum* **82**, 345–352.

Li, Y. and Barber, S.A. (1991) Calculating changes of legume rhizosphere soil pH and soil solution phosphorus from phosphorus uptake. *Communications in Soil Science and Plant Analysis* **22**, 955–973.

Ludwig, F., Dawson, T.E., Kroon, H., Berendse, F. and Prins, H.H.T. (2003) Hydraulic lift in *Acacia tortilis* trees on a East African savanna. *Oecologia* **134**, 293–300.

Luo, Y., Ou Yang, Z., Yuan, G., Tang, D. and Xie, X. (2003) Evaluation of macroscopic root water uptake models using lysimeter data. *Transactions of the American Society of Agricultural Engineers* **46**, 625–634.

Macariola-See, N., Woodard, H.J. and Schumacher, T. (2003) Field verification of the Barber-Cushman mechanistic phosphorus uptake model for maize. *Journal of Plant Nutrition* **26**, 139–158.

McCully, M.E. and Canny, M.J. (1988) Pathways and processes of water and nutrient movement in roots. *Plant and Soil* **111**, 159–170.

McIntyre, B.D., Riha, S.J. and Flower, D.J. (1995) Water uptake by pearl millet in a semiarid environment. *Field Crops Research* **43**, 67–76.

Marschner, H. (1995) *Mineral Nutrition of Higher Plants*, 2nd edn. Academic Press, London.

Mickovski, S.B. and Ennos, A.R. (2003) The effect of unidirectional stem flexing on shoot and root morphology and architecture in young *Pinus sylvestris* trees. *Canadian Journal of Forest Research* **33**, 2202–2209.

Molz, F.J. (1981) Models of water transport in the soil-plant system: a review. *Water Resources Research* **17**, 1245–1260.

Monteith, J.L. (1986) How do crops manipulate water supply and demand? *Philosophical Transactions of the Royal Society, London* **A 316**, 245–259.

Monteith, J.L. and Unsworth, M.H. (1990) *Principles of Environmental Physics*, 2nd edn. Edward Arnold, London.

Nassar, I.N. and Horton, R. (1992) Simultaneous transfer of heat, water and solute in porous media: I. Theoretical development. *Soil Science Society of America Journal* **56**, 1350–1356.

Nassar, I.N., Horton, R. and Globus, A.M. (1992) Simultaneous transfer of heat, water and solute in porous media: II. Experiment and analysis. *Soil Science Society of America Journal* **56**, 1357–1365.

Nicoll, B.C. and Ray, D. (1996) Adaptive growth of tree root systems in response to wind action and site conditions. *Tree Physiology* **16**, 891–898.

Nobel, P.S. (1999) *Physicochemical and Environmental Plant Physiology*, 2nd edn. Academic Press, New York.

Nobel, P.S. and Cui, M. (1992) Hydraulic conductances of the soil, the root-soil air gap and the root: changes for desert succulents in drying soil. *Journal of Experimental Botany* **43**, 319–326.

Nye, P.H. and Marriott, F.H.C. (1969) A theoretical study of the distribution of substances around roots resulting from simultaneous diffusion and mass flow. *Plant and Soil* **30**, 459–472.

Nye, P.H. and Tinker, P.B. (1969) The concept of a root demand coefficient. *Journal of Applied Ecology* **6**, 293–300.

Passioura, J.B. (1983) Roots and drought resistance. *Agricultural Water Management* **7**, 265–280.

Passioura, J.B. (1988a) Root signals control leaf expansion in wheat seedlings growing in dry soil. *Australian Journal of Plant Physiology* **15**, 687–693.

Passioura, J.B. (1988b) Water transport in and to roots. *Annual Review of Plant Physiology and Plant Molecular Biology* **39**, 245–265.

Passioura, J.B. and Cowan, I.R. (1968) On solving the non-linear diffusion equation for the radial flow of solute to roots. *Agricultural Meteorology* **5**, 129–134.

Pinthus, M.J. (1973) Lodging in wheat, barley, and oats: the phenomenon, its causes, and preventative measures. *Advances in Agronomy* **25**, 209–263.

Prenzel, J. (1979) Mass flow to the root system and mineral uptake of a beech stand calculated from 3-year field data. *Plant and Soil* **51**, 39–49.

Raper, C.D., Patterson, D.T., Parsons, L.R. and Kramer, P.J. (1977) Relative growth and nutrient accumulation rates for tobacco. *Plant and Soil* **46**, 473–486.

Ray, D. and Nicoll, B.C. (1998) The effect of soil water-table depth on root-plate development and stability of Sitka spruce. *Forestry* **71**, 169–182.

Reicosky, D.C. and Ritchie, J.T. (1976) Relative importance of soil resistance and root resistance in root water absorption. *Soil Science Society of America Journal* **40**, 293–297.

Rengel, Z. (1993) Mechanistic simulation models of nutrient uptake: a review. *Plant and Soil* **152**, 161–173.

Richards, J.H. and Caldwell, M.M. (1987) Hydraulic lift: substantial nocturnal water transport between soil layers by *Artemisia tridentata* roots. *Oecologia* **73**, 486–489.

Roberts, A.G. and Oparka, K.J. (2003) Plasmodesmata and the control of symplastic transport. *Plant, Cell and Environment* **26**, 103–124.

Robertson, M.J., Fukai, S., Ludlow, M.M. and Hammer, G.L. (1993a) Water extraction by grain sorghum in a sub-humid environment. I. Analysis of the water extraction pattern. *Field Crops Research* **33**, 81–97.

Robertson, M.J., Fukai, S., Ludlow, M.M. and Hammer, G.L. (1993b) Water extraction by grain sorghum in a sub-humid environment. II. Extraction in relation to root growth. *Field Crops Research* **33**, 99–112.

Ross, P.J. (2003) Modeling soil water and solute transport – fast, simplified numerical solutions. *Agronomy Journal* **95**, 1352–1361.

Rowell, D.L., Martin, M.W. and Nye, P.H. (1967) The measurement and mechanism of ion diffusion in soils. III. The effect of moisture content and soil solution concentration on the self-diffusion of ions in soil. *Journal of Soil Science* **18**, 204–222.

Russell, R.S. and Clarkson, D.T. (1976) Ion transport in root systems. In: *Perspectives in Experimental Biology*, Vol. 2 (ed. N. Sunderland), pp. 401–411. Pergamon, Oxford.

Ryel, R.J., Caldwell, M.M., Yoder, C.K., Or, D. and Leffler, A.J. (2002) Hydraulic redistribution in a stand of *Artemisia tridentata*: evaluation of benefits to transpiration assessed with a simulation model. *Oecologia* **130**, 173–184.

Salisbury, F.B. and Ross, C.R. (1992) *Plant Physiology*, 4th edn. Wadsworth Publishing Co., Belmont.

Sattelmacher, B. (2001) The apoplast and its significance for plant mineral nutrition. *New Phytologist* **149**, 167–192.

Sauter, A., Davies, W.J. and Hartung, W. (2001) The long-distance abscisic acid signal in the droughted plant: the fate of the hormone on its way from root to shoot. *Journal of Experimental Botany* **52**, 1991–1997.

Schaetzl, R.J., Johnson, D.L., Burns, S.F. and Small, T.W. (1989) Tree uprooting: review of terminology, process, and environmental implications. *Canadian Journal of Forest Research* **19**, 1–10.

Schreiber, L., Hartmann, K., Skrabs, M. and Zeier, J. (1999) Apoplastic barriers in roots: chemical composition of endodermal and hypodermal cell walls. *Journal of Experimental Botany* **50**, 1267–1280.

Sekiya, N. and Yano, K. (2004) Do pigeon pea and sesbania supply groundwater to intercropped maize through hydraulic lift? – Hydrogen stable isotope investigation of xylem waters. *Field Crops Research.* **86**, 167–173.

Seneviratne, R. and Wild, A. (1985) The effect of mild drying on the mineralization of soil nitrogen. *Plant and Soil* **84**, 175–179.

Seward, P., Barraclough, P.B. and Gregory, P.J. (1990) Modelling potassium uptake by wheat (*Triticum aestivum*) crops. *Plant and Soil* **124**, 303–307.

Silberbush, M. (1996) Simulation of ion uptake from the soil. In: *Plant Roots – The Hidden Half* (eds Y. Waisel, A. Eshel and U. Kafkafi), 2nd edn, pp. 643–658. Marcel Dekker: New York.

Silberbush, M. and Barber, S.A. (1983) Sensitivity analysis of parameters used in simulating K uptake with a mechanistic mathematical model. *Agronomy Journal* **75**, 851–854.

Silberbush, M. and Barber, S.A. (1984) Phosphorus and potassium uptake of field-grown soybean cultivars predicted by a simulation model. *Soil Science Society of America Journal* **48**, 592–596.

Sperry, J.S., Stiller, V. and Hacke, U.G. (2003) Xylem hydraulics and the Soil-Plant-Atmosphere Continuum: opportunities and unresolved issues. *Agronomy Journal* **95**, 1362–1370.

Steudle, E. (1994) Water transport across roots. *Plant and Soil* **167**, 79–90.

Steudle, E. (2000) Water uptake by roots: effects of water deficit. *Journal of Experimental Botany* **51**, 1531–1542.

Steudle, E. (2001) The cohesion-tension mechanism and the acquisition of water by plant roots. *Annual Review of Plant Physiology and Plant Molecular Biology* **52**, 847–875.

Steudle, E. and Heydt, H. (1997) Water transport across tree roots. In: *Trees – Contributions to Modern Tree Physiology* (eds H. Rennenberg, W. Eschrich and H. Zeigler), pp. 239–255. Backhuys Publishers, The Netherlands.

Steudle, E. and Peterson, C.A. (1998) How does water get through roots? *Journal of Experimental Botany* **49**, 775–788.

Steudle, E., Murrmann, M. and Peterson, C.A. (1993) Transport of water and solute across maize roots modified by puncturing the endodermis. *Plant Physiology* **103**, 335–349.

Tabatabaei, S.J., Gregory, P.J. and Hadley, P. (2004) Uneven distribution of nutrients in the root zone affects the incidence of blossom end rot and concentration of calcium and potassium in fruits of tomato. *Plant and Soil* **258**, 169–178.

Taleisnik, E., Peyrano, G., Córdoba, A. and Arias, C. (1999) Water retention capacity in root segments differing in the degree of exodermis development. *Annals of Botany* **83**, 19–27.

Taylor, H.M. and Klepper, B. (1978) The role of rooting characteristics in the supply of water to plants. *Advances in Agronomy* **30**, 99–128.

Tinker, P.B. (1976) Transport of water to plant roots. *Philosophical Transactions of the Royal Society, London* **B 273**, 445–461.

Tinker, P.B. and Nye, P.H. (2000) *Solute Movement in the Rhizosphere*. Oxford University Press, Oxford.

Vaidyanathan, L.V. and Nye, P.H. (1966) The measurement and mechanism of ion diffusion in soils. II. An exchange resin paper method for measurement of the diffusive flux and diffusion coefficient of nutrient ions in soils. *Journal of Soil Science* **17**, 175–183.

Wan, C., Xu, W., Sosebee, R.E., Machado, S. and Archer, T. (2000) Hydraulic lift in drought-tolerant and -susceptible maize hybrids. *Plant and Soil* **219**, 117–126.

Weatherley, P.J. (1982) Uptake and flow in roots. In: *Encyclopedia of Plant Physiology 12B, Physiological Plant Ecology* (eds O.L. Lange, C.B. Osmond and H. Zeigler), pp. 79–109. Springer-Verlag, Heidelberg.

Wegner, L.H. and Raschke, K. (1994) Ion channels in the xylem parenchyma of barley roots: a procedure to isolate protoplasts from this tissue and a patch-clamp exploration of salt passageways into xylem vessels. *Plant Physiology* **105**, 799–813.

Weir, A.H. and Barraclough, P.B. (1986) The effect of drought on the root growth of winter wheat and on its water uptake from a deep loam. *Soil Use and Management* **2**, 91–96.

White, P.J. (2001) The pathways of calcium movement to the xylem. *Journal of Experimental Botany* **52**, 891–899.

White, P.J. and Broadley, M.R. (2000) Mechanisms of caesium uptake by plants. *New Phytologist* **147**, 241–256.

White, P.J. and Broadley, M.R. (2001) Chloride in soils and its uptake and movement within the plant: a review. *Annals of Botany* **88**, 967–988.

White, P.J. and Broadley, M.R. (2003) Calcium in plants. *Annals of Botany* **92**, 487–511.

White, P.J., Whiting, S.N., Baker, A.J.M. and Broadley, M.R. (2002) Does zinc move apoplastically to the xylem in roots of *Thlaspi caerulescens*? *New Phytologist* **153**, 199–211.

Whitehead, D.C. (2000) *Nutrient Elements in Grassland: Soil-Plant-Animal Relationships*. CAB International, Wallingford.

Wild, A. (1988) *Russell's Soil Conditions and Plant Growth*, 11th edn. Longman Scientific and Technical, Harlow.

Wilderotter, O. (2003) An adaptive numerical model for the Richards equation with root growth. *Plant and Soil* **251**, 255–267.

Wilkinson, S. and Davies, W.J. (2002) ABA-based chemical signalling: the co-ordination of responses to stress in plants. *Plant, Cell and Environment* **25**, 195–210.

Willatt, S.T. and Taylor, H.M. (1978) Water uptake by soya-bean roots as affected by their depth and by soil water content. *Journal of Agricultural Science, Cambridge* **90**, 205–213.

Wu, J., Zhang, R. and Gui, S. (1999) Modeling soil water movement with water uptake by roots. *Plant and Soil* **215**, 7–17.

Xu, X. and Bland, W.L. (1993) Reverse water flow in sorghum roots. *Agronomy Journal* **85**, 384–388.

Yamauchi, A., Taylor, H.M., Upchurch, D.R. and McMichael, B.L. (1995) Axial resistance to water flow of intact cotton roots. *Agronomy Journal* **87**, 439–445.

Zhang, J. and Davies, W.J. (1989) Abscisic acid produced in dehydrating roots may enable the plant to measure the water status of the soil. *Plant, Cell and Environment* **12**, 73–81.

Chapter 5

Roots and the Physico-Chemical Environment

The root systems described in Chapter 3 resulted from a mass of interactions of the plant with the aerial and edaphic environments. This chapter describes some of the responses of roots and root systems to physical and chemical components of the environment; biological interactions are considered in Chapter 6. Because soil consists of three physical phases (solid, liquid and gas) changing the volume of any one component in a non-shrinking soil will necessarily change the volume of at least one other component. For example, if water is removed from the pore space, then the volume of air necessarily increases and *vice versa*. So, although the chapter is structured to examine individual environmental factors, many of them interact and these interactions are also considered. The physico-chemical environment of soils is also rarely uniform and there is marked heterogeneity at different scales.

To understand such interactions and the response of roots, it is often necessary to work in controlled conditions, so that much of the work described in this chapter has been undertaken with young root systems grown in containers filled with sieved soil or other media. Such systems allow both better environmental control and ease of access for root measurements but may not reflect the response of older root systems that have acclimated to their particular soil environment. The effects of soil structure at different scales are particularly poorly represented in most experimental studies to date.

5.1 Temperature

As with the shoot, temperature affects both the expansion of the root system through effects on development and growth, and a range of metabolic processes that determine the activity of the root system. Our understanding of these effects has been growing steadily but because of the number and complexity of interacting processes, much remains to be learnt. Reviews by Cooper (1973), Voorhees *et al.* (1981), and Kaspar and Bland (1992) demonstrate the progress that has been possible, although fundamental understanding is still lacking.

The response of a root system to temperature is similar to that of the shoot system with a minimum temperature below which no growth occurs, an optimum at which growth is maximal, and a maximum at which no growth again occurs (Fig. 5.1). The minimum (base) and optimal temperatures depend upon the plant species and are typically in the ranges of 0–12°C and 25–35°C, respectively, while the maximum is almost always around 40–45°C. The optimum is often broad rather than a sharp peak, and there is frequently a broad range of tem-

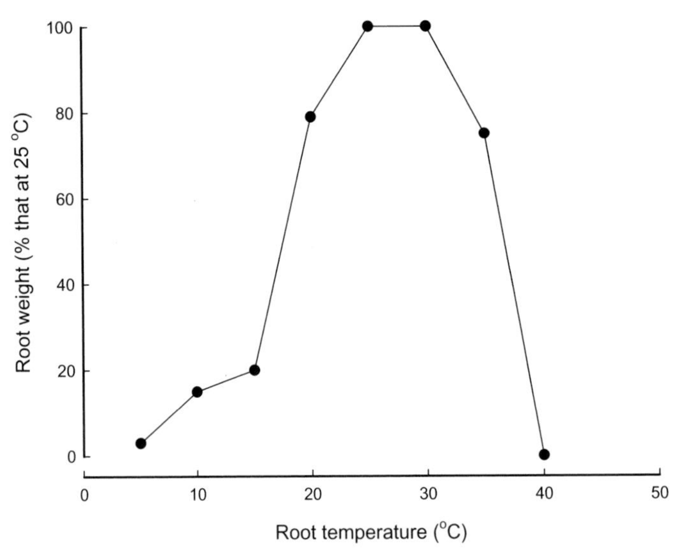

Fig. 5.1 Effect of root temperature on root dry weight of maize at 24 days after germination. (Redrawn from Brouwer, 1962.)

peratures at which root growth rates are ≥50% of their maximum (Table 5.1). Both total root mass and total root length show similar overall responses to temperature although the size of the response to a particular temperature may differ between the two measures. Plants are exposed to temperatures below the minimum (ice and frost in temperate regions) and above the maximum (surface soil temperatures in tropical regions may exceed 45°C), and have evolved mechanisms to cope with these extremes that are beyond the scope of this book.

5.1.1 Root development and growth

The expansion of a root system results from two temperature-dependent processes, namely development and growth. In section 3.2 we saw that the development of roots and shoots was closely linked. This is most obvious in cereals where the nodal roots are linked to the

Table 5.1 Root temperature ranges for root growth rates ≥50% of the maximum

	Temperature range (°C)	
Plant species	Low	High
Flax (*Linum usitatissimum*)	10	31
Pea (*Pisum sativum*)	9	33
Bean (*Phaseolus vulgaris*)	12	33
Maize (*Zea mays*)	17	37
Strawberry (*Fragaria* sp.)	5	31
Faba bean (*Vicia faba*)	12	32
Rape (*Brassica napus*)	16	32
Oat (*Avena sativa*)	9	32

From Klepper, 1987.

number of leaf nodes. Because the interval between the appearance of successive leaves (the phyllochron) in cereals is primarily affected by temperature, the appearance of nodal axes in such plants can be predicted from the thermal time (ε) defined as (Vincent and Gregory, 1989a):

$$\varepsilon = \sum_1^n (T_a - T_b)$$
$$\text{if } T_a < T_b \text{ then } T_a - T_b = 0 \quad (5.1)$$

where T_a is daily mean temperature, T_b is a base temperature below which development ceases, and n is the number of days between each developmental event. Gregory (1983) grew pearl millet (*Pennisetum typhoides* S. & H.) in stands in a glasshouse with air and soil temperatures ranging from 19 to 31°C. As expected, both soil and air temperatures had a marked effect on the development of root axes, with development at a specified air temperature being slower in the cooler soil (Fig. 5.2a). When the number of axes was plotted as a function of the accumulated temperature at the shoot meristem using a T_b of 12°C, a good linear relation was obtained for all soil/air temperature combinations up to 28 days after sowing (Fig. 5.2b). During the study period, lateral roots were produced when axes were 50–80 mm long. Once branching commenced, the number of lateral roots on each plant increased almost exponentially with time as axes produced first order laterals which in turn developed second order laterals. As with root axes, the number of lateral roots was correlated best with the accumulated temperature at the shoot meristem (Fig. 5.2c). Vincent and Gregory (1989a, 1989b) found similar results in controlled glasshouse and field experiments for the appearance of root axes of winter wheat, although low irradiance during early growth of glasshouse-grown plants resulted in the failure of some axes to emerge. Such observations in cereals have formed the basis of quantitative models to predict root development and growth (e.g. Porter *et al*., 1986). Temperature has also been found to af-

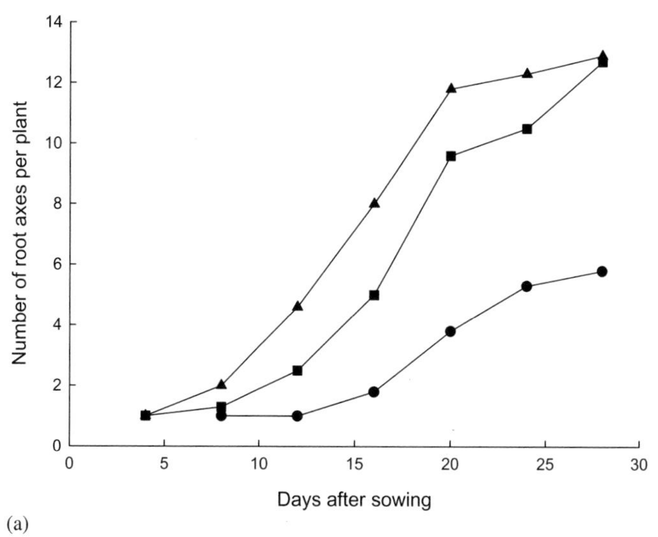

(a)

Fig. 5.2 Effects of soil temperature on the production of roots of pearl millet: (a) production of root axes with an air temperature of 25°C and soil temperatures of 19°C (●), 25°C (■) and 31°C (▲). *Continued.*

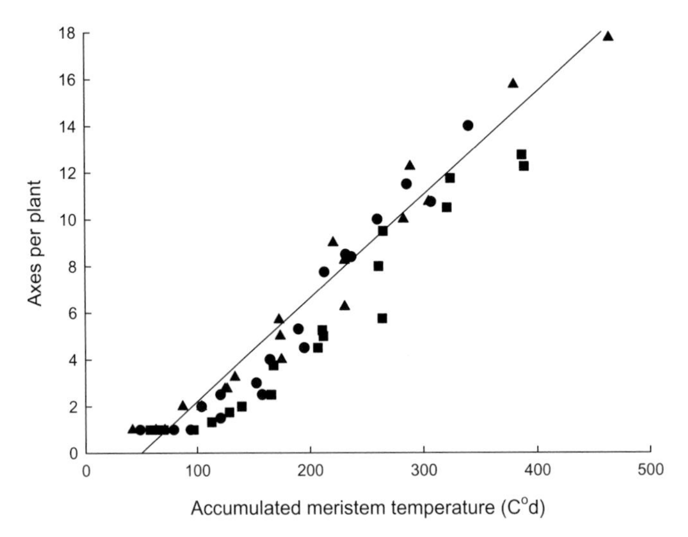

(b)

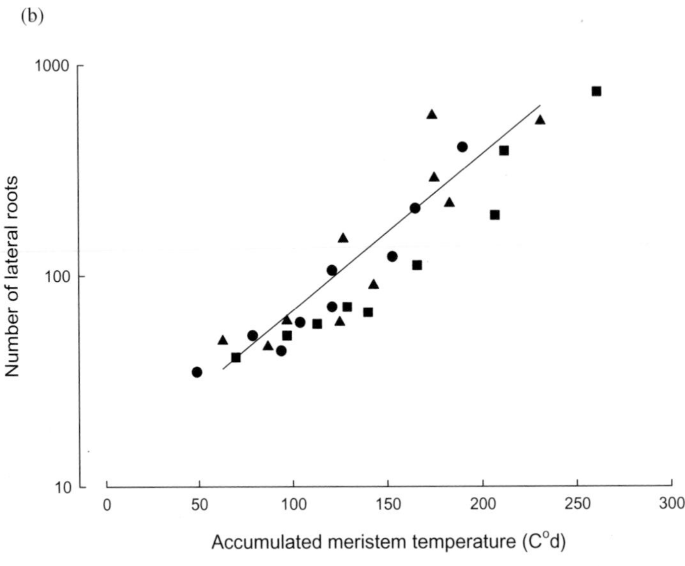

(c)

Fig. 5.2 Effects of soil temperature on the production of roots of pearl millet: (b) the relationship between the number of root axes and thermal time measured at the shoot meristem using a base temperature of 12°C – air temperatures were 19°C (●), 25°C (■) and 31°C (▲); (c) the relationship between the number of lateral roots and thermal time measured at the shoot meristem using a base temperature of 12°C – air temperatures were as for (b). (Reproduced with permission from Gregory, *Journal of Experimental Botany*; Oxford University Press, 1983.)

fect the development of lateral roots in soyabean (Stone and Taylor, 1983), pea (Gladish and Rost, 1993) and sunflower (Seiler, 1998) seedlings, but there is, as yet, no theoretical basis for summarizing the results.

Growth can be specified either in terms of root length or root mass. Many studies (e.g. Gladish and Rost, 1993; Seiler, 1998) have shown that the extension rate of roots is faster as temperature increases up to the optimum and decreases thereafter. However, there is no unique relation between extension rate of a particular root axis and temperature because elongation rates depend on the size and development stage of the shoot, canopy light interception, and the age of the root axis itself (Gregory, 1986; Gladish and Rost, 1993). For example, Gregory (1986) found that while the extension rate of an individual root axis of pearl millet increased with temperature between 20 and 32°C, the second root axis had a faster elongation rate than the first, probably because the shoot was larger when the second axis was elongating (Fig. 5.3). The total length of the root system is also influenced by temperature (Vincent and Gregory, 1989a; Seiler, 1998) but patterns are complex because while expansion is typically exponential initially, it becomes linear later (Vincent and Gregory, 1989b).

Root mass depends on assimilate supply as well as temperature, but there appears to be some difference of opinion in the literature about the relative importance of these two factors in determining the size of the system. Stone and Taylor (1983) grew soyabean roots in containers filled with vermiculite in water baths in a glasshouse and concluded that the rates of tap root and lateral root extension in plants up to 20 days old were limited by the temperature around the roots rather than light intensity. However, temperature affects the rate of leaf expansion and hence the amount of radiation intercepted by a plant, which in turn affects the amount of assimilate available for partitioning to roots. Gregory (1986) found that total root mass of pearl millet across several temperature treatments was linearly related to the cumulative intercepted radiation and Vincent and Gregory (1989b) also showed that effects of sowing date, site and nitrogen applications on total root length could also be substantially accounted for when the amount of light intercepted was taken into account (Fig. 5.4). Aguirrezabal *et al.* (1994) demonstrated a role for intercepted radiation in the rate of elongation of the tap root and laterals of sunflower sown at different times and experiencing different temperature regimes. In their experiments, elongation rate of the tap root and laterals was affected by the temperature only between germination and the two-leaf stage when the carbon used in root growth originated predominantly from the

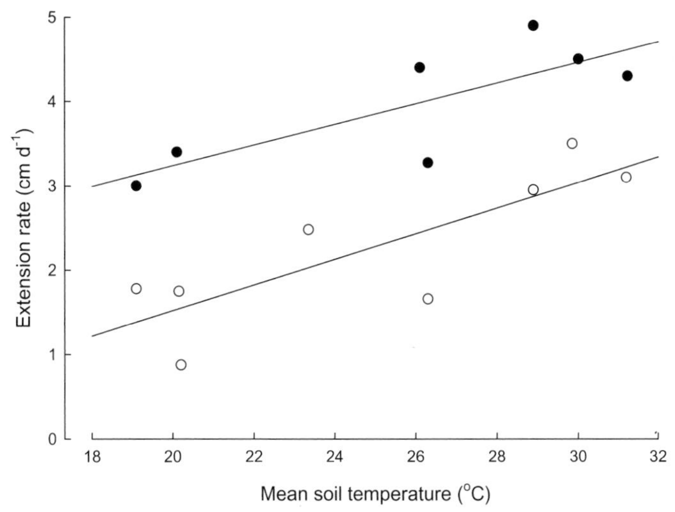

Fig. 5.3 Effects of soil temperature measured at 5 cm depth on rate of extension of axes 1 (○) and 2 (●) over the period 12–20 days after sowing. The lines shown are linear regressions. (Reproduced with permission from Gregory, *Journal of Experimental Botany*; Oxford University Press, 1986.)

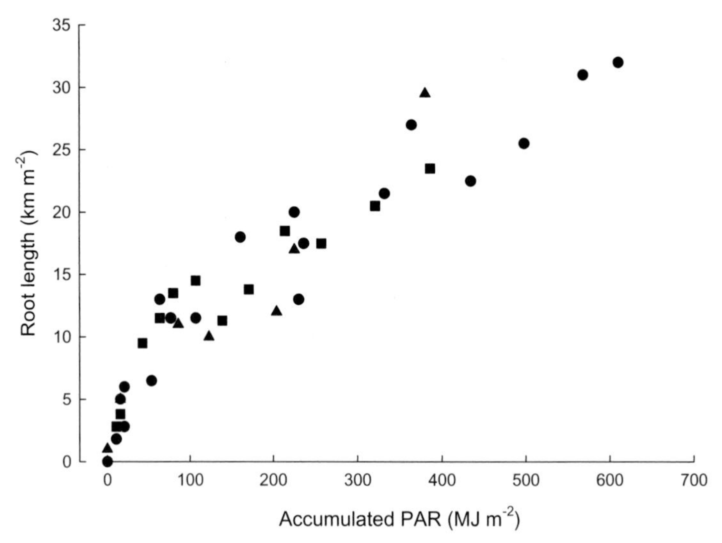

Fig. 5.4 The increase of root length of UK winter wheat crops with accumulated intercepted photosynthetically active radiation (PAR); crops grown at Rothamsted (●), Sutton Bonington (■), and Sonning (▲). (Reproduced with permission from Vincent and Gregory, *Plant and Soil*; Springer Science and Business Media, 1989b.)

seed. Thereafter, the correlation between tap root elongation rate and soil temperature was weak and no relation was observed for laterals; tap root elongation in this second phase was linked to the amount of radiation intercepted (Aguirrezabal and Tardieu, 1996).

Aguirrezabal and Tardieu (1996) suggest that their results reflect a difference between dicot and monocot species, but it is probably more likely to be a reflection of differences in the size/developmental stage of plants used in the various studies; as the canopy increases in size, direct effects of temperature become less important relative to the amounts of radiation intercepted. Some evidence for this suggestion can be detected in studies on native UK grasslands. Fitter *et al.* (1998) measured root growth of three grass species at sites on an altitudinal gradient ranging from 171 to 845 m. They found that while root production (estimated with a minirhizotron) and root biomass showed pronounced seasonal peaks, neither was a simple function of altitude. Increased root accumulation in summer was related to the change in length of the growing season not to soil temperature. Root growth and root respiration rate were most closely correlated with total radiation flux rather than temperature, implying that root growth was determined more by resource availability and source-sink relationships within the plants than by close coupling to temperature. Similar results were obtained in a study of grassland dominated by *Holcus lanatus* L. in which root respiration was insensitive to soil temperature over much of the year but root growth was strongly affected by incident radiation (Edwards *et al.*, 2004).

5.1.2 Root orientation

Root plagiotropism, growth at angles from the vertical, occurs in many species and appears to be temperature-sensitive. Two possible mechanisms for this temperature response were discussed by Kaspar and Bland (1992): (i) two geotropic processes with different temperature responses attain a balance depending on the mean temperature at the root apex; and (ii) temperature gradients in the soil are sensed and roots grow toward the more favourable temperature. Whatever the mechanism, which remains to be determined, the influence of temperature is clear. For example, Onderdonk and Ketcheson (1973) grew maize seedlings in soil for short periods in growth cabinets and found that the angle of growth of the seminal axis was at a minimum relative to the horizontal at a constant 17°C and greater above and below this temperature (range 10–30°C). Cyclic temperatures also affected the angle of growth with the maximum temperature of the cycle apparently determining the angle. Tardieu and Pellerin (1991) assessed the effect of temperature on field-grown maize by changing sowing date, mulching, season and cultivar. In their results, cultivar had no effect but the horizontal component of the trajectory of axes appearing from the first four internodes differed significantly between species, ranging from 93 to 700 mm in roots from the second internode and from 71 to 569 mm for those from the third internode. They determined that the period during which temperature affected trajectory was short, beginning at root appearance and ending between 50 and 100°Cd (i.e. when the root was < 100 mm long).

Despite the similar performance of two cultivars determined by Tardieu and Pellerin (1991), others have found genotypic differences in response. Kaspar *et al.* (1981) found substantial variation in both plagiotropic angle and its temperature sensitivity in 10 soyabean lines with mean growth angles from the horizontal of upper laterals changing from 16.7° to 20.9° as temperature changed from 10 to 26.7°C. Some lines showed no temperature sensitivity, while the most sensitive varied from 12.5° at 10°C to 22.0° at 26.7°C.

5.1.3 Other root functions

Temperature affects the rate and duration of many processes in the root including water and solute transport (Clarkson *et al.*, 1988) and nodulation of leguminous species (e.g. Peltzer *et al.*, 2002). Section 3.5 discusses the important effects that temperature has on root longevity and turnover, but many metabolic processes are similarly affected. For example, McMichael and Burke (1994) found that electron transport in root tips of cotton was affected by temperature and that there was a developmental regulation of the temperature response. This regulation was postulated to be related to the changes in mobilization of seed reserves from the cotyledons. Root metabolism may become more temperature-sensitive as development occurs because the mobilization of cotyledonary reserves decreases during early seedling growth.

Clarkson *et al.* (1988) showed that temperature affects processes related to nutrient acquisition at different organizational levels within the plant, but that acclimatory changes occur following prolonged exposure, especially to low temperature. While lipid composition and carrier activity may change at the membrane level, and the size and morphology may change at the root system level, the overall effect is to reduce the temperature dependence of ion transport and to ensure that nutrient uptake does not limit growth at low temperatures.

5.2 Gravity and other tropistic responses

The gravitropic response of roots has been extensively researched and is a major influence on the shape and architecture of root systems. It has been appreciated for over a century that the perception of gravity occurs in the root cap and 'that this part transmits some influence to the adjoining parts, causing them to curve downwards' (Darwin and Darwin, 1880). The site of perception, then, is located at some distance from the site of growth response which occurs in the zone of elongation. It is now recognized that the gravitropic response is a process that consists of phases of perception, signalling, transmission and growth response (Boonsirichai *et al.*, 2002) and results in the plant growing roots and other organs at a defined angle from the gravity vector, known as the gravitational set point angle. This gravitational set point angle varies from species to species, and organ to organ, and with the developmental stage of the plant, its physiological status, and the environmental conditions (Firn and Digby, 1997).

5.2.1 Gravisensing and the response of roots

The dominant hypothesis of gravity sensing in roots of higher plants is that it occurs via the sedimentation of starch-filled amyloplasts (statoliths) within specialized columella cells of the root cap (statocytes). In contrast to roots, in shoots the statocytes are located in the endodermis that surrounds the vascular system. Amyloplasts present in the root columella cells settle to the bottom side of the cell; in a vertically growing root this would be the distal end of the cell but if the root is reoriented horizontally, the amyloplasts sediment to the bottom longitudinal side of the cell (Plate 5.1). Recent studies using laser ablation techniques to ablate particular cells together with analysis of *Arabidopsis* mutants has allowed the identification of specific root cap cells as important gravity sensors (Boonsirichai *et al.*, 2002; Blancaflor and Masson, 2003). To test whether the position of the amyloplasts might be involved in gravity perception, Kuznetsov and Hasenstein (1996) used high-gradient magnetic fields to move the amyloplasts within the columella cells. The displacement resulted in a curvature at the tip of the root while in starchless *Arabidopsis pgm* mutants this did not occur, strongly suggesting that the movement of amyloplasts is sufficient to promote root curvature. There is also some evidence for a secondary mechanism of gravity sensing, possibly located in the distal zone of elongation, which operates hydrostatically (Boonsirichai *et al.*, 2002).

The relation between the sedimentation of amyloplasts and the resulting direction of growth of roots within a root system is complex and hypothesized to be related to the preferential location of gravireceptors along certain walls. For example, in oil palm (*Elaeis guineensis* Jacq.) field observations show that roots can grow either upwards or downwards (orthogravitropic), horizontally (diagravitropic), or without any specific direction in relation to gravity (agravitropic), and that particular root types maintain the same direction of growth throughout their lives (Jourdan and Rey, 1997). All of the graviresponsive roots, irrespective of their direction of growth, had a statenchyma with amyloplasts sedimenting under the influence of gravity while the agravitropic roots did not. However, the position of the amyloplasts varied between roots with accumulation against distal walls in roots growing vertically downward, against the longitudinal wall in horizontal roots, and against the proximal wall in roots growing vertically upward (Jourdan *et al.*, 2000). Stability in the direction of growth may be a consequence of an absence of receptors along specified walls, so that horizontal growth may be linked to an absence of sensors along the longitudinal wall of the statocyte, while vertical growth may be linked to an absence of sensors along the transverse (proximal or distal) walls (Jourdan *et al.*, 2000).

The way in which the information about the physical position of amyloplasts is converted into a physical signal is not known with certainty. It has been proposed that settled amyloplasts interact with the actin-based cytoskeletal network in the statocytes resulting in: (i) cytosolic Ca^{2+} pulses that trigger downstream signalling events; and/or (ii) disruption of stretch receptors at the plasma membrane (Boonsirichai *et al.*, 2002). Whatever the translation process, there is a range of experimental evidence for a role for Ca^{2+} and/or cytosolic pH as messengers (Blancaflor and Masson, 2003). In 1928 both Cholodny and Went proposed that tropistic stimuli, including gravity, promoted the lateral transport of auxin toward the bottom of plant organs in the elongating zone. Because auxin promotes cell elongation in shoots and inhibits it in roots, the auxin gradient was hypothesized to cause the observed upward curvature in shoots and downward curvature in roots. Recent research (e.g. Ottenschläger *et al.*, 2003) confirms the basic Cholodny-Went theory and demonstrates the linkage between gravity sensing in the root cap and the curvature response that develops further back in the elongating zone. Ottenschläger *et al.* (2003) used a GFP-based auxin biosensor to demonstrate that when roots were gravistimulated, an asymmetric auxin flux occurred from columella cells to the lateral root cap, which was then progressively transmitted from the cap to the elongation zone, where the curvature response developed. These results support the ‘fountain’ model of auxin transport proposed by Konings (1967) in which an auxin (indole-3-acetic acid, IAA) is synthesized in young shoots and transported to the root tip where it adds to another pool of locally synthesized auxin (Plate 5.2). The auxin is then redistributed laterally to more peripheral tissues and transported back towards the elongation zone where it regulates cellular elongation. With gravistimulation, auxin redistribution occurs preferentially downward across the root cap, resulting in the formation of a lateral auxin gradient which is transmitted to the elongation zone where it promotes downward curvature of the tip (Boonsirichai *et al.*, 2002). Both active metabolism and the presence of Ca^{2+} appear to be required for the redistribution process. Boonsirichai *et al.* (2002) concluded that auxin transport is essential for root gravitropism, as demonstrated by the fact that many auxin-response mutants display defects in root gravitropism. They further proposed that the signal transduction pathway controlled the localization of auxin efflux carrier complexes within the columella cells of the root cap, thereby facilitating the lateral transport of auxin during gravistimulation.

5.2.2 Phototropism, hydrotropism and thigmotropism

Darwin and Darwin (1880) recognized that the root tip was sensitive not only to gravity but also to water (hydrotropism) and to contact with hard objects (thigmotropism). In short, ‘These two kinds of sensitiveness conquer for a time the sensitiveness to geotropism (now called gravitropism), which, however, ultimately prevails’. These three environmental factors also interact with light (phototropism) which is also sensed in the root tip. Correll and Kiss (2002) show that while light is sensed by photoreceptors such as phytochromes in contrast to the amyloplasts that sense gravity, the resulting differential growth patterns are similar. Moreover, each stimulus can enhance or reduce the effectiveness of the other, with red and blue light especially effective at enhancing or reducing the gravitropic response depending on plant species.

The response of the root to hard objects and the importance of the root tip in sensing them were demonstrated by Goss and Russell (1980) by observing the response of the seminal axis of maize to small glass ballotini. For roots with an intact root cap, the rate of elongation decreased from about 15 μm min^{-1} in the 20 minutes before contact to about 3.6 μm min^{-1} in the first 5 minutes after contact, before gradually increasing to the initial rate of elongation after 15 minutes. When the root cap was removed, the same initial rate of elongation was measured but no significant change in elongation rate occurred when the ballotini was touched.

The mechanisms of the interactions between the various tropistic responses are slowly being unravelled as *Arabidopsis* root mutants with these characters become available (Eapen *et al.*, 2005). Roots grow towards areas of higher water potential but the processes underlying hydrotrophic sensing and signal transduction remain unknown, although a gradient in apoplastic Ca^{2+} and Ca^{2+} influx through plasma membranes appears to be involved in signal transduction (Takahashi, 1997). Takahashi *et al.* (2003) showed that a severe moisture gradient applied to roots of *Arabidopsis*, leading to subsequent development of a hydrotropic response, was accompanied by the degradation of amyloplasts in the columella and reduced graviresponsiveness (Fig. 5.5). These results suggest that the reduced responsiveness to gravity is, at least in part, attributable to the degradation of amyloplasts in the columella cells. Takahashi (1997) reviewed relevant literature and concluded that roots could sense a water potential gradient as small as 0.5 MPa mm^{-1} so that hydroresponsiveness may contribute to both avoidance of drought stress and modifications to root system architecture.

Similarly, while the mechanism of touch responsiveness is unknown, touch can modulate graviresponsiveness by influencing gravity sensing in the root cap. Massa and Gilroy (2003) grew *Arabidopsis* in gel and induced touch by placing a glass barrier across the direction of growth, resulting in a step-like growth habit. The root axis grew parallel to, but not touching the barrier, while the root cap remained in contact with it. Laser ablation of the columellar cells changed this pattern of growth such that the tip angle became more horizontal with respect to the glass barrier. Similarly when the peripheral cap cells were touched using a glass micropipette, there was a significant reduction in the sedimentation rates of amyloplasts in story 1 and 2 columella cells (Table 5.2). Thus, as with moisture, touch modulates root responses to gravity by affecting gravity sensing within the cap. Darwin and Darwin (1880), Massa and Gilroy (2003), and Blancaflor and Masson (2003) all suggest that this interaction of gravitropic and thigmotropic responses may facilitate the avoidance of obstacles while generally maintaining downward growth. However, the true picture may be even more complex because in soil-grown roots, there is ample evidence that a degree of 'touch' is beneficial to root growth. For example, Stirzaker *et al.* (1996) found that barley grew best in soil packed at an intermediate bulk density, which reflected a compromise between soil which was soft enough to allow good root growth but sufficiently compact to give good root–soil contact. A few large pores (3.2 mm diameter) were detrimental to barley growth and some pea radicles died. Barley grew best in a network of narrow pores made by lucerne and ryegrass roots and responded positively to the larger pores being filled with peat. Similarly, Donald *et al.* (1987) grew maize in a silt loam sieved to four aggregate sizes but with adequate nutrient and water supplies and found that the length of the root system grown in the coarsest aggregates (6.4–12.8 mm diameter) was only 60% that in the finest (<1.6 mm) aggregates. Although

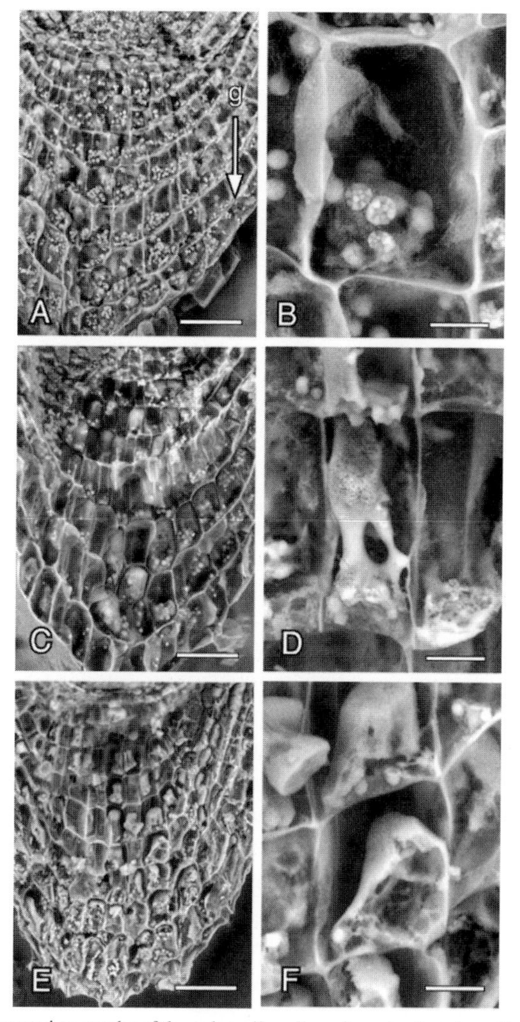

Fig. 5.5 Scanning electron micrographs of the columella cells and amyloplasts in radish (*Raphanus sativus* L.) roots that were hydrotropically stimulated. The left-hand column shows the whole root cap (bar = 50 μm) and the right-hand column is an enlargement of the columella cells with amyloplasts visible (bar = 10 μm). (A and B) Root cap cells not exposed to a moisture gradient; (C and D) root cap after 2 hours exposure to the moisture gradient; and (E and F) root cap after 5 hours exposure to the moisture gradient. The arrow (g) indicates the direction of the gravity vector. (Reproduced with permission from Takahashi *et al.*, *Plant Physiology*; American Society of Plant Biologists, 2003.)

Donald *et al.* (1987) speculated that these findings resulted from ethylene entrapment in roots grown in fine aggregates and to greater penetration resistance of large aggregates, the increased distance between surfaces associated with larger aggregates is also a possible cause. It is noteworthy that farmers with open, fluffy soils (often silty soils with high organic matter content) will roll them to enhance crop growth, although in addition to thigmotropic effects, poor root–soil contact may exacerbate nutrient deficiencies in some soils (Passioura and Leeper, 1963).

Table 5.2 Amyloplast sedimentation rates in columella of root caps of *Arabidopsis* touched with a glass pipette and then rotated through 135° to induce amyloplast sedimentation

	Sedimentation rate (μm min^{-1})	
Region of columella	Control	Touched
Storey 1		
Central	0.9 ± 0.06	0.5 ± 0.08*
Flank	0.5 ± 0.06	0.4 ± 0.07
Storey 2		
Central	1.2 ± 0.06	0.7 ± 0.06*
Flank	0.6 ± 0.06	0.5 ± 0.08
Storey 3		
Central	0.08 ± 0.06	0.1 ± 0.05
Flank	0.07 ± 0.05	0

Results are the mean of 10 observations ± standard error of the mean; *significantly different at $p<0.05$. From Massa and Gilroy, 2003.

5.3 Soil mechanical properties

Soils need to have sufficient mechanical strength to provide anchorage for the plant throughout its life and to sustain the system of pores containing water and air that are essential for plant growth. Dense regions of soil with high strength may limit root growth because they offer a large mechanical resistance to root growth and/or restrict the supply of oxygen to roots. Such dense regions can occur naturally during the formation of soils (for example in the duplex soils of Western Australia; Tennant *et al.*, 1992), and also arise through human activities via compaction by heavy machinery and by the formation of pans at particular depths during cultivation (Soane and van Ouwerkerk, 1994).

Almost all roots growing through soil experience some degree of mechanical impedance, and if continuous pores of appropriate size do not already exist then the root tip region must exert sufficient force to deform the soil. There is not much information about the minimum size of pore into which a root can grow without having to enlarge it, but few plants have roots smaller than 10 μm in diameter and most roots are much larger. This means that roots are often larger than the water-filled pores at field capacity (i.e. the pores filled with water when drainage has ceased; typically diameter <60 μm). Freely draining pores are, then, the main spaces in which roots can grow but growth will be inhibited unless the soil is sufficiently compressible to allow the root tip to make a pore large enough for elongation to proceed. If the soil will not deform, then roots have a limited capacity to modify their anatomy to fit into a pore that is normally smaller than their diameter. For example, Scholefield and Hall (1985) grew perennial ryegrass roots in pots of perlite containing sheets of steel mesh and found that roots were able to penetrate rigid pores as small as one-third of their nominal thickness by reducing their diameter. The degree of constriction was limited by the size of the root cap and that of the stele, which remained unchanged even in severely constricted regions. Under their experimental conditions, seminal axes that were 880 μm in diameter could penetrate rigid pores ≥315 μm. In solution culture, constricted roots growing in narrow capillary tubes elongated more slowly than unconstricted roots but continued to grow if adequately aerated.

5.3.1 Root elongation and mechanical impedance

As described in section 2.3.1, root growth occurs as the cells behind the root tip elongate longitudinally and radially to push the root tip forward. Growth occurs when the turgor pressure of the elongating cells is sufficient to overcome the constraints imposed by the viscoelastic cell walls and the soil matrix. This can be described by the equation:

$$P + W + \sigma = 0 \qquad (5.2)$$

where P is the turgor pressure, W is the wall pressure, and σ is the pressure applied externally by the soil on the root (Greacen and Oh, 1972). The relative simplicity of this expression, though, hides a number of important issues. The first issue relates to the relation between P and σ. In the pioneering experiments of Greacen and Oh (1972), the osmotic pressure was found to be linearly related to the mechanical resistance of the soil with an intercept representing the threshold value of wall pressure for cell elongation. They claimed that adjustments in osmotic pressure within root cells (which they called 'osmoregulation') provided the mechanism whereby roots could grow into strengthening soil. Many other studies, though, have been less conclusive about whether turgor rises with external pressure. For example, Atwell and Newsome (1990) measured turgor pressures in the apical 15 mm of lupin (*Lupinus angustifolius*) roots extracted from a sandy loam and found values ranging from 0.213 to 0.530 MPa when the bulk density was 1.6 Mg m^{-3}, and 0.210 to 0.570 MPa at 1.8 Mg m^{-3}. Mean values were 0.365 MPa at 1.6 Mg m^{-3} and 0.351 MPa at 1.8 Mg m^{-3}. Bengough *et al.* (1997) found that the osmotic pressure in pea roots grown in compressed sand was 0.81 MPa compared to 0.64 MPa in loose sand but that this difference disappeared within 12 hours of the removal of the mechanical impedance. Nevertheless, while the osmotic pressures were similar, the rates of elongation continued to be slower for up to 96 hours afterwards (Fig. 5.6). They suggested that the cell wall properties were more important than turgor in regulating the elongation rate of roots and, moreover, that these were not only functions of the current σ, but also of the stress history. If this is true, then no simple relationship between instantaneous root extension rate and mechanical impedance will exist. Changes in P in response to mechanical impedance may result in some plant species directly through up-regulation of solute import, but it appears increasingly likely that most of the measured changes in P occur indirectly through inhibition of volume expansion. For example, Atwell (1988) found that the ratio of potassium to sugars was preserved in impeded lupin roots and that volume expansion was impaired in such roots to about the same extent as solute accumulation.

A further issue with equation 5.2 is that it implies that, like P, there is a single value of W for any individual cell. However, W is different for the longitudinal and radial walls of an elongating cell so that the axial and radial pressures exerted by roots differ. The maximum axial root pressure of roots differs with species but is typically in the range 0.7–2.5 MPa (Gill and Bolt, 1955; Taylor and Ratliff, 1969), with values of about 1.0 MPa being common (Dexter, 1987). Radial pressures ranged from 0.4 to 0.6 MPa (Gill and Bolt, 1955), although the results of Misra *et al.* (1986) for pea, cotton and sunflower suggested that radial pressure was greater than axial pressure. The differences in maximum root growth pressure obtained by different workers for the same species (Table 5.3) are probably caused by the method of measurement

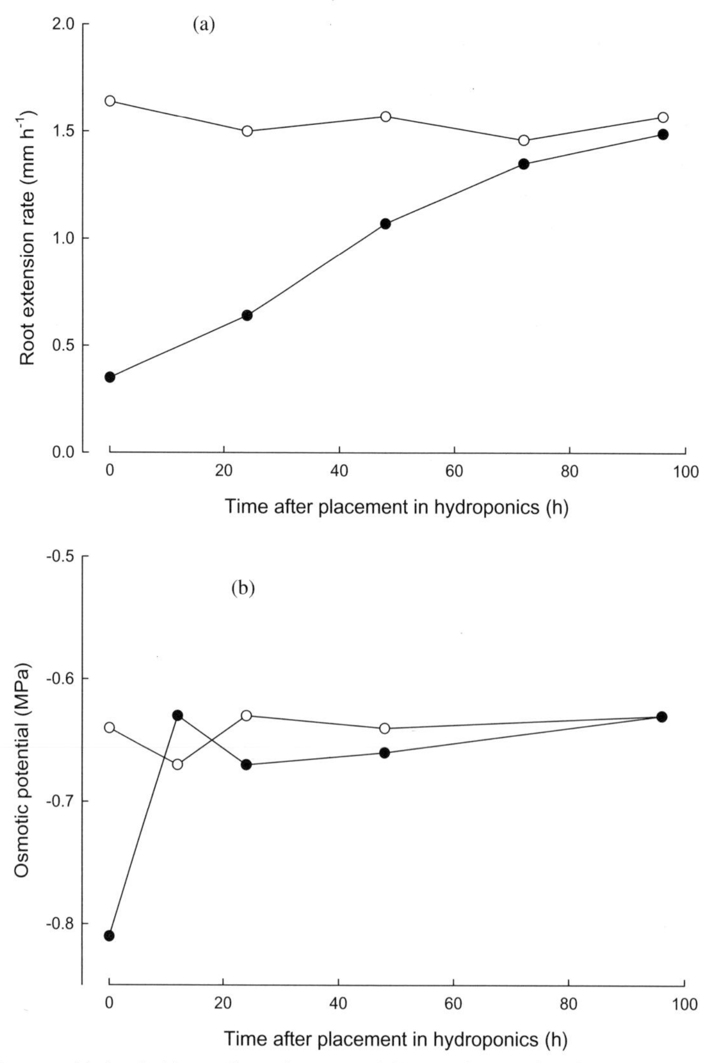

Fig. 5.6 Changes with time in (a) root elongation rate, and (b) osmotic potential of vacuolar sap of pea seedling roots after their removal from either loose sand (○) or compressed sand (●) and transfer to hydroponics. (Reproduced with permission from Bengough *et al.*, *Plant and Soil*; Springer Science and Business Media, 1997.)

Table 5.3 Values of mean maximum root growth pressure (MPa) in some plant species

Species	Taylor and Ratliff, 1969	Eavis *et al.*, 1969	Misra *et al.*, 1986
Cotton	0.95	1.1	0.29
Groundnut	1.16		
Pea	1.31	1.2	0.50
Safflower			0.24

Note that the studies used different experimental techniques.

(Clark *et al*., 2003). But as Atwell (1993) points out, whether or not radial pressures are greater than axial pressures in impeded roots, the key fact is that radial pressures are exerted over a much larger area than the near point impact of the axial pressure. For example, using the experimental results of Pfeffer quoted by Gill and Bolt (1955), Barley and Greacen (1967) calculated that the radial force exerted by a *Vicia faba* root along a 40 mm section of root was about 16 times greater than that exerted axially. Although a significant radial pressure may not be exerted over the whole length of root, even moderate radial pressures are important in causing soil deformation adjacent to the zone of elongation and relieving the resistance to axial growth. The importance of radial forces in overcoming σ and thereby allowing axial elongation was first recognized by Abdalla *et al*. (1969). Swelling of the sub-apical zone of a root (a common feature of roots grown in compacted soil – see section 5.3.2), induces a zone of reduced stress in front of the growing root tip by fracturing the soil in front of the growing apex (Kirby and Bengough, 2002). A zone of reduced stress induced by radial growth was also found in the cylindrical root analogue (finite element model) developed by Richards and Greacen (1986).

Soil mechanical impedance also affects W by altering the properties of cell walls. Relatively little is known about the properties of cell walls of roots compared with those of leaves and stems but Croser *et al*. (2000) demonstrated their importance. Reduced cell extension in mechanically impeded roots of pea was not associated with a reduction in turgor pressure but with what appeared to be a loosening of cell walls in the radial direction and a stiffening of walls in the axial direction.

All of these factors led Bengough *et al*. (1997) to modify the Lockhart equation describing the effects of soil strength on cell expansion as:

$$dl/dt = lm\ (\sigma)\ [P\ (\sigma) - Y\ (\sigma) - \sigma] \tag{5.3}$$

where l is the length of the elongating tissue, t is time, m (σ) is the cell wall extensibility, P (σ) is the turgor pressure in the cell, Y (σ) is the cell wall yield threshold, and σ is the external resisting pressure of the soil. The equation assumes that the permeability of the cell walls to water does not limit the rate of cell expansion, and emphasizes that m, P and Y are physiological properties that depend on the root environment.

5.3.2 Root responses to mechanical impedance

There are two very obvious responses of plant roots to mechanical impedance – slowing of the rate of extension and an increase in root diameter immediately behind the root tip. Figure 5.7a shows that elongation rate decreases as soil strength, measured with a penetrometer, increases but even at penetrometer resistances as high as 6 MPa, elongation of some species continues, albeit at a very slow rate. Figure 5.7a shows that the relative rate of elongation in response to soil strength differs between plant species but that at commonly encountered values of mechanical impedance >2 MPa, the root elongation rate of all species was reduced by at least 50%. The strength of most soils increases as they dry, so that shortage of soil water and hard soils are commonly interlinked. For example, in an oxisol from south-east Queensland, Kirkegaard *et al*. (1992) found that soil strength measured with a small penetrometer ranged from 0.81 to 3.21 MPa in a soil packed to a bulk density of 1.3 Mg m^{-3} as soil water content changed from 0.330 to 0.265 g g^{-1}. In looser soil (1.0

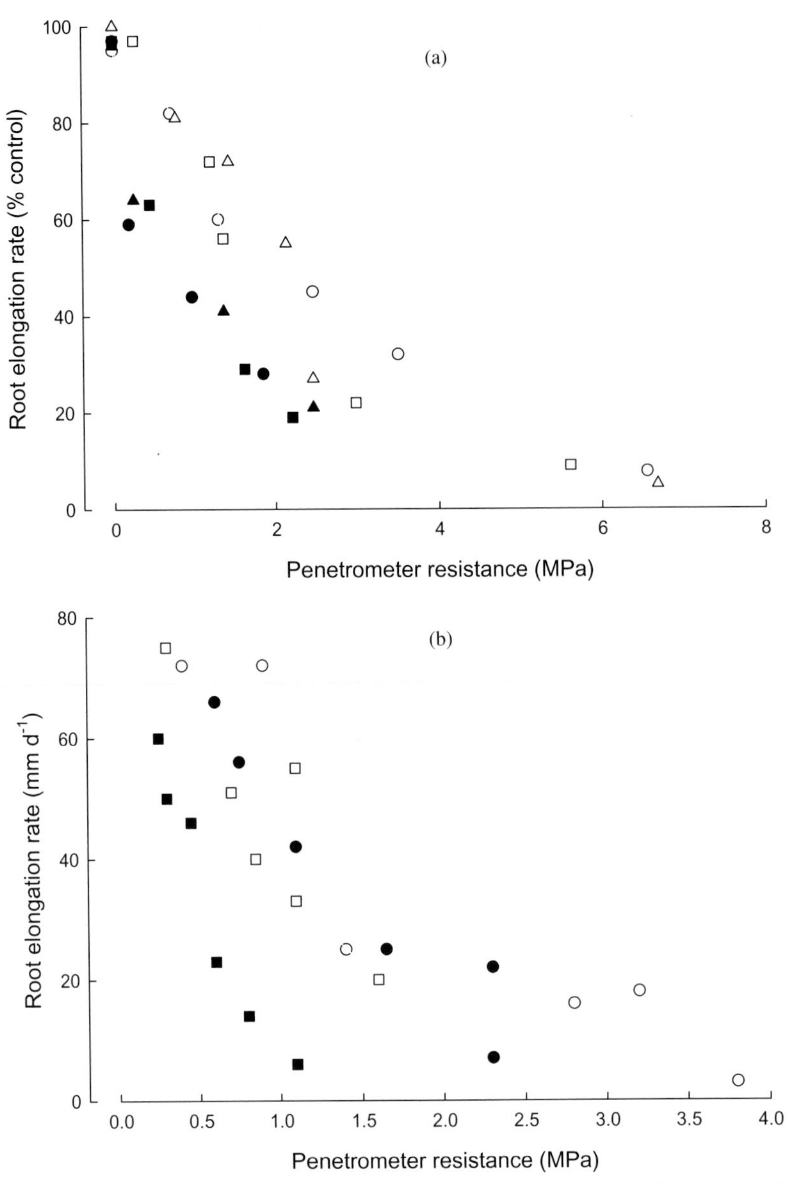

Fig. 5.7 The effect of soil strength on the rate of elongation of root radicles: (a) groundnut (open symbols) and cotton (closed symbols) for the period 40–80 hours after transplanting with approximate soil water contents of 7% (circles), 5% (squares) and 4% (triangles); and (b) pigeonpea for the period 50–75 hours after transplanting with soil gravimetric water contents of 26.5 (○), 28.0 (●), 30.0 (□) and 33.0% (■). ((a) Redrawn and reproduced with permission from Taylor and Ratliff, *Soil Science*; Lippincott, Williams & Wilkins, 1969; and (b) from Kirkegaard *et al.*, *Plant and Soil*; Springer Science and Business Media, 1992.)

Mg m^{-3}), the range of strength was 0.31 to 0.94 MPa for the same range of water content. Tightness of soil particle packing (bulk density), water content and strength therefore interact to affect root growth. Moreover, as water content increases, a fourth factor, the supply of oxygen also becomes a factor. Figure 5.7b shows that root elongation of pigeonpea at a water content of 0.33 g g^{-1} was lower at all values of penetrometer resistance than those at lower water contents. Kirkegaard *et al.* (1992) ascribed this effect, which was evident in all soils when air-filled porosity fell below 0.15, to reduced oxygen availability.

The decrease in elongation rate as mechanical impedance increases is often accompanied by radial expansion of root axes (Atwell, 1988; Materechera *et al.*, 1991). As with elongation rate, the degree of thickening depends on the particular experimental conditions. For example, Atwell (1988) measured radial thickening of lupin in response to soil compaction of about 15% while Materechera *et al.* (1991) found a range of 30–120%. The cells responsible for this thickening appear to be mainly those in the cortex although, again, there is a range of responses in the literature. Wilson *et al.* (1977) found that impedance of barley roots increased their diameter largely due to an increase in the thickness of the cortex along the whole length of the root resulting from an increase in both cell numbers and cell diameter. The stele also increased in diameter by up to 22% within 5 mm of the root apex, although this effect declined with distance and was absent within 40 mm of the root tip. However, Atwell (1988) found no effect of compaction on the diameter of the stele. Root thickening was caused by an increase in the diameter of the cortex which was, in turn, brought about by an increase in the diameter of individual cells rather than an increase in cell number. Cell volumes were up to 50% larger in impeded roots with cell length reduced by 5% in the inner cortex and by 24% in the epidermis.

Lateral root proliferation appears to be a common response to compaction but, again, the precise effects are variable. The loss of apical dominance may cause a proliferation of laterals close to the tip (e.g. Crossett *et al.*, 1975) but this is not a universal response (e.g. Atwell, 1988) and may vary with plant species, the degree of compaction, and the nature of the pore space. Laterals may be initiated near the root apex as a response to compaction and/or as a result of the zone of proliferation being advanced towards the apex.

Morphological and metabolic changes during compaction have rarely been studied simultaneously in root systems. Atwell (1990a, 1990b, 1990c) grew wheat on a deep loamy sand (typic Xeric Psamment) in Western Australia with either a compact soil layer between 10 and 55 cm depth or a deep-tilled (loosened) profile. Table 5.4 summarizes some of the main morphological and metabolic differences found when the plants had five to six leaves and one tiller (46 or 49 days old depending on the season). The number of seminal axes was the same in both treatments, but the elongation rate of the axes was increased from 6 to 17.8 mm d^{-1} by loosening. Cortical cells expanded radially but were 66% shorter to give smaller cell volumes on compacted soil (cf. lupin – Atwell, 1988). Compaction reduced N and K concentrations in the shoot (but not P) and the total concentration of solutes in the roots, but increased the concentrations of soluble sugars in root tissues of the same age. The daily import of solutes into root apices was reduced by 28% on compacted soil, but the amount of recently assimilated dry weight required to produce 10 mm of new root length was twice as high. In roots grown on compacted soil, a greater proportion of the imported substrates went into ethanol-insoluble matter such as cell walls. Radial enlargement, then, involves some cost to the plant in terms of assimilates but allows continued axial elongation into new soil.

Table 5.4 Some morphological and metabolic properties of wheat roots (cv. Eradu) measured at about 46–49 days after sowing for plants grown on a compacted or loosened deep loamy sand

Property	Compact soil	Loosened soil
Seminal axes (no.)	4.5–5.2	4.5–5.2
Rooting depth (cm)	15–30	60–100
Elongation rate of seminal axes (mm d^{-1})	6.0 ± 1.0	17.8 ± 1.5
Cortical cell length 5–15 mm from apex (μm)	89 ± 17	258 ± 42
Root diameter (mm)	1.19	0.86
Root/shoot ratio	3.8	7.7
N concentration (μg g^{-1})	33.7	39.3
P concentration (μg g^{-1})	4.7	4.9
K concentration (μg g^{-1})	28.9	32.8
Sugar concentration for tissue 2 days old (μmol g^{-1} fresh wt)	55	42
Dry matter imported per mm of new axis (μg)	64.2	30.1

The bulk density of both soils was similar (about 1.6 Mg m^{-3}) but the penetrometer resistance of the compacted soil was >2 MPa from 15 to 60 cm whereas that of the loosened soil was <1 MPa. From Atwell, 1990a, 1990b, 1990c.

There has been much interest in the differential ability of crop species and genotypes to penetrate strong soils. Materechera *et al.* (1991) measured the penetration of 22 species into chambers filled with siliceous sand soil with a mean penetration resistance of 4.2 MPa. Generally, the roots of the dicotyledonous plants (with large root diameters) penetrated the soil more than the graminaceous monocotyledons (with smaller diameters) and there was a significant positive correlation between root diameter and elongation over all the species. The initial weight of the seeds was related neither to the radial thickening nor the elongation of the roots. Roots of the dicotyledonous species pea and lupin were also found to penetrate compacted subsoil more readily than wheat and barley (Materechera *et al.*, 1992). In pea and lupin, only 59% of the roots reaching the compact layer penetrated it, whereas in wheat and barley only about 33% penetrated it. Again, the species with thicker roots had better penetration and, more particularly, the extent of thickening relative to the unimpeded control was important. However, despite these results and those of Misra *et al.* (1986) suggesting that differences in maximum root growth pressure (P) might be responsible for differences in root penetration, Clark and Barraclough (1999) found no significant difference in maximum P between four graminaceous species (barley, maize, rice and wheat) and three dicotyledonous species (lupin, pea and sunflower). Dicotyledonous plants had thicker roots than the others (except maize), but the mean maximum P was similar at 0.44 MPa for grasses and 0.41 MPa for dicotyledonous plants.

The ability of roots to penetrate compact soils is, then, not dependent on root thickness or maximum P *per se*, but on the ability of the thickened root both to reduce the impedance ahead of the growing tip and to resist buckling or deflection when encountering strong layers (Barley and Greacen, 1967). Some support for the importance of this latter phenomenon is provided by the work of Clark *et al.* (2000, 2002). They found differences between the abilities of genotypes of lowland rice to penetrate a hard wax layer in columns of sand but no such differences when the roots were grown in columns of uniform, high mechanical impedance. Ability to penetrate strong layers was not correlated with ability to elongate rapidly through uniformly strong media, but the cultivars that penetrated the wax layer best had thicker roots when grown in sand of high impedance. This is consistent with the notion that thicker roots are less likely to deflect or buckle when encountering strong layers.

5.3.3 Roots and soil structure

The strength of a soil is dependent on its cohesive and frictional strengths, both of which vary considerably depending on water content. In many circumstances, roots will first encounter compacted layers when the soil is at field capacity and the soil is at its lowest drained mechanical impedance. When this occurs, mechanical impedance can be approximated by soil dry bulk density. As a general indication (with many exceptions), roots will be severely impeded if bulk densities exceed 1.55, 1.65, 1.80 and 1.85 Mg m^{-3} on clay loams, silt loams, fine sandy loams and loamy fine sands, respectively (Bowen, 1981).

Outside the laboratory, most natural soils have some degree of structure so that the tips of roots will be continually encountering large pores and cracks, and will be affected by them. Under controlled conditions the behaviour of pea, rape and safflower roots growing in cracks was dependent on the strength of the peds, the width of the crack, and the orientation of the crack relative to the preferred geotropic growth direction (Whiteley and Dexter, 1983, 1984). Studies with peds collected from a fine sandy loam soil showed that even when the soil was close to saturation with a penetrometer resistance of about 1.5 MPa, ped surfaces were still a substantial barrier to root penetration. The practical consequence of this is that roots in many subsoils are not uniformly distributed but clumped in cracks (Passioura, 1991). Calculations and models suggest that the flux of water from soil to root, and the root water potential are highly affected by such clumping (Bruckler *et al.*, 1991; Passioura, 1991; Tardieu *et al.*, 1992), with the largest effect in soils with low hydraulic conductivity (i.e. clays) and with high evaporative demand. Dardanelli *et al.* (2003) demonstrated through a revised crop growth model that the main restriction to water uptake in soils with argillic horizons was effected not through effects on rooting depth but by those on root clumping. Cracking of the argillic horizons induced clumping but also decreased the rate of water uptake from lower soil layers. Clumped roots appear to sense dry soil in their immediate vicinity even if the mean soil water potential is high. This finding allows the linking of the effects of soil compaction with those of soil drying, and also suggests that a large part of the effect of soil compaction on plant water behaviour can be accounted for by the effects of root clumping. Further experimental results are needed to substantiate this proposition.

5.4 Soil pores and their contents

Defining all of the characteristics of the pore space that contribute to the growth and functioning of plant roots is very difficult. Size, continuity, tortuosity and quantity all make their contributions in interacting and complex ways. The pore space of most soils is occupied by water and gases in inverse proportions. As a first approximation, many mineral soils are about 50% solids and 50% pores with half of the pore space occupied equally by air and water when drained (Hillel, 1980). Too much water will result in inadequate supplies of O_2 for root respiration and the build up of toxic concentrations of CO_2 and other gases, while too little water may result in reduced plant growth either directly or via increased soil strength. In soils that crack on drying, root systems may also be disrupted by root tearing.

5.4.1 Soil water

Water affects root growth in many ways. Schenk and Jackson (2002), in a survey of water-limited environments, found that the rooting depth of vegetation was more strongly related to mean annual precipitation than to potential evapotranspiration, but with differences between plant growth forms. Maximum rooting depths were positively related to mean annual precipitation for all growth forms except shrubs and trees, with a regression model good enough to explain 62% of the variance observed among rooting depths for arid to sub-humid climates. This relationship is probably a result of restricted infiltration depth limiting the rooting depth of many plant species. In contrast, trees and shrubs had rooting depths ≥2 m in all water-limited environments. As Schenk and Jackson (2002) point out, this result appears to contradict the widely held assumption that root systems become deeper as the environment becomes drier. For a given canopy size, herbaceous plants do, indeed, have deeper maximum rooting depths in drier environments, but because canopy size increases along the rainfall gradient, mean rooting depth also increases. Rainfall also affects shoot and root growth differently so that there are changes in allometric relations with climate. For example, the root:shoot ratio for herbaceous plants increased with aridity while that for shrubs and trees remained unchanged.

Drought and surface drying can reduce the number of nodal axes produced by grasses (Troughton, 1980; Cornish, 1982) and also affect root morphology and root branching. Jupp and Newman (1987) found that soil matric potentials of –2 to –10 MPa caused death of the cortex of perennial ryegrass roots, but death of root tips occurred only at potentials less than –10 MPa. Gradual drying promoted lateral root initiation and growth from the pericycle through the dead cortex, resulting in a three- to five-fold increase in lateral root length. Only a few studies have examined the effects of soil wetting following drought on root growth. In Jupp and Newman's study, re-wetting resulted in root growth by elongation of existing, previously initiated, lateral roots. In contrast, Brady *et al.* (1995) found different responses of seminal and nodal axes in wheat, with re-wetting causing growth of existing lateral roots and proliferation of new root hair zones on seminal axes but initiation and growth of new laterals on nodal axes. Differences in experimental conditions and integrity of the cortex and epidermis may account for these differences in response.

For a given location, numerous publications have confirmed that relatively dry soil conditions can induce plants to develop a more extensive root system. Drying soils can have a range of effects on root systems depending on the exact circumstances including increases in lateral root production, the depth of rooting, and total root length and dry weight when compared with well watered plants. For example, rooting depth of maize grown in columns of potting compost was increased by soil drying but although the distribution of roots was also affected, there was a decrease in total root length and dry weight (Sharp and Davies, 1985). The processes affecting root growth in dry soils are difficult to separate from the simultaneous effects of increased soil strength caused by drying. Sharp *et al.* (1988) grew maize roots in vermiculite (minimizing the influence of strength) maintained at different water potentials. Root extension close to the apex was insensitive to soil water potentials as low as –1.6 MPa,

Plate 2.1. Different types of root: (a) aerial roots of the banyan tree (*Ficus macrophylla* ssp. *columnaris*); (b) aerial roots of the orchid *Ascocenda* cv. Heda; (c) air roots and stilt roots of the grey mangrove tree (*Avicennia marina*); and (d) cluster roots of white lupin (*Lupinus albus*) grown in nutrient solution in the absence of P; the bar is 3 mm. (I am grateful to Dr M. Watt for this photograph.)

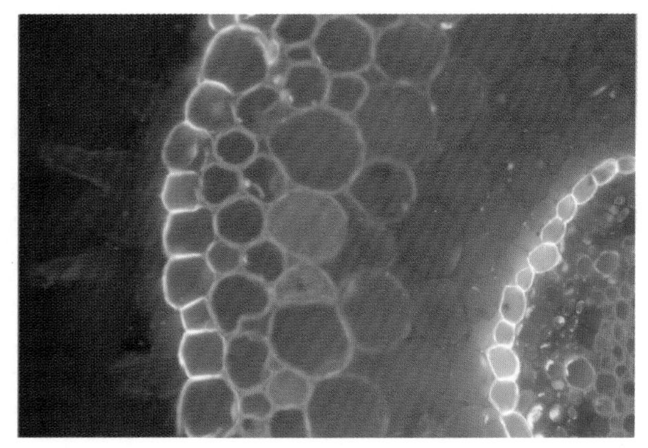

Plate 2.2. Transverse section of a lateral root of maize taken at about 1 mm from a tip that is becoming determinate. The endodermis is mature and the developing hypodermis is already strongly suberized on the anticlinal and outer tangential walls. The epidermis with root hairs is intact. (Reproduced with permission from McCully, *Plant Physiology*; American Society of Plant Biologists, 1995.)

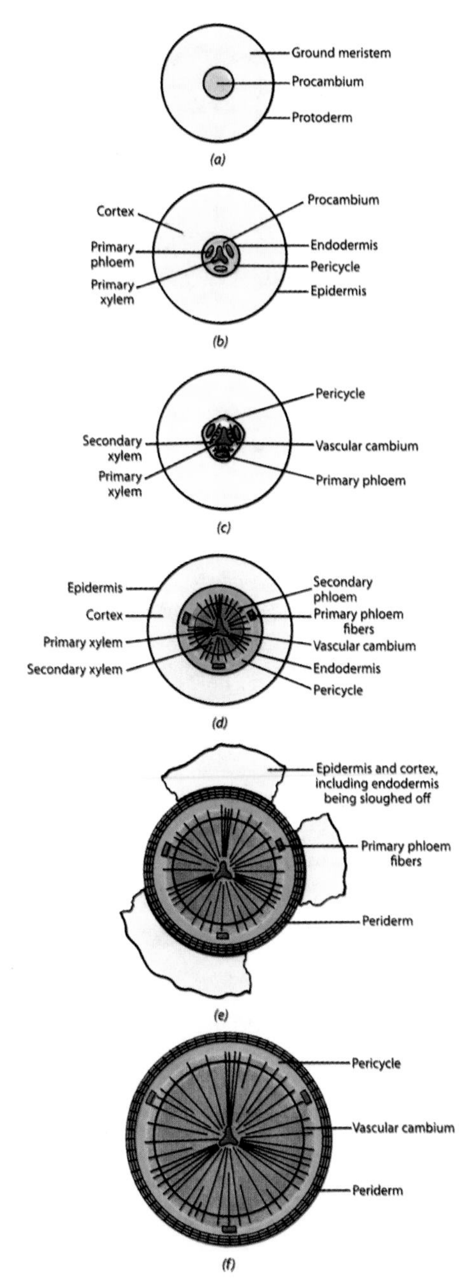

Plate 2.3. Root development in a herbaceous, woody plant showing (a) the primary meristems; (b) the completion of primary root development; (c) the commencement of secondary development with the appearance of vascular cambium; (d) after formation of some secondary phloem and additional secondary xylem; (e) the formation of additional secondary xylem, the appearance of the periderm and the sloughing off of the cortex and epidermis; and (f) the conclusion of secondary development. In (d) to (f), the radiating lines represent rays. (Based on, and reproduced with permission from Esau, *Anatomy of Seed Plants*; John Wiley & Sons Inc., 1977.)

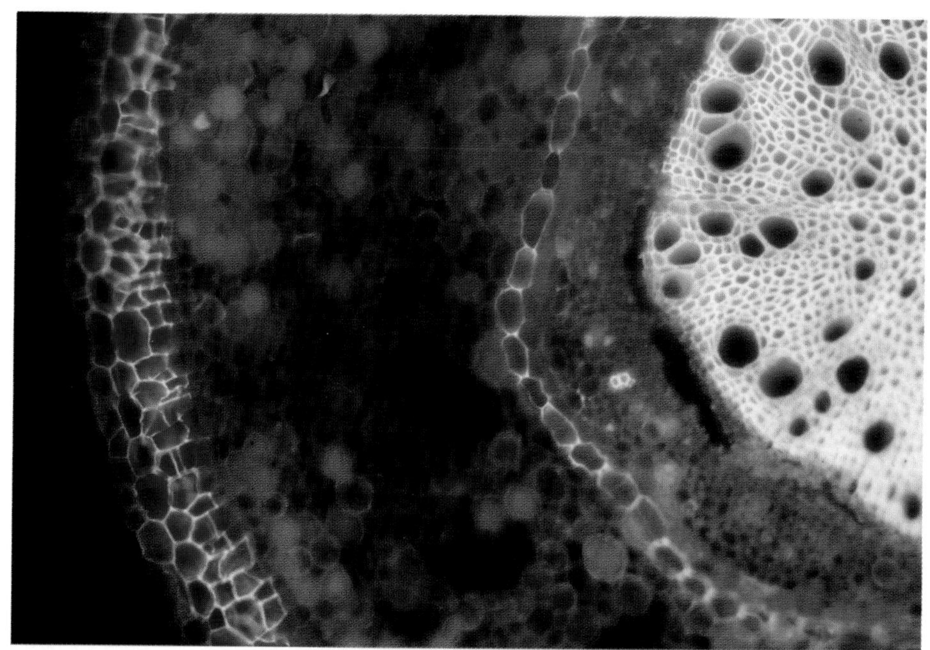

Plate 2.4. Transverse section of a young root of northern catalpa (*Catalpa speciosa*). The considerable secondary growth has displaced the endodermis which has remained intact by new anticlinal divisions. A periderm is forming from cortical cells. (I am grateful to Dr M. McCully for this previously unpublished figure.)

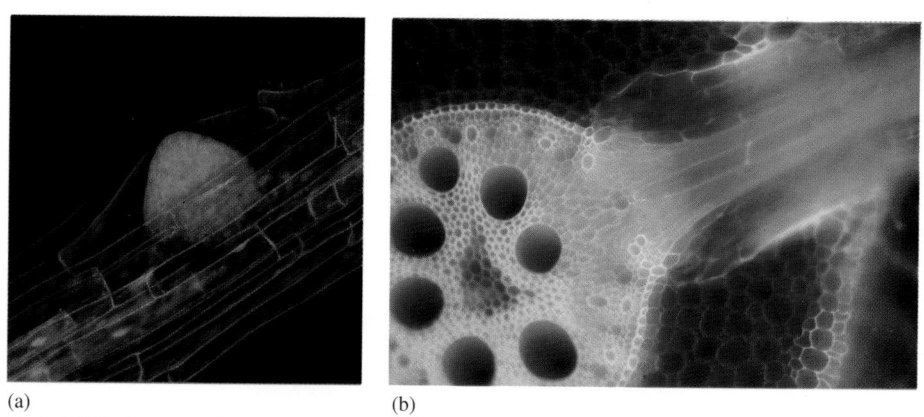

(a) (b)

Plate 2.5. Early stages of lateral root growth. (a) An *Arabidopsis* lateral root emerging through the cortical and epidermal cell layers of the main root axis. Note the buckling of the epidermal cells and gaps in the root surface. The cell walls have been stained red with the fluorochrome propidium iodide, and the green lateral shows phloem unloading of green fluorescent protein (GFP) expressed in response to a sucrose transporter. (I am grateful to Kath Wright, SCRI, Dundee, for this previously unpublished image.) (b) Junction of a young lateral root of field-grown maize with the root axis. In this example, the cortex, hypodermis and epidermis of the root axis press against the lateral to make a tight seal. The large, immature late metaxylem in the lateral adjoins phloem in the root axis and a bed of connecting xylem is developing in the stele of the root axis. (Reproduced with permission from McCully, *Plant Physiology*; American Society of Plant Biologists, 1995.)

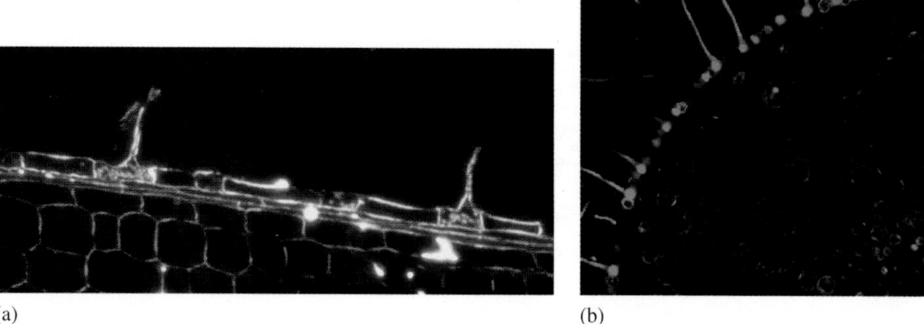

(a) (b)

Plate 2.6. (a) Longitudinal section of a P-deficient nodal root of barley showing red biorefringence of GUS crystals indicating activity of the *Pht1* phosphate transporter gene in the root hairs. (b) Transverse section of a P-deficient nodal root of barley showing GFP expression in green indicating activity of the *Pht1* phosphate transporter gene in the root hairs. (Reproduced with permission from Schünmann *et al.*, *Journal of Experimental Botany*; Oxford University Press, 2004.)

Plate 2.7. Living root cap cells and mucilage on the flanks of a field-grown maize root. The root was sectioned longitudinally and stained with neutral red. Living root cap cells have accumulated this vital stain and mucilage has expanded in the aqueous solution. The dark material is soil. (I am grateful to Dr M. McCully for this previously unpublished figure.)

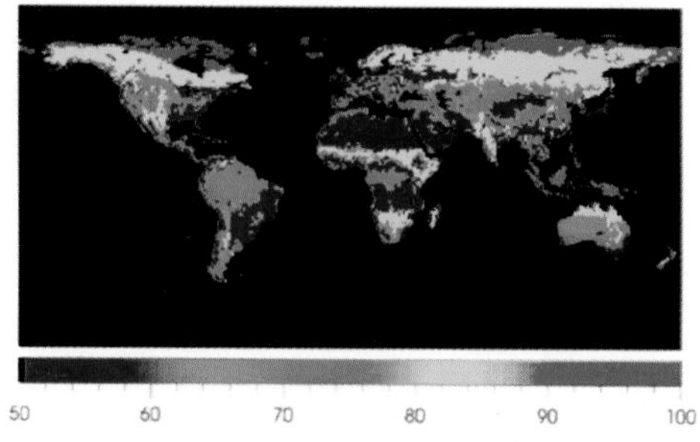

Plate 3.1. A global map of the percentage of root biomass found in the upper 0.3 m of soil. (Reproduced with permission from Jackson *et al.*, *Oecologia*; Springer Science and Business Media, 1996.)

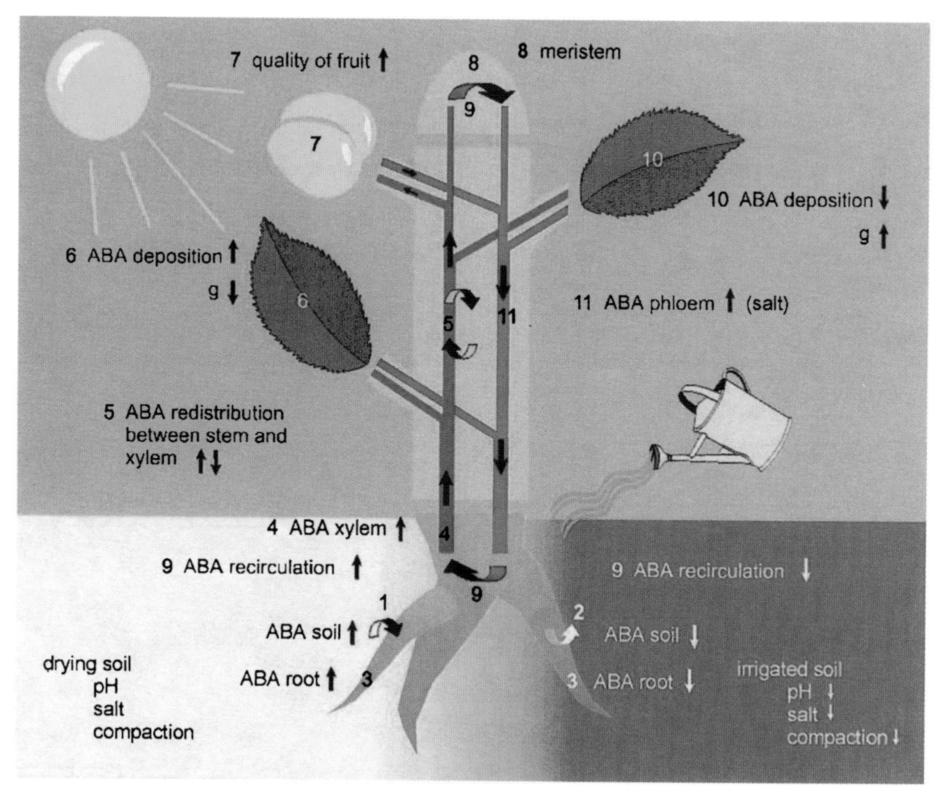

Plate 4.1 Factors influencing the formation and intensity of the ABA long-distance signal. On the left side, plant water shortage is illustrated, while on the right, the plant is well supplied with water. The numbering indicates the number of factors involved. (Reproduced with permission from Sauter *et al.*, *Journal of Experimental Botany*; Oxford University Press, 2001.)

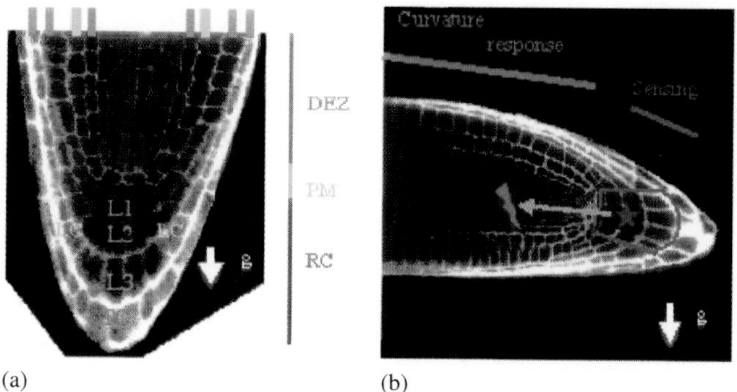

(a) (b)

Plate 5.1. Zones of the *Arabidopsis thaliana* root tip involved in early phases of gravitropism. (a) Confocal image of a propidium iodide-stained root tip showing the root cap (RC), the promeristem (PM), and the distal elongation zone (DEZ). The root cap comprises three layers of columellar cells at the centre (L1, L2 and L3), lateral cells (LC), and tip cells (TC). (b) Confocal image of a propidium iodide-stained gravistimulated root tip showing the site of gravity sensing (columella of the root cap – outlined in blue and marked by a blue star), and the site of graviresponse (the DEZ marked with a lightning sign). The green arrow illustrates the need for signal transmission between the root cap and the elongation zone. The gravity vector (g) is represented by a white arrow. (Reproduced with permission from Boonsirichai *et al.*, *Annual Review of Plant Biology*; Annual Reviews, 2002.)

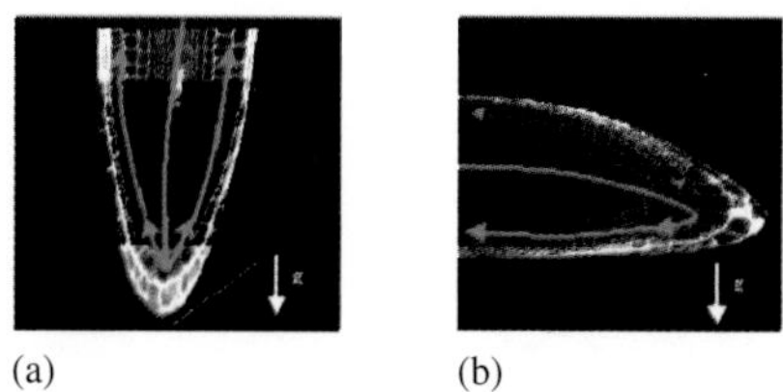

Plate 5.2. The 'fountain' model of auxin transport in roots. Auxin mainly synthesized in young shoot tissues is transported via the vasculature to the young root tip where it is redistributed to more peripheral tissues. It is then transported basipetally to the elongation zone where it regulates cell expansion. (a) When a root grows vertically downward, auxin redistribution to the outer tissues is symmetrical. (b) On gravistimulation, auxin is preferentially transported toward the bottom side of the tip resulting in the formation of a lateral auxin gradient; this gradient is then transmitted to the elongation zone where it is partly responsible for the differential growth that underlies gravitropic curvature. Green arrows represent the direction of auxin transport (width is a relative representation of auxin fluxes), and white arrows represent the gravity vector (g). (Reproduced with permission from Boonsirichai *et al.*, *Annual Review of Plant Biology*; Annual Reviews, 2002.)

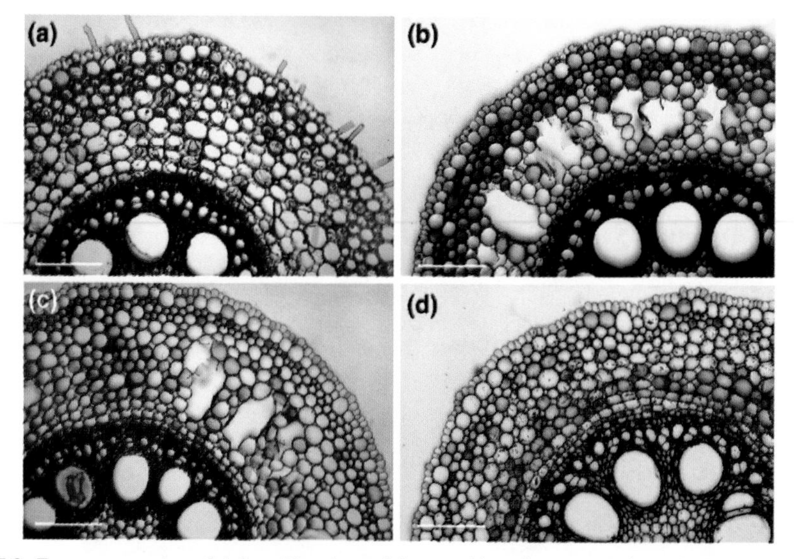

Plate 5.3. Transverse section of 4-day-old maize nodal roots: (a) well-oxygenated root lacking aerenchyma; (b) hypoxic root with lysigenous aerenchyma in the mid cortex; (c) well-oxygenated root treated for 4 days with okadaic acid, an inhibitor of protein phosphatases, with aerenchyma beginning to form; and (d) hypoxic root treated for 4 days with EGTA to complex Ca^{2+}, thereby eliminating aerenchyma formation. Scale bars: 0.25mm. (Reproduced with permission from Drew *et al.*, *Trends in Plant Science*; Elsevier, 2000.)

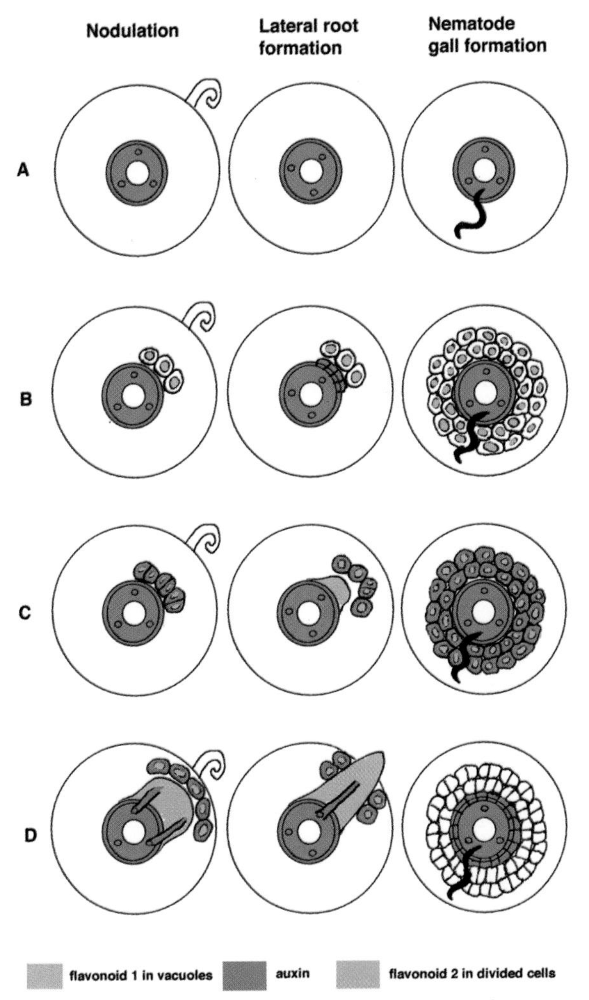

Plate 6.1. Spatial and temporal overlaps and differences in the accumulation of plant compounds induced in response to nodule-inducing rhizobia, plant signals which induce lateral roots, and gall-inducing nematodes. This model suggests that by varying the cell specificity of responses, a plant can form different organs even though the same genes and physiological changes are involved in all cases. (A) The plant perceives a signal from rhizobia (symbolized by curled root hair), from the pericycle (cells from which lateral roots are derived) or from an invading nematode (black worm inside root). (B) Target cells in the cortex which later divide accumulate flavonoid 1 in their vacuoles. (C) Cortex cells targeted for division accumulate auxin probably because flavonoid 1 inhibits auxin breakdown. (D) After division of cells in the forming primordium, a different flavonoid (formononetin) accumulates which stimulates auxin breakdown. The auxin promoter is no longer active. (Reproduced with permission from Mathesius, *Plant and Soil*, Springer Science and Business Media, 2003.)

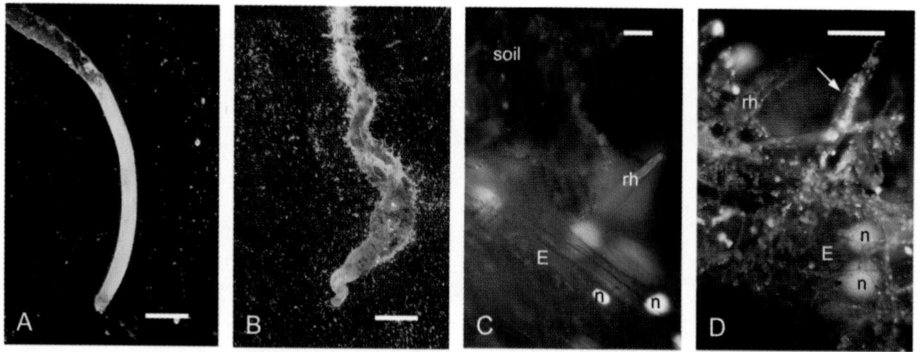

Plate 6.2. Fast-growing (A and C) and slow-growing (B and D) nodal roots of field-grown wheat (cv. Janz). A and B are whole mounts of root apices (bar = 250 μm). The elongation zone is white on the fast-growing root (A), and short and distorted on the slow-growing root (B). C and D are tangential hand sections from approximately 5 mm from the root tip of both root types, stained with PAS reaction and DAPI, and excited with UV epifluorescence to observe bacteria *in situ*. Few bacteria are visible on the fast-growing root (C, bar = 20 μm), while many bacteria (bright spots, some indicated by an arrow) are visible on the slow-growing root (D, bar = 40 μm). E = epidermis, n = nucleus, and rh = root hair. (Reproduced with permission from Watt *et al.*, *Functional Plant Biology*; CSIRO Publishing, 2003.)

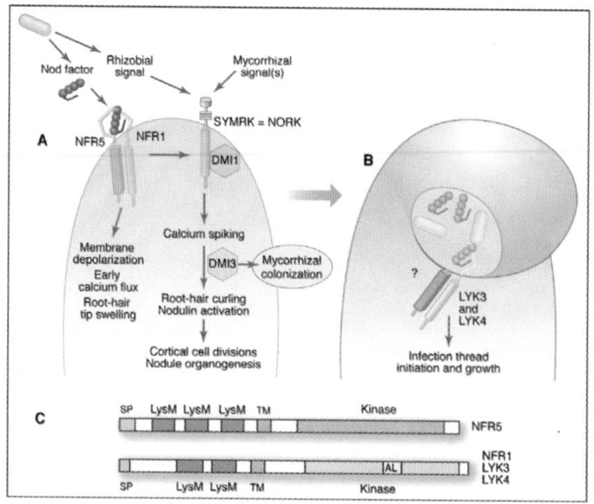

Plate 6.3. The rhizobium infection process. (A) Symbiotic rhizobial bacteria release Nod factors that are perceived by two transmembrane receptors initiating rapid calcium influx and swelling of root hair tips. Simultaneous activation of the NORK/DMI 1 complex results in plant responses to both bacterial and fungal symbionts. (B) Rhizobial bacteria entrapped in a curling root hair. For rhizobia to enter the root hair and to initiate formation of infection threads and nodulation, the Nod factors must be recognized by highly specific plant receptors (e.g. LYK 3 and LYK 4). (C) The chemical domains of some receptors. (Reproduced with permission from Cullimore and Dénarié, *Science*; illustration by Katharine Sutliff, Copyright AAAS, 2003.)

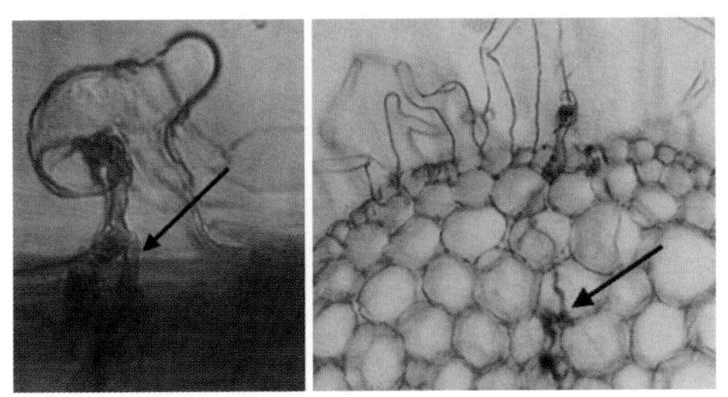

Plate 6.4 Colonization of host cells by rhizobia causes root hair curling (left figure) and the formation of an infection thread (indicated by arrow) which originates as an intrusion of the host cell wall and propagates from cell to cell as a transcellular tunnel in the root cortex (right figure). (Reproduced with permission from Brewin, *Biologist*; Institute of Biology, 2002.)

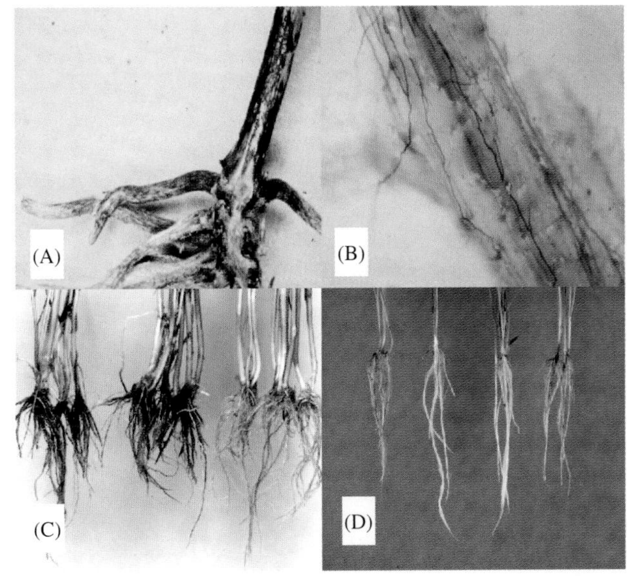

Plate 6.5. Take-all of wheat caused by *Gaeumannomyces graminis* var. *tritici*. (A) Stem base covered by black mycelium and showing blackened roots and internal mottled black and grey tissue. (B) Pigmented runner hyphae on the surface of a fine root. (C) Take-all suppression in one field by transfer of soil from another field that had undergone take-all decline (right) compared with the unamended soil (left) or soil transferred from a non-cropped site near the cropped site (centre). (D) Take-all in cv. Madsen winter wheat (far left) and Penawawa spring wheat (far right) compared to two collections of *Dasapyrum villosum* (centre).(Reproduced with permission from Cook, *Physiological and Molecular Plant Pathology*; Elsevier, 2003.)

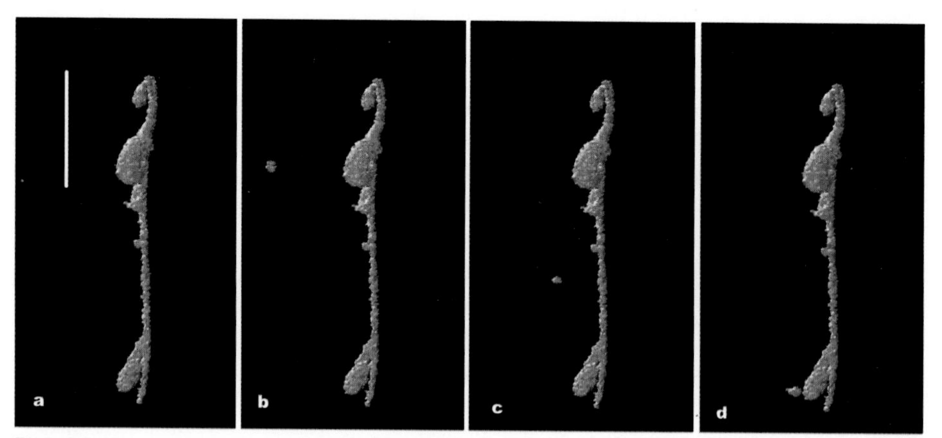

Plate 6.6. X-ray tomographic images of the sequential movement of a neonatal *Sitona lepidus* larva through soil towards the lower clover root nodule. Location at (a) 0 h, (b) 3 h, (c) 6 h, and (d) 9 h. The larva was subsequently recovered from the nodule when the column was dismantled; white bar = 10 mm. (Reproduced with permission from Johnson *et al.*, *Ecological Entomology*; Blackwell Publishing, 2004.)

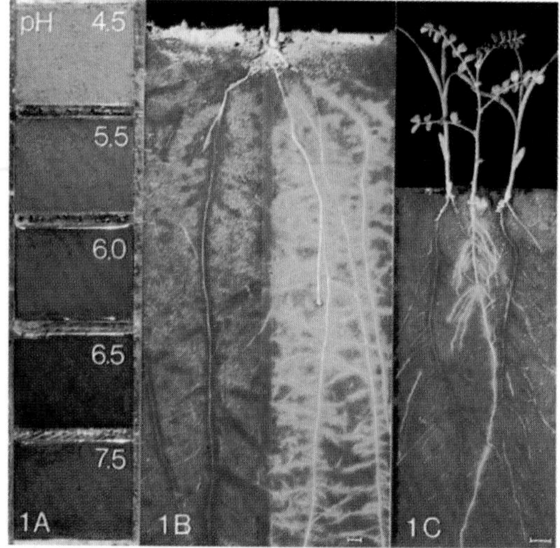

Plate 7.1. Effects of N source and plant species on rhizosphere pH. The soil was infiltrated with agar containing bromocresol purple. (A) pH calibration standards; (B) split root experiment with maize which received 67 mg N kg^{-1} soil as either calcium nitrate (left) or ammonium sulphate (right); and (C) mixed culture of sorghum and chickpea supplied with 83 mg N kg^{-1} soil as calcium nitrate. Bar represents 10 mm. (Reproduced with permission from Marschner *et al.*, *Z. Pflanzenernaehr. Bodenk*; Wiley-VCH Verlag GmbH, 1986.)

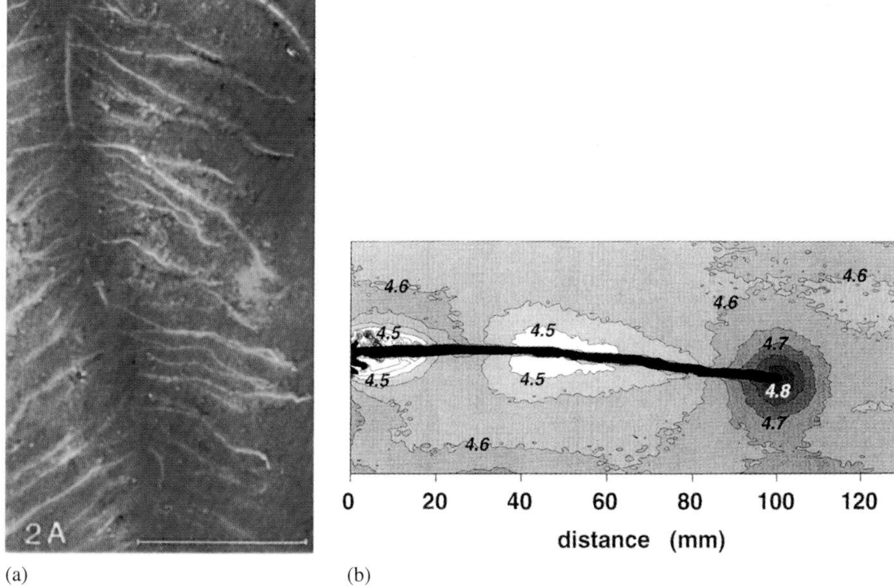

(a) (b)

Plate 7.2. pH changes around roots. (a) Changes in rhizosphere pH of a root axis and associated lateral roots of a 40-day-old maize plant growing in soil of pH 6.0 and supplied with 66 mg N kg^{-1} soil as calcium nitrate. Colours/pH as for Plate 7.1. Bar represents 10 mm. (Reproduced with permission from Marschner *et al.*, *Z. Pflanzenernaehr. Bodenk*; Wiley-VCH Verlag GmbH, 1986.) (b) Map of pH values around the root of a 7-day-old maize seedling after embedding it in an agarose sheet containing bromocresol green at a pH of 4.6 for 120 minutes. The apical region released hydroxyl equivalents while basal regions released protons. (Original research by Jaillard *et al.* (1996); reproduced with permission from Hinsinger, *Advances in Agronomy*; Elsevier, 1998.)

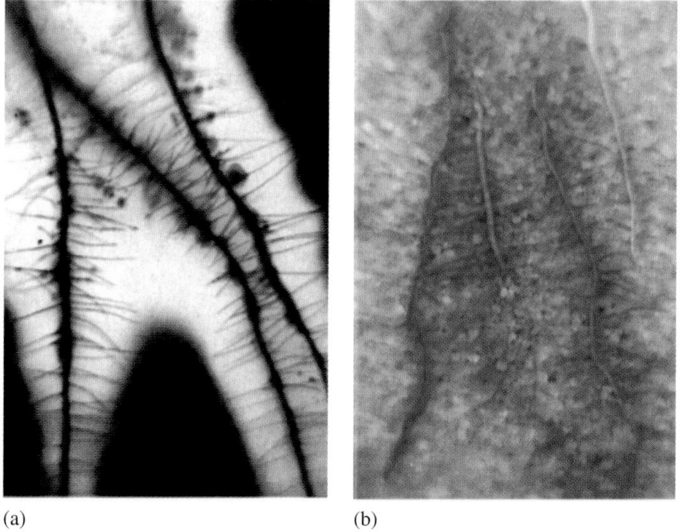

(a) (b)

Plate 7.3. Oxidation of iron around roots of rice. (a) Roots embedded in 0.2% agar containing ferrous sulphate and sodium sulphide giving a black precipitate of ferrous sulphide which has been rendered clear by oxidation close to the root. (b) Roots grown in sand with a ferrous sulphide nutrient medium. (Reproduced with permission from Trolldenier, *Z. Pflanzenernaehr. Bodenk*; Wiley-VCH Verlag GmbH, 1988.)

Plate 9.1. A wheat root growing through the pore created by a lucerne root which is now decaying.

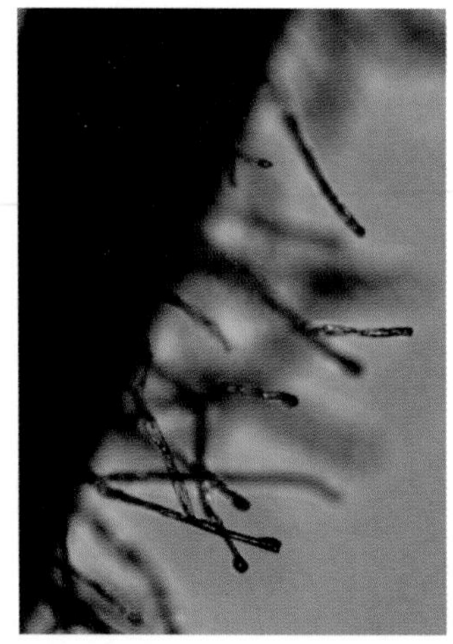

Plate 9.2. Sorghum root hairs exuding yellow drops of sorgleone. (I am grateful to Leslie Weston for this previously unpublished photograph.)

but was inhibited in more basal regions so that the length of the elongation zone decreased as water potential also decreased. Roots grown at low water potential were also thinner; note, though, that increasing strength (which is normally concurrent with soil drying) tends to result in thicker roots (section 5.3.2). The shorter elongation zone was a consequence of the slower growth rate of water-stressed roots, which resulted in shorter cortical cells and a reduction in the rate of cell supply to the cortical files (Fraser *et al.*, 1990).

Irrigation of crop plants induces significant changes in the growth and distribution of root systems which have important consequences for both production and crop quality. Such studies have been particularly important in the horticultural sector where fruit quality is a major issue. Flood irrigation has given way to drip irrigation in much of the industry, with several studies examining consequences for root systems. For example, Araujo *et al.* (1995) measured root distributions of grapevines grown in California that had similar water status and shoot growth patterns when irrigated either by furrow or drippers. The root system of drip irrigated vines was a highly branched mass of fine, fibrous roots with nearly 50% of roots in the upper 0.2 m, and most showing a horizontal growth pattern. In contrast, only a few roots were found in the upper 0.2 m of soil in furrow irrigated vines (about 12%) and most roots were oriented vertically and grew towards deeper layers (Fig. 5.8). The proliferation of roots close to drippers has been observed in many studies, especially as nutrients are also supplied in the water. In Canada, drip irrigated apple trees

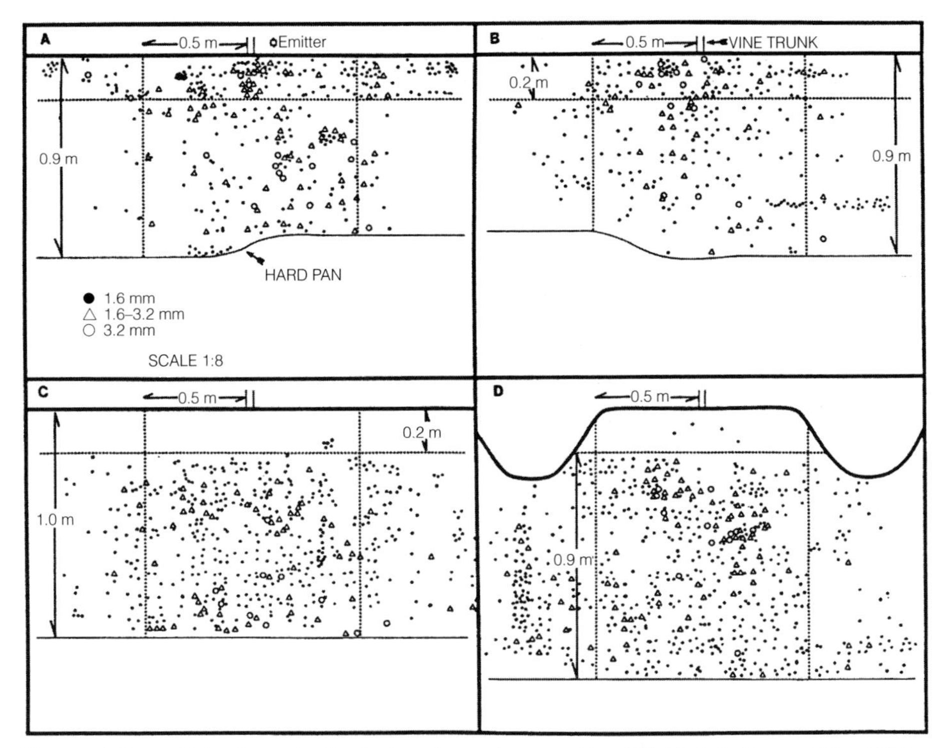

Fig. 5.8 Effects of irrigation on the distribution of grapevine roots: (A and B) drip irrigation; and (C and D) furrow irrigation. Trenches were dug perpendicular to the row (transverse views B and D) and along the row (lateral views A and C) at 15 cm from the trunk. (Reproduced with permission from Arujo *et al.*, *Scientia Horticulturae*; Elsevier, 1995.)

fertigated annually had nearly 50% of their roots within 30 cm depth and lateral distance from the emitter whereas trees irrigated by microjet had roots more widely distributed over a soil volume that extended to 60 cm depth and 90 cm lateral distance from the trees (Neilsen *et al.*, 2000). In some crops, though, the effects are less pronounced. Machado *et al.* (2003) investigated the influence of emitter depth (0, 20 and 40 cm) on the root distribution of field-grown tomatoes in Portugal and found only slight differences between treatments with about two-thirds of roots almost evenly distributed in the upper 50 cm. The non-uniform distribution of water in the root zone can have beneficial effects on fruit yield and quality, and irrigation accompanied by partial root zone drying is now being explored commercially (see section 4.2.2).

5.4.2 Soil aeration

This topic is ignored compared to others, with little research or progress in recent decades. It is well known that for most plants to grow in soil, part of the pore space must be gas-filled. This space allows the supply of O_2 to roots to maintain respiration, and the removal of CO_2 from the root, and if these processes do not occur fast enough then root growth is restricted (e.g. Fig. 5.7b). While there is some evidence that CO_2 at high concentrations (although this is often accompanied by other toxic products) may be deleterious to root growth, the main effect of poor aeration is the lack of O_2. Soil aeration status has been characterized using several indices including gas-filled porosity, the flux of oxygen to a platinum electrode (the so-called 'oxygen diffusion rate', ODR), oxygen flux, and the O_2 concentration or partial pressure. But as Cook and Knight (2003) state 'there has not been a good theoretical basis for investigating these indices' and 'the concept of soil aeration has been based on a correlation between plant performance and these various indices of the O_2 status of the soil'.

Identifying the soil water content at which O_2 supply to plants becomes limiting has been a major preoccupation. For example, Blackwell and Wells (1983) grew seedlings of oats (a plant species that is generally regarded to be tolerant of waterlogging) in columns of waterlogged soil at temperatures similar to those experienced by such crops in a southern UK autumn to determine the oxygen fluxes at which root elongation was slowed and ceased. Elongation of individual roots was slower in waterlogged than freely drained soil at oxygen flux densities <56 ng cm^{-2} min^{-1}, at ODR <121 ng cm^{-2} min^{-1}, and at concentrations of dissolved $O_2 \leq 15\%$ v/v at 10°C (Fig 5.9). Elongation ceased when oxygen flux and dissolved O_2 was close to zero and when ODR was 7.8 ± 1.2 ng cm^{-2} min^{-1}. Such results depend on the temperature which affects the rate of many metabolic processes. Blackwell and Wells (1983) also found that roots whose elongation was reduced by oxygen shortage were thicker (0.66 mm diameter compared with 0.39 mm), possibly related to an increase in endogenous ethylene concentration. Letey (1985) introduced the idea of the non-limiting water range (NLWR) in an attempt to take account of the interactions between water, aeration and mechanical properties; soils with 'good' physical conditions had a high NLWR and 'poor' soils a low NLWR. Such indicators are unlikely to be constants, though, as plant demand for oxygen is temperature-dependent as well as influenced by pore space. Cook and Knight (2003) used a model that coupled factors influencing O_2 diffusion to plant roots at the microscale to O_2 diffusion through soil at the macroscale. Their analysis demonstrated that ODR, O_2 concentration, and air-filled porosity were inter-related and that a

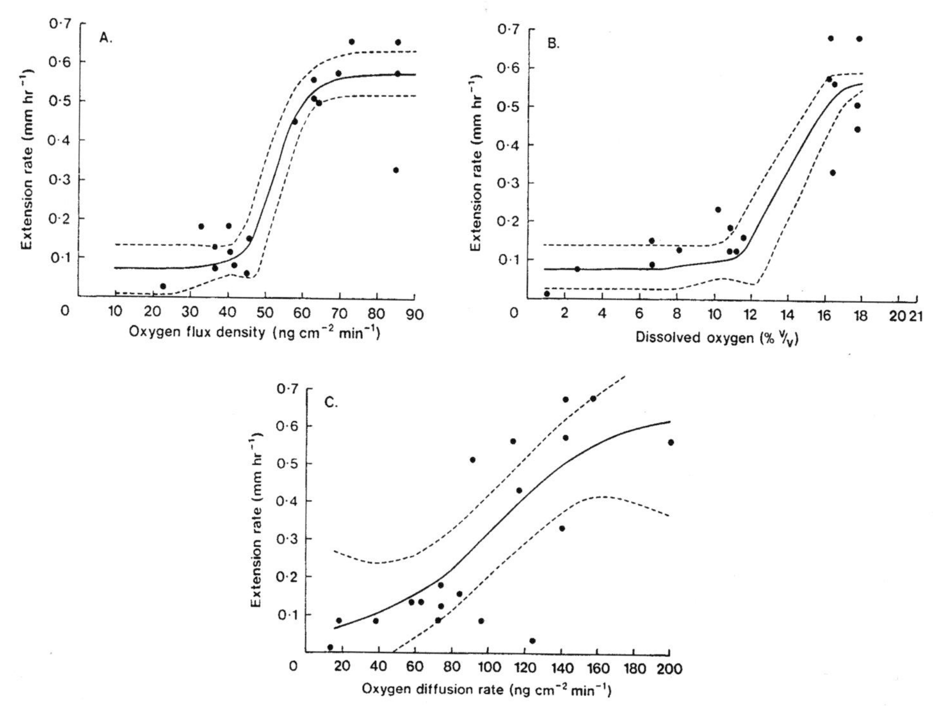

Fig. 5.9 Relationships between the extension rates of individual waterlogged oat roots and (A) oxygen flux density, (B) dissolved oxygen, and (C) oxygen diffusion rate. The dashed lines are 90% confidence intervals for the best fit logistic function. (Reproduced with permission from Blackwell and Wells, *Plant and Soil*; Springer Science and Business Media, 1983.)

single critical value for any of these is unlikely. Indeed, if a constant critical value exists for one of these indices, then it cannot exist for the others.

5.4.3 Waterlogging and aerenchyma

Waterlogging has multiple detrimental effects on plants because of disruption to gas flow below ground. It is a common constraint to the production of agricultural crops, affecting about 10% of the global land area. Aquatic and marsh plants have adapted to cope with such conditions but many plants are very vulnerable. There are, though, differences both between species (e.g. Sitka spruce and Lodgepole pine; Coutts and Philipson, 1978a, 1978b) and within a genus (e.g. *Hordeum*; Garthwaite *et al.*, 2003) in tolerance to waterlogged conditions.

The severity of the effects of waterlogging on roots depends on the developmental stage at which it occurs, the duration of the event, and other environmental factors such as temperature which affect the demand for oxygen by roots and microbes, and the production of chemicals (some of which are toxic to plants) in reduced soils. In cropping regions it is common to introduce drainage systems to alleviate such conditions and ensure aerated soil. Ellis *et al.* (1984) followed the effects of mole drainage on root growth of winter wheat sown on a clay soil (Table 5.5). Root growth beneath the upper 0.1 m was diminished by a water-table at 0.2 m depth, although some root axes continued to grow to 0.75–1.0 m depth. However, these effects were transient and after the water-table disappeared in April, the length of the

Table 5.5 Effect of drainage on root length (cm cm^{-3}) of winter wheat roots grown on a clay soil; the crop was sown on 5–8 October 1979

Depth (cm)	29 January		14 March		8 April		6 May	
	Undrained	Drained	Undrained	Drained	Undrained	Drained	Undrained	Drained
0–10	1.9	2.2	4.7	3.3	7.5	7.5	3.6	3.9
10–20	0.3	0.5	0.5	1.2	0.5	2.4	0.4	0.5
20–30	0.1	0.2	0.1	0.7	0.2	1.5	0.4	0.7
30–40	0.1	0.1	0.1	0.6	0.1	1.3	0.3	0.4
40–50	0.1	0.1	0.1	0.4	0.1	0.6	0.2	0.5
50–75			<0.1	0.1	<0.1	0.2	0.1	0.1
75–100					<0.1	0.1	<0.1	<0.1

From Ellis *et al.*, 1984.

root systems and their distributions with depth became very similar. In contrast, the effect of winter waterlogging persisted in the shoot resulting in 10% decrease in grain yield at maturity. In pot experiments, where intense anaerobic conditions can be generated quickly, even short periods of waterlogging can affect wheat root systems substantially. Malik *et al.* (2002) found that 3 days of waterlogging was sufficient to stop seminal root growth by causing death of root apices and preventing the initiation of new lateral roots, and that these did not recover to control values when waterlogging exceeded 7 days. In contrast, nodal roots resumed elongation after drainage and their dry mass recovered to that of drained controls after 7 days even when they had been in waterlogged conditions for 14 days. Similar differences in the response of seminal and nodal roots have been reported by others (Trought and Drew, 1980; Thomson *et al.*, 1992), but while the nodal roots recover, the severe reduction in the size of the seminal root system has long-lasting effects on overall plant growth (Malik *et al.*, 2002). The different abilities of these two root types to acclimate to waterlogging appear to be related to their abilities to form aerenchyma. In cereal crops, the growth of nodal roots with aerenchyma in the upper, well-aerated zone of the soil profile may substantially reduce the effects of waterlogging on crop yield (Setter and Belford, 1990).

Aerenchyma, plant tissue containing enlarged gas spaces exceeding those found as intracellular spaces, occurs in many plants and is formed either as part of the normal developmental process, or in response to stress, particularly hypoxia as a consequence of waterlogging. It provides the plant with an alternative way for its root tissues to obtain O_2 because the interconnected lacunae provide an internal aeration system that transfers O_2 from the atmosphere to the root rather than relying on the soil atmosphere. Two types of aerenchyma have been identified: (i) schizogenous aerenchyma forms as a result of differential cell growth and subsequent cell separation and is common in wetland species like *Rumex*; and (ii) lysigenous aerenchyma forms after cells in the root cortex die and disappear to leave gas-filled spaces (Evans, 2003). Lysigenous aerenchyma formation is important in many crops plants including maize, wheat, barley and rice but while its formation is induced by flooding in maize and wheat, it forms constitutively (i.e. without any external stimulus) in rice. The end result of aerenchyma formation is to leave a root where the gas spaces are separated by lines of cells bridging the space between the stele and epidermis, like spokes of a wheel (Plate 5.3). Aerenchyma formation increases porosity above that resulting from the usual intercellular spaces to form typically some 15–50% of the root volume (Colmer, 2003a). Table 5.6 shows values

Table 5.6 Porosity (intercellular gas spaces and aerenchyma as a percentage of tissue volume) in roots of selected wetland and non-wetland plant species grown in either drained/aerated or O_2-deficient rooting media

		Porosity (%)	
Species	Root type	Drained/aerated	O_2-deficient
Wetland species			
Oryza sativa	Adventitious roots	15–30	32–45
Typha domingensis	Adventitious roots	10–13	28.34
Phragmites australis	Adventitious roots	43	52
Juncus effusus	Adventitious roots	31–40	36–45
Carex acuta	Adventitious roots	10	22
Rumex palustris	Adventitious roots	15–30	32–45
Plantago maritima	Entire root system	8	22
Ranunculus flammula	Entire root system	9–11	30–37
Non-wetland species			
Triticum aestivum	Adventitious roots	3–6	13–22
Hordeum vulgare	Adventitious roots	7	16
Zea mays	Adventitious roots	4	13
Festuca rubra	Entire root system	1	2
Vicia faba	Entire root system	2	4
Pisum sativum	Entire root system	1	4
Brassica napus	Entire root system	3	3
Trifolium tomentosum	Entire root system	7	11

Adapted from Colmer, 2003a.

for whole root systems or adventitious roots for both wetland and non-wetland species; there is substantial variation both within species, for genotypes within a species, and between root types within a genotype (e.g. seminal compared with nodal roots in cereals). Some plant species (e.g. oilseed rape, *Brassica napus*) do not form aerenchyma (Voesenek *et al.*, 1999), and in crops of oilseed rape subjected to waterlogging, roots may grow upward into better aerated surface soil to reduce the adverse effects (Setter and Belford, 1990).

The processes involved in aerenchyma formation are complex and several different mechanisms may have evolved (Colmer, 2003a; Evans, 2003). Maize has been extensively studied, and in this species hypoxia (and other stresses such as mechanical impedance and N and P shortage) stimulates the production of ethylene which activates a signal transduction pathway involving phosphoinositides and Ca^{2+}, inducing cell death and aerenchyma formation (Drew *et al.*, 2000). It appears that in this and other species, cell death is programmed rather than simply uncontrolled necrosis, but many questions remain about how, for example, dead cell contents and cell walls are removed to leave spaces (Evans, 2003).

The network of spaces is continuous between roots and shoots allowing long-distance transport of gases. Until relatively recently most attention has been directed at the transport of O_2 to the roots but it is now appreciated that the system also transports CO_2, ethylene and methane from waterlogged soils to the shoot and thence the atmosphere (Colmer, 2003a). For example, Butterbach-Bahl *et al.* (1997) estimated that about 90% of methane emitted from two rice cultivars grown in paddy fields in Italy to the atmosphere was plant-mediated and Shannon *et al.* (1996) determined that methane transport through *Scheuchzeria palustris* contributed 64–90% of the net methane efflux from peatland in Michigan, USA.

Oxygen transported to the roots in aerenchyma is either consumed by adjacent cells, or diffuses towards the root apex, or diffuses out of the root into the rhizosphere (radial oxygen loss, ROL; Armstrong, 1979). Losses by ROL can be large, but many wetland plants develop a barrier to ROL in basal zones that enhances longitudinal O_2 diffusion towards the apex. The barrier is induced by growth in stagnant conditions in some species but is constitutive in others (Colmer, 2003a), and results in an increased resistance to O_2 diffusion across the hypodermis/exodermis. In rice, upland, paddy and deep-water genotypes were all found to develop a tight barrier to ROL in basal zones when grown in stagnant solutions in addition to increasing the porosity of their roots (Colmer, 2003b). Colmer (2003a, 2003b) suggested that the plasticity of root form may facilitate rice growth in diverse environments that differ markedly in the intensity of waterlogging, and that lateral roots that remain permeable to O_2 may provide the surfaces for exchange of gases and other substances between roots and the rhizosphere. Some confirmation of the latter suggestion is provided by the calculations of Kirk (2003), which confirm the importance of impermeable but aerenchymatous axes with short, fine, gas-permeable laterals in providing sufficient absorbing surface and loss of O_2 to the rhizosphere to nitrify sufficient NH_4^+ to NO_3^- to allow a plant to absorb 50% of its N as NO_3^-, and to oxidize toxins such as Fe^{2+}.

5.5 The soil chemical environment

5.5.1 Plant nutrients

Fertilizer applications to many soils produce substantial increases in shoot growth and yield while responses below ground are often less marked. For example, Welbank and Williams (1968) found that applications of 50 and 100 kg N ha^{-1} to spring barley at Woburn, UK increased shoot dry weight by 13% and 52%, respectively, relative to the control (0 kg N ha^{-1}) at maturity while root dry weight was increased by only 8% and 16%; there was no further significant response to 150 kg N ha^{-1}. The most obvious effect was that the proportion of total plant weight as roots decreased both with time and increasing nitrogen fertilization (Table 5.7). Similarly, studies with winter and spring wheats, oats and barley found that application of nitrogen fertilizer generally resulted in smaller root systems in the early spring and with root systems that were shorter relative to their dry weight (i.e. specific root length was lower). Later, although N applications produced larger root systems, root mass as a proportion of total plant mass was lower (Table 5.7) (Welbank *et al.*, 1974). On the very P-deficient soils on northern Syria, Brown *et al.* (1987) found that in contrast to Welbank's results, applications of P and N fertilizer increased root mass of two varieties of barley at all samplings, but as with Welbank's results, the effect on shoots was greater so that root weight as a proportion of total plant weight was smaller (Table 5.7). Other studies have suggested that different genotypes of a crop may respond differently to fertilizer applications. Mackay and Barber (1986) grew two genotypes of maize on a silt loam at Lafayette, Indiana, USA and found that N fertilizer (227 kg N ha^{-1}) had no effect on the root length of Pioneer 3732 but increased growth throughout the life cycle of the prolific hybrid B73 × Mo17. At mid-silk, root length was 48% greater in the N-fertilized crop and this,

Table 5.7 Effects of fertilizer applications on root dry weight as a proportion of total plant dry weight

(a) Spring barley

Days after sowing	Fertilizer kg N ha^{-1}				
	0	50	100	150	SE
46	0.417	0.362	0.280	0.300	0.010
60	0.208	0.195	0.171	0.169	0.014
74	0.188	0.151	0.124	0.120	0.010
95	0.133	0.101	0.079	0.081	0.011
130	0.081	0.080	0.064	0.067	0.003

(b) Four cereals (basal = 25 kg N, 38 kg P, 78 kg K and 76 kg Mg ha^{-1} in autumn; +N = basal plus 100 kg N ha^{-1} in spring)

	Winter wheat		Spring wheat		Spring oats		Spring barley		
Date	Basal	+ N	Basal	+ N	Basal	+ N	Basal	+ N	SE
5 May	0.381	0.222	0.349	0.297	0.268	0.218	0.251	0.210	0.024
2 June	0.188	0.115	0.252	0.134	0.297	0.158	0.242	0.109	0.018
30 June	0.129	0.085	0.145	0.094	0.184	0.115	0.125	0.084	0.009

(c) Two varieties of winter barley (–fert = no fertilizer; +fert = 12 kg P plus 40 kg N ha^{-1})

	Arabic abiad		Beecher	
Growth stage	–fert	+fert	–fert	+fert
Stem elongation	0.325	0.288	0.400	0.250
Anthesis	0.206	0.138	0.135	0.163

Data for (a) and (b) from Welbank *et al.* (1974) and data for (c) from Brown *et al.* (1987).

coupled with continued root growth after mid-silk, may contribute to the greater growth of this genotype under a high N fertilizer regime. In contrast, though, Durieux *et al.* (1994) found no consistent differences in either root length or weight between three maize hybrids (Pioneer 3320, I202 × Mo17, and I117 × B73) and while increasing N application (56, 140 and 224 kg N ha^{-1}) decreased root length and weight at maturity, it had no significant effect at 20 days before silking or at silking.

Nutrients are rarely, if ever, uniformly distributed in soils and the response of roots to the heterogeneity of their distribution has been of considerable interest. In an early study of the response, Nobbe (1862) placed 'enriched' soil in various patterns in cylinders and found that the roots tended to follow the enriched soil patterns. Anghinoni and Barber (1980, 1988) studied the effects of $H_2PO_4^-$ and NH_4^+ placement on maize roots. For both nutrients, mixing the same amount of nutrient with different fractional volumes of soil affected neither total root weight nor length, but root lengths were always greater, and root radius smaller, in the fertilized fraction of the soil. The effect of NH_4^+ on root proliferation was described by the simple power function $y = x^{0.70}$ where y is the proportion of the total root length in the NH_4^+-treated soil and x is the proportion of the soil volume treated with NH_4^+. An identical relation was also found by Yao and Barber (1986) with placement of P for maize, soyabean and wheat facilitating calculations of optimal fertilizer placement to maximize the efficiency of fertilizer application (Barber, 1995). Intense nutrient concentrations such as those found in fertilizer granules or bands can either stimulate or decrease root growth depending on the form of fertilizer and time after application. Passioura and Wetselaar (1972) found that wheat roots were absent from zones of soil fertilized with either urea or ammonium sulphate for 2 weeks, presumably due to the low osmotic potential in the fertilized zones. Thereafter roots proliferated in the zone fertilized with ammonium sulphate at 4 and 8 weeks, but while roots proliferated at the periphery of the zone fertilized with urea, they were absent from the centre of the zone.

As indicated above, a common response of roots to a nutrient-rich patch is root proliferation with suppression elsewhere; however, this is by no means a universal response, as confirmed by a review by Robinson (1994) in which one-third of the studies reviewed showed little or no response of roots to localized nutrient supply. Drew and co-workers undertook a series of experiments with barley grown in solution culture and sand irrigated with nutrient solution, with part of seminal axis 1 exposed to either higher or lower concentrations of nutrients than the remainder of the axis (Drew *et al.*, 1973; Drew, 1975; Drew and Saker, 1975, 1978). When exposed to a localized high concentration of nutrient, the root responded by increasing the number and length of first and second order laterals with phosphate, NH_4^+, and NO_3^-, but not K^+ (Fig. 5.10). The reason for the lack of response to localized potassium is unknown. Drew (1975) suggested that it might be because sufficient potassium was translocated within the root from one zone to another to permit optimum branching but there was no evidence to demonstrate such differential mobility. Alternatively it may have been that the low concentrations of potassium in the upper and lower chambers were still sufficient for lateral growth when shoot growth was not limited by overall deficiency of potassium.

The significance of this root proliferation for nutrient acquisition has been assessed by several workers. Robinson (1996) used results from Drew (1975) to calculate the potential exploitation of the individual nutrients and demonstrated that the production of additional lateral roots was highly beneficial in the exploitation of locally available P, principally because of its low diffusion coefficient. For nitrate, though, the necessity for lateral roots, let alone proliferation of laterals, was not demonstrated and their growth appeared to be superfluous for effecting nitrate capture. Why, then, was the proliferation of laterals as great with localized nitrogen supply as for phosphate? Robinson (1996) could provide no definite answer to this question but suggested that in a multi-ion environment such as soil, a non-specific response might allow exploitation of the least mobile nutrients, and that concepts of optimal foraging for non-limiting responses might benefit from re-examination. Hodge (2004) reviewed three experiments in which root proliferation of wheat and two grass species were measured in response to N-enriched organic patches. None of the experiments demonstrated a relationship between root proliferation in, and N capture from, the patches. However, in experiments where two different plant species were grown together and allowed to explore a common enriched patch, there was a direct relation between root proliferation in, and N capture from, the patch (Fig. 5.11). Hodge *et al.* (1999) and Robinson *et al.* (1999) concluded that root proliferation was important for N capture when plants are in interspecific competition for organic patches containing a finite supply of mixed N sources. Remove any one of these factors and the importance of root proliferation for N capture is less obvious. In an agricultural context, then, root proliferation may benefit N capture by

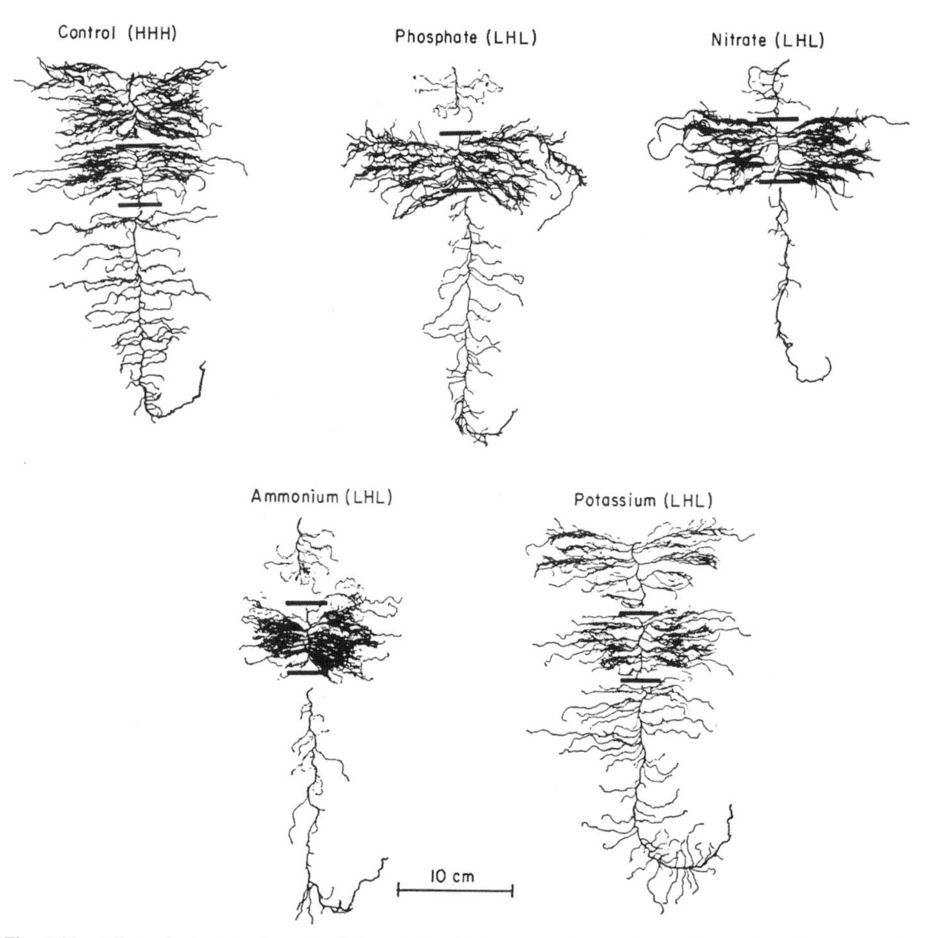

Fig. 5.10 Effect of a localized supply of phosphate, nitrate, ammonium and potassium on lateral root growth on the seminal axis of barley. Control plants (HHH) received a complete nutrient solution to all parts of the root while other plants received the complete solution only in the middle zone with the upper and lower zones supplied with solution deficient in the specified nutrient. (Reproduced with permission from Drew, *New Phytologist*; New Phytologist Trust, 1975.)

plants only when the plant is in a mixed cropping system and when N is released slowly (Hodge, 2004). Whether roots proliferate depends on the plant demand for the nutrient, the mobility of the nutrient within the plant, and the concentration in the patch relative to background (Hodge, 2004). For N, the data in Table 5.8 suggest that, as a generalization, when patches supply >10% of the total plant N, then root proliferation occurs. There are exceptions to this generalization even within the table, indicating that a full understanding of the processes leading to lateral root proliferation remains to be gained.

The plasticity of plant responses to heterogeneous nutrient supplies is evident both in terms of physiological traits and in competitive and symbiotic interactions with microorganisms. Dunbabin *et al.* (2001a) examined growth responses of *Lupinus angustifolius* (dominant tap root and lateral system) and *Lupinus pilosis* (minor tap root and well-developed laterals) to nitrate supplied either uniformly or split between the upper and

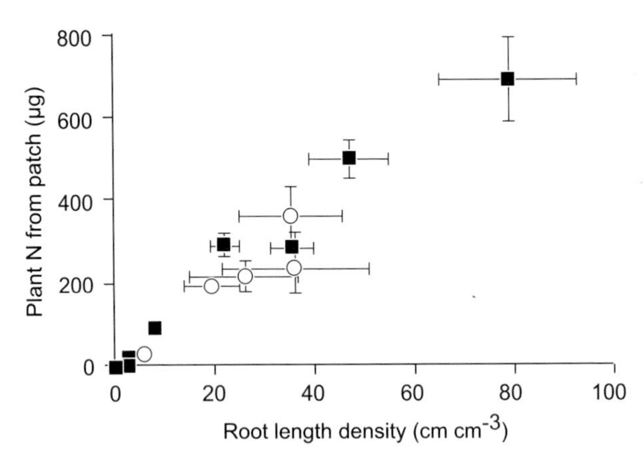

Fig. 5.11 Nitrogen capture by *Lolium perenne* (■) and *Poa pratensis* (○) from a patch of decomposing *L. perenne* shoot material when the two plants were grown in competition. The points are the means of four observations and the bars are ± standard error. (Reproduced with permission from Hodge *et al.*, *Plant, Cell and Environment*; Blackwell Publishing Ltd, 1999.)

Table 5.8 Root proliferation of various plant species in response to patches of organic N (*Lolium perenne* shoots) with different C:N ratio and the percentage of N in the patch that was captured by the plant (%N captured) and the N captured from the patch as a percentage of total plant N (patch/plant %)

Patch C:N	Plant species	Time (days)	Root proliferation	%N captured	Patch/ plant %
31:1	*Festuca arundinacea*	39	Yes	4	7
31:1	*Phleum pratense*	39	Yes	4	16
31:1	*Poa pratensis*	39	No	3	13
31:1	*Dactylis glomerata*	39	Yes	5	11
31:1	*Lolium perenne*	39	Yes	5	10
31:1	*Poa pratensis*	56	Yes	5	18
31:1	*Lolium perenne*	56	Yes	9	16
22:1	*Plantago lanceolata* + *Lolium perenne*	22	Yes	8	2
12:1	*Poa pratensis* + *Lolium perenne*	70	No	26	1

From Hodge, 2004.

lower root system. In both species, increased root proliferation in the high nitrate zone was accompanied by decreased root growth in the low nitrate zone to give about the same total growth as the uniform low nitrate treatment. The growth response of *L. angustifolius* was confined mainly to an increase in number and length of first order laterals, but in *L. pilosus* there was also a substantial increase in number and length of second order laterals (cluster roots) so that its total root system was 1.7 times that of *L. angustifolius*. Dunbabin *et al.* (2001b) further showed that *L. angustifolius* had the ability to increase its rate of nitrate uptake for parts of the root system supplied locally with high nitrate, while *L. pilosus* did not demonstrate this ability and may use its increased root growth to exploit nitrate patches. They concluded that a range of responses may exist within lupin germplasm and that it might be possible to select a lupin type with an enhanced ability to capture nitrate from the soil profile. Root proliferation in response to nutrient inputs usually takes some time so that this does not provide a rapid means to compete with microorganisms (Hodge *et al.*, 2000). Associations with mycorrhizal fungi also provide a means of accessing nutrients to which the plant would not otherwise have direct access. For example, Hodge (2004) provides several examples where hyphae have been shown to proliferate in nutrient-rich zones. However, the limited measurements available show that, generally, arbuscular mycorrhizal fungi have little impact on nutrient capture from patches other than that of phosphate.

The mechanisms by which nutrient availability regulate root architecture are beginning to emerge. *Arabidopsis* seedlings grown on vertical agar plates with a localized zone of NO_3^- increased elongation rates of laterals in the zone though, in contrast to results of Drew (1975), lateral numbers did not increase (Zhang and Forde, 1998). The increased elongation rate was due to enhanced meristematic activity and appeared to arise from a direct signal from the NO_3^- ion rather than from a product of NO_3^- metabolism (a mutant with low nitrate reductase activity showed a similar response to NO_3^--rich zones). This suggests that cells in the lateral root tips have a NO_3^- sensor and a signal transduction pathway to convert the NO_3^- signal into a growth response. Two genes have been identified as playing a role in the signal transduction pathway: (i) *ANR1* is a NO_3^--regulated member of the MADS-box family of transcription factors (Zhang and Forde, 1998); (ii) *AXR4* is an auxin-sensitive gene that may be involved because of the failure of an *axr4* mutant to respond to localized NO_3^- (Zhang *et al.*, 1999; Forde, 2002). While a role for auxin in plant development and in lateral root development is undisputed, a role for auxin in the NO_3^- signal transduction pathway is equivocal (Hodge, 2004). Similarly while Williamson *et al.* (2001) demonstrated that as with responses to localized nitrate, responses to localized phosphate indicated a response to internal phosphate concentration, there was no indication from auxin mutants *axr1*, *aux1* and *axr4* that auxin played any role in the response. In contrast, Al-Ghazi *et al.* (2003) concluded that auxin signalling was involved in the response of root system architecture of *Arabidopsis* to phosphate deprivation, and López-Bucio *et al.* (2003) concluded that the responses of root architecture to nutrients can be modified by several plant growth regulators such as auxins, cytokinins and ethylene, so that nutritional effects on root development may be mediated by changes in hormone synthesis, transport or sensitivity.

5.5.2 Low pH and aluminium

Acid soils occupy almost 50% of all non-irrigated, arable lands and are very common in the tropics where high rainfall for prolonged periods has leached soluble bases. Acid soils can lead to negative effects on plant growth because of toxicity caused by H^+, aluminium or manganese, or through deficiency of calcium or molybdenum. Plants can be separately adapted to H^+ or Al^{3+} toxicity with different effects on root growth and anatomy, and this may have important ecological consequences, especially on very acidic organic soils (Kidd and Proctor, 2001). Aluminium is now widely regarded as the most common limitation to growth on many acid soils because as pH falls to less than about 5.0–5.5, Al-containing minerals become soluble causing phytotoxicity. Even micromolar concentrations of Al^{3+} can inhibit growth in many plant species. Aluminium has similar effects on a wide range of plants with symptoms of toxicity evident first in the roots which appear shortened and swollen, before symptoms appear in the shoot. Typically, the

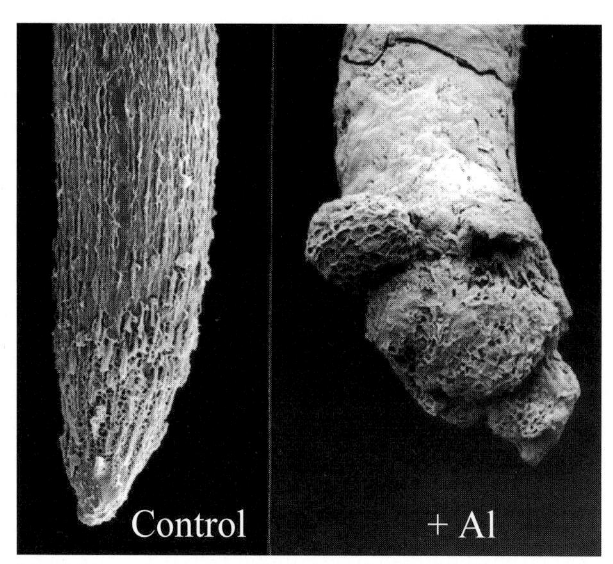

Fig. 5.12 Effects of aluminium on the root apex of near-isogenic wheat seedlings grown for 4 days in a solution containing 5 μM $AlCl_3$ and 200 μM $CaCl_2$ at pH 4.3. The control was a line tolerant of Al while that on the right was Al-sensitive. (Reproduced with permission from Delhaize and Ryan, *Plant Physiology*; American Society of Plant Biologists, 1995.)

root system has a stubby appearance because of swelling to the axes and the inhibition of lateral roots (Fig. 5.12).

The physiological mechanism underlying Al damage to cells is not known with certainty (Kochian, 1995), but because Al bonds strongly with oxygen-donor compounds, it can act with many molecules in the plant root. The root apex has been judged for a long time to be the primary site of action for Al toxicity because the main sites of Al uptake are the peripheral root cap cells and the mucilage surrounding the root. Bennet *et al.* (1984a) showed that Al spread rapidly through the cells of the root cap with much slower entry to the outer cortical cells of the root apex, but there was no evidence for Al reaching actively dividing cells of the primary root meristem fast enough to account for the rapid injury that normally follows exposure to Al. However, rapid disruption to membrane surfaces and interference in membrane transport, and the metabolic activity of cells proximal to the cap/root junction was detected (Bennet *et al.*, 1984b). Miyasaka and Hawes (2001) also demonstrated the importance of the root border cells and associated mucilage in the detection and response to aluminium toxicity.

Plants differ markedly in their response to aluminium toxicity both between and within species, with most of the proposed mechanisms involving either external avoidance or internal tolerance. External avoidance is brought about in several plants through the release of organic acids, especially citrate, oxalate and malate, which chelate aluminium in the rhizosphere (Ma *et al.*, 2001; Mariano and Keltjens, 2003). In wheat and maize Al^{3+} activates an anion channel in the plasma membrane that is permeable to organic acids and chloride, with the activated channel activity restricted to cells localized in a narrow zone within the root apex (Ryan *et al.*, 1997). Because it is the root apex that is most susceptible to Al^{3+}, it is only immediately around this zone that detoxification needs to occur and this minimizes the metabolic cost of this Al-tolerance mechanism. Miyasaka and Hawes (2001) have proposed a second avoidance mechanism in which adsorption of Al by negatively charged root mucilage prevents migration of Al to the root meristem. In studies with two genotypes of *Phaseolus vulgaris* in solutions and agar, they showed that Al killed border cells and that Al induced increased exudation around detached border cells. The genotype known to be more tolerant of soils with high Al had border cells with a thicker layer of mucilage and cell death did not occur, supporting the initial hypothesis that border cells and their associated mucilage help to protect the plant by inhibiting Al uptake into the roots.

Internal tolerance is also associated with processes resulting in the formation of Al–organic acid complexes (Ma *et al.*, 2001). For example, in buckwheat (*Fagopyrum esculentum*) which accumulates Al in its leaves, aluminium enters the root by an unknown mechanism, probably as Al^{3+} which is chelated with oxalate to form a 1:3 Al–oxalate complex once it crosses the plasma membrane. When this is translocated from the root to the shoot in the xylem, ligand exchange occurs to form Al–citrate, which when unloaded from the xylem into leaf cells reforms the Al–oxalate complex. In contrast, in *Brachiaria* species (tropical forage grasses), Al is accumulated at high concentrations in the roots with 70–85% of the Al taken up retained in the root system. In these species the root apices accumulated organic acids (notably citrate and trans-aconitate) as external Al concentration was increased, suggesting a role for organic acids in internal detoxification in root apices (Wenzl *et al.*, 2002). However, the reason for the greater Al tolerance of *Brachiaria decumbens* (Stapf) over *Brachiaria ruziziensis* (Germain & Evrard) was not explainable by this mechanism.

5.5.3 Salinity

Salinity affects about 7% of the world's land area with the area increasing as a consequence of clearing of native, perennial vegetation and the introduction of irrigation schemes without proper drainage. Irrigated plants in arid regions are especially susceptible to salinization, alkalization and waterlogging because waters often contain dissolved Na^+, Ca^{2+} and Mg^{2+} which are excluded by plant roots during water uptake and therefore accumulate in soils. Soil salinity inhibits growth through effects of both Na^+ and Cl^-. If salinity is high and the plant's ability to exclude NaCl is poor, then Na^+ or Cl^- (or both) accumulate in transpiring leaves and eventually exceed the ability of cells to compartmentalize these ions in the vacuole. The ions then build up in either the cytoplasm inhibiting enzyme activity or the cell walls causing them to dehydrate and shrink (Flowers and Yeo, 1986). For some plants (especially woody perennials such as citrus and grapevines), Na^+ is retained in the woody roots and it is the Cl^- accumulated in the leaves that is most damaging, while in other plants (such as cereal crops), Na^+ accumulated in leaves is most damaging (Tester and Davenport, 2003). It is not yet resolved whether the dominant control over plant growth in saline soils is exerted by water status, hormonal regulation or supply of photosynthate (Munns, 2002), but osmotic stress appears to predominate. Salt injury is usually observed first in older leaves where transpiration has occurred for the longest time thereby allowing salt accumulation.

Root growth is often much less affected by salinity than leaf growth, in common with the effect of dry soil, suggesting that the effect is probably due to factors associated with water stress than a salt-specific effect (Munns, 2002). Generally, root cells of non-woody plants have Na^+ and Cl^- concentrations lower than those in the external solution (this rarely occurs in leaves), and do not accumulate Na^+ at concentrations likely to become toxic. For example, a range of five bread and durum wheat genotypes that differed in salt tolerance grown in solutions with from 1 to 150 mM NaCl and at 10 mM Ca^{2+} had concentrations of Na^+ in roots ranging from about 15 to 50 mmol kg^{-1} tissue water (Husain *et al.* 2004). Initial entry of Na^+ from soil solution into the root is via cation channels (Tester and Davenport, 2003), but roots of most plants exclude at least 95% of the salt present in the soil solution and this is the main mechanism of tolerance (Munns, 2005). Differences in salt tolerance within and between plant species are, then, largely a consequence of differences in the degree of exclusion at the root surface (range 92–98% of the salt present in the external solution; Table 5.9) and subsequently another 1–2% at the sites of loading and unloading of the root xylem. If the roots are unable to exclude ≥98% of the salt in the soil solution then, under most conditions, salt will accumulate in leaves and tolerance will require that salt to be compartmentalized in the vacuole. Salt-tolerant cells contribute to salt tolerance, but at the whole plant level of organization, management of Na^+ movement via specific cell types utilizing transporters and signalling elements that are co-ordinated are very important in determining plant adaptation to salinity (Tester and Davenport, 2003).

5.6 Atmospheric CO_2 concentration

With atmospheric CO_2 concentrations now higher than they have ever been in any interglacial period during the last 400,000 years, and increasing, the effects of elevated CO_2

Table 5.9 Salt (Na^+ and Cl^-) concentrations in the xylem and percentage Na^+ and Cl^- exclusion from the xylem for plants grown in solution at 50 mM NaCl for 2 weeks

		Na^+		Cl^-		
Species	Genotype	Xylem Na^+ (mM)	Exclusion (%)	Xylem Cl^- (mM)	Exclusion (%)	Method*
Bread wheat	Janz	0.3	99	3.9	92	Calculated
	Chinese Spring	1.1	98	–	–	Insect feeding
	Kharchia-65	1.4	97	–	–	Calculated
	Punjab-85	1.6	97	–	–	Calculated
Durum wheat	Wollaroi	2.0	96	3.4	93	Calculated
	Tamaroi	2.7	95	2.7	95	Calculated
	Langdon	3.0	94	–	–	Insect feeding
Rice	IR36	2.8	94	–	–	Calculated
Barley	Clipper	3.2	94	4.7	91	Collected sap
Citrus	Rangpur lime	–	–	0.5	99	Calculated
	Etrog citron	–	–	3.4	93	Calculated
	Rangur lime	2.4	95	1.5	97	Collected
	Rough lemon	3.3	93	2.6	95	Collected
Mangrove	*Avicennia marina*	–	–	5.0	90	Calculated
	Aegiceras corniculatum	–	–	5.7	87	Calculated

*Calculated, calculated from rates of Na^+ transport from root to shoot and from transpiration rates; Insect feeding, the xylem-feeding insect *Philaenus spumarius* was used to collect sap; Collected sap, expressed sap by applying pressure to induce flow equivalent to transpiration rate. From Rana Munns, CSIRO, personal communication.

on plant growth have received much attention in the last two decades. Because concentrations in surface soils are typically up to 10 times higher than the atmosphere, any effects on roots are likely to be mediated via shoots rather than as direct effects. Growth in elevated CO_2 generally leads to increased root length and root diameter, and alters patterns of branching (Pritchard *et al.*, 1999). Another major effect of elevated CO_2 is often to increase photosynthesis, resulting in increased growth of both the root and shoot system, and to alter the allocation between the two – often resulting in increased proportions of root dry matter. Rogers *et al.* (1996) reviewed 264 studies of crop plants grown under elevated CO_2 and found a range of responses. Root allocation increased relative to the shoot in almost 60% of studies with 3% with no response and the rest negative; the mean response was an 11% increase in root:shoot ratio. From a literature survey investigating effects on root architecture, Pritchard and Rogers (2000) concluded that, despite highly variable results, elevated CO_2 stimulated the production of lateral roots more than the elongation of root axes, leading to more highly branched, but shallower root systems. However, most of the studies were conducted in pots in growth rooms and glasshouses, with a much smaller number in field conditions.

Free-air CO_2 enrichment studies with crops of cotton in Arizona showed that elevated CO_2 increased total root length and mass, and root mass per unit root length (Prior *et al.*, 1994a). Elevated CO_2 also affected the vertical and lateral distribution of roots with enhancement of both length and mass in the upper 0.6 m (Fig. 5.13), and a smaller proportion of the root system closer to the centre of the rows (Prior *et al.*, 1994b). It also resulted in larger diameter tap roots with higher pulling resistance, and increased the dry weight of lateral roots and their numbers (Prior *et al.*, 1995). Similarly in temperature gradient tunnels, Batts *et al.* (1998) found that elevated CO_2 increased the mass of two cultivars of wheat roots relative to that in shoots. Root mass in elevated CO_2 was 95 g m^{-2} greater (49–186%, coolest to warmest regime) with effects in all layers sampled in the upper 0.4 m, and the root:total plant mass ratio was 56% higher (Table 5.10). In grasses, too, a wide range of effects of elevated CO_2 has been determined including increases in root biomass (Jongen *et al.*, 1995; Fitter *et al.*, 1997) and contrasting responses such as faster turnover (Fitter *et al.*, 1997), and slower decomposition (Jongen *et al.*, 1995). In trees, fine roots play a crucial role in regulating the C balance of the ecosystem so their response to elevated CO_2 has been a focus for several studies. Wan *et al.* (2004) found that both elevated CO_2 and temperature significantly increased production and mortality of fine roots of red maple (*Acer rubrum*) and sugar maple (*Acer saccharum*) when grown in open-top chambers in the field. Such stimulation of fine root activity may provide a mechanism for increased flux of C to soil and for possible sequestration of C in soil organic matter.

This science is still young and many important interactions of elevated CO_2 with other aspects of environmental change (such as altered N cycling and effects on water use) have still to be explored before the apparently different results of experiments can be resolved. Some may be artefacts, with others genuine differences between species, and effects with ecological and economic significance. A major concern is that many studies have been performed with young plants in pots; effects in stands may well be different. For example, Norby *et al.* (1999) reviewed results from experiments with trees and forests and concluded that while the annual increment in wood mass per unit leaf area was

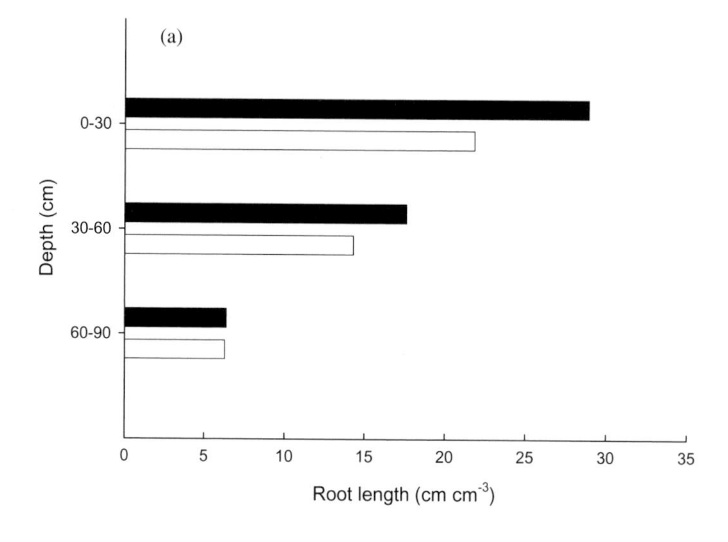

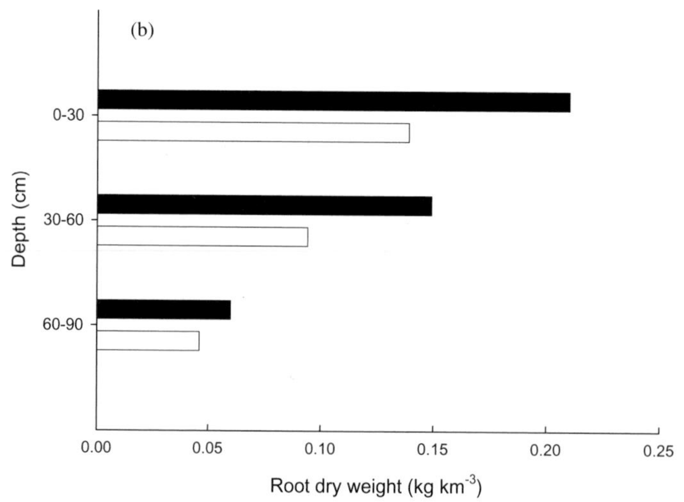

Fig. 5.13 The effect of CO_2 (open 370 μmol mol^{-1}; closed 550 μmol mol^{-1}) on the distribution of (a) root length and (b) root dry weight of cotton grown under field conditions. Significant differences (p>0.05) were found for both measures at 0–30 and 30–60 cm. (Reproduced with permission from Prior *et al.*, *Plant and Soil*; Springer Science and Business Media, 1994b.)

Table 5.10 Effects of elevated atmospheric CO_2 (700 v. 360 μmol CO_2 mol^{-1} air) on total root dry mass, mass in individual soil layers (g m^{-2}) and root mass as a proportion of total plant mass

	Layer (cm)				
CO_2 treatment	0–40	0–10	10–20	20–40	Root mass:plant mass
Ambient	128.8	96.7	11.4	20.7	0.057
Elevated	224.0	166.1	24.9	33.1	0.089
SED	27.8	20.0	3.9	9.2	0.008

Results are the mean of two cultivars of winter wheat at flowering. From Batts *et al.*, 1998.

increased by 27% in elevated CO_2, there was no support for the notion that either the increased root:shoot ratio found in seedlings or the enhanced production of fine roots would persist in a forest.

References

Abdalla, A.M., Hettiaratchi, D.R.P. and Reece, A.R. (1969) The mechanics of root growth in granular media. *Journal of Agricultural Engineering Research* **14**, 236–248.

Aguirrezabal, L.A.N. and Tardieu, F. (1996) An architectural analysis of the elongation of field-grown sunflower root systems. Elements for modelling the effects of temperature and intercepted radiation. *Journal of Experimental Botany* **47**, 411–420.

Aguirrezabal, L.A.N., Deleens, E. and Tardieu, F. (1994) Root elongation rate is accounted for by intercepted PPFD and source-sink relations in field and laboratory-grown sunflower. *Plant, Cell and Environment* **17**, 443–450.

Al-Ghazi, Y., Muller, B., Pinloche, S., Tranbarger, T.J., Nacry, P., Rossignol, M., Tardieu, F. and Doumas, P. (2003) Temporal responses of *Arabidopsis* root architecture to phosphate starvation: evidence for the involvement of auxin signalling. *Plant, Cell and Environment* **26**, 1053–1068.

Anghinoni, I. and Barber, S.A. (1980) Phosphorus application rate and distribution in the soil and phosphorus uptake by corn. *Soil Science Society of America Journal* **44**, 1041–1044.

Anghinoni, I. and Barber, S.A. (1988) Corn root growth and nitrogen uptake as affected by ammonium placement. *Agronomy Journal* **80**, 799–802.

Araujo, F., Williams, L.E., Grimes, D.W. and Matthews, M.A. (1995) A comparative study of young Thompson-Seedless grapevines under drip and furrow irrigation. 1. Root and soil-water distributions. *Scientia Horticulturae* **60**, 235–249.

Armstrong, W. (1979) Aeration in higher plants. *Advances in Botanical Research* **7**, 225–332.

Atwell, B.J. (1988) Physiological responses of lupin roots to soil compaction. *Plant and Soil* **111**, 277–281.

Atwell, B.J. (1990a) The effect of soil compaction on wheat during early tillering. I. Growth, development and root structure. *New Phytologist* **115**, 29–35.

Atwell, B.J. (1990b) The effect of soil compaction on wheat during early tillering. II. Concentrations of cell constituents. *New Phytologist* **115**, 37–41.

Atwell, B.J. (1990c) The effect of soil compaction on wheat during early tillering. III. Fate of carbon transported to the roots. *New Phytologist* **115**, 43–49.

Atwell, B.J. (1993) Response of roots to mechanical impedance. *Environmental and Experimental Botany* **33**, 27–40.

Atwell, B.J. and Newsome, J.C. (1990) Turgor pressure in mechanically impeded lupin roots. *Australian Journal of Plant Physiology* **17**, 49–56.

Barber, S.A. (1995) *Soil Nutrient Bioavailability: A Mechanistic Approach*, 2nd edn. John Wiley & Sons, New York.

Barley, K.P. and Greacen, E.L. (1967) Mechanical resistance as a soil factor influencing the growth of roots and underground shoots. *Advances in Agronomy* **19**, 1–40.

Batts, G.R., Wheeler, T.R., Morison, J.I.L., Ellis, R.H. and Hadley, P. (1998) Developmental and tillering responses of winter wheat (*Triticum aestivum*) crops to CO_2 and temperature. *Journal of Agricultural Science, Cambridge* **127**, 23–35.

Bengough, A.G., Croser, C. and Pritchard, J. (1997) A biophysical analysis of root growth under mechanical stress. *Plant and Soil* **189**, 155–164.

Bennet, R.J., Breen, C.M. and Fey, M.V. (1984a) Aluminium uptake sites in the primary root of *Zea mays* L. *South African Journal of Plant and Soil* **2**, 1–7.

Bennet, R.J., Breen, C.M. and Fey, M.V. (1984b) The primary site of aluminium injury in the root of *Zea mays* L. *South African Journal of Plant and Soil* **2**, 8–17.

Blackwell, P.S. and Wells, E.A. (1983) Limiting oxygen flux densities for oat root extension. *Plant and Soil* **73**, 129–139.

Blancaflor, E.B. and Masson, P.H. (2003) Plant gravitropism. Unraveling the ups and downs of a complex process. *Plant Physiology* **133**, 1677–1690.

Boonsirichai, K., Guan, C., Chen, R. and Masson, P.H. (2002) Root gravitropism: an experimental tool to investigate basic cellular and molecular processes underlying mechanosensing and signal transmission in plants. *Annual Review of Plant Biology* **53**, 421–447.

Bowen, H.D. (1981) Alleviating mechanical impedance. In: *Modifying the Root Environment to Reduce Crop Stress* (eds G.F. Arkin and H.M. Taylor), pp. 21–57. American Society of Agricultural Engineers, Michigan, USA.

Brady, D.J., Wenzel, C.L., Fillery, I.R.P. and Gregory, P.J. (1995) Root growth and nitrate uptake by wheat (*Triticum aestivum* L.) following wetting of dry surface soil. *Journal of Experimental Botany* **46**, 557–564.

Brouwer, R. (1962) Influence of temperature of the root medium on the growth of seedlings of various crop plants. Jaarboek Instituut voor Biologisch en Scheikundig Landbouwgewassen, Wageningen, pp. 11–18.

Brown, S.C., Keatinge, J.D.H., Gregory, P.J. and Cooper, P.J.M. (1987) Effects of fertilizer, variety and location on barley production under rainfed conditions in northern Syria. 1. Root and shoot growth. *Field Crops Research* **16**, 53–66.

Bruckler, L., Lafolie, F. and Tardieu, F. (1991) Modeling root water potential and soil-root water transport: II Field comparisons. *Soil Science Society of America Journal* **55**, 1213–1220.

Butterbach-Bahl, K., Papen, H. and Rennenberg, H. (1997) Impact of gas transport through rice cultivars on methane emission from rice paddy fields. *Plant, Cell and Environment* **20**, 1175–1183.

Clark, L.J. and Barraclough, P.B. (1999) Do dicotyledons generate greater maximum axial root growth pressure than monocotyledons? *Journal of Experimental Botany* **50**, 1263–1266.

Clark, L.J., Aphalé, S.L. and Barraclogh, P.B. (2000) Screening the ability of rice roots to overcome the mechanical impedance of wax layers: importance of test conditions and measurement criteria. *Plant and Soil* **219**, 187–196.

Clark, L.J., Cope, R.E., Whalley, W.R., Barraclough, P.B. and Wade, L.J. (2002) Root penetration of strong soil in rainfed lowland rice: comparison of laboratory screens with field performance. *Field Crops Research* **76**, 189–198.

Clark, L.J., Whalley, W.R. and Barraclough, P.B. (2003) How do roots penetrate strong soil? *Plant and Soil* **255**, 93–104.

Clarkson, D.T., Earnshaw, M.J., White, P.J. and Cooper, H.D. (1988) Temperature dependent factors influencing nutrient uptake: an analysis of responses at different levels of organization. In: *Plants and Temperature* (eds S.P. Long and F.I. Woodward), pp. 281–309. The Company of Biologists, Cambridge, UK.

Colmer, T.D. (2003a) Long-distance transport of gases in plants: a perspective on internal aeration and radial oxygen loss from roots. *Plant, Cell and Environment* **26**, 17–36.

Colmer, T.D. (2003b) Aerenchyma and an inducible barrier to radial oxygen loss facilitate root aeration in upland, paddy and deep-water rice (*Oryza sativa* L.). *Annals of Botany* **91**, 301–309.

Cook, F.J. and Knight, J.H. (2003) Oxygen transport to plant roots: modeling for physical understanding of soil aeration. *Soil Science Society of America Journal* **67**, 20–31.

Cooper, A.J. (1973) *Root Temperature and Plant Growth – A Review*. Research Review No. 4. Commonwealth Bureau of Horticulture and Plantation Crops, Commonwealth Agricultural Bureau, Farnham Royal, UK.

Cornish, P.S. (1982) Root development in seedlings of ryegrass (*Lolium perenne* L.) and phalaris (*Phalaris aquatica* L.) sown onto the soil surface. *Australian Journal of Agricultural Research* **33**, 665–677.

Correll, M.J. and Kiss, J.Z. (2002) Interactions between gravitropism and phototropism in plants. *Journal of Plant Growth Regulation* **21**, 89–101.

Coutts, M.P. and Philipson, J.J. (1978a) Tolerance of tree roots to waterlogging. I. Survival of Sitka spruce and Lodgepole pine. *New Phytologist* **80**, 63–69.

Coutts, M.P. and Philipson, J.J. (1978b) Tolerance of tree roots to waterlogging. II. Adaptation of Sitka spruce and Lodgepole pine to waterlogged soil. *New Phytologist* **80**, 71–77.

Croser, C., Bengough, A.G. and Pritchard, J. (2000) The effect of mechanical impedance on root growth in pea (*Pisum sativum*). II. Cell expansion and wall rheology during recovery. *Physiologia Plantarum* **109**, 150–159.

Crossett, R.N., Campbell, D.J. and Stewart, H.E. (1975) Compensatory growth in cereal root systems. *Plant and Soil* **42**, 673–683.

Dardanelli, J.L., Calmon, M.A., Jones, J.W., Andriani, J.M., Díaz, M.P. and Collino, D.J. (2003) Use of a crop model to evaluate soil impedance and root clumping effects on soil water extraction in three Argentine soils. *Transactions of the American Society of Agricultural Engineers* **46**, 1265–1275.

Darwin, C. and Darwin, F. (1880) *The Power of Movement in Plants*. John Murray, London.

Dexter, A.R. (1987) Mechanics of root growth. *Plant and Soil* **98**, 303–312.
Donald, R.G., Kay, B.D. and Miller, M.H. (1987) The effect of soil aggregate size on early shoot and root growth of maize (*Zea mays* L.). *Plant and Soil* **103**, 251–259.
Drew, M.C. (1975) Comparison of the effects of a localized supply of phosphate, nitrate, ammonium and potassium on the growth of the seminal root system, and the shoot, in barley. *New Phytologist* **75**, 479–490.
Drew, M.C. and Saker, L.R. (1975) Nutrient supply and the growth of the seminal root system in barley. II. Localized, compensatory increases in lateral root growth and rates of nitrate uptake when nitrate supply is restricted to only part of the root system. *Journal of Experimental Botany* **26**, 79–90.
Drew, M.C. and Saker, L.R. (1978) Nutrient supply and the growth of the seminal root system in barley. III. Compensatory increases in growth of lateral roots, and in rates of phosphate uptake, in response to a localized supply of phosphate. *Journal of Experimental Botany* **29**, 435–451.
Drew, M.C., Saker, L.R. and Ashley, T.W. (1973) Nutrient supply and the growth of the seminal root system in barley. I. The effect of nitrate concentration on the growth of axes and laterals. *Journal of Experimental Botany* **24**, 1189–1202.
Drew, M.C., He, C.-J. and Morgan, P.W. (2000) Programmed cell death and aerenchyma formation in roots. *Trends in Plant Science* **5**, 123–127.
Dunbabin, V., Rengel, Z. and Diggle, A. (2001a) The root growth response to heterogeneous nitrate supply differs for *Lupinus angustifolius* and *Lupinus pilosis*. *Australian Journal of Agricultural Research* **52**, 495–503.
Dunbabin, V., Rengel, Z. and Diggle, A. (2001b) *Lupinus angustifolius* has a plastic uptake response to heterogeneously supplied nitrate while *Lupinus pilosis* does not. *Australian Journal of Agricultural Research* **52**, 505–512.
Durieux, R.P., Kamprath, E.J., Jackson, W.A. and Moll, R.H. (1994) Root distribution of corn: the effect of nitrogen fertilization. *Agronomy Journal* **86**, 958–962.
Eapen, D., Barroso, M.L., Ponce, G., Campos, M.E. and Cassab, G.I. (2005) Hydrotropism: root growth responses to water. *Trends in Plant Science* **10**, 44–50.
Eavis, B.W., Ratliff, L. and Taylor, H.M. (1969) Use of a dead-load technique to determine axial root growth pressure. *Agronomy Journal* **61**, 640–643.
Edwards, E.J., Benham, D.G., Marland, L.A. and Fitter, A.H. (2004) Root production is determined by radiation flux in a temperate grassland community. *Global Change Biology* **10**, 209–227.
Ellis, F.B., Christian, D.G., Bragg, P.L., Henderson, F.K.G., Prew, R.D. and Cannell, R.Q. (1984) A study of mole drainage with simplified cultivation for autumn-sown crops on a clay soil. 3. Agronomy, root and shoot growth of winter wheat, 1978–80. *Journal of Agricultural Science, Cambridge* **102**, 583–594.
Evans, D.E. (2003) Aerenchyma formation. *New Phytologist* **161**, 35–49.
Firn, R.D. and Digby, J. (1997) Solving the puzzle of gravitropism – has a lost piece been found? *Planta* **203**, S159–S163.
Fitter, A.H., Graves, J.D., Wolfenden, J., Self, G.K., Brown, T.K., Bogie, D. and Mansfield, T.A. (1997) Root production and turnover and carbon budgets of two contrasting grasslands under ambient and elevated atmospheric carbon dioxide concentrations. *New Phytologist* **137**, 247–255.
Fitter, A.H., Graves, J.D., Self, G.K., Brown, T.K., Bogie, D.S. and Taylor, K. (1998) Root production, turnover and respiration under two grassland types along an altitudinal gradient: influence of temperature and solar radiation. *Oecologia* **114**, 20–30.
Flowers, T.J. and Yeo, A.R. (1986) Ion relations of plants under drought and salinity. *Australian Journal of Plant Physiology* **13**, 75–91.
Forde, B.G. (2002) Local and long-range signalling pathways regulating plant responses to nitrate. *Annual Review of Plant Biology* **53**, 203–224.
Fraser, T.E., Kuhn Silk, W. and Rost, T.L. (1990) Effects of low water potential on cortical cell length in growing regions of maize roots. *Plant Physiology* **93**, 648–651.
Garthwaite, A.J., von Bothmer, R. and Colmer, T.D. (2003) Diversity in root aeration traits associated with waterlogging tolerance in the genus *Hordeum*. *Functional Plant Biology* **30**, 875–889.
Gill, W.R. and Bolt, G.H. (1955) Pfeffer's studies of the root growth pressures exerted by plants. *Agronomy Journal* **47**, 166–168.
Gladish, D.K. and Rost, T.L. (1993) The effects of temperature on primary root growth dynamics and lateral root distribution in garden pea (*Pisum sativum* L., cv. 'Alaska'). *Environmental and Experimental Botany* **33**, 243–258.
Goss, M.J. and Russell, R.S. (1980) Effects of mechanical impedance on root growth in barley (*Hordeum vulgare* L.). III. Observations on the mechanism of response. *Journal of Experimental Botany* **31**, 577–588.
Greacen, E.L. and Oh, J.S. (1972) Physics of root growth. *Nature, New Biology* **235**, 24–25.

Gregory, P.J. (1983) Response to temperature in a stand of pearl millet (*Pennisetum typhoides* S. & H.). III. Root development. *Journal of Experimental Botany* **34**, 744–756.

Gregory, P.J. (1986) Response to temperature in a stand of pearl millet (*Pennisetum typhoides* S. & H.). VIII. Root growth. *Journal of Experimental Botany* **37**, 379–388.

Hillel, D. (1980) *Fundamentals of Soil Physics*. Academic Press, New York.

Hodge, A. (2004) The plastic plant: root responses to heterogeneous supplies of nutrients. *New Phytologist* **162**, 9–24.

Hodge, A., Robinson, D., Griffiths, B.S. and Fitter, A.H. (1999) Why plants bother: root proliferation results in increased nitrogen capture from an organic patch when two grasses compete. *Plant, Cell and Environment* **22**, 811–820.

Hodge, A., Stewart, J., Robinson, D., Griffiths, B.S. and Fitter, A.H. (2000) Competition between roots and soil micro-organisms for nutrients from nitrogen-rich patches of varying complexity. *Journal of Ecology* **88**, 150–164.

Husain, S., von Caemmerer, S. and Munns, R. (2004) Control of salt transport from roots to shoots of wheat in saline soil. *Functional Plant Biology* **31**, 1115–1126.

Jongen, M., Jones, M.B., Hebeisen, T., Blum, H. and Hendrey, G. (1995) The effects of elevated CO_2 concentrations on the root growth of *Lolium perenne* and *Trifolium repens* grown in a FACE system. *Global Change Biology* **1**, 361–371.

Jourdan, C. and Rey, H. (1997) Architecture and development of the oil-palm (*Elaeis guineesis* Jacq.) root system. *Plant and Soil* **189**, 33–48.

Jourdan, C., Michaux-Ferrière, N. and Perbal, G. (2000) Root system architecture and gravitropism in the oil palm. *Annals of Botany* **85**, 861–868.

Jupp, A.P. and Newman, E.I. (1987) Mophological and anatomical effects of severe drought on the roots of *Lolium perenne* L. *New Phytologist* **105**, 393–402.

Kaspar, T.C. and Bland, W.L. (1992) Soil temperature and root growth. *Soil Science* **154**, 290–299.

Kaspar, T.C., Woolley, D.G. and Taylor, H.M. (1981) Temperature effect on the inclination of lateral roots of soybean. *Agronomy Journal* **73**, 383–385.

Kidd, P.S. and Proctor, J. (2001) Why plants grow poorly on very acid soils: are ecologists missing the obvious? *Journal of Experimental Botany* **52**, 791–799.

Kirby, J.M. and Bengough, A.G. (2002) Influence of soil strength on root growth: experiments and analysis using a critical-state model. *European Journal of Soil Science* **53**, 119–128.

Kirk, G.J.D. (2003) Rice root properties for internal aeration and efficient nutrient acquisition in submerged soil. *New Phytologist* **159**, 185–194.

Kirkegaard, J.A., So, H.B. and Troedson, R.J. (1992) The effect of soil strength on the growth of pigeonpea radicles and seedlings. *Plant and Soil* **140**, 65–74.

Klepper, B. (1987) Origin, branching and distribution of root systems. In: *Root Development and Function* (eds P.J. Gregory, J.V. Lake and D.A. Rose), pp. 103–123. Cambridge University Press, Cambridge.

Kochian, L.V. (1995) Cellular mechanisms of aluminium toxicity and resistance in plants. *Annual Review of Plant Physiology and Plant Molecular Biology* **46**, 237–260.

Konings, H. (1967) On the mechanism of the transverse distribution of auxin in geotropically exposed pea roots. *Acta Botanica Neerlandia* **16**, 161–176.

Kuznetsov, O.A. and Hasenstein, K.H. (1996) Intracellular magnetophoresis of amyloplasts and induction of root curvature. *Planta* **198**, 87–94.

Letey, J. (1985) Relationship between soil physical properties and crop production. *Advances in Soil Science* **1**, 277–294.

López-Bucio, J., Cruz-Ramírez, A. and Herrera-Estrella, L. (2003) The role of nutrient availability in regulating root architecture. *Current Opinion in Plant Biology* **6**, 280–287.

Ma, J.F., Ryan, P.R. and Delhaize, E. (2001) Aluminium tolerance in plants and the complexing role of organic acids. *Trends in Plant Science* **6**, 273–278.

Machado, R.M.A., do Rosário, M., Oliveira, G. and Portas, C.A.M. (2003) Tomato root distribution, yield and fruit quality under subsurface drip irrigation. *Plant and Soil* **255**, 333–341.

Mackay, A.D. and Barber, S.A. (1986) Effect of nitrogen on root growth of two corn genotypes in the field. *Agronomy Journal* **78**, 699–703.

McMichael, B.L. and Burke, J.J. (1994) Metabolic activity of cotton roots in response to temperature. *Environmental and Experimental Botany* **34**, 201–206.

Malik, A.I., Colmer, T.D., Lambers, H., Setter, T.L. and Schortmeyer, M. (2002) Short-term waterlogging has long-term effects on the growth and physiology of wheat. *New Phytologist* **153**, 225–236.

Mariano, E.D. and Keltjens, W.G. (2003) Evaluating the role of root citrate exudation as a mechanism of aluminium resistance in maize genotypes. *Plant and Soil* **256**, 469–479.

Massa, G.D. and Gilroy, S. (2003) Touch modulates gravity sensing to regulate the growth of primary roots of *Arabidopsis thaliana*. *The Plant Journal* **33**, 435–445.

Materechera, S.A., Dexter, A.R. and Alston, A.M. (1991) Penetration of very strong soils by seedling roots of different plant species. *Plant and Soil* **135**, 31–41.

Materechera, S.A., Alston, A.M., Kirby, J.M. and Dexter, A.R. (1992) Influence of root diameter on the penetration of seminal roots into a compacted soil. *Plant and Soil* **144**, 297–303.

Miyasaka, S.C. and Hawes, M.C. (2001) Possible role of root border cells in detection and avoidance of aluminium toxicity. *Plant Physiology* **125**, 1978–1987.

Misra, R.K., Dexter, A.R. and Alston, A.M. (1986) Maximum axial and radial growth pressures of plant roots. *Plant and Soil* **95**, 315–326.

Munns, R. (2002) Comparative physiology of salt and water stress. *Plant, Cell and Environment* **25**, 239–250.

Munns, R. (2005) Genes and salt tolerance: bringing them together. *New Phytologist* **167**, 645–663.

Neilsen, G.H., Parchomchuk, P., Neilsen, D. and Zebarth, B.J. (2000) Drip-fertigation of apple trees affects root distribution and development of K deficiency. *Canadian Journal of Soil Science* **80**, 353–361.

Nobbe, F. (1862) Uber die feinere Verastelung der Pflanzenwurzel. Eine Vegetationstudie. *Die landwirtschaftlichen Versuchsstationen* **4**, 212–224.

Norby, R.J., Wullschleger, S.D., Gunderson, C.A., Johnson, D.W. and Ceulemans, R. (1999) Tree response to rising CO_2 in field experiments: implications for the future forest. *Plant, Cell and Environment* **22**, 683–714.

Onderdonk, J.J. and Ketcheson, J.W. (1973) Effect of soil temperature on direction of corn root growth. *Plant and Soil* **39**, 177–186.

Ottenschläger, I., Wolff, P., Wolverton, C., Bhalerao, R.P., Sandberg, G., Ishikawa, H., Evans, M. and Palme, K. (2003) Gravity-regulated differential auxin transport from columella to lateral root cap cells. *Proceedings of the National Academy of Science* **100**, 2987–2991.

Passioura, J.B. (1991) Soil structure and plant growth. *Australian Journal of Soil Research* **29**, 717–728.

Passioura, J.B. and Leeper, G.W. (1963) Soil compaction and manganese deficiency. *Nature* **200**, 29–30.

Passioura, J.B. and Wetselaar, R. (1972) Consequences of banding nitrogen fertilizers in soil. II. Effects on the growth of wheat roots. *Plant and Soil* **36**, 461–473.

Peltzer, S.C., Abbott, L.K. and Atkins, C.A. (2002) Effect of low root-zone temperature on nodule initiation in narrow-leafed lupin (*Lupinus angustifolius* L.). *Australian Journal of Agricultural Research* **53**, 355–365.

Porter, J.R., Klepper, B. and Belford, R.K. (1986) A model (WHTROOT) which synchronizes root growth and development with shoot development for winter wheat. *Plant and Soil* **92**, 133–145.

Prior, S.A., Rogers, H.H., Runion, G.B. and Mauney, J.R. (1994a) Effects of free-air CO_2 enrichment on cotton root growth. *Agricultural and Forest Meteorology* **70**, 69–86.

Prior, S.A., Rogers, H.H., Runion, G.B. and Hendrey, G.R. (1994b) Free-air CO_2 enrichment of cotton: vertical and lateral root distribution patterns. *Plant and Soil* **165**, 33–44.

Prior, S.A., Rogers, H.H., Runion, G.B., Kimball, B.A., Mauney, J.R., Lewin, K.F., Nagy, J. and Hendrey, G.R. (1995) Free-air carbon dioxide enrichment of cotton: root morphological characteristics. *Journal of Environmental Quality* **24**, 678–683.

Pritchard, S.G. and Rogers, H.H. (2000) Spatial and temporal deployment of crop roots in CO_2-enriched environments. *New Phytologist* **147**, 55–71.

Pritchard, S.G., Rogers, H.H., Prior, S.A. and Peterson, C.M. (1999) Elevated CO_2 and plant structure: a review. *Global Change Biology* **5**, 807–837.

Richards, B.G. and Greacen, E.L. (1986) Mechanical stresses on an expanding cylindrical root analogue in granular media. *Australian Journal of Soil Research* **24,** 393–404.

Robinson, D. (1994) The responses of plants to non-uniform supplies of nutrients. *New Phytologist* **127**, 635–674.

Robinson, D. (1996) Resource capture by localized root proliferation: why do plants bother? *Annals of Botany* **77**, 179–185.

Robinson, D., Hodge, A., Griffiths, B.S. and Fitter, A.H. (1999) Plant root proliferation in nitrogen-rich patches confers competitive advantage. *Proceedings of the Royal Society, London* **B266**, 431–435.

Rogers, H.H., Prior, S.A., Runion, G.B. and Mitchell, R.J. (1996) Root to shoot ratio of crops as influenced by CO_2. *Plant and Soil* **187**, 229–248.

Ryan, P.R., Skerrett, M., Findlay, G.P., Delhaize, E. and Tyerman, S.D. (1997) Aluminium activates an anion channel in the apical cells of wheat roots. *Proceedings of the National Academy of Science* **94**, 6547–6552.

Schenk, H.J. and Jackson, R.B. (2002) Rooting depths, lateral root spreads and below-ground/above-ground allometries of plants in water-limited ecosystems. *Journal of Ecology* **90**, 480–494.

Scholefield, D. and Hall, D.M. (1985) Constricted growth of grass roots through rigid pores. *Plant and Soil* **85**, 153–162.

Seiler, G.J. (1998) Influence of temperature on primary and lateral root growth of sunflower seedlings. *Environmental and Experimental Botany* **40**, 135–146.

Setter, T. and Belford, R.K. (1990) Waterlogging: how it reduces plant growth and how plants can overcome its effects. *Western Australia Journal of Agriculture* **31**, 51–55.

Shannon, R.D., White, J.R., Lawson, J.E. and Gilmour, B.S. (1996) Methane efflux from emergent vegetation in peatlands. *Journal of Ecology* **84**, 239–246.

Sharp, R.E. and Davies, W.J. (1985) Root growth and water uptake by maize plants in drying soil. *Journal of Experimental Botany* **36**, 1441–1456.

Sharp, R.E., Kuhn Silk, W. and Rost, T.L. (1988) Growth of the maize primary root at low water potentials. I. Spatial distribution of expansive growth. *Plant Physiology* **87**, 50–57.

Soane, B.D. and van Ouwerkerk, C. (1994) *Soil Compaction in Crop Production*. Elsevier, Amsterdam, The Netherlands.

Stirzaker, R.J., Passioura, J.B. and Wilms, Y. (1996) Soil structure and plant growth: impact of bulk density and biopores. *Plant and Soil* **185**, 151–162.

Stone, D.A. and Taylor, H.M. (1983) Temperature and the development of the taproot and lateral roots of four indeterminate soybean cultivars. *Agronomy Journal* **75**, 613–618.

Takahashi, H. (1997) Hydrotropism: the current state of our knowledge. *Journal of Plant Research* **110**, 163–169.

Takahashi, N., Yamazaki, Y., Kobayashi, A., Higashitani, A. and Takahashi, H. (2003) Hydrotropism interacts with gravitropism by degrading amyloplasts in seedling roots of arabidopsis and radish. *Plant Physiology* **132**, 805–810.

Tardieu, F. and Pellerin, S. (1991) Influence of soil temperature during root appearance on the trajectory of nodal roots of field grown maize. *Plant and Soil* **131**, 207–214.

Tardieu, F., Bruckler, L. and Lafolie, F. (1992) Root clumping may affect the root water potential and the resistance to soil-root water transport. *Plant and Soil* **140**, 291–301.

Taylor, H.M. and Ratliff, L.F. (1969) Root elongation rates of cotton and peanuts as a function of soil strength and soil water content. *Soil Science* **108**, 113–119.

Tennant, D., Scholz, G., Dixon, J. and Purdie, B. (1992) Physical and chemical characteristics of duplex soils and their distribution in the south-west of Western Australia. *Australian Journal of Experimental Agriculture* **32**, 827–843.

Tester, M. and Davenport, R. (2003) Na^+ tolerance and Na^+ transport in higher plants. *Annals of Botany* **91**, 1–25.

Thomson, C.J., Colmer, T.D., Watkin, E.L.J. and Greenway, H. (1992) Tolerance of wheat (*Triticum aestivum* cvs. Gemenya and Kite) and triticale (*Triticosecale* cv. Muir) to waterlogging. *New Phytologist* **120**, 335–344.

Trought, M.C.T. and Drew, M.C. (1980) The development of waterlogging damage in wheat seedlings (*Triticum aestivum* L.). I. Shoot and root growth in relation to changes in the concentrations of dissolved gases and solutes in the soil solution. *Plant and Soil* **54**, 77–94.

Troughton, A. (1980) Production of root axes and leaf elongation in perennial ryegrass in relation to dryness of the upper soil layer. *Journal of Agricultural Science, Cambridge* **95**, 533–538.

Vincent, C.D. and Gregory, P.J. (1989a) Effects of temperature on the development and growth of winter wheat roots. I. Controlled glasshouse studies of temperature, nitrogen and irradiance. *Plant and Soil* **119**, 87–97.

Vincent, C.D. and Gregory, P.J. (1989b) Effects of temperature on the development and growth of winter wheat roots. II. Field studies of temperature, nitrogen and irradiance. *Plant and Soil* **119**, 99–110.

Voesenek, L.A.C.J., Armstrong, W., Bögemann, G.M., McDonald, M.P. and Colmer, T.D. (1999) A lack of aerenchyma and high rates of radial oxygen loss from the root base contribute to the waterlogging intolerance of *Brassica napus*. *Australian Journal of Plant Physiology* **26**, 87–93.

Voorhees, W.B., Allmaras, R.R. and Johnson, C.E. (1981) Alleviating temperature stress. In: *Modifying the Root Environment to Reduce Crop Stress* (eds G.F. Arkin and H.M. Taylor), pp. 217–266. American Society of Agricultural Engineers, Michigan, USA.

Wan, S., Norby, R.J., Pregitzer, K.S., Ledford, J. and O'Neill, E.G. (2004) CO_2 enrichment and warming of the atmosphere enhance both productivity and mortality of maple tree fine roots. *New Phytologist* **162**, 437–446.

Welbank, P.J. and Williams, E.D. (1968) Root growth of a barley crop estimated by sampling with portable powered soil-coring equipment. *Journal of Applied Ecology* **5**, 477–481.

Welbank, P.J., Gibb, M.J., Taylor, P.J. and Williams, E.D. (1974) Root growth of cereal crops. In: *Report for Rothamsted Experimental Station for 1973,* Part 2, pp. 26–66. Rothamsted Experimental Station, UK.

Wenzl, P., Chaves, A.L., Patiño, G.M., Mayer, J.E. and Rao, I.M. (2002) Aluminium stress stimulates the accumulation of organic acids in root apices of *Brachiaria* species. *Journal of Plant Nutrition and Soil Science* **165**, 582–588.

Whiteley, G.M. and Dexter, A.R. (1983) Behaviour of roots in cracks between soil peds. *Plant and Soil* **74**, 153–162.

Whiteley, G.R. and Dexter, A.R. (1984) The behaviour of roots encountering cracks in soil. I. Experimental methods and results. *Plant and Soil* **77**, 141–149.

Williamson, L.C., Ribrioux, S., Fitter, A.H. and Leyser, O. (2001) Phosphate availability regulates root system architecture in *Arabidopsis. Plant Physiology* **126**, 875–882.

Wilson, A.J., Robards, A.W. and Goss, M.J. (1977) Effects of mechanical impedance on root growth in barley, *Hordeum vulgare* L. II. Effects on cell development in seminal roots. *Journal of Experimental Botany* **28**, 1216–1227.

Yao, J. and Barber, S.A. (1986) Effect of one phosphorus rate placed in different soil volumes on P uptake and growth of wheat. *Communications in Soil Science and Plant Analysis* **17**, 819–827.

Zhang, H.M. and Forde, B.G. (1998) An Arabidopsis MADS box gene that controls nutrient-induced changes in root architecture. *Science* **279**, 407–409.

Zhang, H., Jennings, A., Barlow, P.W. and Forde, B.G. (1999) Dual pathways for regulation of root branching by nitrate. *Proceedings of the National Academy of Science* **96**, 6529–6534.

Chapter 6

Roots and the Biological Environment

The stimulation of microbial activity in the rhizosphere described in section 1.3.1 arises from the release of exudates from plant roots. The quantities of exudates released may be substantial (see section 7.1), with fast-growing bacteria, actinomycetes and fungi stimulated to grow by the carbon pulses provided as exudates. The increased amounts of microbial biomass and activity in the rhizosphere lead, in turn, to increases in numbers of microfaunal grazers such as protozoa and nematodes, which have effects on the mineralization and availability of many plant nutrients. The effects of the plant, then, on the microbial community are far-reaching, but the microbes that colonize the rhizosphere also have an impact on plants via their interaction with roots. Some microbes are plant pathogens (see section 6.3), whereas others contribute beneficial effects. The beneficial effects of mycorrhizal fungi and of bacteria belonging to the genera *Rhizobium* and *Frankia* have been widely studied (see section 6.2) and result in symbioses that produce modifications to the morphology and architecture of root systems. Other bacteria, termed plant growth-promoting bacteria, have beneficial effects on plants without the development of such evident symbiotic structures.

Roots are eaten by a wide range of herbivorous insects and parasitized by other plants. These relationships are economically important in managed production systems resulting in considerable crop losses, but there is also a growing interest in how such interactions affect inter-plant competition and the spatial and temporal patterns of natural plant communities.

6.1 Interactions of roots with soil organisms

6.1.1 Root–rhizosphere communication

As described in section 1.3.1, the rhizosphere is a zone that is densely populated with soil organisms, including bacteria, yeasts, fungi, protozoa and insects, feeding on a wide range of substrates. Much research has now demonstrated that compounds released from roots may act as messengers that communicate and initiate interactions between roots and a wide range of soil-dwelling organisms (Walker *et al.*, 2003). Root–microbe and root–insect communication can be either positive (symbiotic) to the plant (e.g. via associations with mycorrhizal fungi, and N-fixing bacteria) or negative to the plant (e.g. interactions with parasitic plants, pathogenic microbes and herbivorous insects), and involves signal exchange and

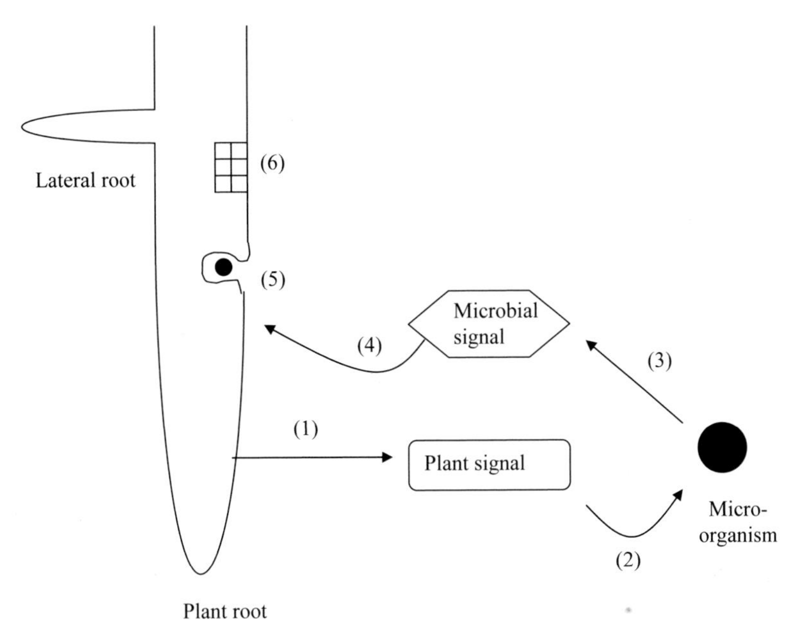

Fig. 6.1 Schematic depiction of different stages of plant–microbe interactions. (1) Exudation of a plant signal into the soil, (2) perception of the plant signal by the microorganism, (3) response of the microorganism with a new signal, (4) perception by the plant of the microbial signal, (5) invasion of the plant by the microorganism, and (6) initiation of plant cell division and/or differentiation. (Reproduced with permission from Mathesius, *Plant and Soil*, Springer Science and Business Media, 2003.)

perception, followed by invasion of the plant by the microbe/insect and concomitant structural changes (Fig. 6.1). In its interactions with other organisms, then, the plant has to control two processes simultaneously: (i) regulating the invasion of the microbe; and (ii) initiating developmental changes resulting from the interaction (Mathesius, 2003).

Mathesius (2003) examined the interactions between plants and the microbes involved in *Rhizobium*–legume symbiosis, mycorrhizas and nematode-induced galls, which, together with lateral root formation, produce a wide range of outcomes. She demonstrated that there are substantial overlaps in the signalling pathways underlying these interactions and concluded that these endophytic microorganisms have 'co-opted plant developmental pathways in their evolution', and that 'the plant is actively orchestrating its interactions with microorganisms'. Results from studies with mutants defective in root developmental pathways show that similar genes are involved in nodulation, root development and mycorrhizal formation. During invasion (stage 5 of Fig. 6.1), microbes often form infection threads or cell wall invaginations requiring changes in cell wall flexibility and re-arrangement of the cytoskeleton, and post-invasion, nutrient exchange systems between host and invader develop. Similar genes control these processes irrespective of the invading organism. Lateral root, nodule and gall formation all result from initiation of cortical or pericycle cell divisions and differentiation (arbuscular mycorrhiza cause only differentiation not division), regulated by plant hormones, especially auxin and cytokinins. Similarities in the perturbance of auxin balance have been established for lateral root, nodule and gall formation, and mycorrhizal symbiosis, possibly mediated by specific plant flavonoids. Mathesius (2003) proposed a schema that, by varying the cell specificity of responses, allowed a plant to form different organs despite the same genes and physiological changes being involved in all cases (Plate 6.1).

Several hundred secondary metabolites from plants and microbes have been identified and implicated in plant–microbe signalling. Werner (2001) identified four groups of signal molecules that have been shown to influence the interaction between plants and microbes:

(1) Flavonoids and isoflavonoids – flavonoids are produced by a wide range of plants but identified as specific molecules in the rhizobia/legume host plant communication. For example, 3,5,7,3'-tetrahydroxy-4'-methoxyflavanone induces the nodulation (*nod*) gene in *Rhizobium leguminosarum* bv. *viciae*, while 3',4',5,7-tetrahydroxyflavone is the inducer of *Sinorhizobium meliloti*. In contrast to the flavonoids, isoflavonoids have a more limited distribution almost confined to the subfamily Papilionoidae. Similar compounds may also play a role in the colonization of root by vesicular-arbuscular mycorrhiza (Walker *et al.*, 2003).
(2) Nod factors – the signal molecules above that activate the nod genes, result in the bacteria producing nod factors (lipochitooligosaccharides) which trigger reactions in the plant such as root-hair curling and meristem induction.
(3) Antibiotic substances – are produced by a wide range of microbes including *Pseudomonas*, *Agrobacterium*, *Rhizobium*, *Erwinia*, *Burkholderia* and *Bacillus*. The fluorescent pseudomonads have been most studied and found to produce up to seven different compounds. Plants, too can produce antimicrobial compounds. For example, basil (*Ocimum basilicum*) released rosmarinic acid (a constitutive antimicrobial compound) in response to the pathogenic fungi *Phytophthora cinnamoni* and *Pythium ultimum* which gave potent antimicrobial activity against an array of soil-borne microbes including *Pseudomonas aeruginosa* and *Aspergillus niger* (Bais *et al.*, 2002).
(4) Vitamins – the effect of vitamins such as thiamine (vitamin B_1), biotin (vitamin H) and pyridoxine (vitamin B_6) in promoting bacterial growth in the rhizosphere has been demonstrated with several plants.

Plant roots also release compounds that are able to disrupt communication between bacteria thereby reducing their susceptibility to infection (Fray, 2002). Many Gram-negative and Gram-positive bacteria, including several important plant pathogens such as *Agrobacterium* spp., *Erwinia* spp. and *Pseudomonas* spp., possess an ability to sense the number and growth status of fellow members within their vicinity. This ability, known as quorum-sensing, enables the individuals in a population to co-ordinate their growth and activities to achieve ends that would not be successful if performed independently (e.g. successful invasion of roots by a pathogen often requires a threshold level of pathogen to be exceeded). Quorum-sensing is achieved by cell-to-cell communication between bacteria using signalling molecules (typically N-acylhomoserine lactones, AHLs, for Gram-negative bacteria, and peptide-signalling molecules for Gram-positive bacteria) which, on achieving a threshold concentration, activate proteins that induce specific genes (Whitehead *et al.*, 2001; Walker *et al.*, 2003). AHLs are found in pathogenic, symbiotic and biological control strains of bacteria, with a higher proportion of AHL-producing bacteria present in the rhizosphere than the bulk soil, suggesting a general role in rhizosphere colonization (Newton and Fray, 2004). Some plants have evolved strategies to interfere with the AHL signalling system including the production of signal mimics, signal blockers, signal-degrading enzymes, or compounds that block the activity of AHL-producing enzymes (Fray, 2002). AHL mimics have been detected in a range of higher plants (e.g. Teplitski *et al.*, 2000), suggesting active interference of eukaryotes with bacterial communication. Mathesius *et al.* (2003) found that exudates from the legume *Medicago truncatula* exposed to either 3-oxo-C_{12}-HL or C_6-HL (the principal AHL produced by certain *Chromobacterium*, *Yersinia* and *Rhizobium* strains) contained signal-mimic compounds that were quantitatively, and perhaps qualitatively, different after exposure to the two AHLs. Their results also suggest that AHLs may provide plants with important information about the presence of bacteria 'that sets in motion an extensive and appropriate set of responses'.

6.1.2 Interactions with bacteria

The symbiotic association of roots and N-fixing bacteria will be described in section 6.2.1, so this section will describe other interactions. The distribution of bacteria along roots is often assumed to reflect patterns of exudation with high numbers near the tip and higher populations further back associated with exudation due to lateral root formation (van Vuurde and Schippers, 1980). This essentially static concept of populations was questioned by Semenov *et al.* (1999) who measured temporal oscillations of both copiotrophic (i.e. fast growing) and oligotrophic (i.e. able to live at low substrate levels) bacteria along wheat roots up to 0.9 m in length (Fig. 6.2). The patterns varied with time and were not consistently correlated with total organic carbon concentrations or with zones of lateral root formation. Semenov *et al.* (1999) suggested that the wavelike patterns were a consequence of cycles of bacterial growth and death and that if the root tip is considered as a moving source of nutrients, then such waves should result. The composition of the rhizosphere bacterial community is likely to be influenced, then, by the nature of the exudates released (i.e. plant species-dependent), location on the root (influencing the quantity and composition of exudates released) and soil type (influencing carbon availability and root–soil adhesion).

Marschner *et al.* (2001) investigated the effects of these factors on the bacterial community composition of three plants (chickpea, rape and Sudan grass) grown for 7.5 weeks in undisturbed cores of three soils (a sand, sandy loam and clay) using polymerase chain reaction-denaturing gradient gel electrophoresis (PCR-DGGE) to analyse the microbial community at either the root tip or the zone of lateral root production. Analysis of 16S rDNA band profiles showed that the three plants had similar indices of bacterial diversity when compared with respect to root zone and soil type. Diversity was significantly greater in the mature root zone than at the root tip in the sandy soil and the clay, but not in the loamy sand. Canonical correspondence analysis showed significant effects of soil type, plant species and root zone on the bacterial community structure (Fig. 6.3). Soil and root zone effects cumulatively explained 84.7, 83.0 and 81.6% of the total variance in chickpea, rape and Sudan grass, respectively. In chickpea, the community composition of the sandy soil differed from that of the other two soils, but in the other two plants the root zone affected composition more than soil type. In the sand and sandy loam, community composition was influenced more by the root zone than the plant species, whereas in the clay soil

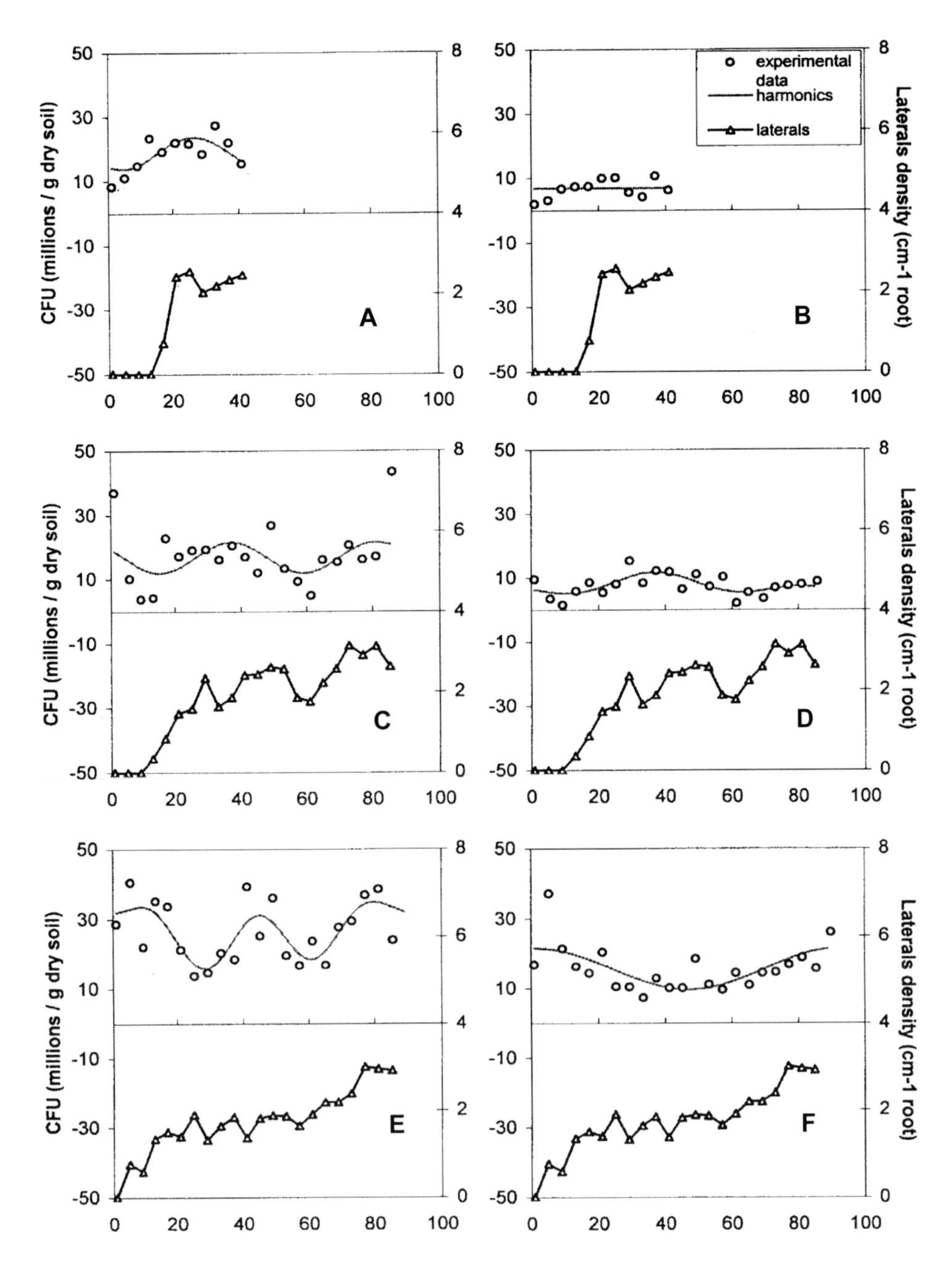

Fig. 6.2 Colony-forming units (CFUs) of copiotrophic (A, C, E) and oligotrophic (B, D, F) bacteria, and concentration of total water-soluble organic carbon (TOC – G, H, I) in the rhizosphere with distance from the root tip of wheat at 2 (A, B, G), 3 (C, D, H) and 4 (E, F, I) weeks after planting. Points are differences of geometric means ($n = 3$) between rhizosphere and bulk soil; lines are predicted values from harmonic analysis. The numbers of laterals per centimetre of root axis are shown in the lower graphs. (Reproduced with permission from Semenov *et al.*, *Microbial Ecology*; Springer Science and Business Media, 1999.) *Continued.*

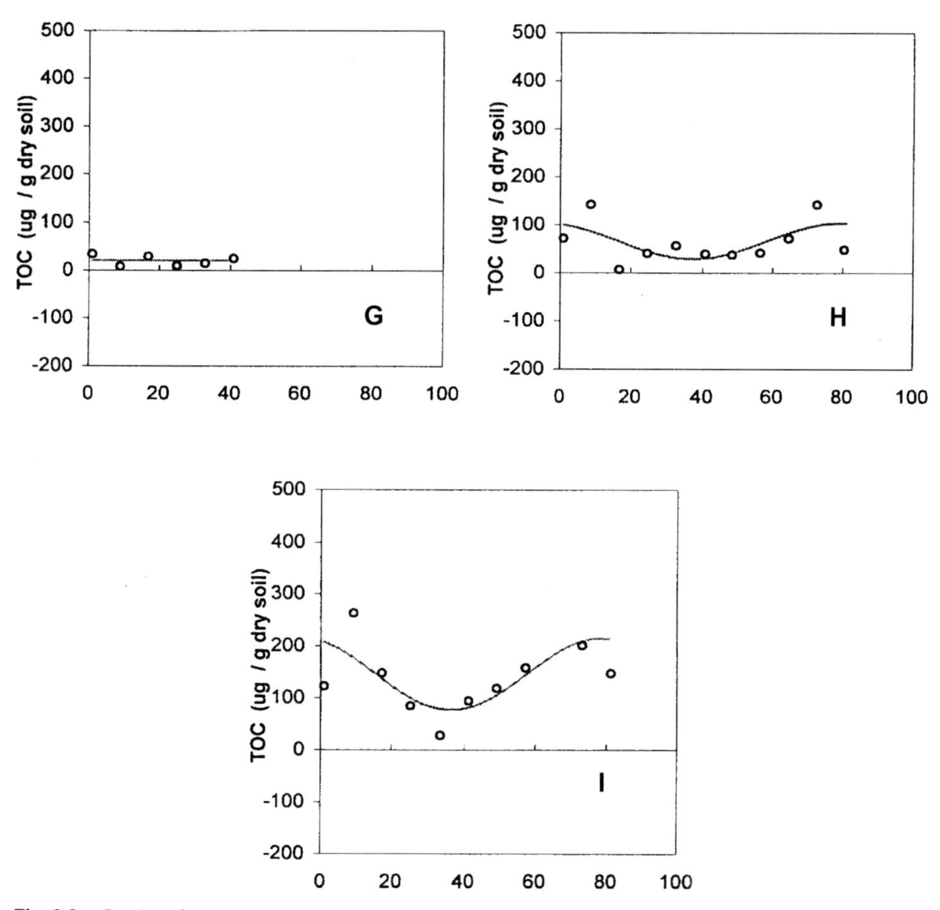

Fig. 6.2 *Continued.*

the community associated with chickpea differed considerably from that of the other plant species. These complex interactions mean that the relative importance of any single factor, for example soil type, on bacterial community structure cannot yet be specified.

Similar PCR-DGGE analysis has also demonstrated the changes in bacterial composition and functional diversity with distance from roots. Kandeler *et al.* (2002) measured spatial gradients in the bacterial community and enzyme activity from a planar root surface of maize seedlings. Communities in unplanted controls were unaffected by distance, but with plants there was a gradient in composition within the first 2.2 mm from the root surface and that composition was also different from those further away from the root. The gradients of invertase, and alkaline and acid phosphatase activities were much closer to the root surface (0.2–0.8 mm); there was no gradient of xylanase activity. Mycorrhizal colonization of maize roots, too, changed the bacterial community structure on the root surface and in the non-rhizosphere soil (Marschner and Baumann, 2003). This change was, at least partially, plant-mediated because the effects of mycorrhizas were not confined to the colonized parts of the root system. The importance, if any, of these community changes in bacteria for the rhizosphere and for plant growth has yet to be deduced.

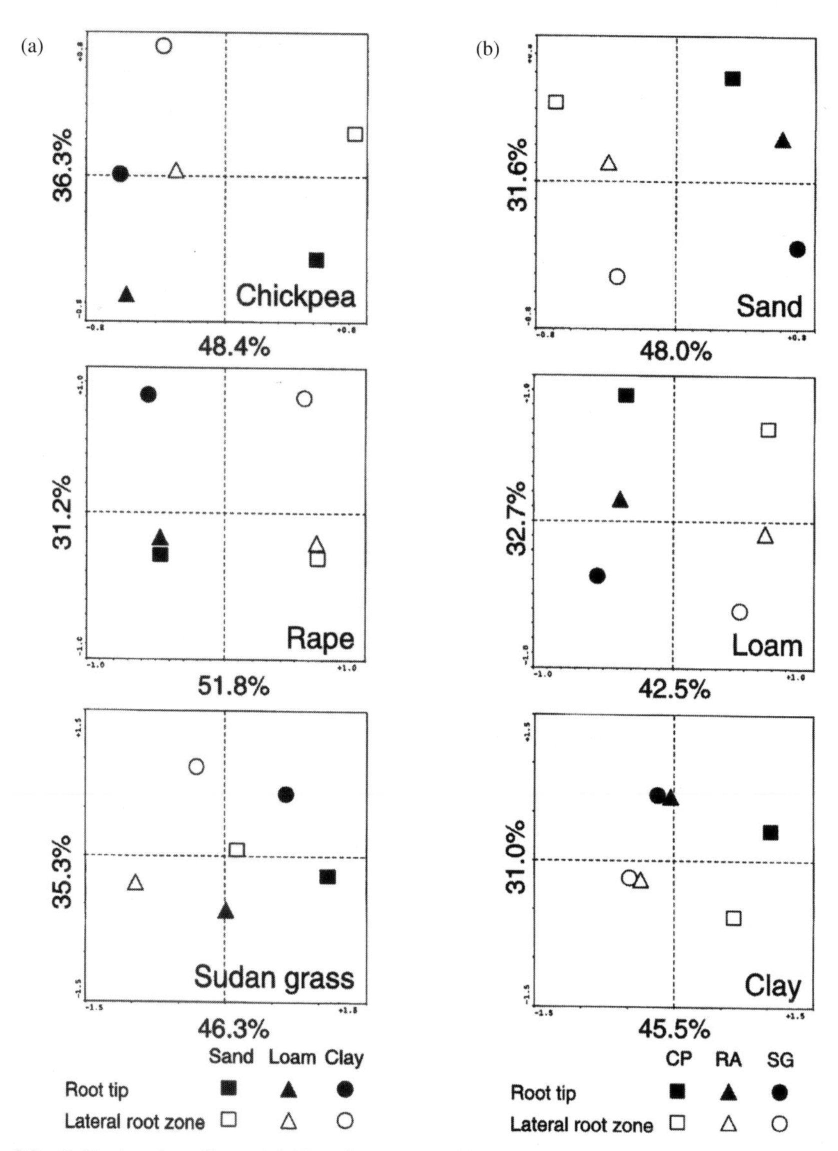

Fig. 6.3 Ordination plots of bacterial rhizosphere communities at the root tip and zones of lateral root emergence generated by canonical correspondence analysis of 16S rDNA profiles for chickpea, rape and Sudan grass growing in three soils: (a) effects of plant species; (b) effects of soil type. (Reproduced with permission from Marschner *et al.*, *Soil Biology and Biochemistry*; Elsevier, 2001.)

Some bacteria promote plant growth (plant growth-promoting bacteria, PGPB) while others, such as *Pseudomonas* spp., have been implicated as inhibitory organisms. Mantelin and Touraine (2004) reviewed the contributions of the two features of PGPB that contribute most to plant performance: (i) many PGPB are diazomorphs that fix N_2 but is any of this N transferred to the host plant? and (ii) PGPBs produce phytohormones that affect root development and growth. Overall, while the nitrogen-fixation genes of rhizobacteria such as *Azospirillum* and *Azoarcus* are expressed in the presence of roots of plants such as wheat and rice, there is little evidence for a transfer of N to the plant so that a direct nutritional role seems unlikely (Bremer *et al.*, 1995). There may, though, be an indirect signalling role for the NH_4^+ ions released by the PGPB in their interactions with plants. However, PGPB induce substantial alterations in root development through increased lateral root length and root hair number and length. These effects on root system development are thought to be the results of phytohormones such as auxins, cytokinins and giberellins produced by the bacteria, although it is not entirely clear whether one or several compounds are responsible for the different changes observed (Mantelin and Touraine, 2004). The increased lateral and root hair growth increases the surface area available for nutrient uptake and this may contribute to the increased nutrient uptake found in plants grown with PGPB (Okon and Kapulnik, 1986). It remains to be determined whether PGPB stimulate NO_3^- uptake directly through stimulation of the plant transport system and/or to what extent the changes in root development contribute to the greater nutrient uptake.

Some bacteria have been demonstrated to have deleterious effects on plant growth, although the effects of individual isolates of rhizobacteria can fluctuate from inhibitory to promotive depending on environmental conditions, the host genotype and mycorrhizal status (Nehl *et al.*, 1996). In field studies, rhizosphere bacteria have been implicated in yield decline in agricultural systems, especially in monocultures (Rovira *et al.*, 1990). The mechanisms resulting in the deleterious effects on growth are several and complex, and may include the production of phytotoxins, phytohormone production, competition for nutrients and inhibition of mycorrhizal function. In their review, Nehl *et al.* (1996) demonstrate a wide range of deleterious effects for several *Pseudomonas* isolates, and, in a later study, fast root elongation of wheat roots in eastern Australia was associated with the avoidance of accumulation of *Pseudomonas* spp. (Watt *et al.*, 2003). Nodal axes of wheat from the upper 20 cm of a red earth (22% clay) in a field experiment elongated in loose (cultivated) soil at three times the rate (20 mm d^{-1}) of those in compacted (direct drilled) soil, had longer elongation zones (6.9 mm vs 1.6 mm), and had about 20 times less *Pseudomonas* spp. and 8 times less total bacteria per unit length than those grown in compacted field soil. The differences in bacterial populations were greatest in the 10 mm nearest the root tips (Plate 6.2). It is not known with certainty whether the *Pseudomonas* spp. cultured in this study were the cause of the inhibited root growth but *Pseudomonas* spp. cultured from the rhizosphere of direct drilled wheat reduced root extension when re-applied to wheat roots in a bioassay (Simpfendorfer *et al.*, 2002). However, not all effects of *Pseudomonas* are deleterious. For example, Weller *et al.* (1988) showed that fluorescent pseudomonads decreased the incidence of the take-all fungus (*Gaeumannomyces graminis*) on soils used for long-term production of wheat and that the suppression was mediated in part by the production of antibiotics and/or siderophores.

Crops grown as continuous monocrops frequently show declining yields with time. Various reasons have been suggested for this phenomenon including depleted nutrients and soil organic matter contents, and degraded soil structure, but changes in biological populations may also play a part. For example, in the Mallee region of southern Australia, yields of second and subsequent wheat crops are typically 5–20% less than first wheats even when practices to control take-all, *Rhizoctonia* and cereal cyst nematodes are adopted. Gupta *et al.* (2004) showed that these yield reductions were a result of biological activity and further demonstrated that the most poorly performing wheat variety had a higher rhizosphere population of copiotrophic and lower number of oligotrophic bacteria than the best performing variety.

6.1.3 Interactions with fungi

Some mycorrhizal and pathogenic fungal associations with roots will be detailed in sections 6.2.2 and 6.3.1, but there is a variety of other fungal associations with roots that do not fit these categories. For example, dark septate root endophytes are a heterogeneous group of conidial or sterile fungi that are distributed throughout the world and associated with roots of many plant species (Jumpponen and Trappe, 1998). Their ecology and function are largely unexplored, and despite their ubiquity, studies to date have been inconsistent regarding their interaction with host plants. Most such fungi colonize roots via the presence of sparse superficial mycelium growing along depressions between epidermal cells, which then form an appressorium before penetration of the cell walls of the cortical layer via a penetration tube; a microsclerotia of rounded, thick-walled cells subsequently forms within the cortical cells of the host root (Jumpponen and Trappe, 1998).

6.1.4 Interactions with protozoa

Protozoa feed mainly on bacteria, yeasts or fungi, and because the rhizosphere is rich in such organisms, protozoa also accumulate in this region. Zwart *et al.* (1994) reviewed several field and pot experiments and found that the biomass of protozoa in the rhizosphere was, on average, six times that in the bulk soil although the variability was large. All comparisons demonstrated an increase in protozoan numbers or biomass in the rhizosphere compared with the bulk soil. The interactions between plant roots and protozoa are, then, largely indirect with carbon and other substrate fluxes promoting the growth of bacteria which are then consumed by microfaunal grazers such as protozoa. Griffiths (1994) demonstrated that this consumption of bacterial biomass by protozoa enhanced flows of nutrients in soil both directly via the excretion of consumed nutrients and mineralization of nutrients on death, and indirectly via changes to the composition of the microbial community and enhancement of microbial mineralization activity in those components remaining. Without this 'microbial loop' whereby consumption of microbes by protozoa (and nematodes) constantly remobilizes nutrients for plant (and microbial) uptake, nutrients would remain sequestered in bacterial biomass. Because of the small difference in C:N ratios of the protozoan predator and bacterial prey and a relatively low assimilation efficiency, typically about 60% of the ingested nutrients are excreted and available for plant or microbial uptake (Griffiths, 1994; Bonkowski, 2004).

Root architecture is also affected by protozoan grazing, at least in laboratory experiments. On agar plates, plants develop an extensive and highly branched root system in the presence of protozoa. Release of ammonium from protozoa, which is rapidly oxidized to nitrate by nitrifying bacteria, may partially explain the increased production of laterals (see section 5.5.1; Forde, 2002), but this effect has also been shown to be due to hormonal effects on root growth rather than to nutrient effects alone (Bonkowski, 2004). Some workers have suggested a direct release of plant hormones by protozoa, but Bonkowski and Brandt (2002) showed that enhanced lateral root growth in water cress (*Lepidium sativum*) was a consequence of stimulation of auxin-producing bacteria. Bonkowski (2004) suggests that the additional root surface produced will allow greater nutrient absorption from the soil and increased rates of exudation which will further stimulate the bacterial–protozoan interaction. So, in addition to stimulating nutrient flows, protozoa promote a mutually beneficial relationship between plant roots and bacteria in the rhizosphere (Fig. 6.4).

6.1.5 Interactions with nematodes and mesofauna

In addition to the microflora and microfauna, plant roots share the soil with a wide range of mesofauna including nematodes, insects and earthworms. Plant-parasitic nematodes will be considered in more detail in section 6.3.2, but non-parasitic nematodes also play a role in root–soil interactions. As with protozoa, many nematodes predate on bacteria and fungi in the rhizosphere and make a significant contribution to N cycling in soils. For example, Ferris *et al.* (1997) measured C:N ratios of eight bacterial-feeding nematodes isolated from soils used to grow agricultural crops and measured values of 5.16–6.83 (mean 5.89) after 48 hours of starvation. This ratio was greater than that of the six bacteria used in the study (mean 4.12). While the population and proportions of individual nematode species changed with time in the field, Ferris *et al.* (1997) estimated that the nematode community contributed 0.28, 0.98 and 1.38 kg N ha^{-1} in April, May and June, respectively.

Among the arthropods in soils, collembola (springtails) are among the most common (Hopkin, 1997). Collembola show definite preferences for certain types of food, with fungal hyphae (including mycorrhizal fungi and root pathogens) preferred over most other food sources. They also consume decomposing leaf litter, and many will feed on fine roots. Interactions between springtails, fungi and roots have not been studied much but the selective feeding on mycorrhizal symbionts and pest fungi can significantly affect the perform-

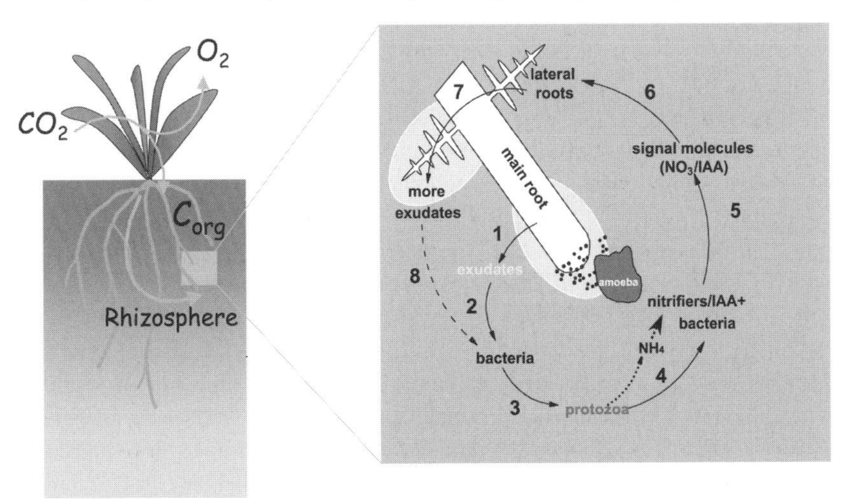

Fig. 6.4 A conceptual model illustrating microfaunal-induced hormonal effects on root growth. Root exudation (1) stimulates growth of a diverse bacterial community (2) and subsequently of bacterial-feeders such as protozoa (3). Ammonia is excreted by protozoa and selective grazing favours nitrifiers and IAA-producing bacteria (4). The release of signal molecules (5), such as nitrate and IAA, induces lateral root growth (6), leading to release of more exudates (7), and more bacterial growth (8), etc. (Reproduced with permission from Bonkowski, *New Phytologist*; New Phytologist Trust, 2004.)

ance of the host plant. For example, Boerner and Harris (1991) studied the competition between mycorrhizal grass *Panicum virgatum* and the non-mycorrhizal black mustard (*Brassica nigra*) in the presence and absence of *Folsomia candida*, a collembolan known to feed preferentially on arbuscular mycorrhizal fungi. Grazing of the mycorrhizal fungi by the collembolan increased N availability which was taken up by the more extensive root system of *B. nigra* when plants were sown simultaneously, so that shoot and root mass of *P. virgatum* was reduced relative to plants not subject to competition. When *B. nigra* was sown later than *P. virgatum*, its smaller root system was unable to take advantage of the additional N and the grass benefited.

Earthworms, too, may also feed on plant roots as well as root detritus and dead organic matter. Lee (1985) summarized studies that showed that the gut contents of species such as *Aporrectodea rosea* and *Lumbricus rubellus* contained roots but that some species switched from leaf to root materials depending on the type of vegetation. Cortez and Bouche (1992) used ^{14}C-labelled plants in columns to show that diets of *Lumbricus terrestris* contained significant quantities of young roots and their associated microflora, and recent root residues, at least under their experimental conditions. More generally, though, earthworms have been shown to stimulate plant and root growth through a number of indirect mechanisms. Scheu (2003) reviewed 67 studies of the response of plants to the presence of earthworms, most with crop plants and European species of earthworms. Shoot biomass of plants was significantly increased in the presence of earthworms in 79% of the studies, and decreased in 9%. The effects on root biomass were less consistent than those on shoots, with only half of the studies in which it was measured (45% of the studies) showing an increase. The processes by which earthworms affect root growth are poorly understood but are thought to include: (i) changed soil structure; (ii) increased N and P availability; (iii) the production of growth-promoting compounds such as IAA; and (iv) dispersal of growth-stimulating and anti-pathogenic microbes. Earthworms also interact with other elements of the soil fauna and flora to influence root herbivore activity (Scheu, 2003) and nutrient transformations and transfer. For example, Tuffen *et al.* (2002) in a study involving leek (*Allium porrum* L.) plants, mycorrhizal fungi and earthworms (*Aporrectodea caliginosa*), found that ^{32}P transfer from a dying root system to unlabelled plants was increased with earthworms present, whereas mechanical disruption eliminated the effect of the mycorrhiza on ^{32}P transfer between mycorrhizal plants. They concluded that mobilization of ^{32}P, partly contained within mycorrhizal hyphae of the donor plant, was mediated by earthworms and led to the enhanced ^{32}P availability to the hyphae of the living plant.

Some species of earthworm such as *Eisenia foetida* produce humic substances with hormonal activity that can affect plant growth. The mechanism of these processes is uncertain, but Canellas *et al.* (2002) found that humic acids isolated from *Eisenia foetida* compost with cattle manure as substrate contained auxin groups. The humic acids stimulated root growth (total root length and surface area was increased by up to 50%) and enhanced the number (by 7–12-fold) and length of lateral roots of 7-day-old maize seedlings when grown in nutrient solution. Humic acids also stimulated H^+-ATPase activity of the plasma membrane, which is probably associated with the induction of mitotic sites that are the precursors of lateral roots.

The burrowing behaviour of earthworms and the distribution of roots also interact. Springett and Gray (1997) investigated the burrowing behaviour of four earthworm species in pure swards of ryegrass, chicory, white clover and lucerne, in a field experiment on a silt loam in which different quantities of roots were achieved by cutting the swards to different heights. When earthworms were introduced into the established swards, the number and distribution of burrows was affected by both the quantity of roots and the plant species (Table 6.1). There were significantly less burrows under the cut sward compared with the uncut sward in 7 of the 10 months of the study, and there were significantly less burrows under ryegrass than the other plant species. Burrows of *Aporrectodea caliginosa* and *Lumbricus rubellus* were more superficial (*L. rubellus* had no burrows below 0.2 m), while those of *A. longa* and *L. terrestris* were deeper. As burrows were established, a small percentage (about 2% of the total after 10 months) of roots grew into them, so that after 10 months 20% of burrows had no roots in them, 42% one root, 34% two roots, and 4% had three or more roots.

6.2 Symbiotic associations

Symbiotic relationships are often established by plants with specific bacteria and fungi which aid the acquisition of nutrients from soil. The most studied relationships are those involving rhizobial bacteria and mycorrhizal fungi, both of which trigger the host genetic programme to allow localized infection and controlled growth. Some of the genes involved in transduction of signals from rhizobia involved in nodule formation are shared with mycorrhizal fungi with a receptor-like kinase required for the transduction of both bacterial and fungal symbiotic signals (Riely *et al.*, 2004). This suggests at least some common steps in seemingly different host/microbe symbioses, which is most likely due to similar infection processes of the invading symbionts (Kistner and Parniske, 2002).

6.2.1 Rhizobia and N fixation

Many books have been written on the symbiotic fixation of molecular nitrogen by rhizobia hosted in nodules by legume roots, and by the actinomycete, *Frankia*, which forms actinorhizal root nodules in eight different plant families that are mainly trees or woody shrubs mostly representing pioneer plants (e.g. Postgate, 1998; Giller, 2001; Leigh, 2002). This section will describe the main features of the root/rhizobium interaction and give some examples of the activity of nodulated root systems in fixing nitrogen.

Many, but by no means all, species of legumes have an ability to form root nodules, which are specialized plant organs, capable of the biological fixation of atmospheric

Table 6.1 The effects of type of vegetation and of cutting on the number and distribution of earthworm channels: the numbers are the mean number of intersections recorded at various depths with a glass tube

	Ryegrass		White clover		Chicory		Lucerne	
Depth (m)	Uncut	Cut	Uncut	Cut	Uncut	Cut	Uncut	Cut
0–0.1	0.87	0.18	1.54	0.86	2.15	1.28	1.18	0.69
0.1–0.2	0.81	0.10	1.46	0.56	1.49	0.87	0.12	0.12
0.2–0.3	0	0	1.05	0.30	1.43	0.50	0.26	0.17
0.3–0.4	0	0	0.83	0.44	1.33	0.50	0.30	0.20

Data adapted from Springett and Gray, 1997.

nitrogen. Nodule formation is induced by soil bacteria of the genera *Rhizobium*, *Sinorhizobium*, *Bradyrhizobium*, *Azorhizobium* and *Mesorhizobium*, generically called rhizobia. It has been known for a long time that some rhizobia are specific to a particular host plant while others are adapted to a range of hosts; the signalling mechanism that underlies these various responses is now clearer. Legume roots release flavonoids, with structures that depend on the plant host, which act as both chemoattractants and activators of *nod* gene products in potentially infective *Rhizobium* (Redmond *et al.*, 1986). These *nod* genes trigger the synthesis and secretion of specific signal molecules, called nod factors (lipo-chitin-oligosaccharides), which initiate the twin processes of nodule initiation and cell colonization (Plate 6.3). The receptors of these signals in the legume host have recently been identified and belong to a family of plant transmembrane receptor-like serine/threonine kinases whose extracellular regions contain Lys M domains with similarity to peptidoglycan-binding proteins (Cullimore and Dénarié, 2003; Madsen *et al.*, 2003).

In response to the flavonoid chemoattractant released from the root, rhizobial bacteria move towards the root, especially the root hair regions, and release Nod factors which are species-specific in their structure. Following recognition of these factors several processes are initiated by host plant cells including rapid influx of calcium ions and calcium spiking, expression of specific nodulin genes, alterations in the epidermal and root hair cells resulting in root hair curling and the formation of an infection thread, and division of cortical root cells (Oldroyd and Downie, 2004). The bacterium is enveloped by the curling root hair and enters the plant via an infection thread which develops as an intrusion in the hair cell wall (Plate 6.4). Eventually the infection thread grows from cell to cell in the root cortex as a transcellular tunnel and the bacterium is delivered to the developing nodule.

Simultaneous with the process of bacterium recognition and entry to the host is that of nodule formation. In most legumes, root nodules are formed by the re-initiation of cell division in the root cortex typically opposite protoxylem poles (Hirsch, 1992). A small group of dividing cells forms a nodule primordium that differentiates into a nodule that is subsequently invaded by bacteria. Exactly how Nod factors stimulate cell division is not known, but research points towards an indirect effect on plant hormone balance (Hirsch, 1992), especially the role of auxin in the stimulation of cell division and of root differentiation. Mathesius *et al.* (1998) demonstrated that Nod factors can induce specific flavonoids during nodulation of white clover which perturb auxin flow in the root during the early stages of nodule formation. It was subsequently shown (Mathesius, 2001) that flavonoids regulated the breakdown of auxin by peroxidases in a manner similar to that which occurs during lateral root formation (see Plate 6.1). In both lateral root and nodule formation, auxin accumulates in the first dividing cells, but this decreases later as cell differentiation and elongation proceed. IAA oxidases are thought to bring about this reduction of auxin concentrations with their activity locally regulated by flavonoid compounds such as the isoflavonoid formononetin.

In many legumes (e.g. clovers, beans and soyabean), root nodules are preferentially initiated just behind the root tip in the root hair zone, while in other legumes (e.g. groundnut) and woody plants lacking root hairs, the sites of lateral root emergence are favoured. Cortical cells at sites of lateral root emergence undergo similar changes in gene expression to those in the cortex of the young root zone, and it has therefore been suggested that rhizobia could 'hijack' cortical cells next to sites of lateral root emergence that have already been activated to divide by the plant (Mathesius *et al.*, 2000). The number of lateral root-associated nodules is often correlated with the number of laterals on the tap root and the nodule that evolves can be connected to the vascular system of either the tap root or the lateral root.

All nodules have an outer, uninfected cortical region surrounding vascular tissue which is connected to the host plant; this is separated from an inner, infected zone by a nodule endodermis (Fig. 6.5). Nodules are of two types: (i) determinate nodules grow for a fixed period and have a finite lifespan; whereas (ii) indeterminate nodules have an apical meristem which continues to be active for a prolonged period and results in infection zones of different ages; perennial legumes have indeterminate nodules allowing new activity each growing season. The inner infected zone usually contains a mixture of plant cells that either contain or do not contain rhizobial bacteria, although in some legumes (e.g. groundnut) all cells are infected. The bacteria are contained within a plant membrane (an invagination of the plasmalemma called the 'peribacteroid membrane') within the original host cortical cells (called a bacteroid) in which they may multiply. The role, if any, of the uninfected cells is only partially known. In plants that assimilate fixed N in the form of ureides (Table 6.2), the uninfected cells are the sites of ureide synthesis, but in legumes that assimilate fixed N as amides, their role is unknown.

The process of nitrogen fixation involves the reduction of atmospheric dinitrogen gas (N_2) and of two protons (Giller, 2001), and is effected by nitrogenase enzymes (Fisher and Newton, 2002):

$$N_2 + 8H^+ + 8e^- + 16ATP \rightarrow 2NH_3 + H_2 + 16ADP + 16P_i \quad (6.1)$$

Much of the resulting H_2 is oxidized within the nodule by hydrogenases but some may be lost as hydrogen gas. The ammonia gas is generally protonated to form ammonium ions which are then converted to glutamate and glutamine; this fixed N, though, is not synthesized by the bacteroid but by the host plant. Lodwig *et al.* (2003) showed that ammonium assimilation in the bacteroids of pea nodules was shut down and that the plant supplied amino acids to the bacteroids in exchange for amino acids returned to the plant for asparagines synthesis. This amino acid cycling reinforces the mutualistic relationship of host and rhizobium. Legumes can be divided into two groups depending on the form in which the

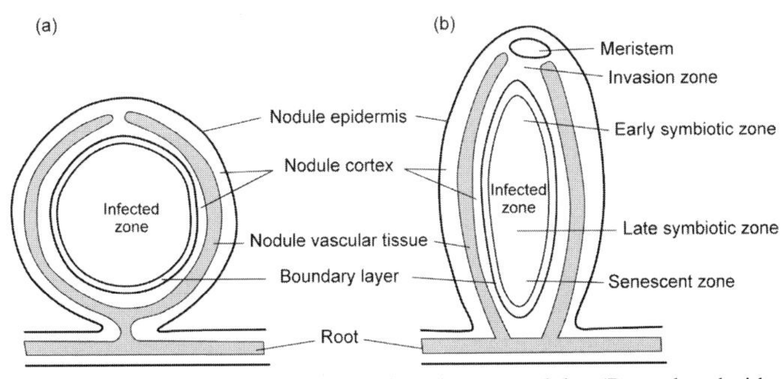

Fig. 6.5 Structure of (a) determinate, and (b) indeterminate legume nodules. (Reproduced with permission from Giller, *Nitrogen Fixation in Tropical Cropping Systems*, CAB International, 2001.)

fixed N is exported from the nodule (Table 6.2). Amide producers export either glutamine or asparagines synthesized in the infected cells, while ureide producers export either allantoin or allantoic acid synthesized as a consequence of uric acid synthesis in infected cells which is then transferred to adjacent uninfected cells where the final synthesis occurs. Most temperate legumes are amide exporters (with the notable exception of common bean, *Phaseolus vulgaris*), whereas most tropical legumes are ureide exporters (with the notable exception of groundnut). The proximity of the bacteroids to vascular tissues ensures the ready transfer of amino acids from rhizobium to host and *vice versa*.

The amount of N fixed by rhizobia is substantial and exceeds that of the world's fertilizer industry. Vitousek *et al.* (1997) reported that nitrogen fixation in terrestrial ecosystems worldwide prior to extensive human activity was 90–140 million t N a^{-1} which compares with about 80 million t of N applied in inorganic fertilizers and about 40 million t of N fixed by leguminous crops. Measuring the amounts of nitrogen fixed by plants is not an easy task and each method has its own advantages and limitations (Peoples *et al.*, 2002). The ^{15}N natural abundance and enrichment techniques are likely to produce the most certain estimates of fixation although even these are vulnerable to errors. N fixation has been measured most commonly in crop plants because of its economic importance in many agricultural systems worldwide. It is difficult, though, to generalize about the amounts of N fixed by particular crops because growth and N uptake vary considerably from season to season. Unkovich and Pate (2000) summarized literature for a range of annual legume crops and found a range of between 0 and 450 kg N ha^{-1} (Table 6.3). The values in Table 6.3 do not include any estimate of the amount of N contained in the root system or transferred to the soil; this N has been shown to represent a substantial proportion of the plant's total N content in several studies with fixed N present below ground in roots, nodules and root exudates. For example, in chickpea and faba bean crops grown on a vertisol in New South Wales, Australia, Khan *et al.* (2003) found that 25% and 77%, respectively, of total plant N was below ground in roots and as root-derived N in soil (the comparable value for barley was 36%). Similarly, for pasture legumes, McNeill *et al.* (1998) found in column-grown plants that 31–33% of N was below ground in subterranean clover (*Trifolium subterraneum* L.) and 25–29% in serradella (*Ornithopus compressus* L.), with <60% of the below-ground N in both species recovered in the washed roots.

Table 6.2 Examples of legumes that transport either amides or ureides in xylem sap as the dominant products of N_2 fixation exported from nodules

Amides (asparagines, glutamine)	Ureides (allanoin, allantoic acid)
Chickpea (*Cicer arietinum*)	Soyabean (*Glycine max*)
Lentil (*Lens culinaris*)	Pigeonpea (*Cajanus cajan*)
Pea (*Pisum sativum*)	Mung bean (*Vigna radiata*)
Faba bean (*Vicia faba*)	Black gram (*Vigna mungo*)
Lupin (*Lupinus angustifolius*)	Cowpea (*Vigna unguiculata*)
Groundnut (*Arachis hypogaea*)	Rice bean (*Vigna umbellata*)
	Common bean (*Phaseolus vulgaris*)

From Peoples *et al.*, 2002.

Table 6.3 Amounts of nitrogen fixed for several annual crop legumes

Crop	Area harvested (million ha)	%N in crop derived from the atmosphere	N fixed in shoot (kg ha^{-1})
Soyabean	67.6	0–95	0–450
Groundnut	23.7	22–92	32–206
Common bean	13.0	0–70	0–165
Chickpea	11.3	0–82	0–141
Pea	6.5	5–95	4–244
Lentil	3.3	28–87	5–191
Faba bean	2.2	19–97	12–330
Lupin	1.3	20–97	19–237

From Unkovich and Pate, 2000.

6.2.2 Mycorrhizas

As with rhizobial symbioses, there is a very large literature on this topic (see, for example, books by Harley and Smith, 1983; Smith and Read, 1997), but in contrast to the limited group of plants that form root/rhizobial associations, about 90% of seed-bearing plants belong to families in which most species form mycorrhizal associations, as do the roots of most pteridophytes. Similar associations are also found in bryophytes which do not have true roots. An important family to plant scientists and agriculturists that is non-mycorrhizal is the Brassicaceae (Cruciferae), so that neither *Arabidopsis* nor rape is mycorrhizal. Mycorrhizas ('fungus-roots') are, then, a normal part of the root system and its functioning for the vast majority of terrestrial plants and there is substantial evidence in the fossil record that the fungi and roots have co-evolved into the many forms that exist today. Mycorrhizas are important because they usually function as mutualistic symbioses in which the fungus supplies nutrients (especially ions that diffuse slowly in soils such as phosphate and zinc) to the plant in exchange for sugars from recent photosynthesis. This reciprocal transfer of resources is often prolonged, although it does not appear to be either directly or indirectly coupled, so that mycorrhizas do not always give positive growth responses and sometimes the fungus can appear functionally parasitic, at least under experimental conditions (Johnson *et al.*, 1997; Ryan and Graham, 2002).

Many plant species cannot complete their life cycle in their usual habitats without forming mycorrhizas (called obligately mycorrhizal plants), while others form mycorrhizas but can complete their life cycle without them (facultatively mycorrhizal). To some extent this division depends upon the medium in which the plants are grown because most obligately mycorrhizal plants can complete their life cycles when grown in fertilized soils (Smith *et al.*, 2003a). Seven types of mycorrhizas are recognized although several of these are very similar (Table 6.4) with many developmental stages in common (Smith *et al.*, 2003a):

(1) Growth of fungal hyphae from germinating spores or other sources of inoculum such as previously colonized plant roots.
(2) Recognition of a plant root as a potential partner by the fungus, and *vice versa*.
(3) Evasion or inhibition of attack/defence mechanisms of the partners.
(4) Extensive colonization of the root.

Table 6.4 Characteristics of the major types of mycorrhizas

	Type of mycorrhiza						
Feature	Arbuscular (AM)	Ecto-mycorrhiza (ECM)	Ectendo-mycorrhiza	Ericoid (ERM)	Arbutoid	Mono-tropoid	Orchid
Fungal taxa	Zygo	Basidio Asco (Zygo)	Basidio Asco	Asco	Basidio	Basidio	Basidio
Plant taxa	Bryo Pterido Angio Gymno	Angio Gymno	Angio Gymno	Angio: Ericales Bryo	Angio: Ericales	Angio: Ericales	Angio: Orchidaceae
Intracellular interfaces:							
Hyphal coils or pegs	+ or –	–	+	+	+	+	+
Arbuscules	+ or –	–	–	–	–	–	–
Extracellular interfaces:							
Intercellular hyphae	+ or –	+	+	–	+	+	–
Fungal sheath	–	+	+ or –	–	+ or –	+	–

Based on Smith and Read, 1997 and Smith *et al.* 2003b. Fungal taxa are abbreviated from Zygomycetes, Ascomycetes and Basidiomycetes, and the plant taxa from Bryophyta, Pteridophyta, Angiospermae and Gymnospermae. Presence or absence of structures in the mature state is indicated by + or –.

(5) Development of fungal–plant interfaces that are stable over periods of days, weeks or longer.
(6) Development of external fungal hyphae that spread extensively into the soil.

Three types of mycorrhizas are commonly encountered:

(1) *Arbuscular mycorrhizas* (AM) – are the most common mycorrhizal class occurring in about 80% of terrestrial plants including trees, shrubs, forbs and grasses. Originally known as vesicular-arbuscular mycorrhiza and placed in the zygomycete order Glomales, it has recently been proposed that the fungi belong in a new phylum, the Glomeromycota. There are two morphological types, *Arum* and *Paris*, with the latter the most common (Fig. 6.6). In *Arum* types, fungal hyphae in the root grow between cortical cells (intercellularly) and then penetrate the cortical cell walls to form profusely branched arbuscules, whereas in *Paris* types, the hyphae spread from cortical cell to cortical cell (i.e. intracellularly) forming well developed coils from which small, branched arbuscules develop. Both the coils and arbuscules are surrounded by the plant plasma membrane and remain within the apoplastic compartment of the root. Both types may form storage structures (vesicles) containing lipids and proteins but they are not formed by all AM fungi. Colonization of roots by AM fungi not only changes the internal morphology of the plant but can also change the architecture of the system by affecting the number and length of lateral roots thereby altering the ability of host

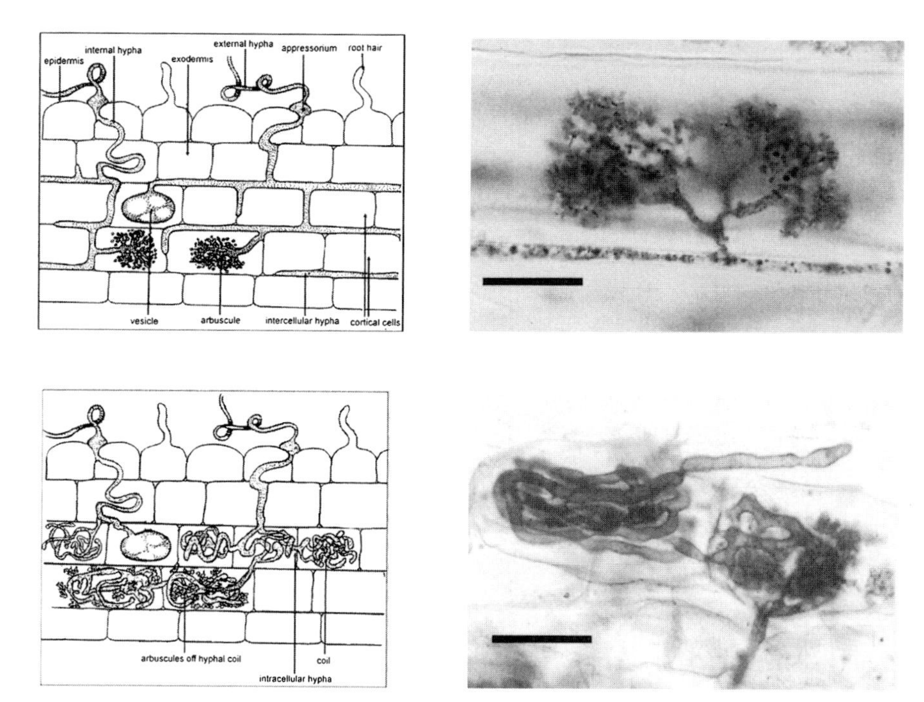

Fig. 6.6 Arbuscular mycorrhizal structures: top – diagram of *Arum*-type (left) with arbuscule and intercellular hyphae in *Allium porrum* colonized by *Glomus coronatum* (right), bar = 20 μm; bottom – diagram of *Paris*-type (left) with intracellular hyphal coils with arbuscules in *Panax quinquefolius* colonized by an unknown glomalean fungus (right), bar = 75 μm. (Reproduced with permission from Smith *et al.*, *Root Ecology*; Springer Science and Business Media, 2003a.)

plants to acquire nutrients independently of the nutrient acquisition by the AM fungus (Hetrick, 1991; Yano *et al.*, 1996).

(2) *Ectomycorrhizas* (ECM) – form on a wide range of forest trees including many economically important species, and consist mainly of basidiomycetes with some ascomycetes. ECM fungal hyphae proliferate between the outer cells of the root to form a sheath (the Hartig net) which covers the root surface (Fig. 6.7). Usually there is no intracellular penetration by the fungus but ingrowths occur in the cell walls of epidermal cells (Massicotte *et al.*, 1986). As with AM fungi, ECM have major effects on root architecture with shortening of lateral roots, swelling of root tips, and inhibition of root hair growth (Smith and Read, 1997; Ditengou *et al.*, 2000).

(3) *Ericoid mycorrhizas* (ERM) – occur in heathland plants of the Ericaceae, Epacridaceae and Empetraceae which typically have very fine roots with only one or two layers of cortical cells and no root hairs. The fungi (mainly ascomycetes with some basidiomycetes) form a loose surface covering of hyphae which penetrate the epidermal cells to form very dense hyphal complexes (Read, 1996). Unlike the AM and ECM fungi, these fungi are ecologically specialized and do not seem to cross-colonize other plants.

Most plant ecotypes contain plants that can form mycorrhizal associations with more than one type of fungus so, for example, temperate forests with mixed tree, shrub and herb

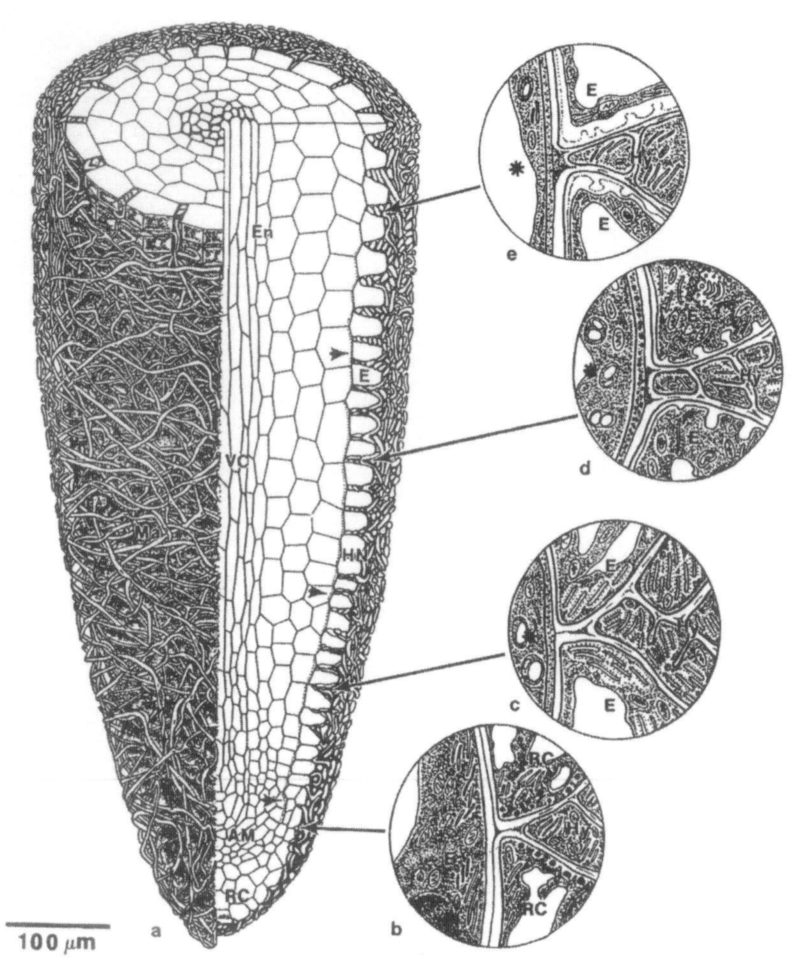

Fig. 6.7 Schematic drawing of an ectomycorrhizal root of *Alnus crispa–Alpova diplophloeus* emphasizing the development and the structure of the epidermis–fungus interface. (a) A mantle consisting of intertwining hyphae covers the root surface. A paraepidermal Hartig net (HN) is present in the region proximal to the apical meristem (AM). Concurrent modifications of the outer tangential walls (arrows) of the exodermis occur. Epidermal cells (E) are progressively more radially enlarged from the apex back. The root cap (RC) shows limited development. The vascular cylinder (VC) and endodermis (En) are also shown. (b) to (e) Cytological features showing the gradual penetration of the fungal hyphae (Hy) between the walls of the epidermal cells and concurrent changes in cell wall structure – see Massicotte *et al.* (1986) for further details. (Reproduced with permission from Massicotte *et al.*, *Canadian Journal of Botany*; National Research Council of Canada, 1986.)

species contain a range of AM, ECM and other types of mycorrhiza. Some individual plants, too, can be colonized at the same time by more than one fungus, with some trees forming simultaneous AM and ECM associations (Smith and Read, 1997).

At present, there is little information available about the signal compounds or receptors in mycorrhizal plants or fungi (Smith and Read, 1997), but this situation may change quickly. In their review, Kistner and Parniske (2002) suggest that the fact that a common set of genes appear to be involved in AM and rhizobial symbioses is evidence of a link between the infection processes, and that this link relates to the development and growth of the intracellular infection thread. Perotto *et al.* (1994) demonstrated that the matrix surrounding the infection thread of the rhizobium and that surrounding AM hyphae was structurally and biochemically similar in pea roots, suggesting that common genetic determinants control both infection processes. It is now appreciated that the way AM fungi develop in host roots is highly variable, with a continuum of types between well-known Arum types and Paris types (Dickson, 2004). In *Arum* types hyphae grow intercellularly in the root cortex; lateral branches penetrate the cortical cells and branch extensively to form terminal arbuscules which invaginate, but do not penetrate the plant plasma membrane (Fig. 6. 6). *Paris* types are characterized by lack of an intercellular phase. The hyphae grow from cell to cell in the cortex, forming intracellular coils which may bear arbusculate branches. Evidence is mounting that both arbuscules and coils provide sites of P (and perhaps other nutrient) transfer to the plant. The type of AM formed depends on the identity of both the fungus and the plant (Cavagnaro *et al.*, 2004).

Mycorrhizas and plant nutrition

The nutritional benefits to host plants arising from mycorrhiza have received considerable attention and this aspect of the symbiosis tends to dominate the literature. Many experimental studies have shown that colonization by AM fungi increases P uptake and plant growth compared with non-mycorrhizal controls. For example, Stribley *et al.* (1980a, 1980b) grew leek seedlings on 10 different soils each amended with five different levels of added P and found that both shoot growth and tissue P concentrations were increased by mycorrhiza (Fig. 6.8a). Moreover, the mycorrhizal and non-mycorrhizal plants responded differently to the quantity of bicarbonate-soluble P (a measure of the plant extractable inorganic P) in the soil (Fig. 6.8b). AM infection decreased the critical level of soil P for maximum yield from 150 to 50 mg P kg^{-1} soil, but the shape of the response curve suggests that response to infection may be small at both very low and high soil P concentrations, with maximum response at intermediate soil P concentrations. An interesting feature of Stribley *et al.*'s (1980b) results was that the responses obtained in Fig. 6.8 were only obtained with artificially infected plants. In naturally infected plants, there was no clearly defined response curve to soil P, and the plants showed a poor relationship between shoot P concentration and yield.

Four possible explanations have been put forward for the increased uptake of P by mycorrhizal roots (Smith and Read, 1997):

(1) The main theory is that fungal hyphae outside the root are able to access supplies of P at some distance from the root that would otherwise take considerable time to diffuse to the root surface. The fungus takes up the ions external to the plant, translocates them to the internal mycelium, and transfers them to the host, thereby bypassing the slow diffusion process in the soil. This mechanism is similar to that of root hairs, but the hyphae can extend many centimetres from the root surface. Growth of hyphae involves a smaller expenditure of C per unit of absorbing area than that of roots, and their smaller diameter allows access to pores unavailable to roots, thereby increasing the volume of accessible soil. There is good evidence that AM infection extends the phosphate depletion zone beyond the root surface (Li *et al.*, 1991; Fig. 6.9) and that a consequence of

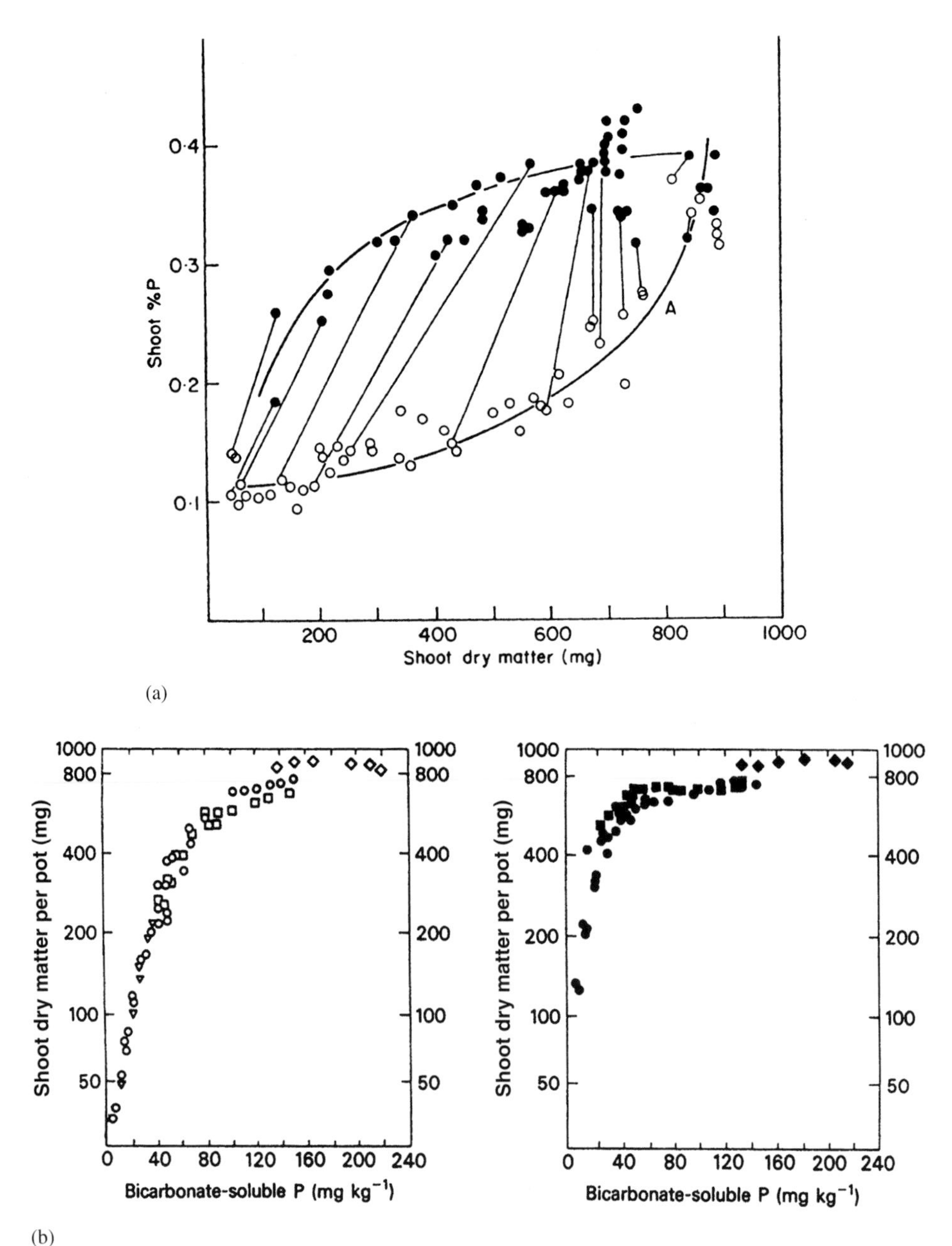

(b)

Fig. 6.8 (a) Different concentrations of P in leaves of leeks grown with (●) and without (○) arbuscular mycorrhizal infection. Lines connect plants growing in the same soils at the same phosphorus concentrations. (Reproduced with permission from Stribley *et al.*, *New Phytologist*; New Phytologist Trust, 1980a.) (b) Effect of soil phosphate concentration on yield of leeks when grown with (solid symbols) or without (open symbols) arbuscular mycorrhizal infection in 10 soils all given 5 rates of P fertilizer. (Reproduced with permission from Stribley *et al.*, *Journal of Soil Science*; Blackwell Publishing, 1980b.)

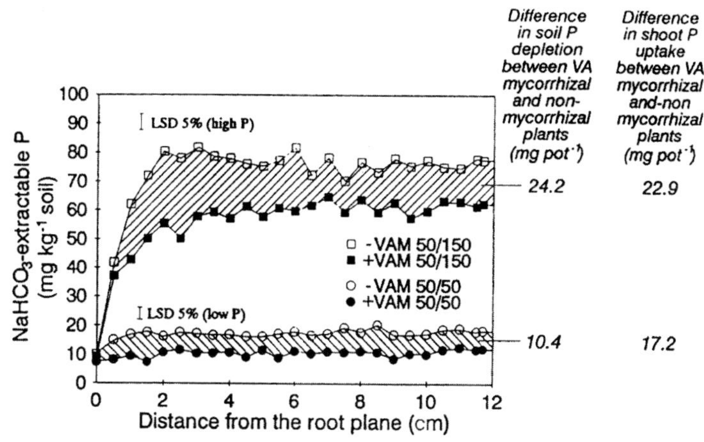

Fig. 6.9 Depletion of extractable P in soil adjacent to a clover root plane, with two levels of soil P, and with (solid symbols) and without (open symbols) arbuscular mycorrhizal infection. (Redrawn and reproduced with permission from Li *et al.*, *Plant and Soil*; Springer Science and Business Media, 1991.)

this is that inflow of P is typically two to five times greater than into non-mycorrhizal plants (Tinker and Nye, 2000). Again, this varies and will depend on changes in root and fungal uptake pathways.

(2) The hyphae may be more effective than roots in competing with other soil microbes for P and thus circumvent the immobilization and/or adsorption of P in soils. There is some evidence that AM fungi can intercept inorganic P released during mineralization and thereby prevent immobilization in the microbial biomass or adsorption by minerals (e.g. Joner and Jakobsen, 1994) but more work is needed to confirm and quantify this mechanism.

(3) The kinetics of P uptake into hyphae may differ from that of roots. There is some evidence that hyphae have a higher affinity (lower K_m) than roots, leading to greater uptake at low soil solution concentrations.

(4) The fungus can access sources of P that are unavailable to plant roots either by solubilizing inorganic P or by hydrolysis of organic P. Results to date are equivocal but suggest that this is not a major mechanism for increased uptake in mycorrhizal plants except, perhaps, those with ECM. There are, though, results demonstrating increased phosphatase activity around mycorrhizal roots (Dodd *et al.*, 1987), that AM can increase uptake of P from organic sources (Jayachandran *et al.*, 1992), and that applications of rock phosphate to soils low in P can increase the growth of mycorrhizal plants relative to non-mycorrhizal plants (Ba and Guissou, 1996; Barea *et al.*, 2002).

For ECM, the fungal sheath and hyphae extending into the soil can become the entire absorbing surface for water and nutrients of the whole root system (Smith and Read, 1997). The increased diameter of the root brought about by the sheath is of limited advantage in increasing the uptake of ions transported in the soil by diffusion, but the hyphae extending into the soil confer similar advantages to AM fungi. Increasingly, though, there has been more attention paid to broad role of ECM in securing a wide range of plant nutrients (Harley and Smith, 1983) and, particularly, the role of ECM in accessing nutrients in organic forms because the temperate forest trees that have most ECM often have roots in or close to the litter layer where many nutrients are in that form. There is now evidence that ECM can release proteolytic and other enzymes, and take up amino acids but the significance of this for forest stands has yet to be fully quantified. For example, Bending and Read (1995) grew *Pinus sylvestris* colonized by *Suillus bovinus* or *Thelephora terrestris* on non-sterile peat in containers and introduced litter from the fermentation horizon of a pine forest. The mycelium of the fungi proliferated over the litter substrate to form dense patches. After 120 days it was found that there was no significant loss of N or P from the controls but that N and P contents of the litter were depleted by 23% by *S. bovinus* and that N content was depleted by 13% by *T. terrestris* with no significant effect on P (Table 6.5). Similarly, Dinkelater and Marschner (1992) demonstrated that acid phosphatase activity was greater in ectomycorrhizal roots of Norway spruce (*Picea abies*) than in non-mycorrhizal roots, suggesting that sources of organic P might be utilized.

In ERM, too, there is evidence for the utilization of organic sources of nutrients, with an increasing challenge to the conventional view that all plants are dependent for supplies of N and P on mineralization (Read and Perez-Moreno, 2003). In upland heath environments, and in some forests, it is suggested that ERM and ECM symbioses may facilitate the acquisition of nutrients, thereby pre-empting both the immobilization by other microbes, and their incorporation into humic compounds. However, the mechanisms by which ERM and ECM intervene to acquire nutrients are yet to be worked out; whether the plants take up N and P in organic forms and/or somehow facilitate mineralization before uptake of inorganic nutrients has yet to be determined.

Transfer of nutrients between the fungus and the host is still not well understood (Smith and Read, 1997; Tinker and Nye, 2000). Ryan *et al.* (2003) examined the indigenous AM of field- and glasshouse-grown plants of subterranean clover, white clover, leek and pea roots by cryo-analytical scanning electron microscopy and found that P concentration in hyphae within the root was generally 60–170 mM (maximum 600 mM) and 30–50 mM in turgid branches of young arbuscules. In the hyphae, P concentration was linearly related with K concentration up to 350 mM and with Mg concentration up to 175 mM. The concentration of P was low or undetectable in 96% of uncolonized cortical cell vacuoles, but normally increased in the vacuoles surrounding arbuscules and in the liquid and/or gel surrounding the hyphae in the intercellular spaces. Ryan *et al.* (2003) suggest that nutrient transfer to the host and C transfer to the fungus occur both in young arbuscules and in intercellular hyphae, and that Mg and K ions are probably balanc-

Table 6.5 Concentrations of N, P and K (mg g^{-1} litter dry weight) in litter obtained from the fermented horizon of a pine forest that was either uncolonized with ECM or colonized with either *Suillus bovinus* or *Thelephora terrestris* for about 120 days

	Uncolonized	T. terrestris	*S. bovinus*
N	13.2 a	11.9 b	10.6 c
P	0.72 a	0.70 a	0.56 b
K	0.70 a	0.57 b	0.51 b

Treatments possessing different letters within a row are significantly different, $p<0.01$. Data from Bending and Read, 1995.

ing cations for P transfer. This suggestion of more than one possible site for transfer is consistent with other reports of growth responses of the host before arbuscules are fully formed (Smith and Read, 1997).

In the same way that different elements of a root system may not function identically, so different mycorrhizal fungi do not all behave in identical manner. The result is that different fungal–plant combinations are functionally diverse. Some fungi grow only close to a root while others extend to considerable distance. Smith *et al.* (2003b, 2004) grew combinations of flax (*Linum usitatissimum*), medic and tomato with three AM fungi (*Glomus caledonium*, *Glomus intraradices* and *Gigaspora rosea*) in pots with root + hyphal, and hyphal only compartments. The growth medium was a 1:1 mixture of sand and irradiated soil low in extractable P; ^{33}P was added to the hyphal only compartment and used to separate the contributions of roots and hyphae to P uptake. Dry weight of flax was increased by all fungi relative to a non-mycorrhizal control, while medic increased with the two *Glomus* spp. but not *Gigaspora rosea*, and tomato showed no positive responses with any species. External hyphal development with tomato was less than with flax or medic, and external hyphae of *Gi. rosea* also grew poorly such that values were often equal to or lower than those in non-mycorrhizal pots (Fig. 6.10). The only exceptions were in the root + hyphal compartments of tomato. In contrast, the external hyphae of both *Glomus* spp. grew extensively in all three host plants. There was, then, no simple relation between plant growth and external hyphal growth in the soil. When P and ^{33}P uptake by the plants was analysed, total P uptake showed a similar pattern to growth in response to AM colonization, but specific activity of ^{33}P showed that the *G. intraradices* delivered close to 100% of the P to flax and tomato, and 60–80% of P to medic. The contribution of *G. caledonium* at week 6 was 25–40% (Table 6.6), with small contributions of *Gi. rosea* throughout. Mycorrhizal uptake of P can, then, replace uptake by roots and root hairs, even in plants such as tomato which showed no growth response to AM colonization. Lack of a growth response to an AM fungus does not mean that the fungus makes no contribution to P uptake; this observation has significant ecological and agricultural implications if found more widely in field conditions (Smith *et al.*, 2003b).

The results of Smith *et al.* (2004) are interesting, too, because they provide the only example where the relative contributions of fungus and root to the nutrition of the plant have been quantified. We still know very little about the importance of the mycorrhizas for scarce nutrients such as Zn and Cu, or of their role over the whole life cycle of a plant. Most work has been performed with young plants grown in media that are not natural soils where nutrients may be in uncommon forms and without the microbes that would normally be part of the mineralization processes.

Mycorrhizas – benefits and costs

Smith *et al.* (2003a) summarized the benefits of mycorrhiza for the acquisition of soil nutrients as depending greatly on the fungal hyphae that extend well beyond the usual rhizosphere of the root to form a large 'mycorhizosphere' (Fig. 6.11). AM hyphae can spread more than 250 mm from a root and ECM hyphae can grow several metres, thereby exploiting soil nutrients substantial distances from plant roots. Coupled with this is evidence that hyphae can link the roots of more than one plant to form an extensive underground network, thereby possibly allowing the transfer of C and nutrients from one plant to another, or at least from fungal mycelium in one plant to mycelium in another. For

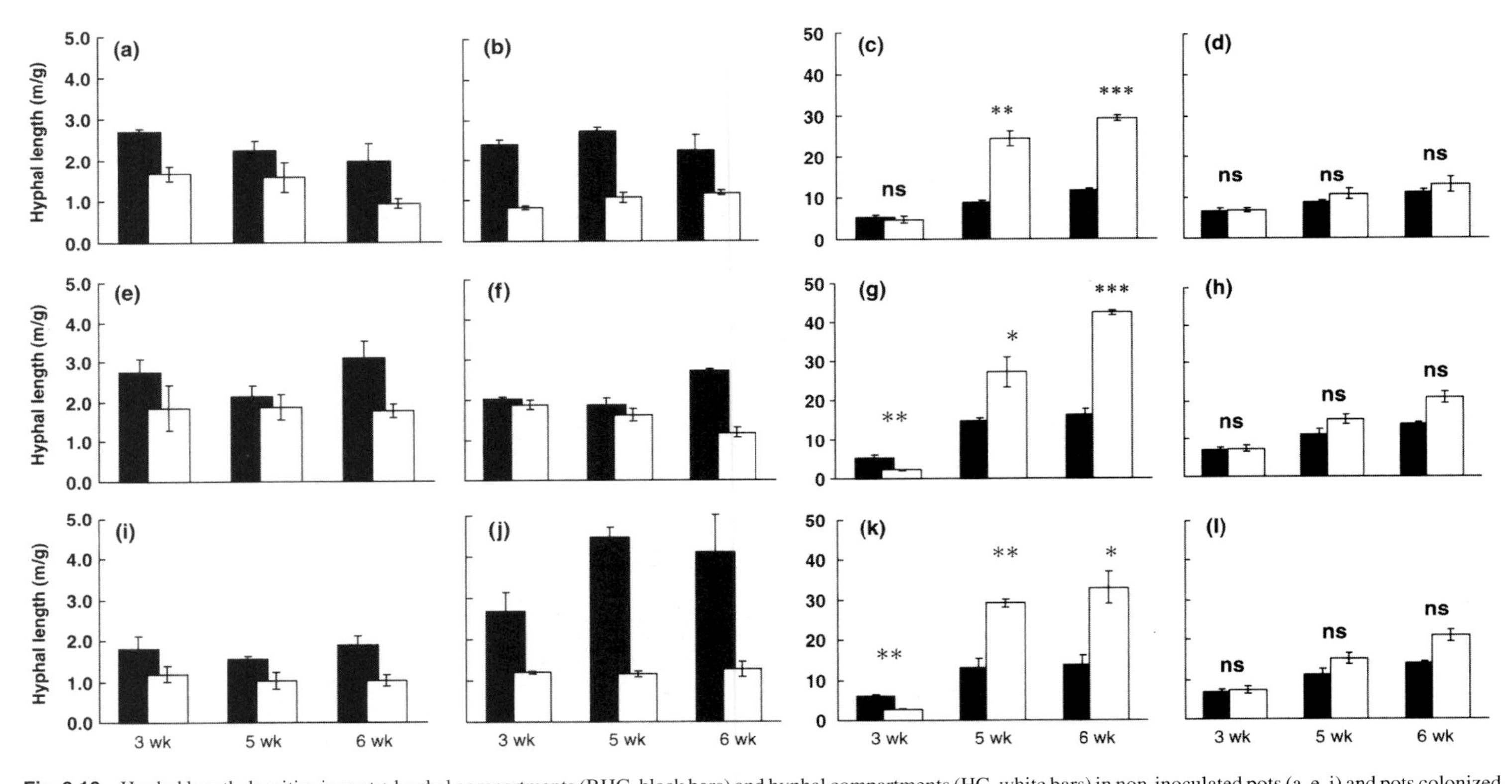

Fig. 6.10 Hyphal length densities in root + hyphal compartments (RHC, black bars) and hyphal compartments (HC, white bars) in non-inoculated pots (a, e, i) and pots colonized by *Gigaspora rosea* (b, f, j), *Glomus caledonium* (c, g, k) and *Glomus intraradices* (d, h, l) in symbiosis with flax (*Linum usitatissimum* – top row), medic (*Medicago truncatula* – centre row), and tomato (*Lycopersicon esculentum* – bottom row). For non-inoculated pots, hyphal length density in RHC > HC ($p<0.005$) and HLD in medic > flax > tomato ($p<0.005$). Pairwise comparisons shown for treatments involving *G. caledonium* and *G. intraradices*: *$p<0.05$; **$p<0.01$; ***$p<0.001$; ns, not significant. (Reproduced with permission from Smith *et al.*, *New Phytologist*; New Phytologist Trust, 2004.)

Table 6.6 Contribution (%) of the mycorrhizal pathway to P uptake in plants of flax, medic and tomato after 6 weeks when colonized by either *Glomus caledonium* or *Glomus intraradices*

	G. caledonium	*G. intraradices*
Flax	42 (3)	94 (15)
Medic	38 (5)	62 (5)
Tomato	25 (8)	83 (12)

The standard error is in brackets. Data from Smith *et al.*, 2004.

Fig. 6.11 Seedling of *Pinus* colonized by *Suillus bovinus* and grown on soil from a forest in an observation chamber about 20 cm across. The upper arrow shows mycelial strands and the lower one shows the advancing hyphal front. (Original photo by D.J. Read. Reproduced with permission from Smith *et al.*, *Root Ecology*; Springer Science and Business Media, 2003a.)

example, Simard *et al.* (1997) used reciprocal labelling with ^{13}C and ^{14}C to demonstrate a bidirectional carbon transfer between ECM tree species resulting in a 3–10% net transfer of carbon from *Betula papyrifera* trees to *Pseudotsuga menziesii* through the direct hyphal pathway. Such a 'wood-wide web', if shown more widely, would have important implications for the productivity, stability and sustainability of ecosystems. Similarly, an 'in-turf net' has been demonstrated for AM fungi in grassland ecosystems (Grime *et al.*, 1987; Watkins et al., 1996) with apparent transfer of C from one species to another. Watkins *et al.* (1996) estimated that 10% of the total C in roots of *Cynodon dactylon* was transferred from *Plantago lanceolata* over a 10-week period; a quantity similar to the estimated net C cost to a plant of a mycorrhizal symbiosis. In the latter case, though, Fitter *et al.* (1998) showed that the transferred C remained in the roots and was not transported to the shoots; it probably remained inside the fungal structures in the root and was unavailable to the plants to which it was apparently transferred. Robinson and Fitter (1999) suggest that the C sharing that occurs is most likely an essential aspect of the carbon economy of the fungus with C moving according to its own carbon demands, but remaining within the fungal mycelium. This retention of C within the mycelium was demonstrated experimentally for a *Glomus intraradices–Daucus carota* association by isotopic labelling (Pfeffer *et al.*, 2004). No inter-plant transfer of P has yet been demonstrated, and more research is required to establish the true beneficiaries of C movement.

As described earlier, there are many examples of mycorrhiza increasing rates of plant growth in soils low in nutrients (especially P) and of introduced tree species that will not grow until an appropriate ECM fungus is also introduced (Malajczuk *et al.*, 1982; Smith and Read, 1997). However, the fungal/plant association is dynamic and can move from mutualistic to parasitic when the net cost of the symbiosis exceeds the net benefit (Johnson *et al.*, 1997). In managed production systems where fertilizers and manures are applied, the C cost of maintaining the fungus may exceed the benefits of its activities in P uptake and there has been little evidence to date for a beneficial role of AM fungi in improving productivity. For example, Ryan *et al.* (2002) conducted field experiments on a range of crop rotations in south-eastern Australia and found that while the degree of AM fungal colonization in wheat crops was lower following non-mycorrhizal brassica crops, there was no relationship between AM colonization and early wheat growth, pre-anthesis P and Zn uptake, or grain yield (Table 6.7). Moreover, high levels of AM colonization did not protect crop roots from damage by root pathogens. Ryan and Graham (2002) reviewed literature of field-based studies and concluded that AM fungi 'do not play a vital role in the nutrition and growth of plants in many production-orientated agricultural systems'. Even when P availability was

Table 6.7 Effects of crop rotation (crops in year 1 shown in first column) on AM fungal colonization, growth and nutrition of subsequent crops of wheat grown (values in columns 4–8) at Morangarell, southern NSW, Australia

	Year 1		Year 2 – wheat crops				
Year 1 crop	AM colonization at tillering (% of root length)	Grain yield (t ha^{-1})	AM colonization at tillering (% of root length)	Shoot biomass at tillering (t ha^{-1})	Shoot P at tillering (mg g^{-1})	Grain yield (t ha^{-1})	Grain P (mg g^{-1})
Wheat	50	3.2	62	0.39	2.4	2.1	2.2
Linola	40	1.1	55	0.46	2.4	2.1	2.1
Canola	0	0.6	43	0.40	2.4	2.2	2.3
Mustard	0	0.7	41	0.45	2.3	2.3	2.3
Fallow	–	–	21	0.61	2.4	2.5	2.4

From Ryan *et al.*, 2002.

low and AM colonization levels were high (as on some long-term organic farms), there was no obvious net benefit from the symbiosis. Such firm conclusions, though, appear to be at odds with the results in controlled conditions of Smith *et al.* (2003b, 2004) summarized earlier, and those of Karasawa *et al.* (2001) who demonstrated substantial effects of non-mycorrhizal crops within rotations on growth and P uptake during the vegetative stages of maize growth. They grew maize in pots in a glasshouse after either sunflower or non-mycorrhizal mustard on 17 soils and found higher shoot weight and P uptake on 14 of them after sunflower; the effects of the preceding crop were eliminated if the soils were sterilized, suggesting that AM fungi played a key role in the growth of the maize crop. Until such results can be replicated in field conditions, the role, if any, for AM fungi in sustainable crop production systems is a long way from being incorporated into economically viable management systems (see Harrier and Watson [2003] for an account).

6.3 Root pathogens and parasitic associations

6.3.1 Fungal diseases

A wide range of fungal pathogens are known to enter the plant via roots producing a variety of effects on the host plant (Table 6.8). While fungi that infect through the leaf have to breach the tough, waxy cuticle – often by generating considerable turgor pressures (up to 8 MPa; Howard and Valent, 1996) – root-infecting fungi face less of a mechanical barrier but must compete effectively with other rhizosphere microbes and the antimicrobial compounds released by plant roots (Deacon, 1996). Soil-borne fungal pathogens have been studied much less than those invading via leaves because of the difficulty of study in a heterogeneous and opaque medium. Consequently, relatively little is known about signalling pathways promoting host plant recognition or the infection process except for a few fungal species. Deacon (1996) provides examples of host specificity of fungal pathogens ranging from the non-specific, tissue-rotting *Sclerotium rolfsii* regulated largely by environmental conditions to the soil-borne pathogen *Phytophthora sojae* with gene-for-gene relationships with its host, soyabean. Within these limits, two levels of specificity can be recognized: (i) to a particular family or major taxon of plants; and (ii) to crop-specific strains.

Host specificity is accompanied by host recognition in the pre-infection stages including: (i) host-specific triggering of propagule germination or activation of inoculum; (ii)

Table 6.8 Some major soil-borne fungal diseases that infect through roots and their effects

Common name	Latin name	Plants affected	Effect on plant
Take-all	*Gaeumannomyces graminis*	Wheat, oats and some grasses	Root death, reduced tillering, white ears
Damping off disease	*Pythium* spp.	Broad host range	Root rot and damping off
Root rot	*Phytophthora* spp.	Many trees and shrubs	Root rot and damping off
Wilt	*Fusarium* spp.	Many horticultural crops, trees and shrubs	Root rot, wilting, death of leaves, plant death
Wilt	*Verticillium* spp.	Many horticultural crops, trees and shrubs	Wilting, death of leaves, plant death
Bare patch, crater disease, root rot	*Rhizoctonia* spp.	Many annual plants and perennial grasses, trees	Damping off and root rot

orientation or movement to a suitable host (in the case of motile zoospores such as *Pythium* and *Phytophthora*, there seems to be both general and specific receptor-based recognition for chemotaxis to root exudates); and (iii) adhesion to the host (for zoospores this occurs during encystment often involving interaction with root-produced mucilage; Deacon, 1996). Fungi can enter plants in several ways ranging from entry through natural plant openings, to the creation of lesions, to the generation of structures able to penetrate cells. In the latter group, several fungi in contact with plant tissues swell to form a single cell structure containing melanin, known as an appressorium, from which develops a penetration peg. This peg allows the pathogen to enter the plant cell when an infection hypha starts to develop which subsequently differentiates into a branched hypha enabling proliferation throughout the host tissue (for an example with the rice blast leaf pathogen, *Magnaporthe grisea* see Howard and Valent, 1996). In *Rhizoctonia*, the fungus forms an 'infection cushion' consisting of a bundle of hyphae on the root surface which then develop an infection peg similar to that developed by appressoria; infectious hyphae are typically pigmented.

The fungal disease take-all, caused by *Gaeumannomyces graminis* var. *tritici*, an ascomycete, is one of the most studied and most important diseases of wheat worldwide (Cook, 2003). It grows as dark pigmented hyphae on the surface of roots and lower stem and produces short, melanized appressoria-like structures (hyphopodia) through which a hyphal peg grows into the root. This facilitates the penetration of thin, transparent hyphae through the root cortex and endodermis into the stele. The hyphae on the root surface continue to grow and can eventually form a complete coverage of the upper roots and lower stem (Plate 6.5); they may also grow from plant to plant if soil moisture conditions permit. The invasion of the root tissues results in their death and wheat plants with severe take-all can easily be pulled from soil because of their weakened root systems. The number of tillers is often substantially reduced in infected plants, and the ears senesce early and appear white in an otherwise green crop, and have small or no grain. The fungus survives saprophytically in the dead roots and stem bases of the host to infect the next crop. An interesting feature of this fungus is its interactions with other soil microbes and the partial control that can be affected thereby. A feature of take-all is that there is a spontaneous decline in its severity after a few years of severe disease, although yields do not reach those of crops grown in appropriate rotations. Because the hyphae grow at the root/soil interface they are susceptible to the antagonistic effects of other soil microbes, and appear particularly sensitive to antibiotics produced by fluorescent *Pseudomonas* species. Root-colonizing strains of *Pseudomonas fluorescens* producing antibiotics have been associated with the phenomenon of take-all decline in several countries (Cook, 2003). The most important antibiotic appears to be 2,4-diacetylphloroglucinol, with *P. fluorescens* populations of 10^5 to 10^6 CFU g^{-1} of root sufficient to fully suppress the fungus (Raaijmakers *et al.*, 1997, 1999). However, attempts to introduce *Pseudomonas fluorescens* as a seed coating to control take-all in wheat crops have met with mixed success because many soil properties, including organic matter content, affect the interaction between the two microbes and the host and, therefore, the incidence and severity of disease in the field (Ownley *et al.*, 2003).

6.3.2 Nematodes

The majority of nematodes in the rhizosphere feed on bacteria (see section 6.1.5), but most plant-parasitic nematodes are soil-dwelling and infect plants via the roots. They cause major damage to many plant types but it is their impact on economically important crops that has received most attention. Most of the damage is caused by a small number of the many nematode genera that infect crops, especially the sedentary root-knot (*Meloidogyne* spp.) and cyst (*Globodera* and *Heterodera* spp.) nematodes, and several migratory nematodes including *Pratylenchus* and *Radopholus* spp. (Bird and Kaloshian, 2003). Most plant-parasitic nematodes are migratory and feed as ectoparasites or endoparasites of roots: the most successful have sedentary female stages that induce specialized feeding cells in their hosts and then produce 100–1000 eggs per female. Such nematodes include the semi-endoparasitic *Tylenchulus* and *Rotylenchulus* spp., in which the females only partially invade the root, and the endoparasitic *Heterodera* and *Globodera* spp., in which the females swell so much that they can rupture the root cortex and epidermis, and *Nacobbus* and *Meloidogyne* spp., which remain embedded in galls so that only the egg masses produced by the females are on the root surface (Kerry, 2000).

In the widely studied root-knot nematode, *Meloidogyne* spp., the females release their eggs into a proteinaceous substance on the root surface from which second-stage larvae hatch in the soil and typically re-infect the same plant. These larvae do not feed but are long-lived, surviving for long periods (several months) on stored lipid reserves; this stage in the life cycle allows for dispersal of the organism. Exit from this stage is controlled by signals from food sources and nematode population density based on quorum-sensing (see section 6.1.1). Larvae of root-knot nematodes penetrate the root, preferentially in the zone of elongation or at the site of lateral root emergence, and migrate intercellularly to the vascular cylinder where they select a feeding site (Bird and Kaloshian, 2003). The basis of site selection is unknown. All plant-parasitic nematodes possess a hollow, protrusible spear in the oral cavity, known as the stylet, which is used to pierce plant tissues, inject secretions and withdraw host nutrients (Fig. 6.12) (Davis *et al.*, 2004). During migration of the second-stage larva into the plant, extensive secretion of proteins occurs from glands that play numerous parts in the host–parasite interaction. Different glands are more active at different stages in the migratory process so that, for example, in root-knot and cyst nematodes the subventral gland cells are more active prior to host penetration, whereas the dorsal gland enlarges and becomes more active as parasitism progresses (Bird and Kaloshian, 2003; Davis *et al.*, 2004). Various enzymatic functions for the secretions have been proposed with production of cellulase (β-1,4-endoglucanase), pectinases, polygalacturonase and phenol oxidase identified (Smant *et al.*, 1998). The form of cellulase produced is noteworthy because it is not found in any other animals and is believed to have been acquired via horizontal gene transfer from prokaryotes. The cellulose and pectinase enzymes secreted result in depolymerization of cell wall compounds and the dismantling of the cell walls facilitating entry of the nematode. Enzymes capable of modifying host cell walls (endoglucanases and xylanase) appear to be active only in the subventral gland cells and only during larval migration, whereas plant endoglucanases are up-regulated in the formation of feeding sites (Davis *et al.*, 2004).

Root-knot and cyst nematodes induce root cells in the vascular cylinder close to the nematode head to become substantially enlarged feeding sites. Synchronous nuclear division without cell division accompanied by cell wall modification and a reduction in plasmodesmatal connections occurs in a group of neighbouring cells to form 'giant cells' in root-knot nematodes and 'syncytia' in cyst nematodes (Fig. 6.13). These cells function as carbon and assimilate sinks transferring host products to the nematode. The precise

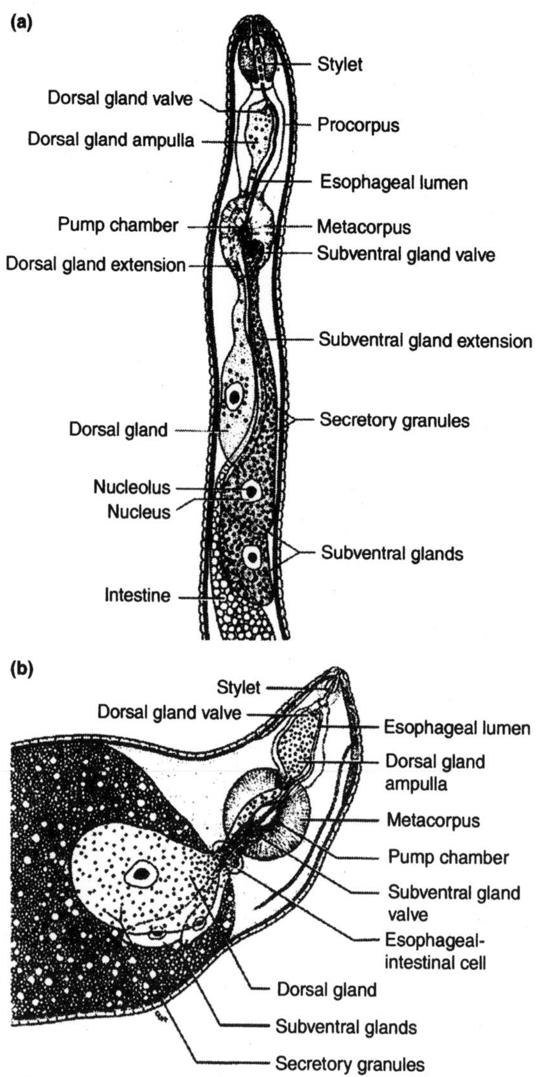

Fig. 6.12 Illustration of the single dorsal and two subventral oesophageal gland cells that produce secretions that are introduced into host tissues through the stylet (feeding spear) of plant-parasitic nematodes. (a) A pre-infective second-stage juvenile with numerous secretions packaged within granules in the subventral gland cells. (b) A late parasitic stage female with enlarged dorsal gland cell packed with secretory granules and reduced subventral gland cells. (Reproduced with permission from Davis *et al.*, *Trends in Parasitology*; Elsevier, 2004.)

signal leading to the development of these cells is unknown but auxin and peptide signal molecules have been implicated (Goverse *et al.*, 2000; Olsen and Skriver, 2003). Whatever the mechanism, changes in cell form are closely coupled with the development of the nematode and in many hosts the cortical and pericycle cells also divide and expand to form a gall or knot around the nematode and feeding cells. It is probable that phytohormones are involved in gall formation, with auxin playing a major role in the expansion of the cells, and the inhibition of auxin transport in the presence of the nematode mediated through activa-

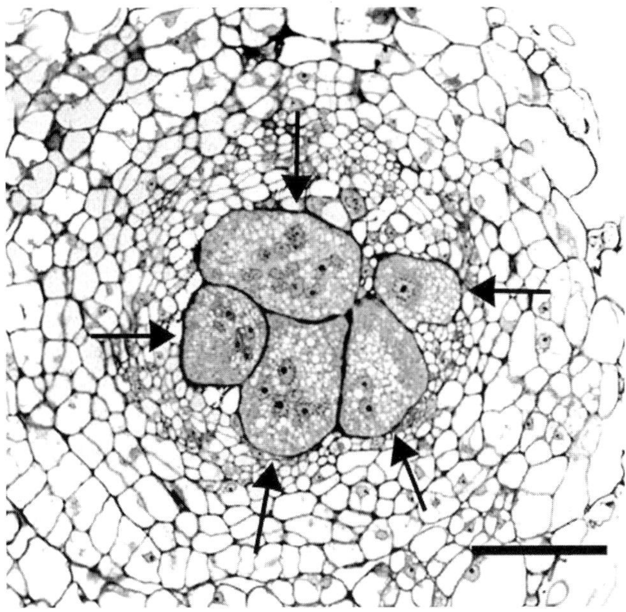

Fig. 6.13 Giant cells induced in tomato roots by the nematode *Meloidogyne incognita*. The five cells are indicated with arrows; bar = 100 μm. (Reproduced with permission from Bird and Kaloshian, *Physiological and Molecular Plant Pathology*; Elsevier, 2003.)

tion of a flavonoid pathway (Hutangura *et al.*, 1999). A similar induction pathway has been observed in lateral root and rhizobial nodule formation (Mathesius, 2003) (Plate 6.1).

Nematodes can have significant detrimental effects on the yield of crops but interactions with the root system can be more complex. For example, Smit and Vamerali (1998) showed in field experiments in The Netherlands that potato cyst nematode reduced the total root length only during early growth, with compensatory changes in later rates of production and death resulting in similar-sized systems later in the season. However, the date at which maximum root length was achieved was retarded by at least 20 days depending on the level of infection. Similar results were also obtained by De Ruijter and Haverkort (1999) but they also found that the compensatory growth was confined to the upper 0.3 m and that nematodes reduced rooting depth.

Resistance genes to nematodes have been isolated and a range of different mechanisms appear to operate. For example, Seah *et al.* (2000) studied resistance to cereal cyst nematode in both wheat and barley and found that the syncytia in infected barley with the *Ha2* or *Ha4* genes both developed and degenerated faster than in the resistant wheats with the *Cre1* or *Cre3* genes. In the resistant wheat plants with the *Cre1* or *Cre3* genes, the syncytia remained extensively vacuolated and less metabolically active throughout.

6.3.3 Parasitic weeds

The parasitic weeds *Striga* and *Orobanche* species are major pests of crops in several parts of the world, resulting in lower yields and, in some cases, complete crop failure. A thorough account of these parasitic associations and of possible means of controlling them is given by Parker and Riches (1993), but this section focuses on interactions with roots.

Striga

Striga species, commonly called 'witchweed', are a serious problem in cereal crop production for many millions of farmers in semi-arid areas of Africa and parts of Asia. Cereal crops affected include maize, sorghum, pearl millet, finger millet, rice and sugar cane, with maize and sorghum in Africa particularly affected by two species of *Striga*, *S. hermonthica* and *S. asiatica*. One species, *S. gesnerioides*, infects species of Fabaceae, especially cowpea.

Germination of *Striga* seed is complex but, following a pre-conditioning period, is stimulated by a range of chemicals exuded from host and non-host roots (Sato *et al.*, 2003). It is likely that several substances are involved in each species and that different populations within a species may differ in the range of substances to which they will respond (Stewart and Press, 1990) Upon germination, contact with the host root is achieved in response to as yet unidentified chemical signals, and elongation of the radicle stops and development of a haustorium starts. Dörr (1997) observed that infection of maize by *S. hermonthica* and of sorghum by *S. asiatica* followed a similar course. The haustorium consists of upper hyaline tissue and a lower endophytic part which penetrates the host by splitting the cortex longitudinally or compressing the cortical cells laterally (Fig. 6.14). Haustorial cells did not grow between or into the cortical cells of the host but penetrated the vascular tissue primarily through the pits. Once inside the

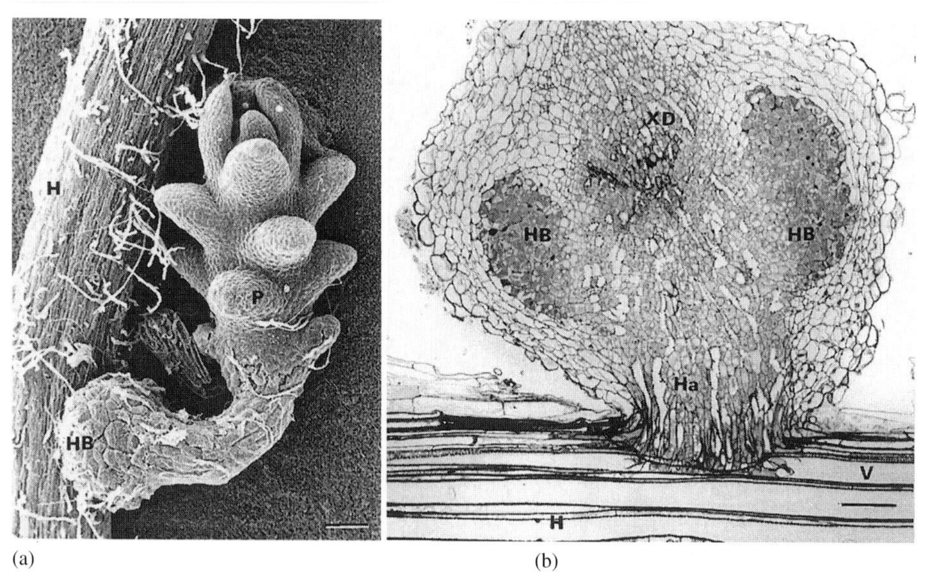

(a) (b)

Fig. 6.14 (a) Scanning electron micrograph of a young *Striga* seeding (P) growing on a host root (H) with the hyaline body (HB); bar = 100 μm. (b) Light micrograph of a longitudinal section through the hyaline body (HB) of the haustorium (Ha). In the region of the haustorium the cortex of the root is reduced. Parasitic cells have spread in the vessels (V) of the host and xylem elements of the parasite have traversed the haustorium fusing in the xylem disc (XD); bar = 100 μm. (Reproduced with permission from Dörr, *Annals of Botany*; Oxford University Press, 1997.)

vessels, the parasite cells spread in clusters. The invading part of the contact cell (often cup-like in shape) is named the 'osculum' and it is through these structures that water and nutrients are absorbed. Several oscula are directly connected to the xylem of the haustorium allowing ready transfer of water and solutes from host to parasite. In Dörr's (1997) study, no phloem elements were found in the haustorial complex although carbohydrate is known to be transferred from the host (Graves *et al.*, 1989). Typically this primary haustorium remains small (1–2 mm in diameter), but as the parasite grows, adventitious roots develop from the elongating stem. These roots can form secondary haustoria when they come into contact with host roots. Okonkwo (1966) demonstrated that although these secondary haustoria develop as lateral rather than terminal organs on the root, they have similar internal structure and organization.

The consequences of infestation by *Striga* are generally greater in soils of low fertility, especially those low in N (Farina *et al.*, 1985; Boukar *et al.*, 1996), although the effects of N fertilizer application are equivocal with variable results (Aflakpui *et al.*, 2002). Many workers, though, have observed that the main effect of *Striga* infection is on shoot growth of the host rather than root growth, so the proportion of total biomass allocated to roots is greater in infected plants than uninfected plants. For example, Aflakpui *et al.* (2002) found that maize infected with *S. hermonthica* significantly decreased shoot mass but had no effect on root mass so that the proportion of total biomass in roots was increased from about 0.3 in uninfected plants to 0.5 in one experiment and 0.4 in another (Table 6.9); there was no significant *Striga* × nitrogen interaction on the responses measured. Similar results were also recorded by Graves *et al.* (1989) for a *S. hermonthica* association with sorghum, but Taylor *et al.* (1996) found higher root biomass in infected maize. The severity of infection and other conditions affecting growth appear to influence the growth responses observed.

Although some progress has been made in breeding maize and sorghum with some tolerance to *Striga*, no completely resistant variety has yet been identified. Arnaud *et al.* (1999) compared the partially resistant variety Framida with the susceptible CK-60B and found no difference between them at the germination or attachment stages. Differences appeared during the establishment of the functional haustorium with reduced metabolite uptake and growth rate of the parasite on Framida roots. Later stages of the infection were characterized by significant accumulation of a coloured phenolic-like material in and

Table 6.9 The effect of *Striga hermonthica* infection on root mass as a proportion of total plant mass in two experiments with maize

Experiment 1						
Days after planting	12	29	61	90	104	
Infected	0.46	0.52	0.52	0.53	0.54	
Uninfected	0.43	0.42	0.40	0.34	0.32	
SE	0.02	0.04	0.02	0.03	0.03	
Experiment 2						
Days after planting	29	43	57	78	85	99
Infected	0.31	0.43	0.44	0.43	0.42	0.40
Uninfected	0.26	0.030	0.28	0.27	0.29	0.28
SE	0.01	0.02	0.02	0.02	0.02	0.03

From Aflakpui *et al.*, 2002.

around Framida vascular tissues but not CK-60B. The significance of this latter development was not known with certainty but inhibition of uptake by the parasite was suspected.

Orobanche

The broomrapes include about 14 genera, all holo-parasites lacking chlorophyll, of which 5 are serious parasites of crops. *Orobancha ramosa*, *O. aegyptiaca* and *O. crenata* are important crop pests around the Mediterranean. Their effects are not generally as devastating on crop yields as *Striga* but a wide range of dicotyledonous crops is affected including rape, mustard, chickpea, faba bean, aubergine, carrot, tobacco and tomato.

There are many similarities in the development of the host–*Orobanche* relationship with that of the host–*Striga* relationship. Germination of *Orobanche* requires a moist, conditioning period and follows response to the reception of a chemical stimulant from host roots (Yokota *et al.*, 1998; Sato *et al.*, 2003). After germination, the procaulome of the *Orobanche* attaches itself to the surface of the host root, and the tip forms a haustorium, which pentrates the host root, eventually establishing connections with the vascular system of the host. The young parasite then develops a tubercle with adventitious roots, and a non-photosynthesizing shoot (Fig. 6.15). Unlike *Striga*, the adventitious roots do not form secondary attachments to the host but they may take up some nutrients (Parker and Riches, 1993).

Soil nutrient conditions appear to be less important in the degree of infection of *Orobanch* than of *Striga*, but nitrogen fertilizer in the ammonium form was more inhibitory than nitrate (Westwood and Foy, 1999). The effect was on the elongation of the seedling procaulome rather than on seed germination, with *Orobanche* spp. much more sensitive to ammonium toxicity than the host species.

Resistance to *Orobanche* exists, especially in chickpea, but its exact nature has yet to be fully established. Zehhar *et al.* (2003) examined the interaction between *O. ramosa* and

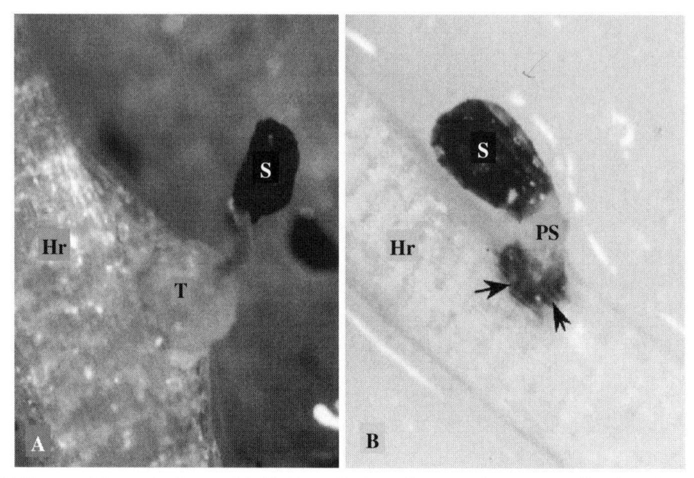

Fig. 6.15 (A) Successful penetration and initial installation of crenate broomrape (*Orobanche crenata* Forsk.) on pea (*Pisum sativum* L.). (B) Unsuccessful haustorial penetration on chickpea (*Cicer arietinum* L.) showing the darkening of the surrounding tissues (arrows). S, parasitic seed; Hr, host root; T, parasite tubercle; PS, parasite seedling. (Reproduced with permission from Rubiales *et al.*, *Weed Science*; Weed Science Society of America, 2003.)

susceptible and resistant varieties of carrot. In the susceptible carrot, no defence reactions were observed and the developing parasite inhibited tap root formation, whereas in the resistant varieties, the parasite germinated, attached to the host, but became necrotic. Histological studies showed that the resistance mechanisms in the two resistant varieties were different. In one variety, the parasite connected to the host xylem but there was thickening of the host xylem vessel cell walls and, sometimes, complete occlusion of these vessels by phenolic compounds, gel or gum-like substances at the interface of the host and parasite xylem. Hence the xylem connections were not functional. In the other variety, entry of the haustorium into the host initiated the death of inner cortical cells several layers thick. The dead cells contained globular structures and formed a mechanical barrier to the parasite and inhibited nutrient transfer. In chickpea, the primary mechanism of resistance was found to be low rates of stimulant production resulting in lower populations of the germinated parasite (Rubiales *et al.*, 2003). The secondary resistance mechanism was evident by a darkening of host cell tissue in contact with the procaulome (Fig. 6.15). Histological studies showed no necrosis of the host cells but darkening of the haustorium cells seemingly as a result of an accumulation of a dark secretion between parasite and host cells. Eventually the whole parasite became dark and died.

6.4 Root herbivory by insects

A wide range of insects have stages in their life cycle that feed on roots including such well-known pests as corn rootworm (*Diabrotica virgifera virgifera*), cabbage root fly (*Delia radicum*), cranefly (*Tipula* spp.) and clover root weevil (*Sitona lepidus*), but relative to shoot feeders, little is known about their consumption and the consequences for plant growth (Hunter, 2001). It is recognized that below-ground insect herbivory is likely to influence the diversity of plant communities, the rate and direction of succession, competitive interactions among plants, the susceptibility of plants to other herbivores and pathogens, and the yields of crops and forests, but such interactions are difficult to study (Brown and Gange, 1989; Hunter, 2001). There is also very little information about the chemicals that roots deploy to defend themselves against insect attack, although the presence of toxins in roots has been recorded for a few plant species (Hunter, 2001; Van der Putten, 2003).

From the limited measurements available, it appears that herbivory on roots facilitates population growth of insect herbivores above ground. Masters *et al.* (1993) suggested that root herbivory might lead to an accumulation of soluble amino acids and carbohydrates in foliage, resulting in increased performance of foliar feeders. Conversely, foliar feeders lead to less assimilate being translocated to the root system, resulting in reduced root growth and limited food availability for the root feeder. This ‘plus-minus’ model in which root herbivory has a positive effect on performance of foliar feeders while foliar herbivory has a detrimental effect on the growth of root feeders was based on field and laboratory observations demonstrating that interactions between shoot and root herbivores appear to be mediated by increases in soluble N and sugars in shoots and by decreases in root biomass. However, more studies are needed because not all insects and plants respond in this way (Masters *et al.*, 1993; Hunter, 2001).

Sitona lepidus is a significant pest of white clover (*Trifolium repens* L.) with newly hatched larvae feeding within the N-fixing nodules, and later stages feeding on progressively larger roots. Estimates of the damage caused in field pastures are sparse but in growth chambers, plants infested with larvae had significantly lower biomass and a lower number of nodules compared with uninfested controls (both about 60% of control values), although root mass as a proportion of total biomass (0.41 and 0.43, respectively) was similar (Murray *et al.*, 2002). The association between the insect and the host is specific and newly hatched larvae move preferentially to the nodules of the host rather than to associated grass roots or the roots of other legumes in the sward, although the signalling molecule(s) facilitating this has yet to be identified (Johnson *et al.*, 2004). Plate 6.6 shows the movement of a larva toward a nodule recorded using X-ray micro-tomography which allowed the rate and direction of movement in soil to be measured for use in subsequent modelling of the interaction (Johnson *et al.*, 2004; Zhang *et al.*, 2005).

References

Aflakpui, G.K.S., Gregory, P.J. and Froud-Williams, R.J. (2002) Growth and biomass partitioning of maize during vegetative growth in response to *Striga hermonthica* infection and nitrogen supply. *Experimental Agriculture* **38**, 265–276.

Arnaud, M.-C., Véronési, C. and Thalouarn, P. (1999) Physiology and histology of resistance to *Striga hermonthica* in *Sorghum bicolor* var. Framida. *Australian Journal of Plant Physiology* **26**, 63–70.

Ba, A.M. and Guissou, T. (1996) Rock phosphate and vesicular-vesicular mycorrhiza effects on growth and nutrient uptake of *Faidherbia albida* (Del) seedlings in an alkaline sandy soil. *Agroforestry Systems* **34**, 129–137.

Bais, H.P., Walker, T.S., Schweizer, H.P. and Vivanco, J.M. (2002) Root specific elicitation and antimicrobial activity of rosmarinic acid in hairy root cultures of *Ocimum basilicum*. *Plant Physiology and Biochemistry* **40**, 983–995.

Barea, J.M., Toro, M., Orozco, M.O., Campos, E. and Azcon, R. (2002) The application of isotopic (^{32}P and ^{15}N) dilution techniques to evaluate the interactive effect of phosphate-solubilizing rhizobacteria, mycorrhizal fungi and *Rhizobium* to improve the agronomic efficiency of rock phosphate for legume crops. *Nutrient Cycling in Agroecosystems* **63**, 35–62.

Bending, G.D. and Read, D.J. (1995) The structure and function of the vegetative mycelium of ectomycorrhizal plants. V. Foraging behaviour and translocation of nutrients from exploited litter. *New Phytologist* **130**, 401–409.

Bird, D.M. and Kaloshian, I. (2003) Are roots special? Nematodes have their say. *Physiological and Molecular Plant Pathology* **62**, 115–123.

Boerner, R.E.J. and Harris, K.K. (1991) Effects of collembola (Arthropoda) and relative germination date on competition between mycorrhizal *Panicum virgatum* (Poaceae) and non-mycorrhizal *Brassica nigra* (Brassicaceae). *Plant and Soil* **136**, 121–129.

Bonkowski, M. (2004) Protozoa and plant growth: the microbial loop in soil revisited. *New Phytologist* **162**, 617–631.

Bonkowski, M. and Brandt, F. (2002) Do soil protozoa enhance plant growth by hormonal effects? *Soil Biology and Biochemistry* **34**, 1709–1715.

Boukar, I., Hess, D.E. and Payne, W.A. (1996) Dynamics of moisture, nitrogen, and *Striga* infestation on pearl millet transpiration and growth. *Agronomy Journal* **88**, 545–549.

Bremer, E., Janzen, H.H. and Gilbertson, C. (1995) Evidence against associative N_2 fixation as a significant N source in long-term wheat plots. *Plant and Soil* **175**, 13–19.

Brown, V.K. and Gange, A.C. (1989) Differential effects of above- and below-ground insect herbivory during early plant succession. *Oikos* **54**, 67–76.

Canellas, L.P., Olivares, F.L., Okorokova-Façanha, A.L. and Façanha, A.R. (2002) Humic acids isolated from earthworm compost enhance root elongation, lateral root emergence, and plasma membrane H^+-ATPase activity in maize roots. *Plant Physiology* **130**, 1951–1957.

Cavagnaro, T.R., Smith, F.A., Hay, G., Carne-Cavagnaro, V.L. and Smith, S.E. (2004) Inoculum type does not affect overall resistance of an arbuscular mycorrhiza-defective tomato mutant to colonisation but inoculation does change competitive interactions with wild-type tomato. *New Phytologist* **161**, 485–494,

Cook, R.J. (2003) Take-all of wheat. *Physiological and Molecular Plant Pathology* **62**, 73–86.
Cortez, J. and Bouche, M.B. (1992) Do earthworms eat living roots? *Soil Biology and Biochemistry* **24**, 913–915.
Cullimore, J. and Dénarié, J. (2003) How legumes select their sweet talking symbionts. *Science* **302**, 575–577.
Davis, E.L., Hussey, R.S. and Baum, T.J. (2004) Getting to the roots of parasitism by nematodes. *Trends in Parasitology* **20**, 134–141.
De Ruijter, F.J. and Haverkort, A.J. (1999) Effects of potato cyst nematodes (*Globodera pallida*) and soil pH on root growth, nutrient uptake and crop growth of potato. *Journal of Plant Pathology* **105**, 61–76.
Deacon, J.W. (1996) Ecological implications of recognition events in the pre-infection stages of root pathogens. *New Phytologist* **133**, 135–145.
Dickson, S. (2004) The *Arum-Paris* continuum of mycorrhizal symbioses. *New Phytologist* **163**, 187–200.
Dinkelaker, B. and Marschner, H. (1992) *In vivo* demonstration of acid phosphatase activity in the rhizosphere of soil-grown plants. *Plant and Soil* **144**, 199–205.
Ditengou, F.A., Béguiristain, T. and Lapeyrie, F. (2000) Root hair elongation is inhibited by hypaphorine, the indole alkaloid from the ectomycorrhizal fungus *Pisolithus tinctorius*, and restored by indole-3-acetic acid. *Planta* **211**, 722–728.
Dodd, J.C., Burton, C.C., Burns, R.G. and Jeffries, P. (1987) Phosphatase activity associated with the roots and the rhizosphere of plants infected with vesicular-arbuscular mycorrhizal fungi. *New Phytologist* **107**, 163–172.
Dörr, I. (1997) How *Striga* parasitizes its host: a TEM and SEM study. *Annals of Botany* **79**, 463–472.
Farina, M.P.W., Thomas, P.E.L. and Channon, P. (1985) Nitrogen, phosphorus and potassium effects on the incidence of *Striga asiatica* (L.) Kuntze in maize. *Weed Research* **25**, 443–447.
Ferris, H., Venette, R.C. and Lau, S.S. (1997) Population energetics of bacterial-feeding nematodes: carbon and nitrogen budgets. *Soil Biology and Biochemistry* **29**, 1183–1194.
Fisher, K. and Newton, W.E. (2002) Nitrogen fixation – a general overview. In: *Nitrogen Fixation at the Millenium* (ed. G.J. Leigh), pp. 1–34. Elsevier, Amsterdam.
Fitter, A.H., Graves, J.D., Watkins, N.K., Robinson, D. and Scrimgeour, C. (1998) Carbon transfer between plants and its control in networks of arbuscular mycorrhizas. *Functional Ecology* **12**, 406–412.
Forde, B.G. (2002) Local and long-range signalling pathways regulating plant responses to nitrate. *Annual Review of Plant Biology* **53**, 203–224.
Fray, R.G. (2002) Altering plant-microbe interactions through artificially manipulating bacterial quorum sensing. *Annals of Botany* **89**, 245–253.
Giller, KE. (2001) *Nitrogen Fixation in Tropical Cropping Systems*, 2nd edn. CAB International, Wallingford.
Goverse, A., Overmars, H., Engelbertink, J., Schots, A., Bakker, J. and Helder, J. (2000) Roth induction and morphogenesis of cyst nematode feeding cells are mediated by auxin. *Molecular Plant-Microbe Interactions* **13**, 1121–1129.
Graves, J.D., Press, M.C. and Stewart, G.R. (1989) A carbon balance model of the sorghum-*Striga hermonthica* host-parasite association. *Plant, Cell and Environment* **12**, 101–107.
Griffiths, B.S. (1994) Soil nutrient flow. In: *Soil Protozoa* (ed. J.F. Darbyshire), pp. 65–91. CAB International, Wallingford.
Grime, J.P., Mackey, J.M.L., Hillier, S.H. and Read, D.J. (1987) Floristic diversity in a model system using experimental microcosms. *Nature* **328**, 420–422.
Gupta. V.V.S.R., Roget, D.K. and Coppi, J.A. (2004) Identification of a previously unrecognized constraint to yield in sequential wheat crops. In: *Proceedings of the 3rd Australasian Soilborne Disease Symposium* (eds K. Ophel Keller and B. Hall), pp. 13–14. SARDI, Adelaide.
Harley, J.L. and Smith, S.E. (1983) *Mycorrhizal Symbiosis*. Academic Press, London.
Harrier, L.A. and Watson, C.A. (2003) The role of arbuscular mycorrhizal fungi in sustainable cropping systems. *Advances in Agronomy* **79**, 185–225.
Hetrick, B.A.D. (1991) Mycorrhizas and root architecture. *Experientia* **47**, 355–362.
Hirsch, A.M. (1992) Developmental biology of legume nodulation. *New Phytologist* **122**, 211–237.
Hopkin, S.P. (1997) *Biology of the Springtails (Insecta: Collembola)*. Oxford University Press, Oxford.
Howard, R.J. and Valent, B. (1996) Breaking and entering: host penetration by the rice blast pathogen *Magnaporthe grisea*. *Annual Review of Microbiology* **50**, 491–512.
Hunter, M.D. (2001) Out of sight, out of mind: the impacts of root-feeding insects in natural and managed systems. *Agricultural and Forest Entomology* **3**, 3–9.
Hutangura, P., Mathesius, U., Jones, M.G.K. and Rolfe, B.G. (1999) Auxin induction is a trigger for root gall formation caused by root-knot nematodes in white clover and is associated with the activation of the flavenoid pathway. *Australian Journal of Plant Physiology* **26**, 221–231.

Jayachandran, K., Schwab, A.P. and Hetrick, B.A.D. (1992) Mineralization of organic phosphorus by vesicular-arbuscular mycorrhizal fungi. *Soil Biology and Biochemistry* **24**, 897–903.

Johnson, N.C., Graham, J.H. and Smith, F.A. (1997) Functioning of mycorrhizal associations along the mutualism-parasitism continuum. *New Phytologist* **135**, 575–585.

Johnson, S.N., Read, D.B. and Gregory, P.J. (2004) Tracking larval insect movement within soil using high resolution X-ray microtomography. *Ecological Entomology* **29**, 117–122.

Joner, E.J. and Jakobsen, I. (1994) Contribution by two arbuscular mycorrhizal fungi to P uptake by cucumber (*Cucumis sativus* L.) from ^{32}P-labelled organic matter during mineralization in soil. *Plant and Soil* **163**, 203–209.

Jumpponen, A. and Trappe, J.M. (1998) Dark septate endophytes: a review of facultative biotrophic root-colonizing fungi. *New Phytologist* **140**, 295–310.

Kandeler, E., Marschner, P., Tscherko, D., Gahoonia, T.S. and Nielsen, N.E. (2002) Microbial community composition and functional diversity in the rhizosphere of maize. *Plant and Soil* **238**, 301–312.

Karasawa, T., Kasahara, Y. and Takebe, M. (2001) Variable response of growth and arbuscular mycorrhizal colonization of maize plants to preceding crops in various types of soil. *Biology and Fertility of Soils* **33**, 286–293.

Kerry, B.R. (2000) Rhizosphere interactions and the exploitation of microbial agents for the biological control of plant-parasitic nematodes. *Annual Review of Phytopathology* **38**, 423–441.

Khan, D.F., Peoples, M.B., Schwenke, G.D., Felton, W.L., Chen, D. and Herridge, D.F. (2003) Effects of below-ground nitrogen on N balances of field-grown fababean, chickpea, and barley. *Australian Journal of Agricultural Research* **54**, 333–340.

Kistner, C. and Parniske, M. (2002) Evolution of signal transduction in intracellular symbiosis. *Trends in Plant Science* **7**, 511–518.

Lee, K.E. (1985) *Earthworms: Their Ecology and Relationships with Soils and Land Use*. Academic Press, Sydney.

Leigh, G.J. (ed.) (2002) *Nitrogen Fixation at the Millenium*. Elsevier, Amsterdam.

Li, X.-L., George, E. and Marschner, H. (1991) Extension of the phosphorus depletion zone in VA-mycorrhizal white clover in a calcareous soil. *Plant and Soil* **136**, 41–48.

Lodwig, E.M., Hosie, A.H.F., Bourdès, A., Findlay, K., Allaway, D., Karunakaran, R., Downie, J.A. and Poole, P.S. (2003) Amino-acid cycling drives nitrogen fixation in the legume-*Rhizobium* symbiosis. *Nature* **422**, 722–726.

Madsen, E.B., Madsen, L.H., Radutoiu, S., Olbryt, M., Rakwalska, M., Szczyglowski, K., Sato, S., Kaneko, T., Sandal, N. and Stougaard, J. (2003) A receptor kinase gene of the LysM type is involved in legume perception of rhizobial signals. *Nature* **425**, 637–640.

Malajczuk, N., Molina, R. and Trappe, J.M. (1982) Ectomycorrhiza formation in *Eucalyptus*. I. Pure culture synthesis, host specificity and mycorrhizal compatibility with *Pinus radiata*. *New Phytologist* **91**, 467–482.

McNeill, A.M., Zhu, C. and Fillery, I.R.P. (1998) A new approach to quantifying the N benefit from pasture legumes to succeeding wheat. *Australian Journal of Agricultural Research* **49**, 427–436.

Mantelin, S. and Touraine, B. (2004) Plant growth-promoting bacteria and nitrate availability: impacts on root development and nitrate uptake. *Journal of Experimental Botany* **55**, 27–34.

Marschner, P. and Baumann, K. (2003) Changes in bacterial community structure induced by mycorrhizal colonisation in split-root maize. *Plant and Soil* **251**, 279–289.

Marschner, P., Yang, C.-H., Lieberei, R. and Crowley, D.E. (2001) Soil and plant specific effects on bacterial community composition in the rhizosphere. *Soil Biology and Biochemistry* **33**, 1437–1445.

Massicotte, H.B., Peterson, R.L., Ackerley, C.A. and Piché, Y. (1986) Structure and ontogeny of *Alnus crispa – Alpova diplophloeus* ectomycorrhizae. *Canadian Journal of Botany* **64**, 177–192.

Masters, G.J., Brown, V.K. and Gange, A.C. (1993) Plant mediated interactions between above- and below-ground insect harbivores. *Oikos* **66**, 148–151.

Mathesius, U. (2001) Flavenoids induced in cells undergoing nodule organogenesis in white clover are regulators of auxin breakdown by peroxidase. *Journal of Experimental Botany* **52**, 419–426.

Mathesius, U. (2003) Conservation and divergence of signalling pathways between roots and soil microbes – the *Rhizobium*-legume symbiosis compared to the development of lateral roots, mycorrhizal interactions and nematode-induced galls. *Plant and Soil* **255**, 105–119.

Mathesius, U., Schlaman, H.R.M., Spaink, H.P., Sautter, C., Rolfr, B.G. and Djordjevic, M.A. (1998) Auxin transport inhibition preceded root nodule formation in white clover roots and is regulated by flavenoids and derivatives of chitin oligosaccharides. *The Plant Journal* **14**, 23–34.

Mathesius, U., Weinman, J.J., Rolfe, B.G. and Djordjevic, M.A. (2000) Rhizobia can induce nodules in white clover by 'hijacking' mature cortical cells activated during lateral root development. *Molecular Plant-Microbe Interactions* **13**, 170–182.

Mathesius, U., Mulders, S., Gao, M., Teplitski, M., Caetano-Anollés, G., Rolfe, B.G. and Bauer, W.D. (2003) Extensive and specific responses of a eukaryote to bacterial quorum-sensing signals. *Proceedings of the National Academy of Science* **100**, 1444–1449.

Murray, P.J., Dawson, L.A. and Grayston, S.J. (2002) Influence of root herbivory on growth response and carbon assimilation by white clover plants. *Applied Soil Ecology* **20**, 97–105.

Nehl, D.B., Allen, S.J. and Brown, J.F. (1996) Deleterious rhizosphere bacteria: an integrating perspective. *Applied Soil Ecology* **5**, 1–20.

Newton, J.A. and Fray, R.G. (2004) Integration of environmental and host-derived signals with quorum sensing during plant-microbe interactions. *Cellular Microbiology* **6**, 213–224.

Okon, Y. and Kapulnik, Y. (1986) Development and function of *Azospirillum*-inoculated roots. *Plant and Soil* **90**, 3–16.

Okonkwo, S.N.C. (1966) Studies on *Striga senegalensis* Benth. I. Mode of host-parasite union and haustorial structure. *Phytomorphology* **16**, 453–463.

Oldroyd, G.E.D. and Downie, J.A. (2004) Calcium, kinases and nodulation signalling in legumes. *Nature Reviews Molecular Cell Biology* **5**, 566–576.

Olsen, A.N. and Skriver, K. (2003) Ligand mimicry? Plant-parasitic nematode polypeptide with similarity to CLAVATA3. *Trends in Plant Science* **8**, 55–57.

Ownley, B.H., Duffy, B.K. and Weller, D.M. (2003) Identification and manipulation of soil properties to improve the biological control performance of phenazine-producing *Pseudomonas fluorescens*. *Applied and Environmental Microbiology* **69**, 3333–3343.

Parker, C. and Riches, C.R. (1993) *Parasitic Weeds of The World: Biology and Control*. CAB International, Wallingford.

Peoples, M.B., Boddey, R.M. and Herridge, D.F. (2002) Quantification of nitrogen fixation. In: *Nitrogen Fixation at the Millenium* (ed. G.J. Leigh), pp. 357–389. Amsterdam, Elsevier.

Perotto, S., Brewin, N.J. and Bonfante, P. (1994) Colonization of pea roots by the mycorrhiza fungus *Glomus versiforme* and by *Rhizobium* bacteria: immunological comparison using monoclonal antibodies as probes for plant cell surface components. *Molecular Plant-Microbe Interactions* **7**, 91–98.

Pfeffer, P.E., Douds D.D., Jr, Bücking, H., Schwartz, D.P. and Shachar-Hill, Y. (2004) The fungus does not transfer carbon to or between roots in an arbuscular mycorrhizal symbiosis. *New Phytologist* **163**, 617–627.

Postgate, J.R. (1998) *Nitrogen Fixation*, 3rd edn. Cambridge University Press, Cambridge.

Raaijmakers, J.M., Weller, D.M. and Thomashow, L.S. (1997) Frequency of antibiotic-producing *Pseudomonas* spp. in natural environments. *Applied and Environmental Microbiology* **63**, 881–887.

Read, D.J. (1996) The structure and function of the ericoid mycorrhizal root. *Annals of Botany* **77**, 365–374.

Raaijmakers, J.M., Bonsall, R.F. and Weller, D.M. (1999) Effect of population density of *Pseudomonas fluorescens* on production of 2,4-diacetylphloroglucinol in the rhizosphere of wheat. *Phytopathology* **89**, 470–475.

Read, D.J. and Perez-Moreno, J. (2003) Mycorrhizas and nutrient cycling in ecosystems – a journey towards relevance? *New Phytologist* **157**, 475–492.

Redmond, J.R., Batley, M., Djordjevic, M.A., Innes, R.W., Keumpel, P.L. and Rolfe, B.G. (1986) Flavones induce expression of *nod* genes in *Rhizobium*. *Nature* **323**, 632–635.

Riely, B.K., Ane, J.M., Penmetsa, R.V. and Cook, D.R. (2004) Genetic and genomic analysis in model legumes bring Nod-factor signalling to center stage. *Current Opinion in Plant Biology* **7**, 408–413.

Robinson, D. and Fitter, A.H. (1999) The magnitude and control of carbon transfer between plants linked by a common mycorrhizal network. *Journal of Experimental Botany* **50**, 9–13.

Rovira, A.D., Elliott, L.F. and Cook, R.J. (1990) The impact of cropping systems on rhizosphere organisms affecting plant health. In: *The Rhizosphere* (ed. J.M. Lynch), pp. 389–436. John Wiley & Sons, Chichester, UK.

Rubiales, D., Pérez-de-Luque, A., Joel, D.M., Alcántara, C. and Sillero, J.C. (2003) Characterization of resistance in chickpea to crenate broomrape (*Orobanche crenata*). *Weed Science* **51**, 702–707.

Ryan, M.H. and Graham, J.H. (2002) Is there a role for arbuscular mycorrhizal fungi in production agriculture? *Plant and Soil* **244**, 263–271.

Ryan, M.H., Norton, R.M., Kirkegaard, J.A., McCormick, K.M., Knights, S.E. and Angus, J.F. (2002) Increasing mycorrhizal colonisation does not improve growth and nutrition of wheat on Vertosols in south-eastern Australia. *Australian Journal of Agricultural Research* **53**, 1173–1181.

Ryan, M.H., McCully, M.E. and Huang, C.X. (2003) Location and quantification of phosphorus and other elements in fully hydrated, soil-grown arbuscular mycorrhizas: a cryo-analytical scanning electron microscopy study. *New Phytologist* **160**, 429–441.

Sato, D., Awad, A.A., Chae, S.H., Yokota, T., Sugimoto, Y., Takeuchi, Y. and Yoneyama, K. (2003) Analysis of strigolactones, germination stimulants for *Striga* and *Orobanche*, by high performance liquid chromatography/tandem mass spectrometry. *Journal of Agricultural and Food Chemistry* **51**, 1162–1168.

Scheu, S. (2003) Effects of earthworms on plant growth: patterns and perspectives. *Pedobiologia* **47**, 846–856.

Seah, S., Miller, C., Sivasithamparam, K. and Lagudah, E.S. (2000) Root responses to cereal cyst nematode (*Heterodera avenae*) in hosts with different resistance genes. *New Phytologist* **146**, 527–533.

Semenov, A.M., van Bruggen, A.H.C. and Zelenev, V.V. (1999) Moving waves of bacterial populations and total organic carbon along roots of wheat. *Microbial Ecology* **37**, 116–128.

Simard, S.W., Perry, D.A., Jones, M.D., Myrold, D.D., Durall, D.M. and Molina, R. (1997) Net transfer of carbon between ectomycorrhizal tree species in the field. *Nature* **388**, 579–582.

Simpfendorfer, S., Kirkegaard, J.A., Heenan, D.P. and Wong, P.T.W. (2002) Reduced early growth of direct drilled wheat in southern New South Wales – role of root inhibitory pseudomonas. *Australian Journal of Agricultural Research* **53**, 323–331.

Smant, G., Stokkermans, J.P.W.G., Yan, Y., de Boer, J.M., Baum, T.J., Wang, X.L., Hussey, R.S., Gommers, F.J., Henrissat, B., Cavis, E.L., Helder, J., Schots, A. and Bakker, J. (1998) Endogenous cellulases in animals: isolation of β-1,4-endoglucanase genes from two species of plant-parasitic cyst nematodes. *Proceedings of the National Academy of Science* **95**, 4906–4911.

Smit, A.L. and Vamerali, T. (1998) The influence of potato cyst nematodes (*Globodera pallida*) and drought on rooting dynamics of potato (*Solanum tuberosum* L.). *European Journal of Agronomy* **9**, 137–146.

Smith, F.A., Smith, S.E. and Timonen, S. (2003a) Mycorrhizas. In: *Root Ecology* (eds H. De Kroon and E.J.W. Visser), pp. 257–295. Springer-Verlag, Berlin.

Smith, S.E. and Read, D.J. (1997) *Mycorrhizal Symbiosis*, 2nd edn. Academic Press, London.

Smith, S.E., Smith, F.A. and Jakobsen, I. (2003b) Mycorrhizal fungi can dominate phosphate supply to plants irrespective of growth responses. *Plant Physiology* **133**, 16–20.

Smith, S.E., Smith, F.A. and Jakobsen, I. (2004) Functional diversity in arbuscular mycorrhizal (AM) symbioses: the contribution of the mycorrhizal P uptake pathway is not correlated with mycorrhizal responses in growth or total P uptake. *New Phytologist* **162**, 511–524.

Springett, J. and Gray, R. (1997) The interaction between plant roots and earthworm burrows in pasture. *Soil Biology and Biochemistry* **29**, 621–624.

Stewart, G.R. and Press, M.C. (1990) The physiology and biochemistry of parasitic angiosperms. *Annual Review of Plant Physiology and Plant Molecular Biology* **41**, 127–151.

Stribley, D.P., Tinker, P.B. and Rayner, J.H. (1980a) Relation of internal phosphorus concentration and plant weight in plants infected by vesicular-arbuscular mycorrhizas. *New Phytologist* **86**, 261–266.

Stribley, D.P., Tinker, P.B. and Snellgrove, R.C. (1980b) Effect of vesicular-arbuscular mycorrhizal fungi on the relations of plant growth, internal phosphorus concentration and soil phosphate analysis. *Journal of Soil Science* **31**, 655–672.

Taylor, A., Martin, J. and Seel, W.E. (1996) Physiology of the parasitic association between maize and witchweed (*Striga hermonthica*): is ABA involved? *Journal of Experimental Botany* **47**, 1057–1065.

Teplitski, M., Robinson, J.B. and Bauer, W.D. (2000) Plants secrete substances that mimic bacterial N-acyl homoserine lactone signal activities and affect population density-dependent behaviours in associated bacteria. *Molecular Plant-Microbe Interactions* **13**, 637–648.

Tinker, P.B. and Nye, P.H. (2000) *Solute Movement in the Rhizosphere*. Oxford University Press, Oxford.

Tuffen, F., Eason, W.R. and Scullion, J. (2002) The effect of earthworms and arbuscular mycorrhizal fungi on growth of and ^{32}P transfer between *Allium porrum* plants. *Soil Biology and Biochemistry* **34**, 1027–1036.

Unkovich, M.J. and Pate, J.S. (2000) An appraisal of recent field measurements of symbiotic N_2 fixation by annual legumes. *Field Crops Research.* **65**, 211–228.

Van der Putten, W.H. (2003) Plant defense belowground and spatiotemporal processes in natural vegetation. *Ecology* **84**, 2269–2280.

van Vuurde, J.W.L. and Schippers, B. (1980) Bacterial colonization of seminal wheat roots. *Soil Biology and Biochemistry* **12**, 559–565.

Vitousek, P.M., Aber, J.D., Howarth, R.W., Likens, G.E., Matson, P.A., Schindler, D.W., Schlesinger, W.H. and Tilman, D.G. (1997) Human alteration of the global nitrogen cycle: sources and consequences. *Ecological Applications* **7**, 737–750.

Walker, T.S., Bais, H.P., Grotewold, E. and Vivanco, J.M. (2003) Root exudation and rhizosphere biology. *Plant Physiology* **132**, 44–51.

Watkins, N.K., Fitter, A.H., Graves, J.D. and Robinson, D. (1996) Carbon transfer between C_3 and C_4 plants linked by a common mycorrhizal network, quantified using stable carbon isotopes. *Soil Biology and Biochemistry* **28**, 471–477.

Watt, M., McCully, M.E. and Kirkegaard, J.A. (2003) Soil strength and rate of root elongation alter the accumulation of *Pseudomonas* spp. and other bacteria in the rhizosphere of wheat. *Functional Plant Biology* **30**, 483–491.

Weller, D.M., Howie, W.J. and Cook, R.J. (1988) Relationship between in vitro inhibition of *Gaeumannomyces graminis* var. *tritici* and suppression of take-all of wheat by fluorescent pseudomonads. *Phytopathology* **78**, 1094–1100.

Werner, D. (2001) Organic signals between plants and microorganisms. In: *The Rhizosphere: Biochemistry and Organic Substances at the Soil-Plant Interface* (eds R. Pinton, Z. Varanini and P. Nannipieri), pp. 197–222. Marcel Dekker, New York.

Westwood, J.H. and Foy, C.L. (1999) Influence of nitrogen on germination and early development of broomrape (*Orobanche* spp.). *Weed Science* **47**, 2–7.

Whitehead, N.A., Barnard, A.M.L., Slater, H., Simpson, N.J.L. and Salmond, G.P.C. (2001) Quorum-sensing in Gram-negative bacteria. *FEMS Microbiology Reviews* **25**, 365–404.

Yano, K., Yamauchi, A. and Kono, Y. (1996) Localized alteration in lateral root development in roots colonized by an arbuscular mycorrhizal fungus. *Mycorrhiza* **6**, 409–415.

Yokota, T., Sakai, H., Okung, K., Yoneyama, K. and Takeuchi, Y. (1998) Alectrol and orobanchol, germination stimulants for *Orobanche minor*, from its host red clover. *Phytochemistry* **49**, 1967–1973.

Zehhar, N., Labrousse, P., Arnaud, M.-C., Boulet, C., Bouya, D. and Fer, A. (2003) Study of resistance to *Orobanche ramosa* in host (oilseed rape and carrot) and non-host (maize) plants. *European Journal of Plant Pathology* **109**, 75–82.

Zhang, X., Johnson, S.N., Gregory, P.J., Crawford, J.W., Young, I.M., Murray, P.J. and Jarvis, S.C. (2005) Modelling the movement and survival of the root-feeding clover weevil, *Sitona lepidus*, in the rhizosphere of white clover. *Ecological Modelling*. In press.

Zwart, K.B., Kuikman, P.J. and Van Veen, J.A. (1994) Rhizosphere protozoa: their significance in nutrient dynamics. In: *Soil Protozoa* (ed. J.F. Darbyshire), pp. 93–121. CAB International, Wallingford.

Chapter 7

The Rhizosphere

The rhizosphere, as originally conceived by Hiltner (1904), was a zone of soil close to the root in which the activity of the microbial community was enhanced with consequent effects on nutrient (especially nitrogen) availability to the plant (legumes in Hiltner's studies). Nowadays, the rhizosphere is more broadly defined as the volume of soil affected by the presence of roots of growing plants. The changes to the soil may be biological, chemical or physical in nature and this broader definition has meant that the size of the rhizosphere has also increased in breadth from a narrow zone extending some <1–2 mm from the root surface, to one extending some >10–20 mm in the case of some mobile nutrients and water, or to even greater distances for volatile compounds released from roots.

Many compounds are released into the rhizosphere but for most them our current understanding of their real significance, in for example plant nutrition, is very limited. Most of the experiments reported in this chapter have been conducted with young plants, often grown in nutrient solution or to form dense root mats against blocks of soil, with expression of the chemical of interest stimulated by nutrient deficiency in the growing medium. The quantitative relevance of such results to root systems growing in soils is difficult to extrapolate from such observations. There is little doubt that in most cases release of carbon compounds by roots stimulates at least a temporary increase in microbial activity close to the root (see section 1.3.1), but at present it appears prudent to weigh carefully the evidence for other claimed effects for compounds released. As Jones *et al.* (2004) state 'Root exudates have been hypothesized to be involved in the enhanced mobilization and acquisition of many nutrients from soil or the external detoxification of metals. With few exceptions, there is little mechanistic evidence from soil-based systems to support these propositions'.

7.1 Rhizodeposition

A wide variety of carbon compounds (see section 7.1.2) is released from living roots to the soil via several mechanisms including:

(1) Exudation of low molecular weight, water-soluble compounds, such as glucose, which are lost passively without the involvement of plant metabolic activity.
(2) Secretion of higher molecular weight compounds, such as polysaccharide mucilage and enzymes, involving root metabolic processes.
(3) Lysates released from sloughed off root cells and, with time, whole roots.

(4) Gases such as CO_2, ethylene and hydrogen cyanide.

In practical terms, it is impossible to distinguish exudates from secretions and the two are usually considered together as exuded materials. There is, though, some confusion in the literature over the sources of C to be included in the term 'rhizodeposition'. Some researchers such as Lynch and Whipps (1990) and Grayston *et al.* (1996) include all four sources of C, while others such as Swinnen *et al.* (1995) and Paterson and Sim (1999) do not include gaseous losses. There is some sense in separating gaseous from other losses because, while the gases mainly diffuse to the atmosphere, all the other losses constitute substrates for microorganisms in the rhizosphere and sustain their increased growth and activity relative to the bulk soil.

In graminaceous species (which have been the focus for most studies) photosynthetically fixed C is transported very rapidly below ground (Gregory and Atwell, 1991; Kuzyakov and Cheng, 2001) and can be detected in the root and its external environment within 1 hour when non-soil growing media are used. For example, Dilkes *et al.* (2004) found that exudation from wheat grown in nutrient solution following a pulse of ^{14}C was maximal after 2–3 hours and declined to one-third of its maximum after 5 hours. Cumulative ^{14}C exudation after 20 hours was about 3% of the ^{14}C fixed. This rapid response indicates a close link between current photosynthesis and exudation, although the rate of exudation was more closely related to the rate of carbon import into the root which is, in turn, related to the concentrations of soluble sugars in the root. The rhizodeposits are an important substrate for the soil microbial community and there is a complex interplay between this community and the quantity and type of compounds released. Not only are substances released from roots rapidly utilized by microbes but additions of rhizodeposits to soils can speed up ('prime') the decomposition of native soil organic matter (Paterson, 2003). Ryan *et al.* (2001) demonstrated that when soluble C is added to the rhizosphere at realistic concentrations, then rapid mineralization occurs with a half-life for most sugars, and amino and organic acids of 0.5–2 hours. This means that photosynthate produced in a shoot may have a half-life of only about 3–6 hours in the highly dynamic plant/soil system. Although most of the exuded materials are respired (Kuzyakov and Cheng, 2001), some C will be incorporated into the microbial biomass which has a slower turnover time (typically 30–90 days).

Addition of rhizodeposits to soil can enhance the decomposition rate of other organic substrates in the soil, although this effect is dependent on plant species and the C:N ratio of the deposits. For example, Cheng *et al.* (2003) found that the rate of soil organic matter decomposition in a pot study was from zero to almost four times greater when soyabean and wheat plants were grown than in a control with no plants. This rhizosphere priming effect was responsible for a major component of the total soil C efflux. The effect was greater in soyabean than wheat, with soil C loss for soyabean of 1.44 mg C g^{-1} soil (averaged across all fertilizer treatments) compared with 1.07 mg C g^{-1} soil for wheat and 0.54 mg C g^{-1} soil for the control. The bigger effect of soyabean than wheat on decomposition was ascribed to the higher substrate quality of soyabean rhizodeposits and their likely N-rich compounds.

7.1.1 Quantities of rhizodeposits

Estimates of rhizodeposition vary quite markedly with values of up to 40% of the dry matter produced by plants reported; such values include root respiration (Lynch and Whipps,

1990). The difference between estimates is largely a result of the different methodologies used to measure rhizodeposition and the consequent bias in the interpretation of the results. Plants have commonly been grown in nutrient solution or sand flushed with nutrient solution (Shepherd and Davies, 1993) to enable characterization of exudate composition, or labelled with ^{14}C or ^{13}C to facilitate C budgets of plants grown in soil. Labelling with ^{14}C has involved either continuous labelling (e.g. Whipps and Lynch, 1983; Martin and Merckx, 1992) or pulse-chase labelling (e.g. Gregory and Atwell, 1991), and each technique has its own benefits and limitations (Meharg, 1994). Growing plants in a continuously maintained ^{14}C atmosphere labels the plant homogenously and allows total C inputs to the soil to be calculated. An advantage is that the soil microbial biomass also becomes labelled, facilitating quantitative assessment of the C flux through the biomass. A disadvantage, though, is that this technique does not allow discrimination between root exudation and root turnover (Meharg, 1994). Pulse-labelling on the other hand, is necessary to determine changes in carbon allocation in response to plant development or environmental conditions; this is impossible with continuous labelling because recently assimilated carbon cannot be distinguished from that assimilated previously. A practical difficulty with this method lies in the choice of equilibration period before determining the fate of the label, and the method also fails to account for root turnover and sloughing of cells. Thornton *et al.* (2004) conclude that while pulse-labelling is an effective method for tracking inputs of C to soil from recent assimilation, this input is unlikely to constitute the most abundant source of substrate to microbes in the rhizosphere. They consider that older assimilates released into the rhizosphere are likely to be more complex organic materials than recently assimilated materials and consequently processed at different rates and by different microbial communities.

Measurements on field crops are few, reflecting the practical difficulties of labelling plants with isotopes in field conditions. Martin and Kemp (1986) pulse-labelled wheat plants at early and late tillering stages and measured the fate of the ^{14}C until anthesis. They found that a smaller proportion of the ^{14}C was translocated to the roots at the later sampling (about 0.1 compared with 0.4) and that of the C translocated below ground, 65% was released as CO_2 with the early labelling and 45% at the later labelling. Similarly, Keith *et al.* (1986) found that about 50% of photosynthate was translocated below ground in wheat plants at early tillering of which about one-half was respired and one-quarter each was recovered in the soil and in the roots. However, after anthesis <5% of the ^{14}C was translocated below ground and the estimated total seasonal rhizodeposition (including respiration) was 130 g C m^{-2}. Gregory and Atwell (1991) compared allocation to field-grown wheat and barley crops and found similar patterns of partitioning with time even though sampling was conducted within 24 hours of the labelling period. At early tillering, a substantial percentage (33% wheat and 54% barley) of the label recovered was in below-ground components with respiration accounting for almost one-half of this, whereas from anthesis onwards a much smaller percentage was allocated below ground (9% wheat and 5% barley). The estimated inputs of C into the soil over the whole season from the crops were 48 g C m^{-2} for wheat and 58 g C m^{-2} for barley (shoot dry matter was 494 and 735 g m^{-2}, respectively) with exudates comprising only 1.6 g C m^{-2} for wheat and 2.4 g C m^{-2} for barley. The short equilibration period following labelling undoubtedly underestimated rhizodeposition because significant re-allocation of assimilates may occur to roots after 24 hours, and root decay and sloughing of cells occur over longer periods.

Swinnen *et al.* (1995) pulse-labelled crops of winter wheat and spring barley at regular intervals and allowed the assimilated ^{14}C to be allocated within the plant–soil system for 21 days. A model was used to allocate the C assimilated to various fractions based on time-dependent specific activity of the daily fixed assimilates (Swinnen *et al.*, 1994) and on the measured rates of respiration of rhizodeposits (Swinnen, 1994). For crops that were grown with conventional agricultural practice, some 18.2% of net assimilation was allocated below ground in winter wheat and 33.3% in spring barley over the whole growing season (shoot dry matter at maturity was about 1800 and 1080 g m^{-2}, respectively; Table 7.1). Accumulated over the season, the quantity of C transferred to the roots was 155–225 g C m^{-2} and total rhizodeposition (omitting root respiration) was 65–99 g C m^{-2} (some 7–15% of net assimilation), representing twice the quantity of roots left at crop maturity (Table 7.1). A similar seasonal balance was also obtained by Jensen (1994) for a winter barley crop grown on a sandy loam in Denmark. Total seasonal below-ground transfer of C was 237 g C m^{-2}, of which rhizosphere respired C was 39.5% and root C at maturity 23.7–43.8%. Rhizodeposition (excluding respiration) was 175 g m^{-2}, of which 67% was retained below ground at maturity.

Pastures have marked seasonal patterns of growth and, because they are perennial, periods of marked root decay. Saggar and Hedley (2001) used ^{14}C pulse-labelling on pastures grown on fine sandy loams in New Zealand to assess below-ground C fluxes. Annual pasture production was 1600 g m^{-2} with marked seasonal differences in the rates of growth (7.5–7.9 g C m^{-2} d^{-1} in spring and 1.8–2.0 g C m^{-2} d^{-1} in winter), rates of root decomposition (half-life 111 days for autumn roots and 64 days for spring roots) and respiratory losses (66–70% of assimilated ^{14}C in summer, autumn and winter, but 37–39% during spring). Over the season 649 g C m^{-2} remained in shoots, 682 g C m^{-2} was translocated to roots, and 132 g C m^{-2} was transferred to soil as rhizodeposits. Most other studies with pastures have been in controlled conditions. For *Lolium perenne*, Kuzyakov *et al.*

Table 7.1 Seasonal accumulated C fluxes in crops of winter wheat and spring barley grown under field conditions on a calcareous silt loam in The Netherlands under conventional management

	Winter wheat		Spring barley	
	(g C m^{-2})	% net assimilation	(g C m^{-2})	% net assimilation
Net assimilation	879 (5)		649 (9)	
Shoot growth	719 (0)	81.8 (0.3)	433 (0)	66.7 (0.5)
Transfer to roots	160 (5)	18.2 (0.3)	216 (9)	33.3 (0.5)
Root growth	63 (3)	7.1 (0.3)	94 (5)	14.1 (0.8)
– Root decay	27 (1)	3.1 (0.1)	51 (3)	7.7 (0.4)
– Net root increase	35 (2)	4.0 (0.2)	43 (2)	6.4 (0.3)
Root release of young photosynthate	100 (5)	11.3 (0.4)	128 (9)	19.6 (0.8)
– Root respiration*	57 (7)	6.4 (0.8)	94 (10)	14.2 (1.3)
– Rhizodeposition*	47 (8)	5.3 (0.8)	40 (11)	6.0 (1.5)
Total rhizodeposition*	73 (8)	8.2 (0.8)	88 (11)	13.4 (1.4)
Below-ground organic input*	107 (8)	12.1 (0.8)	128 (11)	19.6 (1.3)

The values are means of a range (half the range is shown in brackets). The range is a consequence of different extrapolation and, for those indicated with an asterisk, rhizodeposit modelling procedures. Adapted from Swinnen *et al.*, 1995.

(2001) found that the C input into the soil was about 50% of the easily available organic substances and that the root exudates stimulated CO_2 efflux relative to soil with no plant roots present. This priming effect of root exudates on microbial activity led to additional soil organic matter decomposition of about 6 g C m^{-2} d^{-1}. In mixed pastures, amounts and types of C released may differ between species, resulting in different rates of incorporation of the rhizodeposits into soil organic matter. For example, de Neergaard and Gorissen (2004) found that white clover released less of the ^{14}C assimilated than perennial ryegrass as rhizodeposits and incorporated it sooner into stable compounds either in the plant or the soil, but that the root-derived C compounds of clover decomposed faster than those from grass. The effects of defoliation on rhizodeposition by pasture plants have not been widely studied and, as with root growth (see section 3.3.1), have led to contrasting conclusions. Paterson and Sim (1999) defoliated perennial ryegrass grown axenically in nutrient solution and found that while the response was highly variable, defoliation significantly increased exudation of soluble compounds for a period of 3–5 days. However, Crawford *et al.* (2000) found that when barrel medic (*Medicago truncatula*) grown in pots of soil was defoliated, the rate of rhizosphere respiration decreased, probably because of a decrease in the below-ground flux of newly assimilated carbon. Interestingly there was no net translocation of assimilate from roots to new shoots after defoliation, indicating that new shoot growth occurred by re-mobilization of above-ground stores and newly assimilated carbon.

Differences in C allocation below ground have been found between plant species, with an annual grass species losing more C in the rhizosphere by respiration and exudation than a slower-growing perennial species (Warembourg and Estelrich, 2001). Similarly, in a glasshouse study of 12 Mediterranean species belonging to different functional groups (grasses, legumes and non-legume forbs), a functional hierarchy of rhizosphere activity was determined (Warembourg *et al.*, 2003). Measurements 6 days after labelling with $^{14}CO_2$ showed that rhizosphere respiration as a percentage of below-ground respiration was significantly smaller for non-legume forbs (42%) than for grasses (46%) and legumes (51%), and consequently there was more ^{14}C incorporated into root biomass of the non-legume forbs. The non-legume forbs, then, were the least rhizosphere-active in terms of respiration and exudation, and invested more in long-term C storage. These differences were partially related to traits such as root nitrogen concentration, but root architecture may also have influenced below-ground C allocation.

7.1.2 Composition of rhizodeposits

As indicated in the previous section, a wide range of compounds is released into the rhizosphere. Some are the almost immediate products of photosynthesis while others are more complex products of plant assimilation and metabolism (Table 7.2). Uren (2001) suggests that, except for chlorophyll and some compounds associated with the photosynthetic process, root products probably contain every type of compound that exists in plants. As analytical methods have developed it has been possible to identify a wider range of products. Chromatographic methods, which require some prior knowledge of what might be present, are gradually being overtaken by other techniques. For example, Fan *et al.* (2001) employed nuclear magnetic resonance with gas chromatography-mass spectrometry and mass

Table 7.2 Organic compounds released from plant roots

Group	Compounds
Sugars and polysaccharides	Arabinose, fructose, galactose, glucose, maltose, mucilages of various compositions, oligosaccharides, raffinose, ribose, sucrose, xylose
Amino acids	α-Alanine, β-alanine, γ-aminobutyric, arginine, aspartic, citrulline, cystathionine, cysteine, cystine, deoxymugineic, 3-epihydroxymugineic, glutamine, glutamic, glycine, homoserine, isoleucine, leucine, lysine, methionine, mugineic, ornithine, phenylalanine, praline, serine, threonine, tryptophane, tyrosine, valine
Organic acids	Acetic, aconitic, ascorbic, benzoic, butyric, caffeic, citric, *p*-coumaric, ferulic, fumaric, glutaric, glycolic, glyoxilic, malic, malonic, oxalacetic, oxalic, *p*-hydroxy-benzoic, propionic, succinin, syringic, tartaric, valeric, vanillic
Fatty acids	Linoleic, linolenic, oleic, palmitic, stearic
Sterols	Campesterol, cholesterol, sitosterol, stigmasterol
Growth factors	*p*-Amino benzoic acid, biotin, choline, *N*-methyl nicotinic acid, niacin, pantothenic, vitamins B_1 (thiamine), B_2 (riboflavin) and B_6 (pyridoxine)
Enzymes	Amylase, invertase, peroxidase, phenolase, phosphatases, polygalacturonase, protease
Flavonones	Adenine, flavonone, guanine, uridine/cytidine
Miscellaneous	Auxins, scopoletin, hydrocyanic acid, glucosides, unidentified ninhydrin-positive compounds, unidentified soluble proteins, reducing compounds, ethanol, glycinebetaine, inositol and myo-inositol-like compounds, Al-induced polypeptides, dihydroquinone, sorgoleone

From Uren, 2001.

spectrometry to identify 27 compounds in crude exudates of barley, wheat and rice plants without sample fractionation.

7.1.3 Nitrogen rhizodeposits

The influence of rhizodeposit quality and especially of C:N composition on soil organic matter decomposition has been referred to in section 7.1, but there has been interest, too, in the possible transfer of N via rhizodeposits from a N-rich legume to a N-poor non-legume, especially in mixed swards and grassland systems. In laboratory studies, Paynel *et al.* (2001) used a ^{15}N dilution technique to demonstrate transfer of N from roots of white clover to perennial ryegrass in pots and micro-lysimeters. Comparison of the two studies suggested that the primary route of N transfer to the grass was via the exudation of N compounds (especially NH_4^+ with some amino acids) into the soil by the legume followed by uptake by the companion grass; the contribution of N from the death of legume roots and nodules was estimated to be small. Høgh-Jensen and Schjoerring (2001) labelled plants with ^{15}N in a field study in Denmark and followed the fate of the N for two seasons. Rhizodeposition of N by unfertilized red clover, white clover and perennial ryegrass grown in pure stands was 64, 71 and 9 g N m^{-2}, respectively, for the two growing seasons and 89 and 32 g N m^{-2} for red clover and white clover, respectively, grown with ryegrass. These rhizodeposits, including fine roots, were >80% of the total plant-derived N in the soil and exceeded the amounts present in stubble and larger roots (Fig. 7.1). Rhizodeposited N from legumes, then, may play a substantial role in the N economy of grasslands, exceeding even the amount of N

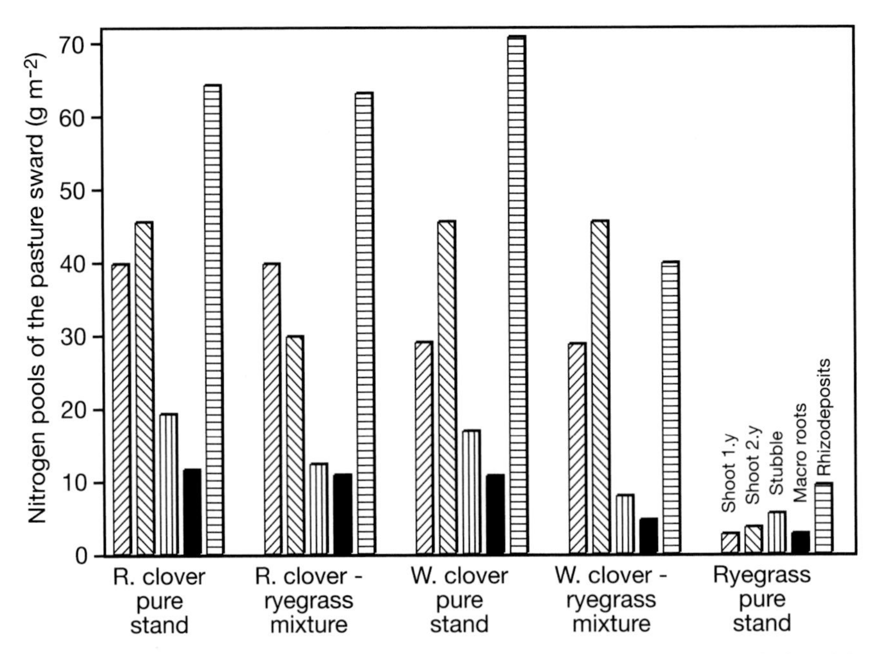

Fig. 7.1 Total nitrogen content of harvested shoots, stubble, large (macro) roots and rhizodeposits for red clover, white clover and ryegrass grown in mixtures or pure stands over two complete growing seasons. (Reproduced with permission from Høgh-Jensen and Schjoerring, *Soil Biology and Biochemistry*; Elsevier Science Ltd, 2001.)

harvested in shoot material (Fig. 7.1). It is noteworthy that in this study even N-deficient ryegrass grown in a pure stand had substantial rhizodeposited N relative to the N content of the shoots, stubble and larger roots.

In grain legumes, too, rhizodeposits may be a significant N pool in crop rotations, although there are few quantitative assessments. Mayer *et al.* (2003) grew faba bean, pea and white lupin in pots of soil to maturity and using ^{15}N found that rhizodeposition accounted for 13% of total plant N for faba bean and pea, and 16% for white lupin. Of these rhizodeposits, 7% (white lupin) to 31% (pea) were recovered as fine roots, 14–18% was recovered in the microbial biomass, and 3–7% was found in the mineral N fraction. Some 48% (pea) and 72% (white lupin) of rhizodeposited N was not recoverable in any of these fractions, suggesting rapid immobilization of this N as microbial residues. Whether such materials are then available to subsequent crops remains to be ascertained.

7.2 Chemical changes affecting nutrient acquisition

In addition to the considerable changes in microbiology induced by roots in the rhizosphere, there are also significant changes in the chemical environment which affect the ability of roots to acquire nutrients (Darrah, 1993). Some of these changes are a direct consequence of the selectivity of nutrient uptake by roots detailed in section 4.3.3, while others are a consequence of plant metabolism seeking to maintain anion/cation balance or to solubilize sources of nutrients that might otherwise remain unavailable. Root activity, then, has a direct effect on soil chemistry and on the ability of the plant to acquire nutrients.

Hinsinger (1998) details four major root-induced changes to the chemical environment of the rhizosphere, to which a fifth has been added:

(1) Ionic concentrations
(2) pH
(3) Reduction/oxidation conditions
(4) Complexation of metals
(5) Enzyme activity.

These five changes will be considered in turn, although it should be understood that several such changes may contribute to determining the bioavailablity of particular ions, and particularly to phosphorus acquisition.

7.2.1 Rhizosolution composition and replenishment

Section 4.3.2 has already detailed the selection of ions that occurs at the root surface resulting in the frequent accumulation of some ions (e.g. calcium) in the rhizosphere and the depletion of others (e.g. nitrogen and phosphorus). Many factors influence the concentration of ions in solution with distance from the root including the buffer capacity for an ion (related to the number and type of sorption sites), the solubility of other salts, the chemical equilibrium in soil solution, the soil water content, and the uptake capacity of the root. There are many studies that have modelled the changing concentration of ions with distance from the root (e.g. Darrah, 1993; Barber, 1995) but measurements of rhizosphere solutions are uncommon because of the technical difficulties of measuring concentrations in small volumes at a point and the disturbance to ionic movement that occurs whenever a sample is withdrawn. For example, Lorenz *et al.* (1994) found that while the pH of rhizosphere soil determined by displacing soil solution with water was less than that of non-rhizosphere soil, the pH of soil centrifugates was similar because of buffering by colloids during sampling. Displacement solutions demonstrated the accumulation of calcium and magnesium and the depletion of potassium, zinc and cadmium in the rhizosphere of radish (*Raphanus sativus*) roots more clearly than soil centrifugates. Lorenz *et al.* (1994) concluded that soil centrifugation had many limitations when used to sample soil solutions. Microsuction cups have been used in boxes with roots grown as a mat to measure gradients of ion concentrations with distance from the root surface, but the exact zone of sampling is unknown and temporal resolution is low (Vetterlein and Jahn, 2004).

When concentrations of some ions fall to low levels at the root surface then far-reaching effects on some soil constituents can be engendered. For example, Claassen and Jungk (1982) estimated that the concentration of K close to the root surface of maize seedlings grown in two different soils decreased to about 2–3 μM compared with the bulk soil concentration of several hundred μM. Such a large concentration gradient not only leads to the diffusion of K to the root surface but also shifts the equilibrium of adsorption/desorption of K towards enhanced desorption and to the depletion of non-exchangeable K as well as of exchangeable K close to the root (Fig. 7.2). Kuchenbuch and Jungk (1984) found that root-induced release of non-exchangeable K contributed up to 80% of the plant uptake in soils in which it might have been anticipated from the bulk soil K concentration in solution

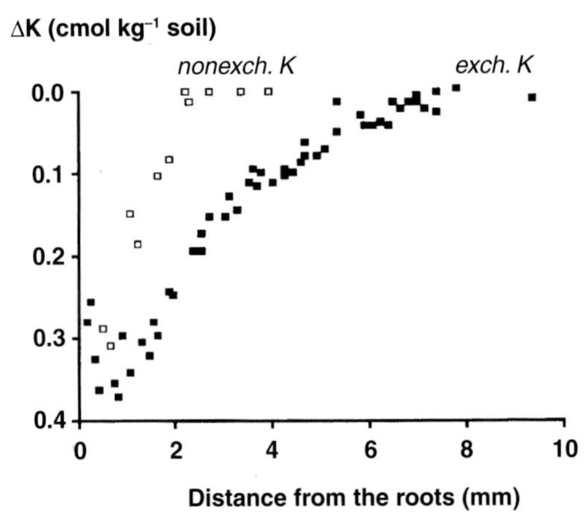

Fig. 7.2 Profiles of exchangeable and non-exchangeable K in the rhizosphere of rape grown for 7 days on a loess-derived soil. (Original research by Kuchenbuch and Jungk (1984). Reproduced with permission from Hinsinger, *Advances in Agronomy*; Elsevier, 1998.)

that its contribution would be insignificant. By decreasing the K concentration in solution below that which allows K release from phyllosilicates to occur, significant release of non-exchangeable K can occur and concurrent transformation (weathering) of minerals results. For example, in laboratory experiments in which ryegrass and rape were grown in containers allowing a mat of roots to form against agar containing the K-rich mica phlogopite, root-induced release of interlayer K resulted in transformation of the K-bearing phyllosilicate to vermiculite (Fig. 7.3). Transformation of the mica started within a few days of contact with the root mat by an exchange of interlayer K with cations such as Ca and Mg with a larger hydration energy, leading to an expansion of the lattice. In ryegrass, the process of vermiculite formation was evident to a distance of 1.5 mm from the root surface within 4 days (Hinsinger and Jaillard, 1993). The intensity and extent to which such transformations occur in the field are likely to be far less under normal conditions but become significant over time. Nevertheless, similar mineralogical transformation of mica to vermiculite and interstratified clay minerals was measured by Kodama *et al.* (1994) by comparing bulk soil mineralogy with that of rhizosphere soil sampled from maize roots. Bulk and rhizosphere soil samples collected from three sites near Ottawa, Canada after 4–5 weeks of maize growth (plants at the two-leaf stage) showed that while the content of non-phyllosilicates and mica was about 5% lower in bulk soil than rhizosphere soil, that of vermiculite and interstratified clay minerals was about 6% higher.

Ammonium ions also occur in soils in exchangeable and non-exchangeable forms and, as with potassium, there is evidence for use of non-exchangeable sources in the rhizosphere. Scherer and Ahrens (1996) grew rape, ryegrass and red clover on soils containing different clay minerals in containers allowing a root mat to develop against a volume of soil. In soils containing smectites and vermiculite as the dominant clay mineral, the depletion of non-exchangeable and extractable ammonium extended to about 3 mm from the root surface, and for rape the concentration of non-exchangeable ammonium N at 0.2 mm from the root surface declined to about 300 mg kg^{-1} compared with the bulk soil value of 410 mg kg^{-1} (i.e. a reduction of 27%). For soils containing illite as the dominant mineral the amount of depletion was

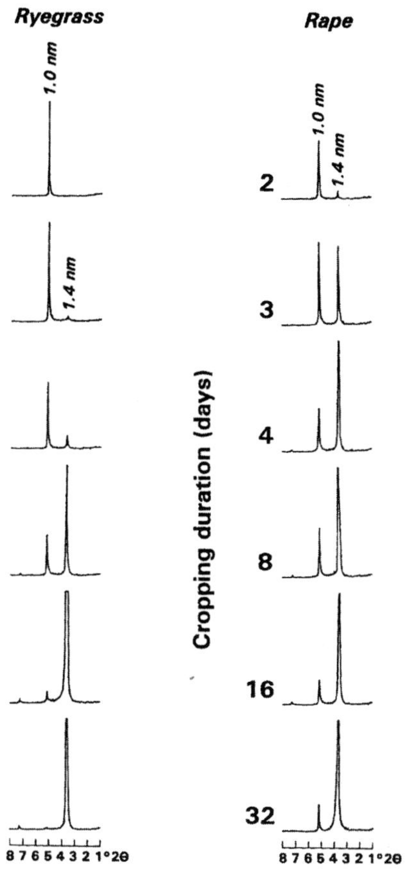

Fig. 7.3 Changes in the X-ray diffraction patterns for minerals in the rhizospheres of Italian ryegrass and rape with time demonstrating the alteration of phlogopite to vermiculite. The only source of K for the plants was phlogopite which has a typical peak at 1.0 nm. The appearance of a second peak at 1.4 nm after 2–3 days for both plant species indicates the vermiculitization of phlogopite accompanying the release of interlayer K to the plants. (Reproduced with permission from Hinsinger, *Advances in Agronomy*; Elsevier, 1998.)

less (the authors claim it was 'negligible'), with the concentration of non-exchangeable ammonium N at 0.2 mm from the root surface for rape declining to about 110 mg kg^{-1} compared with the bulk soil value of 140 mg kg^{-1} (i.e. a reduction of 21%). Depletion by ryegrass in soils containing smectites and vermiculite at 0.2 mm from the root surface (concentration 320 mg kg^{-1}) was slightly less than that by rape, but depletion by red clover was only about 15 mg kg^{-1}. The effect of the roots acting as a sink is one factor contributing to the change of cation exchange equilibrium resulting in the release of non-exchangeable ammonium, but the difference between the plant species points to other possible factors. Scherer and Ahrens (1996) compared their results with those of other workers and suggested that the greater root density of ryegrass than red clover probably contributed to the greater depletion of non-exchangeable ammonium (and potassium) by the former. Moreover, rape is a plant that is known for the release of H^+ into its rhizosphere (see later in this chapter). The exchange of H^+ for NH_4^+ in the clay crystal lattice leaves the lattice in a more expanded state, thereby leaving the non-exchangeable ammonium more accessible to replacing ions. As with potassium, then, non-exchangeable ammonium may be released in the rhizosphere, contributing to the nutrition of plants and accounting for a significant proportion of the measured seasonal depletion found beneath crop plants (Mengel and Scherer, 1981; Mengel *et al.*, 1990).

The depletion of P in the rhizosphere of plants is a well-known phenomenon (see Tinker and Nye, 2000, for many examples) but because of the low solubility of many P compounds in soils, very low concentrations have to be achieved before there are significant changes to the adsorption/desorption and dissolution/precipitation equilibria in most soils (Hinsinger, 1998). Such low concentrations are almost certain to be too low for many plants to grow optimally. Under some laboratory conditions, it has been possible to demonstrate accumulation of phosphate in the rhizosphere rather than depletion, especially if roots induce dissolution of a P compound at a faster rate than it is taken up. Hinsinger and Gilkes (1996) grew ryegrass and subterranean clover with either ammonium or nitrate as the source of N and rock phosphate or alumina-sorbed phosphate as the source of P. When P was supplied as rock phosphate there was an increase in NaOH-extractable P in the rhizosphere of both plant species, but larger for ryegrass. The largest increase was obtained for ryegrass supplied with ammonium (Fig. 7.4). When P was supplied sorbed to alumina, extractable P decreased in the rhizosphere of ryegrass for both sources of N but increased for subterranean clover, especially when supplied with ammonium. These results were consistent with the measured

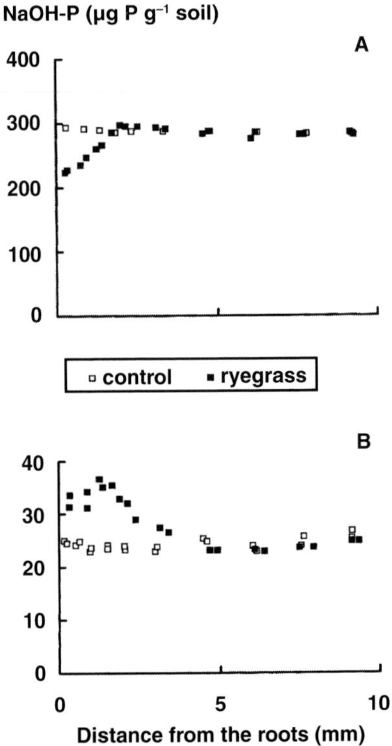

Fig. 7.4 Profiles of NaOH-extractable P in the rhizosphere of ryegrass grown for 14 days on an alumina sand with P supplied either as P sorbed onto alumina (A), or as phosphate rock (B). The control contained no plants. (Reproduced with permission from Hinsinger, *Advances in Agronomy*; Elsevier, 1998.)

changes in rhizosphere pH; pH decreased when ammonium was supplied, was larger for ryegrass than subterranean clover, and was 1.5 units lower when P was supplied as alumina-sorbed phosphate and 0.8 units lower with rock phosphate. H^+ excretion by roots of ryegrass induced dissolution of the rock phosphate, which occurred at a faster rate than P uptake. The significance, if any, of such processes for field-grown plants has not been explored.

7.2.2 Changes in pH

Changes of pH in the rhizosphere have been widely demonstrated and their causes and consequences have been reviewed in several publications (Nye, 1986; Marschner, 1995; Hinsinger *et al.*, 2003). Changes of 0.5–1 unit have frequently been reported within 1–2 mm of the root surface, although the size of the change is often very dependent on other components of the soil chemical environment such as P, Fe and Al status. The origin of the root-mediated pH changes includes:

(1) Imbalance of cation/anion uptake by plants
(2) Release of organic anions
(3) Root respiration
(4) Microbial production of acids from root exudates.

Cation/anion imbalance

The differential uptake of cations and anions by plant roots is a major contributor to the release of H^+ and HCO_3^- ions into the rhizosphere. The release is a necessary consequence of the regulation of cellular electrical charge and pH which is maintained within a narrow range of about 7.3 in the cell cytoplasm (see Marschner, 1995, for details of the mechanism of regulation). When a plant takes up more cations than anions, protons are released from the cytoplasm to the apoplasm to maintain electrical neutrality; these protons are then released into the rhizosphere contributing to its acidification. Conversely, when anion uptake into the cells exceeds cation uptake, the excess negative charge is balanced by the release of hydroxyl or bicarbonate ions (the former are usually rapidly carbonated by respired CO_2) leading to alkalinization of the rhizosphere. These effects have been demonstrated many times in solution culture experiments where it is relatively easy to change the nutrients bathing the roots and to observe the consequent alteration of pH. For example, Hiatt (1967) incubated excised barley roots in nutrient solutions and demonstrated that when plants were grown in K_2SO_4 this led to greater uptake of K^+ than SO_4^- and consequent acidification of the nutrient solution. With a $CaCl_2$ solution, plant uptake of Cl^- was greater than that of Ca^{2+}, resulting in alkalinization of the solution, while with KCl solution, K^+ and Cl^- were taken up at similar rates so that there was little change in pH of the nutrient solution. The result is that there is a tight coupling between H^+ and HCO_3^- release and cation/anion balance as demonstrated by the study by Rufyikiri et al. (2001) with banana seedlings grown in solution culture with different Al, NO_3^- and NH_4^+ concentrations (Fig. 7.5).

Nitrogen nutrition plays an important role in the cation/anion balance because it is often the nutrient present in highest concentrations in plant tissues and because it can be taken up as a cation (NH_4^+), as an anion (NO_3^-) or, in legumes in symbiotic association

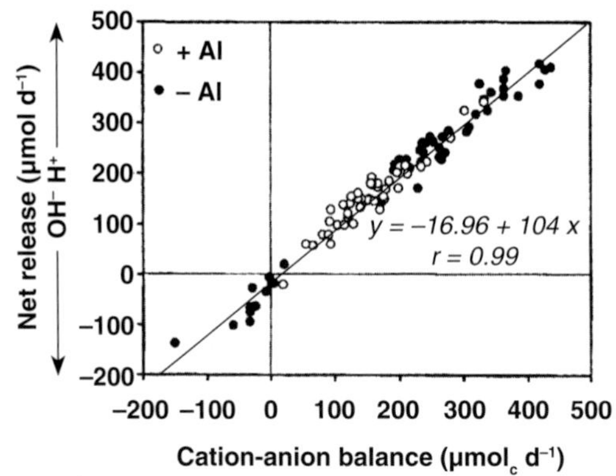

Fig. 7.5 Relationship between the amount of H^+ released and cation/anion balance of seedlings of banana grown in solution culture with complete nutrients (N as ammonium and nitrate) with (○) or without (●) 78.5 μM Al. (Original research by Rufyikiri *et al.* (2001). Reproduced with permission from Hinsinger *et al.*, *Plant and Soil*; Springer Science and Business Media, 2003.)

with *Rhizobia*, as a molecule (N_2). Many studies in both solutions and soils have shown that plants supplied with NO_3^- as their source of N will release OH^- or HCO_3^- into the rhizosphere thereby increasing the pH, while plants supplied with NH_4^+ as their principal source of N will release H^+ thereby acidifying the rhizosphere (e.g. Marschner *et al.*, 1986; Nye, 1986) (Plate 7.1). Legumes fixing atmospheric N_2 take up more cations than anions and acidify their rhizosphere in consequence (Jarvis and Robson, 1983). For example, Tang *et al.* (1997) investigated the proton excretion of 12 pasture legumes grown in solution culture and found that more cations than anions were taken up in all species. There was a strong linear relation ($r^2 = 0.94$) between the quantities of H^+ released (mmol plant^{-1}) and the excess of cations over anions released with a slope of 1.06.

The foregoing description is a useful generalization but, nevertheless, a simplification. Some plants such as rape, chickpea and lupin frequently, although not always, acidify their rhizosphere even when supplied with NO_3^- as their source of N (Grinsted *et al.*, 1982; Marschner and Römheld, 1983). There is also spatial variation along roots so that different portions of a root or root system may respond differently. For example, Marschner *et al.* (1986) showed that maize plants supplied with NO_3^- increased the rhizosphere pH along the main axes but pH along the laterals either remained unaltered or decreased slightly (Plate 7.2a). Similarly Jaillard *et al.* (1996) grew roots on agar and demonstrated that while the apical regions of a maize root alkalinized the rhizosphere pH, the basal portions, especially in the zones of lateral root emergence and elongation, were acidifying the rhizosphere (Plate 7.2b). Such measurements suggest that the cation/anion balance varies along a root, probably as a consequence of differential uptake of cations and anions along it. There is, however, little experimental evidence to support or refute this conclusion.

Another factor that may affect cation/anion balance of cells is the release of organic anions (Hinsinger *et al.*, 2003). Jones and Darrah (1994) demonstrated that amino acids can be both released as exudates and taken up, thereby contributing to the net anion balance and the nitrogen nutrition of the plant. Their results and calculations with maize indicated that when inorganic N concentrations in soil were limiting (≤0.1 μmoles cm^{-3} soil), the uptake of amino-N accounted for up to 90% of the total N taken up by the roots. When inorganic N levels were high, simulations indicated that amino acid N contributed ≤30% of total N uptake. Results such as those shown in Fig. 7.5 suggest that this may not be a major source of imbalance in many circumstances, although the full significance of the phenomenon has yet to be quantified for field-grown plants. Phosphate ions, too, can have different charges dependent on the ambient pH of the soil ($H_2PO_4^-$ is the dominant ion for soil pH <7.2 and HPO_4^{2-} for pH >7.2), thereby affecting the cation/anion balance differently depending on the ionic species taken up. However, in practice the acidic pH of the apoplast usually ensures that the major form of P taken up is $H_2PO_4^-$ even in soils with pH >7.2.

The measured change of pH in the rhizosphere is not a good indicator of the release of H^+ or OH^-/HCO_3^- released by roots when comparing studies because soils are differently buffered for pH depending on the nature and type of clay minerals, and calcium carbonate and organic matter contents (Wild, 1988). Table 7.3 compares results from a pot experiment using eight different soils in which faba bean was grown (Schubert *et al.*, 1990). In the three soils containing 2–6% $CaCO_3$ (i.e. with a large buffering capacity), there was no measurable decrease in pH, while in the others the decrease in pH ranged from 0.73 to 1.49. The calculated H^+ release determined on the basis of the pH change and buffer capacity was very similar in all soils, so that the lack of pH change in the calcareous soils does not mean that there was no H^+ release but, most likely, that the H^+ was consumed in dissolution reactions with $CaCO_3$ (Hinsinger *et al.*, 2003). Similar dissolution reactions between rock phosphate and H^+ also result in pH changes that are less than the H^+ efflux because H^+ is consumed during the reaction (Hinsinger and Gilkes, 1996).

The importance of root-induced pH changes in the rhizosphere lies in the influence that these may have on the bioavailability of many nutrients and toxic elements together with physiological effects on roots and microbes. In particular, the dynamics of the many forms of sparingly soluble inorganic P in soils are strongly pH-dependent, with phosphate ions readily precipitating to form Fe and Al phosphates in acidic conditions and Ca phosphates in alkaline conditions (Hinsinger, 2001). In soils of about pH 6–8.5, P reacts with Ca to form various calcium phosphates such as octocalcium phosphate and hydroxyapatite (Wild, 1988). If a supply of protons is made available (i.e. the medium becomes more acidic) then the phosphate mineral dissolves and P becomes available to the plant root:

$$Ca_5(PO_4)_3OH + 7H_3O^+ \leftrightarrow 3H_2PO_4^- + 5Ca^{2+} + 8H_2O \qquad (7.1)$$

Table 7.3 Effect of $CaCO_3$ concentration, proton buffering capacity and soil pH on plant-induced pH change by faba beans grown in a pot experiment for their complete life cycle

Soil	$CaCO_3$ (g kg^{-1})	Buffering capacity	Initial pH	Final pH	H+ release (mmol H^+ g^{-1} plant dry weight)
Calcaric regosol	69.0	15 430	7.60	7.56	–
Calcaric fluvisol	42.9	7644	7.50	7.50	–
Fluvial colluvisol	21.0	5477	7.45	7.45	–
Calcic cambisol	9.8	37	7.25	6.52	1.09
Rhodic acrisol	0.0	38	7.20	6.35	1.12
Eroded orthic luvisol	13.0	23	7.00	6.25	0.92
Orthic luvisol	4.7	12	7.35	6.25	0.92
Dystric cambisol	2.9	6	7.30	5.81	1.14

Data from Schubert *et al.* (1990); adapted by Hinsinger (2001).

These reactions are more complex than the simple equation suggests (for more details see Wild, 1988; Hinsinger, 2001), but there are many examples in the literature of plant roots increasing P availability and uptake by protonation of the rhizosphere. For example, Grinsted *et al.* (1982) grew rape in thin layers of a phosphate-deficient soil supplied with phosphate-free nutrient solution containing N as nitrate and measured a decrease in pH as large as from 6.5 to 4.1. The pH decrease was associated with an increase in solution P concentration in the rhizosphere soil compared with that obtained by shaking unplanted, control, soil with extractant at the initial soil pH of 6.1 (Fig. 7.6). Calculations based on the deviation of solution P concentration from the standard desorption isotherm indicated that a decrease in pH from 6.5 to 4.1 in this sandy loam soil could result in at least a 10-fold increase in P released into solution, which could be available for plant uptake. This study is also interesting because pH initially rose, as expected when N is supplied as nitrate, but then fell in response to the P-deficient conditions.

Several studies have indicated that the proton release occurring as a result of ammonium fertilizer application may improve the P nutrition of crop plants by enhancing P dissolution. For example, Riley and Barber (1971) grew soyabean in pots of a silt loam soil to which lime was added to create a range of pH values and the plants were supplied with either NH_4^+ or NO_3^- fertilizers. After 21 days, the pH of rhizosphere soil ranged from 4.7 to 7.2 for plants given NH_4^+ and from 6.6 to 7.4 for plants given NO_3^-, and there were linear relations between P uptake and rhizosphere pH (uptake mg P pot^{-1} = 19.46–2.38pH; r^2 = 0.96) and the P concentration in the shoot and rhizosphere pH (P [mg g^{-1}] = 3.68–0.34pH; r^2

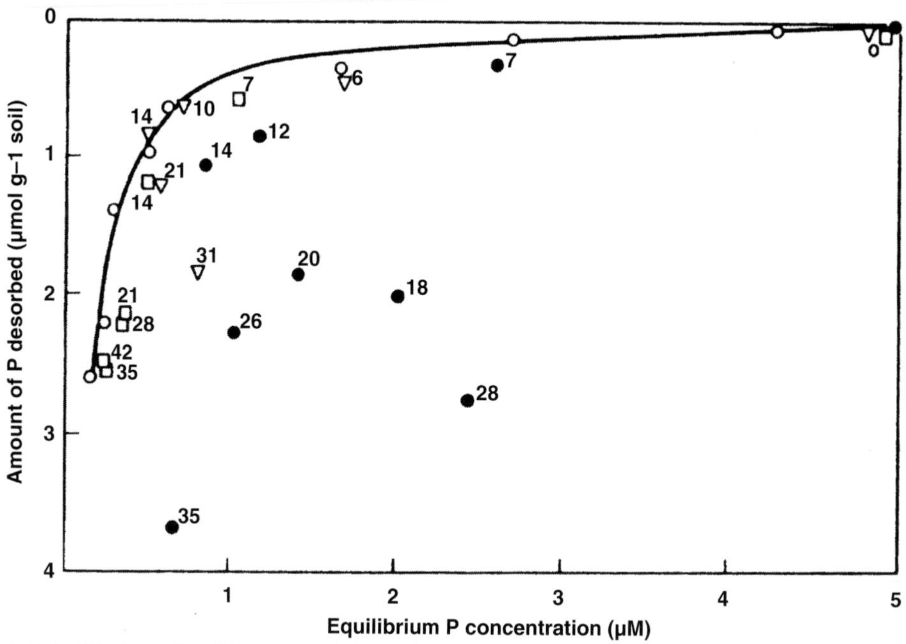

Fig. 7.6 The desorption of P from a sandy loam soil by dilution in 0.01 M calcium nitrate (○) and by uptake by rape in three repeated experiments (the three symbols ●, □ and ▽). The age in days of the rape plants is given with each data point. The increase in solution P concentration after 14 days suggests solubilization of P occurring in the rhizosphere soil. (Reproduced with permission from Grinsted *et al.*, *New Phytologist*; New Phytologist Trust, 1982.)

= 0.94). These results suggest enhanced bioavailability of soil P by root-induced acidification. Similar results were obtained by Gahoonia and Nielsen (1992) with ryegrass grown in pots containing a luvisol; plants fed with NH_4^+ took up more P, acidified the rhizosphere more, and depleted HCl-extractable P more than those fed with NO_3^-.

Dissolution of rock phosphate can be considerably speeded up by H^+ release from plant roots and the relatively good performance of plants such as buckwheat and some legumes such as pigeonpea and lupin has been ascribed to their ability to release protons, thereby increasing the bioavailability of P (Hinsinger, 2001). For example, Haynes (1992) compared the abilities of wheat, barley, rape, kale, narrow-leaf lupin (*Lupinus angustifolius*) and buckwheat (*Fagopyrum esculentum*) to grow and utilize two sources of rock phosphate mixed with a low-P silt loam soil (pH 5.5) in pot experiments. When N was supplied as NO_3^-, all crops (including lupin) except buckwheat took up more anions than cations and raised soil pH, limiting the use made of rock phosphate. However, N-fixing lupin and buckwheat took up more cations than anions, acidified the soil (by 0.4 and 0.2 units, respectively) and utilized the phosphate rocks. Similarly when tea (*Camellia sinensis* L.) was supplied with NH_4^+, the cation/anion imbalance resulted in acidification of the rhizosphere relative to the bulk soil and the enhanced use of rock phosphate (Zoysa *et al.*, 1998).

The solubility of ions such as Fe and Al is also markedly affected by pH, decreasing as pH increases. Many plant species have been shown to respond to Fe deficiency by acidifying their rhizosphere (e.g. Marschner *et al.*, 1989), although the concentration of Fe in solution is almost always less than the 10^{-6} M required for plants to meet their Fe requirements in the pH range commonly found in soils. The contribution of proton fluxes to Fe nutrition of plants is, then, unresolved (Hinsinger, 1998).

Alkalinization of the rhizosphere also occurs but it has not been as widely reported despite the widespread use of nitrate fertilizers in many crop production systems which would suggest that its frequency should be common (Nye, 1986). Gahoonia and Nielsen (1992) showed alkalinization of the rhizosphere of ryegrass grown in pots of a luvisol or oxisol when supplied with NO_3^- and that this resulted in less depletion of total-P and bicarbonate extractable-P but more depletion of NaOH-P (i.e. P bound to Fe and Al oxides) compared with rhizosphere acidification. An explanation of this observation is that the OH^-/HCO_3^- released by roots supplied with NO_3^- desorbed P from the Fe and Al oxides by ligand exchange reactions. Again, the importance of such reactions in field-grown plants is unknown. It may be important in an ecological context, but is probably not important for crop plants because in very acidic soils, where such forms of phosphate predominate, the high dissociation constant of carbonic acid means that a substantial rise of pH would be needed for bicarbonate ions to occur in soil solution at the concentrations necessary to affect the desorption of P ions from the oxides (Hinsinger, 2001).

Despite the many examples of rhizosphere acidification reported in laboratory studies, the contribution that such changes may make to field-grown plants is uncertain and largely unquantified (Darrah, 1993; Hinsinger, 2001). Many of the large changes reported are induced by nutrient-deficient conditions, so the extent of root-mediated pH changes as an adaptive strategy for nutrient acquisition has still to be resolved. Nevertheless, because of the limited zone of soil from which nutrients such as P are taken up, it is clear that the pH of the rhizosphere is of greater importance than that of the bulk soil in determining the bioavailability of some nutrients.

Organic anion release and cluster roots

The ability of roots to release malate and citrate as a means of reducing the toxic effects of Al in soil solution has already been described (see section 5.5.2), but plants release a wide range of organic anions which may contribute to changes in rhizosphere chemical properties. Kirk (1999) outlined the processes whereby organic anions contribute to the solubilization of phosphate as: (i) excretion accompanied by excretion of H^+ may acidify the rhizosphere thereby solubilizing phosphate; (ii) displacement of phosphate from adsorption sites; (iii) chelation of metal ions that would otherwise immobilize phosphate; and (iv) formation of soluble metal–chelate complexes with P. The anions released include carboxylic acids that are important metabolites in plant cells such as oxalate, oxalo-acetate, α-cetoglutarate, malate, fumarate, succinate, isocitrate and citrate, as well as compounds such as aconitate, formate, lactate and piscidate that are released in some root exudates. Because the dissociation constant of most organic acids is low compared with the neutral pH of plant cell contents, they are dissociated within the cell and released as organic anions rather than as acids (Hinsinger, 2001). The release of organic anions may contribute to acidification of the rhizosphere via compensation for the release of net negative charge that they represent (i.e. organic anion efflux should be counterbalanced by an equivalent influx of OH^- or efflux of H^+) but, in general their effect on pH is small relative to the effect of the inorganic cation/anion balance of the plant (Kirk, 1999; Hinsinger *et al.*, 2003).

The quantities and types of organic anions released vary considerably between plant species and on the environmental conditions, and their effects operate through multiple mechanisms (Jones, 1998). Deficiency of nutrients such as P and Fe appears to promote anion release in some species. For example, P deficiency in rape doubled citrate concentration in leaves and the citrate:sugar ratios of the phloem compared with plants in which P was adequately supplied (Hoffland *et al.*, 1992). Citrate produced in the leaves was exported to roots where it accumulated in the region of excretion before its release. Malate was synthesized in the excreting zone of the root and was the major component of the organic anion exudate. Interestingly, when no P was supplied to the root zone, the zone of excretion was located 10–20 mm behind root tips but when rock phosphate was applied locally, the zone of excretion moved to that part of the root in direct contact with the rock phosphate particles. This observation suggests that plants may have a capacity to determine the optimal position in the root system for exudation of anions to ensure the maximum benefit to P nutrition. Citrate, malate and oxalate are the organic anions most commonly released, with each tending to dominate in particular species (Table 7.4). There is, though, considerable variation in the amounts released even between genotypes of a single species. For example, in chickpea, a legume known to have high concentrations of malonate in its roots and nodules (Li and Copeland, 2000), Wouterlood *et al.* (2004) found that while carboxylate concentrations in the rhizosphere of lateral roots steadily increased with time for cv. Heera, those for cv. Tyson did not at about 33 days after sowing. In this study carboxylates (malonate, malate and citrate) were released even when the soil was well supplied with P, with carboxylate concentration increasing with distance from the root apex.

There have been relatively few attempts to measure the release of organic anions and assess quantitatively their contribution to plant nutrition. Rice is generally known as a plant adapted to acid, P-deficient upland soils. It was found to take up P from alkali-soluble in-

Table 7.4 Rates of exudation of organic anions from various parts of roots or whole root systems (wrs)

Plant	Exudate location	Treatment	Rate of exudation (pmol g^{-1} root fw s^{-1})		
			Malate	Citrate	Oxalate
Rape	Tip	+P, –P	10–59	4–18	–
Rape	Base	+P, –P	2–14	0.5–9	–
Lupin	Proteoid roots	+P, –P	8–141	6–158	–
Lucerne	wrs	+P, –P	0.8–1.7	0.2–0.8	–
Sorghum	wrs	+P, –P	< detection	0.5–5.8	–
Maize	Tip	–Al, +Al	1–8	0.1–36	–
Wheat	Tip/wrs	–Al, +Al	14–338	1.3–3.1	–
Maize	wrs	+/– nutrients	5–165	3.3–38	–
Barley	wrs	+Fe, –Fe	1–40	–	–
Maize	wrs	+K, –K	0.25–0.42	0.02–0.04	–
Agropyron cristatum	wrs	+ nutrients	0.25	–	–
Agropyron smithii	wrs	+ nutrients	0.17	–	–
Bouteloua gracilis	wrs	+ nutrients	<0.1	–	–
Calcifuge plants (nine species)	wrs	+ nutrients	0.08		0.44
Calcicole plants (nine species)	wrs	+ nutrients	0.08	0.38	1.22

From Jones (1998), in which original references are cited.

organic forms when grown on P-deficient soil, possibly as a result of organic anion release (Hedley *et al.*, 1994), and also to release citrate when grown in nutrient solution (Kirk *et al.*, 1999a). Kirk *et al.* (1999b) grew upland rice in thin layers of a clay soil supplied with P-free nutrient solution and sampled the soil and plants at weekly intervals over a 6-week period. The amounts of organic anions in the soil increased in the presence of plants, with citrate the main anion together with smaller quantities of oxalate, malate, lactate and fumarate. Rates of release were similar for plants fed NH_4^+ and NO_3^-, but pH decreased by 0.6 units and increased by 0.4 units, respectively, mainly due to inorganic cation-anion balance in the plant rather than to organic anion excretion. Subsidiary studies showed that the half-life of citrate in soil solution was <5 hours, being a function of both rate of decomposition by rhizosphere microbes and rate of *de novo* synthesis from other released substances. Apparent rates of citrate release ranged from 337 to 155 nmol g^{-1} root fresh weight h^{-1} over the course of plant growth equivalent to 2–3% of plant dry weight. Using these measurements together with Kirk's (1999) model of phosphate solubilization by organic anions, which allows for the diffusion of the anion away from a root, its reaction and decomposition in the soil, and diffusion of the solubilized P to and from the root, Kirk *et al.* (1999b) demonstrated that the measured P uptake by the plants was a consequence of P solubilization by citrate (Fig. 7.7). The main mechanism of solubilization involved the chelation of metal ions that would otherwise have immobilized P or formation of soluble citrate–metal–P complexes, or both. Solubilization by displacement of P from adsorption sites was unimportant.

Plants that form proteoid or cluster roots are a group of particular interest with regard to organic anion excretion. Such roots are almost universal in the family Proteaceae (1600 species) and in fewer species in another seven families (Lamont, 2003). There may be 10–1000 rootlets cm^{-1} of parent root, each rootlet of limited length and often densely covered with root hairs. The whole looks like a bottlebrush (see Plate 2.1d). A 'working definition' of a cluster root is that it is a lateral root with defined clusters of 10 or more rootlets

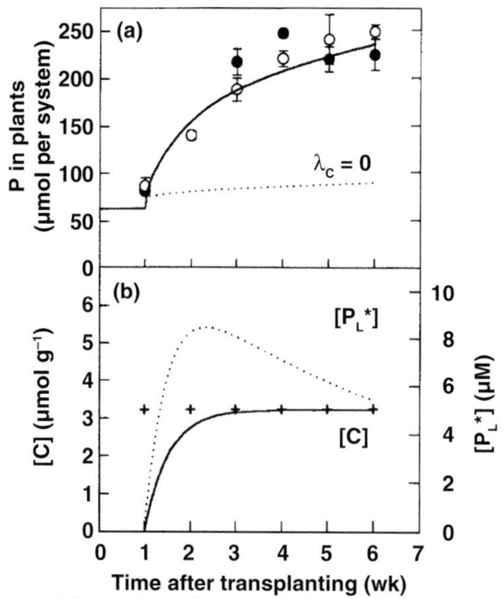

Fig. 7.7 P uptake and citrate. (a) Comparison of the measured and calculated time courses of P uptake with and without solubilization by citrate. The points are measured data (NH_4^+ (●); NO_3^-(○)) and lines are the calculated values with (solid line) and without (dotted line) citrate solubilization. (b) The corresponding calculated concentrations of citrate in the whole soil and P in the soil solution (lines), and the measured mean citrate concentrations in the soil (+). (Reproduced with permission from Kirk *et al.*, *New Phytologist*; New Phytologist Trust, 1999.)

cm^{-1} of lateral root (Johnson *et al.*, 1996; Watt and Evans, 1999a). Cluster root formation is common among slow-growing sclerophyllous shrubs and trees in Western Australia and South Africa which are adapted to habitats of very low chemical fertility and with P as a major limiting nutrient (Pate and Watt, 2002). They are relatively unimportant among crop plants except in some species of the genus *Lupinus*. Cluster roots arise from the pericycle opposite the protoxylem poles along lateral roots with formation stimulated by a shortage of P and, in some species, deficiency of Fe. Low internal P status of the plant is one trigger for cluster root formation but external factors may also be involved because formation can also be stimulated in nutrient-rich patches and in upper soil layers rich in organic matter (Neumann and Martinoia, 2002).

White lupin (*Lupinus albus*) has been used as a model plant for the study of the physiology and functioning of cluster roots because of its ease of growth for controlled experiments. Under conditions of P deficiency, white lupin forms cluster roots which may be up to 60% of the total root biomass (Keerthisinghe *et al.*, 1998). These release large amounts of carboxylates (mainly citrate and malate, but also others such as malonate and *trans*-aconitate) which mobilize P from Fe- and Al-P fractions, and from acid-soluble Ca-P fractions, with concurrent solubilization of Fe, Mn, Zn, Al and P bound to Fe- and Al-humic acid complexes (Gardner *et al.*, 1983; Dinkelaker *et al.*, 1989). Acid phosphatases that may assist in the release of organic P have also been measured coincident with the exudation of organic anions (Neumann *et al.* 1999). Several processes, then, contribute to the increased ability of the plant to acquire P (Fig. 7.8). The factors regulating cluster root formation and carboxylate formation are still being determined but appear to be related to the P concentration in the shoots. For example, Shen *et al.* (2003) grew white lupin plants in pots of sand

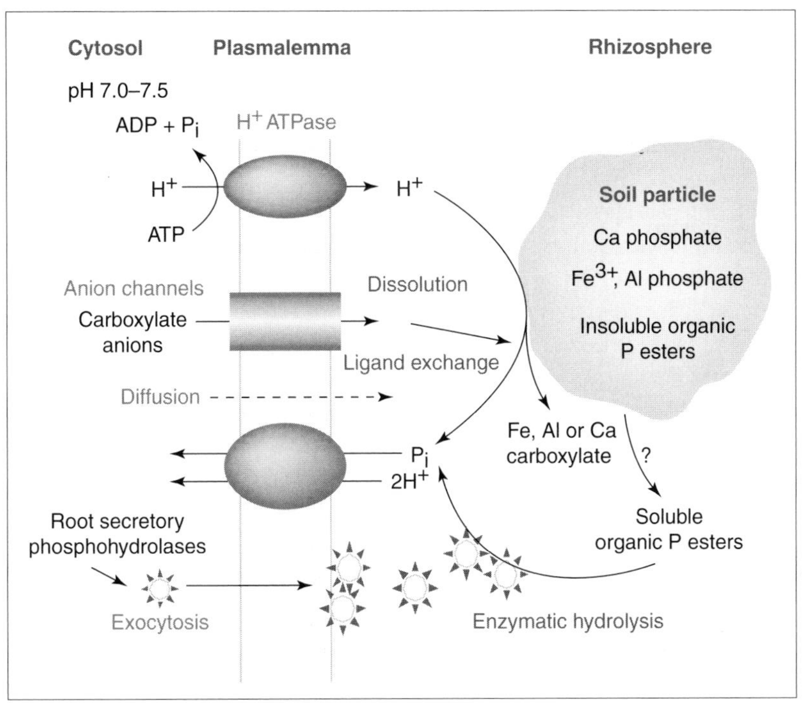

Fig. 7.8 A model for root-induced chemical mobilization of phosphate in the rhizosphere by exudation of carboxylates, protons and phosphohydrolases. (Reproduced with permission from Neumann and Martinoia, *Trends in Plant Science*; Elsevier, 2002.)

to which P was added to give either deficient or adequate P for growth and found that the number of cluster roots was inversely related to the concentration of P in either roots or shoots. Similarly application of P fertilizer to a loam soil decreased cluster root length, giving an inverse relationship between cluster root production and shoot P concentration in the range 2–8.5 mg g^{-1}. Total root length and total shoot weight were increased by only about 20% as P application was increased from 0 to 1 g P kg^{-1} soil (Peek *et al.*, 2003). Exudation of organic anions from the roots is accompanied by increased activity in the roots of malate dehydrogenase (MDH) and phosphoenolpyruvate carboxylase (PEPC) with the latter fixing some 25–34% of the C excreted as citrate and malate by P-deficient roots of white lupin (Johnson *et al.*, 1996). Using cloned white lupin PEPC and MDH cDNAs, Uhde-Stone *et al.* (2003) demonstrated that under P-deficient conditions, PEPC and MDH were both expressed in the cortex of emerging and mature cluster rootlets, with PEPC (but not MDH) also expressed in the meristem. Their results indicate that acclimation of white lupin to P deficiency involves modification to the expression of genes involved in carbon metabolism and that the expression of PEPC and MDH in different tissues may reflect changes in the composition of excreted organic anions during cluster root development.

The altered carbon metabolism of the roots is linked to their development, and there appears to be a tightly regulated sequence of events that triggers the initiation of clusters, limits rootlet growth, alters metabolism, and activates then deactivates exudate release (Watt and Evans, 1999b). Once cluster roots have been initiated, they grow for about 3–4 days after which no further meristematic activity is detectable in the rootlet. These determinate second order laterals differentiate root hairs and vascular tissues along their length during this period. Watt and Evans (1999a) followed the development of a cluster and showed that while PEPC activity increased during the period of rootlet growth, citrate exudation did not start until 1 day after the rootlets had reached their maximum length, 4 days after emergence (Fig. 7.9). Release of citrate was transient with a diurnal pulse during daylight for a period of 2–3 days before returning to trace levels. Interestingly, growth under elevated CO_2 caused both final rootlet length and citrate efflux to be advanced by 1 day with no discernible effect on the synchrony of rootlet activity (Watt and Evans, 1999a). The mechanism of exudation is not yet known, but the composition of the organic anions exuded does not reflect that of the plant tissues nor do rates of exudation reflect tissue concentration, indicating that the mechanism is specific and not simply driven by a concentration gradient between root and soil (Keerthisinghe *et al.*, 1998; Neumann *et al.*, 1999). Export may be partly via anion channels (Fig. 7.8) since efflux was reduced by 50% when channel inhibitors were used (Neumann *et al.*, 1999), but packing of citrate in vesicles and release by exocytosis is also possible (Watt and Evans, 1999b; see Skene *et al.*, 1996, for a description of exocytosis in cluster roots of *Grevillea robusta*).

Extrapolation of results with white lupin to other species must be undertaken with care because although there are some visual similarities between cluster roots of different species, other features may differ. For example *Banksia* spp. and white lupin differ in the following: (i) the soils on which banksias grow usually have very much lower P concentrations than those on which white lupin is grown; (ii) banksias and many other Proteaceae form distinctive mats of proteoid root clusters just below or within the litter at the soil surface which are not formed by white lupin; (iii) banksias have extensively branched proteoid root clusters that are more complex morphologically than those of white lupin and many other Proteaceae (e.g. *Hakea* spp.); (iv) banksias differ from white clover in the spectra of organic anion exudates; and (v) banksias are woody species that can live for many years and can generate successive sets of proteoid roots throughout a season unlike the annual white lupin (Pate and Watt, 2002). The range of carboxylates exuded by roots is still being explored but Roelofs *et al.* (2001) demonstrated that malate, malonate, lactate, citrate and *trans*-aconitate were invariably present as a

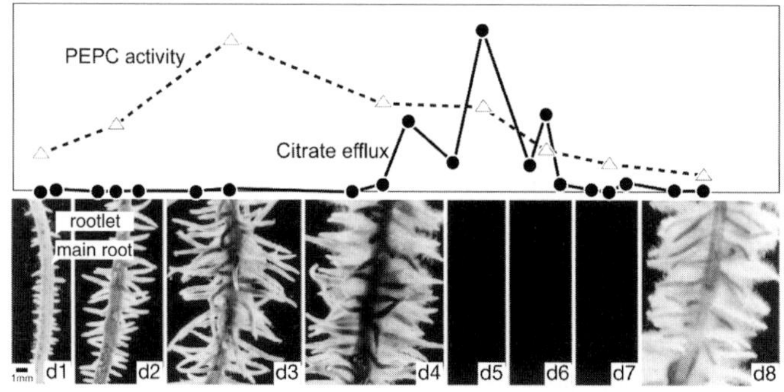

Fig. 7.9 Time course of cluster rootlet growth, PEPC activity (triangle – scale 0 to 1.6 μmol NADH m^{-1} root axis min^{-1}), and citrate efflux (● – scale 0 to 0.11 μmol m^{-1} root axis min^{-1}). Note that the time scale is not linear but scaled to fit the size of the micrographs. PEPC peaked as rootlets reached their final length (d3) and no longer have meristems. (Reproduced with permission from Watt and Evans, *Plant Physiology*; American Society of Biologists, 1999b.)

large proportion of the total carboxylate exudation in a range of Australian Proteaceae when grown in nutrient solution (Table 7.5). Another contrast with white lupin was that while white lupin typically exudes carboxylates accompanied by protons that acidify the rhizosphere, no concomitant proton release was measured in the Proteaceae studied.

A wide range of *Lupinus* spp. is known to form cluster roots but, until recently, the ability to form such roots was believed to be confined to Old World species, although not all such species produce such roots (Skene, 2000). However, Hocking and Jeffery (2004) induced the Old World species *L. hispanicus* and *L. luteus*, and the New World species *L. mutabilis*, to form cluster roots that secreted organic anions when grown in P-deficient sand. Citrate was the dominant organic anion exuded, although *L. luteus* also exuded large amounts of succinate. Rates of citrate efflux from *L. hispanicus* and *L. luteus* were comparable to those reported for P-deficient white lupin (up to 1.09 nmol g^{-1} root fresh weight s^{-1}) but about 30 times higher than the rate from *L. mutabilis* (0.036 nmol g^{-1} fresh weight s^{-1}). Citrate efflux showed a diurnal pattern that was maximal during daylight with no appreciable efflux from first order lateral roots or root tips; both these features are as reported for white lupin. In contrast, *L. angustifolius* did not produce cluster roots under P-deficient conditions but did under conditions of high nitrate supply; these roots also exuded citrate. These results suggest that many more *Lupinus* spp. than appreciated previously have an inherent capacity to form cluster roots.

Root respiration

Significant quantities of CO_2 are released into the rhizosphere by the respiration of roots (typically some 12–25% of daily photosynthesis; Lambers *et al.*, 1996) and the breakdown of root exudates and other materials by rhizosphere microbes (see section 7.1). This CO_2 has sometimes been assumed to have an acidifying effect but there are few measurements to sustain or refute such claims (Hinsinger *et al.*, 2003). Fluxes of CO_2 from roots of 10–120 nmol g^{-1} root dry weight s^{-1} were reported for a range of 12 plant species by Lambers *et al.*

Table 7.5 Composition of the exudates (% of total carboxylates exuded) collected from cluster roots of seven Proteaceae species

Carboxylate	*Banksia occidentalis*	*Banksia prionotes*	*Hakea baxteri*	*Hakea petiolaris*	*Hakea prostrata*	*Hakea undulata*	*Dryandra sessilis*
Tartarate	0	0.5	0	0	2.6	0.8	0
Formate	0	0	7.3	5.7	2.9	2.6	0
Malate	15.1	11.4	14.5	19.2	9.4	9.9	10.1
Malonate	16.9	8.3	31.7	19.2	28.1	18.7	19.7
Lactate	7.8	37.4	0.5	20.0	9.9	25.4	11.2
Acetate	1.6	7.8	3.1	2.3	9.4	0.5	4.2
Maleate	0	1.0	0	0.5	0.3	0.5	0.5
Citrate	35.8	22.9	30.1	23.9	34.8	18.7	26.5
Succinate	1.3	0.5	0.5	0	0.5	2.1	7.3
Fumarate	1.6	2.6	4.2	2.6	0.5	2.6	1.3
cis-Aconitate	3.6	1.3	1.6	1.6	0.5	2.6	2.1
trans-Aconitate	15.8	7.8	6.2	5.5	1.3	16.1	16.6

From Roelofs *et al.*, 2001.

(1996) but, except in very wet soils, this can be expected to diffuse rapidly from the root surface and bulk soil concentrations are typically 11–32 mmol mol^{-1} (i.e. some 30–100-fold higher than atmospheric concentration [0.36 mmol mol^{-1}]; Wild, 1988). Nye (1981) calculated that the difference in partial pressure of CO_2 between soil at the root surface and the bulk soil under typically aerated conditions would be only 2×10^{-4} kPa, insufficient to cause any significant difference in equilibrium pH. Smooth gradients of CO_2 with distance from the root surface of hybrid *Sorghum bicolor*/*Sorghum sudanese* grown in two calcareous soils were measured by Gollany *et al.* (1993) using a microelectrode indicating the anticipated diffusion from the root surface. Few other measurements of CO_2 gradients around roots are available because of the difficulties of making such measurements, although recent developments in mass spectrometry have allowed small gradients around grass and clover roots to be measured (Fig. 7.10).

In acid soils, theoretical considerations show that CO_2 will make a negligible contribution to changes in rhizosphere pH – both because of diffusion and because the dissociation constant of H_2CO_3 is such that it remains undissociated at acidic pH values. In calcareous soils, Hinsinger *et al.* (2003) suggest that a soil with a pH of 8.3 at ambient atmospheric CO_2 partial pressure will decrease to a pH of 6.7 if the partial pressure of CO_2 increases to 100 mmol mol^{-1} (i.e. to values typically found in bulk soil occupied by roots). So, the CO_2 partial pressures measured by Gollany *et al.* (1993) of 80–110 mmol mol^{-1} may result in pH values of 6.70–6.78 in the rhizosphere. More quantitative data are required to ascertain whether root and microbial respiration represents a major source of acidification in calcareous soils. Given the rapid diffusion of CO_2 in most aerobic soils, the impact is likely to be slight.

7.2.3 Changes in redox conditions

The balance of reduction/oxidation in a soil is expressed in terms of the redox potential (the potential of a platinum electrode relative to that of a hydrogen electrode, Eh [mV]) with

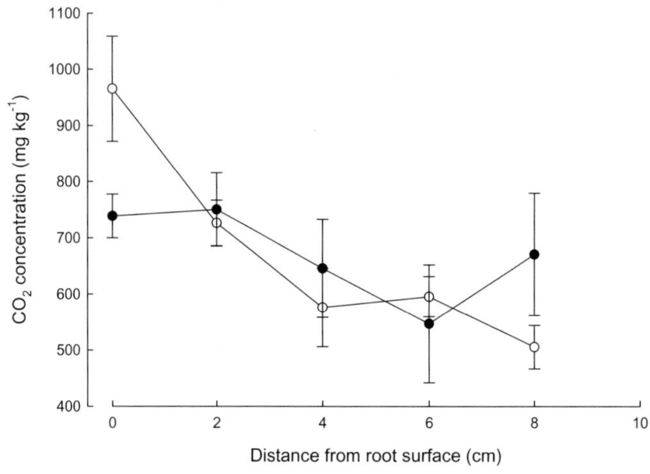

Fig. 7.10 Profile of CO_2 concentration with distance from roots of white clover (○) and ryegrass (●) measured with a mass spectrometer assuming atmospheric concentration was 525 mg kg^{-1}. (I am grateful to Dr P. Murray, IGER, North Wyke, UK for these unpublished data.)

values that are typically 400–600 mV for aerated, oxidized soils and +100 to –250 mV for anaerobic, reduced soils (see Wild, 1988, for more details). Eh is important in the dynamics of those elements that can occur in different oxidation states in soils because it affects their solubility and hence availability to plants. Availability of the essential nutrients iron and manganese is influenced in this way.

Redox processes in the rhizosphere are intimately coupled with pH changes (Ponnamperuma, 1972; Hinsinger *et al.*, 2003) because changes in oxidation states of elements imply the consumption or production of protons. For example, in an anaerobic soil, if electrons are available then reduction of Fe^{3+} to Fe^{2+} together with the dissolution of Fe^{3+}-bearing minerals such as goethite occurs according to the following equation:

$$FeOOH + 3H_3O^+ + e^- \leftrightarrow Fe^{2+} + 5H_2O \quad (7.2)$$

Electrons are most often supplied by the microbial oxidation of organic matter according to the following generalized equation:

$$CHO_2^- + H_2O \leftrightarrow CO_2 + H_3O^+ + 2e^- \quad (7.3)$$

Combining these two equations gives the observed reaction in soils which shows that reduction of Fe^{3+} to the more soluble Fe^{2+} is accompanied by oxidation of other compounds such as organic matter and by a consumption of protons that will result in an increase in pH. Conversely, the oxidation of Fe^{2+} to Fe^{3+} will be accompanied by a decrease in soil pH:

$$2FeOOH + CHO_2^- + 5H_3O^+ \leftrightarrow 2Fe^{2+} + CO_2 + 9H_2O \quad (7.4)$$

The result is that in a flooded, acidic soil, the pH of the bulk soil typically increases and stabilizes at about 6.0–6.5; in calcareous and alkaline soils, changes to the solubility of the dominant compounds ($CaCO_3$ and $NaHCO_3$, respectively) also result in a soil with pH close to 7. Flooded soils used to grow rice, then, typically have a bulk soil pH close to 7 (Ponnamperuma, 1972).

As indicated by the equations, reduction processes increase the activity of Fe^{2+} and might enable plants to meet their nutritional demand for Fe. However, the partial pressure of oxygen has to fall to very low levels before a significant increase occurs in the solubility of iron oxides and this is normally deleterious to other aspects of plant growth. However, several plant species respond to Fe deficiency by both enhancing proton excretion and increasing the reducing capabilities of their roots via induction of a plasma membrane-bound reductase (Römheld and Marschner, 1983; Marschner and Römheld, 1994).

Of course, the roots of most plants cannot function properly in a reduced environment, so the increased solubility of iron and manganese is of limited utility and they may even become toxic. However, as described in section 5.4.3, many plant species that grow in wet soils typically develop aerenchyma in their roots and release (leak) O_2 into the rhizosphere. Plants such as rice, then, have roots in a very complex soil medium with gradients of O_2, CO_2, pH, iron and nitrogen species (among others) between rhizosphere and bulk soil (Fig. 7.11) (Begg *et al.*, 1994). In the oxidized rhizosphere of rice, Fe^{2+} is oxidized to $Fe(OH)_3$ releasing H^+ and causing brown coatings near the roots (Trolldenier, 1988) (Plate 7.3), and

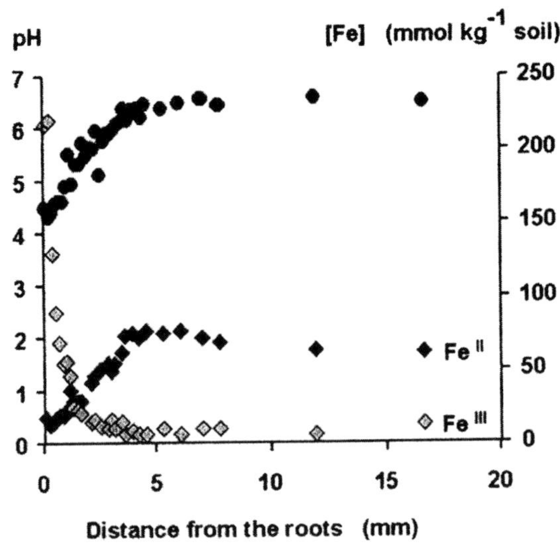

Fig. 7.11 Concentrations of Fe^{2+} and Fe^{3+} (diamonds) and of pH (●) with distance from the surface of rice roots grown in reduced soil conditions. (Original research by Begg *et al.* (1994), reproduced with permission from Hinsinger *et al.*, *Plant and Soil*; Springer Science and Business Media, 2003.)

the diffusion of Fe^{2+} from the bulk soil to the rhizosphere. In reduced soil, nitrogen exists mainly as NH_4^+ and its diffusion to the root and subsequent uptake results in an excess of cations over anions in the plant with the release of H^+ to maintain electrical neutrality. The two sources of H^+ (oxidation of Fe^{2+} and excess cation uptake) lead to a reduction of rhizosphere pH and tend to solubilize P and micronutrients such as Zn and Mn (Kirk and Bajita, 1995; Kirk and Saleque, 1995). Because the air-filled porosity of anaerobic soils is low, the ability of CO_2 to diffuse away from the root is limited and it may return to the atmosphere via the root's aerenchyma (Higuchi, 1982). This net uptake of CO_2 with decreased concentration of carbonic acid near the root may partially offset the acidity produced by oxidation and excess cation uptake.

In such a complex environment that varies spatially and temporally, mathematical models provide a good means of exploring the many concurrent chemical reactions. Saleque and Kirk (1995) measured the phosphate solubilized with distance from a plane of rice roots growing in a flooded ultisol and determined that about 90% of this was released from acid-soluble pools. Using these results, they developed a model that described diffusion of acid from the roots, the solubilization of P from ferrous hydroxides and carbonates, and the diffusion of the solubilized P to the roots. Measured and modelled P and pH profiles agreed well and the model demonstrated that over 80% of the P taken up by the plant was solubilized by the acid produced (Kirk and Saleque, 1995). Calculations showed that rice plants grown on many different soils would depend on solubilization for the majority of their P. It is noteworthy, too that the plants took up only about 50% of the P solubilized; the remainder diffused to the bulk soil. Nitrogen uptake by rice may also be more complex than has sometimes been suggested, with appreciable oxidation of ammonium in the rhizosphere to nitrate depending on the rates of O_2 flux from roots. Kirk (2001) modelled this process over a 10-day period taking account of O_2 diffusion from the root, transport to the root and uptake of NH_4^+, nitrification of NH_4^+ to NO_3^-, transport to the root and uptake of NO_3^-, and denitrification of NO_3^- to N_2, and calculated that up to 33% of the total N taken up may be as NO_3^-. Calculated concentration profiles are shown in Fig. 7.12.

7.2.4 Root exudates and phytosiderophores

As described earlier (see section 7.1), plants release a wide range of exudates from the roots, some of which exhibit complexing or chelating properties with metal ions. Complexes can be formed with a wide range of plant nutrients such as Fe, Cu, Mn and Zn and with potentially toxic metals such as Pb and Cd among others (Mench *et al*., 1987). The identity of exudates capable of fulfilling this role in significant quantities is not yet certain, but some of the simple organic anions described in section 7.2.2 are known to fulfil this function as well as improve the availability of P. For example, Dinkelaker *et al*. (1989) found that the concentration of DTPA-extractable Fe was increased from 34 $\mu mol\ kg^{-1}$ soil in bulk soil to 251 $\mu mol\ kg^{-1}$ in rhizosphere soil of white lupin; corresponding values for Mn were 44 to 222 $\mu mol\ kg^{-1}$ and for Zn from 2.8 to 16.8 $\mu mol\ kg^{-1}$. In acidic conditions, anions such as citrate and malate may contribute substantially to the dissolution of iron hydroxides and thereby to the Fe nutrition of plants (Jones *et al*., 1996), but such ions may also be rapidly metabolized by microbes in the rhizosphere – diminishing their effectiveness in the process of complexation (Darrah, 1993).

Roots of the Gramineae respond to Fe deficiency by the release of non-proteinogenic amino acids, such as mugineic acid and its derivatives, which are effective in mobilizing Fe^{3+} by chelation from sparingly soluble inorganic Fe^{3+} compounds (Takagi *et al*.,1984). Such exudates (known as phytosiderophores) have been shown to be released in increasing amounts as the degree of Fe deficiency increases. Fan *et al*. (1997) found that as roots of barley plants were exposed to increasingly Fe-deficient nutrient solutions, so the concentration of a range of organic and amino acids changed as well as that of the mugineic acid derivative 3-epihydroxymugineic acid (Table 7.6). Under conditions of moderate

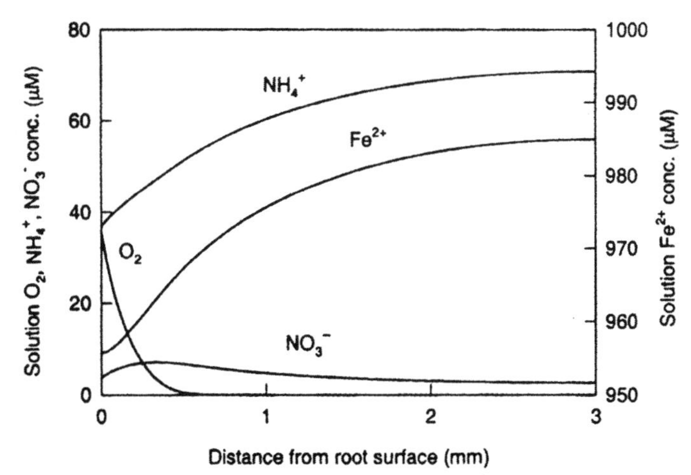

Fig. 7.12 Calculated concentration profiles of nitrate, ammonium, Fe^{2+} and O_2 around rice roots after 10 days in a flooded soil. The parameter values used in the calculations were typical of a healthy root growing in a soil used for growing lowland rice. (Reproduced with permission from Kirk, *Plant and Soil*; Springer Science and Business Media, 2001.)

Table 7.6 Quantities (μmol g^{-1} root dry weight) of various compounds in exudates from barley roots grown with different amounts of iron in solution: FeMS is the total iron-mobilizing substances and MAs is the total of the three forms of mugineic acid

	Fe treatment			
Compound	More than sufficient	Sufficient	Deficient	Very deficient
Lactate	1.11	4.09	5.00	4.82
Alanine	0.31	4.37	1.54	0.66
Glycine	0.07	0.43	0.19	0.16
Val	0.04	0.58	0.34	0.19
Leucine	0.02	0.19	0.07	0.04
Ile	0.00	0.13	0.03	0.00
Succinate	0.04	0.98	0.16	0.14
γ-Aminobutyrate	0.02	2.50	0.40	0.27
Pro	0.01	0.17	0.12	0.00
Fumarate	0.05	0.29	0.17	0.18
Serine	0.01	0.39	0.14	0.03
Thr	0.02	0.74	0.11	0.04
Malate	0.00	2.79	0.23	0.10
Asparagine	0.12	0.41	0.36	0.26
Glu	0.02	0.16	0.05	0.00
Glycinebetaine	0.53	5.91	1.66	2.74
Acetate	1.58	4.68	1.70	2.25
3-Epihydroxymugineic acid	1.12	8.09	16.14	17.37
Mugineic acid + 2-deoxymugineic acid	–	0.16	1.39	3.59
FeMS	1.5	4.64	5.99	7.27
MAs:FeMS	0.75	1.78	2.93	2.88

Adapted from Fan *et al.*, 1977.

Fe deficiency, 3-epihydroxymugineic acid represented about 22% of the total exudates released, but as the degree of deficiency increased to severe, its percentage increased to 50% of the total, although the total quantity of exudates per unit weight of root remained unchanged. Different species differ in their quantitative production of phytosiderophores. The qualitative importance of differences in phytosiderophore release was demonstrated by Römheld and Marschner (1990), who showed that when a range of cereal species was grown in nutrient solution, the relative resistance of the different species to Fe deficiency was directly related to their relative ability to release phytosiderophores (Fig. 7.13). Similarly, differences exist within a plant species, although in a study of 12 maize hybrids grown in solution culture the quantities of phytosiderophore released were not uniquely related to perceived susceptibility to iron chlorosis (Bernards *et al.* 2002). Takagi *et al.* (1984) showed that release of phytosiderophores occurred during daylight hours in a defined diurnal pattern (as with citrate from cluster roots; section 7.2.2) and was particularly intensive from apical root zones. Bernards *et al.*, (2002) found that the rate of release per unit dry mass of nodal roots of maize hybrids was about twice that from the tap root (primary axis), suggesting that they are more important in Fe uptake than the seminal roots and that basal zones of axes, especially with laterals, may also contribute significantly to Fe uptake.

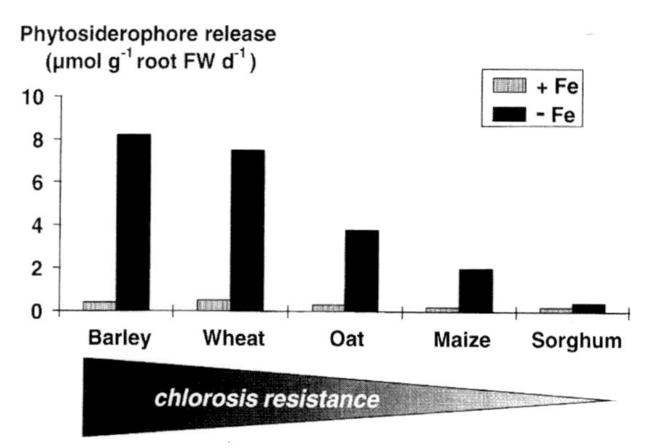

Fig. 7.13 Phytosiderophore release from roots of Fe-sufficient (+Fe) and Fe-deficient (–Fe) seedlings of Graminaceae species differing in their response to lime-induced chlorosis. Species such as barley and wheat are the most chlorosis-resistant and also have the highest rate of phytosiderophore release and the largest response to Fe deficiency. (Original research by Römheld and Marschner (1990), reproduced with permission from Hinsinger, *Advances in Agronomy*; Elsevier, 1998.)

Phytosiderophores also form complexes with other metal ions, notably Zn and Cu. For example, Chaignon *et al.* (2002) found that when two wheat cultivars grown in Fe-deficient conditions were transferred to a calcareous soil containing a high concentration of copper, the acquisition of soil Cu increased three- to four-fold in response to Fe deficiency compared with a non-deficient control. Enhanced release of phytosiderophores in the Fe-deficient plants was associated with significantly higher concentrations of Cu in both roots and shoots. Zhang *et al.* (1989) showed that the enhanced release of phytosiderophores by roots of wheat was not specific to Fe deficiency, but also occurred under Zn-deficient conditions. Some researchers have demonstrated that Zn-efficient genotypes released more phytosiderophore than Zn-inefficient types (e.g. Cakmak *et al.*, 1996, for durum wheat) but it is, as yet, uncertain whether or not phytosiderophore interaction with soil Zn plays a major role in Zn nutrition compared to uptake of Zn^{2+} (Hacisalihoglu and Kochian, 2003). Similarly, while phytosiderophores released from Fe-deficient wheat and barley plants grown in solution formed chelates with soil Cd, there was no evidence for enhanced Cd uptake by the plants (Shenker *et al.*, 2001).

7.2.5 Enzyme activity

Roots and microbes release a range of enzymes into soil to aid the acquisition of nutrients. Most attention has been given to the synthesis and release of acid phosphatase enzymes by roots because such synthesis is a universal response of plants to P deficiency. Phytase and other acid phosphatase enzymes are released in quite large quantities by many microbes to access soil P in organic forms such as the inositol phosphates (phytate) which can be up to 50% of total soil P. Until recently the role of phytase in the nutrition of plants has been little investigated, mainly because it was thought that P uptake by plants occurred only from inorganic forms (Wild, 1988).

Extracellular phosphatases have been measured in the rhizosphere of several plant species grown in soils of low P status (Vance *et al.*, 2003) and it is clear that there are differences between species in their ability to utilize P from organic sources. For example, George *et al.* (2002a, 2002b) found that while white lupin, *Tithonia diversifolia* and *Tephrosia vogelii* were able to deplete sodium hydroxide-extractable organic P from their rhizospheres, maize was not. The differential depletion of organic P by the species was not entirely explained by phosphatase activity alone and other changes, such as the depletion of phosphate monoesters around roots of species such as *Tithonia* may also play a part. Efforts to increase the release of phytase from root systems are described in section 8.2.2, but as with other root-released products, their susceptibility to rapid decomposition by the rhizosphere microbial community or their inactivation due to sorption by soil colloids mean that their quantitative significance for plant nutrition is currently uncertain.

7.3 Physical changes in the rhizosphere

Physical changes induced by roots have been much less studied than biological or chemical changes, yet differences in soil aggregation (section 1.3.2) and the formation of coherent rhizosheaths (section 2.4.2) around roots have been clearly demonstrated and might be expected to have effects on the response of roots to soil and the flux of solution and gases to and from the root surface. The biological and physical interactions occurring in the rhizosphere, especially as a consequence of rhizodeposition and microbial activity, and the repeated wetting and drying of soil at the root–soil interface result in a heterogeneous soil matrix with properties that are different to soil at some distance from the root. For example, Czarnes *et al.* (2000a) found that in a silty soil, soil adhering to maize roots had higher aggregate strength (450–500 kPa) than that not adhering (410–420 kPa) and that the friability (defined as the coefficient of variation of the aggregate strength) was lower (67% compared with 49%, respectively, at a soil water tension of 30 kPa). Young (1998) concluded that soil microbes and carbon substrates act together to stabilize the structural framework and thereby affect the development of heterogeneity of soil structure within the soil profile. Moreover, the biophysical processes that occur at the root surface on a small scale, directly affect processes such as structure formation and water movement at soil profile scale.

Besides aggregation, the two physical properties in the rhizosphere that have been most studied are changes in bulk density or porosity and various aspects of water relations. Water flow to roots was described in section 4.2.2, but an unresolved issue has been the possibility that the rhizosphere could be a major resistance in the flow of water to plants. If soil were to dry out locally around roots then this would limit the flow of water (and nutrients) by decreasing the hydraulic conductivity. In the laboratory, it is possible to induce such drying in the rhizosphere (Dunham and Nye, 1973), but the inflow of water required to produce the effect was much greater than that usually required of field-grown plants. Sparsely rooted roots may experience localized rhizosphere drying but other issues such as poor root–soil contact, and/or restriction of roots to a few pores are probably more common causes of restricted water flow to roots.

7.3.1 Bulk density and porosity

As described in section 5.3.1, when a root grows it deforms soil by expanding radially, and the volume occupied by the root is matched by an equivalent loss of pore space from the surrounding soil (Cockcroft *et al.*, 1969; Dexter, 1987). Dexter's model (1987) predicts an exponential decrease in density with distance from the root–soil interface within a homogeneous soil matrix according to the equation:

$$\eta = \eta_{min} + (\eta_i - \eta_{min})\{1 - \exp[-k(r - r_0)/(r_0)]\} \tag{7.5}$$

where η is porosity, η_i is the initial porosity before root growth, η_{min} is the porosity at the root:soil interface, r_0 is the radial distance to which the root compresses the soil to η_{min} (i.e. the radius of the root), and r is the radial distance.

Using pea roots growing in remoulded clay soil, Cockcroft *et al.* (1969) showed that the bulk density increased from 1.21 Mg m^{-3} (voids ratio of 1.193) initially to 1.26 Mg m^{-3} (voids ratio 1.099) for soil immediately around the root. This represents a change in porosity of only 0.02 and was, therefore, largely discounted as having any effect on root functioning and gas and liquid transport. In contrast to studies with remoulded soil materials, studies of natural soils using thin sections have suggested bigger effects, especially in losses of medium-sized pores (mesopores). Bruand *et al.* (1996) used backscattered electron scattering images of maize roots growing in silty clay loam and clay loam materials to quantify changes in porosity with a resolution of 60 μm and found that porosity was 22–24% less within the soil adjacent to the root than in the surrounding soil. Bulk density increased to 1.80 Mg m^{-3} at the root–soil interface compared with 1.54 Mg m^{-3} at distance (Fig. 7.14). An interesting feature of these results was that the zone over which the change in porosity occurred was only about 800 μm, a distance that is much less than that recorded in the studies of Cockcroft *et al.* (1969) and Dexter (1987). The difference is probably a consequence of the sizes of pores in natural and remoulded soils. In the study by Bruand *et al.* (1996), the reduction in porosity resulted from a decrease in micropores and removal of macropores which were numerous in the natural soils but may have been absent in remoulded soils. However, the limited distance over which changes occurred in

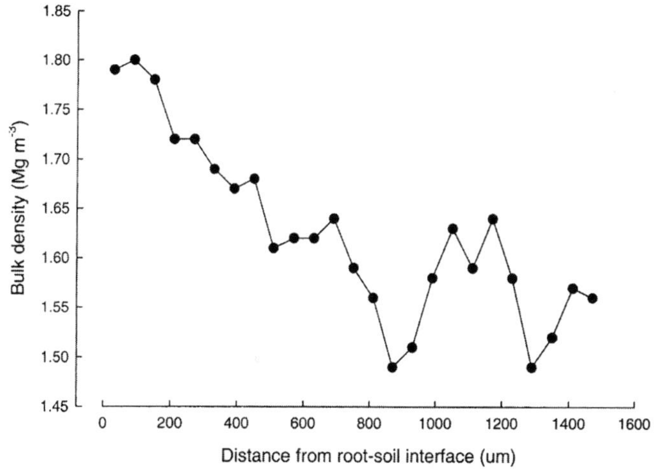

Fig. 7.14 Changes in bulk density with distance from the root surface measured by backscattered electron emission of thin sections. (Reproduced with permission from Bruand *et al.*, *Soil Science Society of America Journal*; Soil Science Society of America, 1996.)

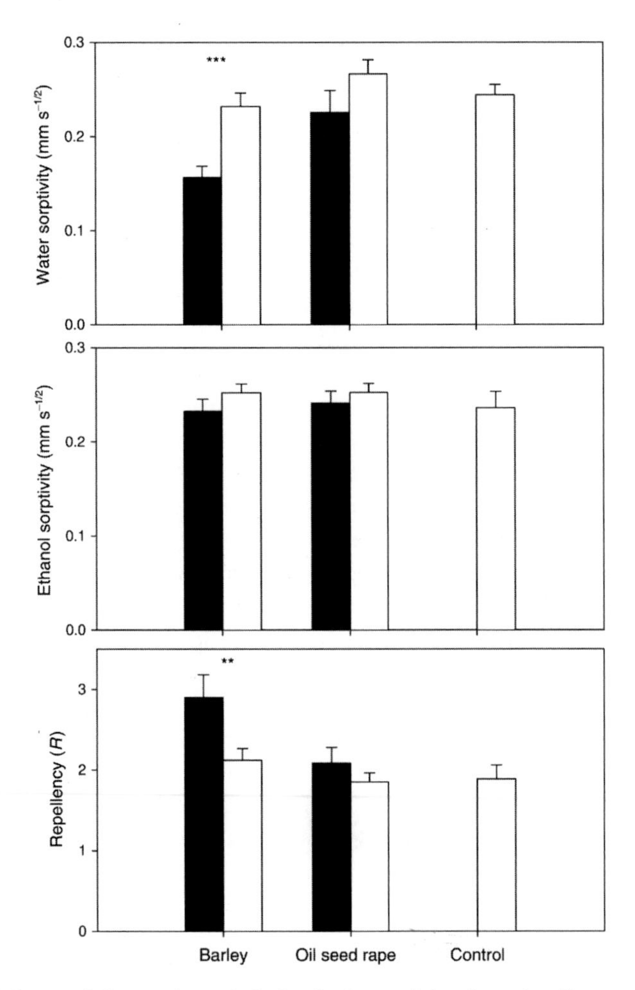

Fig. 7.15 The influence of plant species on the hydraulic characteristics of a sandy soil measured in a laboratory experiment at ambient water content. Closed bars show the rhizosphere and open bars the bulk soil. Significant differences between rhizosphere and bulk soil are indicated by $^{**}p<0.05$ and $^{***}p<0.001$. (Reproduced with permission from Hallett *et al.*, *New Phytologist*; New Phytologist Trust, 2003.)

this study contrasts with the effects over 4–5 mm summarized by Young (1998). Clearly, more studies of naturally structured soils at fine resolution are required to resolve both the size of changes that occur and their potential consequences for liquid and gaseous flow at the root–soil interface.

7.3.2 Water

Water relations at the root–soil interface are complex, not only because of issues of poor root-soil contact leading to an interfacial resistance to water flow (see section 4.2.2) and solute accumulation leading to large osmotic pressures at the interface (Stirzaker and Passioura, 1996), but also because of the release of organic substances which may have contrasting effects on soil water behaviour. Surfactants present in root mucilage may alter the relationship between water content and soil matric potential (the moisture characteristic curve) making the soil drier at a given value of soil matric potential (Read *et al.*, 2003). Passioura (1988) suggested that this may be particularly beneficial for water uptake at lower values of matric potential (drier soils), although Read *et al.* (2003) measured a bigger effect on the moisture characteristic curve at high matric potentials. This effect together with the coating of soil particles by complex organic compounds (DeBano, 2000) will tend to make the soil more hydrophilic and enhance wetting. Conversely, some organic compounds produced by roots and microbes bind to soil particles to make the soil more hydrophobic (Czarnes *et al.*, 2000b) and, ultimately, water-repellent. There have been very few direct studies of rhizosphere hydraulic properties *in situ* but Hallett *et al.* (2003) used a miniaturized infiltrometer to study the hydraulic characteristics of barley, rape, potato and grass rhizospheres. In excavated blocks of field soil there were significant differences between plant species on sorptivity and water repellency, but there was no difference between soil within 1 mm of the root surface (rhizosphere soil) and that 20 mm from a root. In laboratory studies with sieved soil, water sorptivity in the rhizosphere was less than that in soil 20 mm from the root surface for barley but not for rape (Fig. 7.15). The measured values of water repellency would have negligible effect on the ability of plants to extract water from soil but might provide a buffer against desiccation in drier soils.

Whalley *et al.* (2004) used microcosms containing a packed, sieved sandy loam to demonstrate the effect of growing maize roots on hydraulic properties. The rate of infiltration of water was less in soil that had been deformed by roots than non-deformed soil, and this was suggested to be a consequence of an increase in bulk density of the rhizosphere. In this study there was no evidence for effects of organic additions on water repellence. Again, more research is required to resolve the size and direction of any changes in rhizosphere hydraulic properties and to quantify their significance for field-grown plants.

References

Barber S.A. (1995) *Soil Nutrient Bioavailability: A Mechanistic Approach*, 2nd edn. John Wiley & Sons, New York.

Begg, C.B.M., Kirk, G.J.D., MacKenzie, A.F. and Neue, H.-U. (1994) Root-induced iron oxidation and pH changes in the lowland rice rhizosphere. *New Phytologist* **128**, 469–477.

Bernards, M.L., Jolley, V.D., Stevens, W.B. and Hergert, G.W. (2002) Phytosiderophore release from nodal, primary, and complete root systems in maize. *Plant and Soil* **241**, 105–113.

Bruand, A., Cousin, I., Nicoullaud, B., Duval, O. and Begon, J.C. (1996) Backscattered electron scanning images of soil porosity for analyzing soil compaction around roots. *Soil Science Society of America Journal* **60**, 895–901.

Cakmak, I., Sari, N., Marschner, H., Ekiz, H., Kalayci, M., Yilmaz, A. and Braun, H.J. (1996) Phytosiderophore release in bread and durum wheat genotypes differing in zinc efficiency. *Plant and Soil* **180**, 183–189.

Chaignon, V., Di Malta, D. and Hinsinger, P. (2002) Fe-deficiency increases Cu acquisition by wheat cropped in a Cu-contaminated vineyard soil. *New Phytologist* **154**, 121–130.

Cheng, W., Johnson, D.W. and Fu, S. (2003) Rhizosphere effects on decomposition: controls of plant species, phenology, and fertilization. *Soil Science Society of America Journal* **67**, 1418–1427.

Claassen, N. and Jungk, A. (1982) Kaliumdynamik im wurzelnahen Boden in Beziehung zur Kaliumaufnahme von Maispflanzen. *Zeitschrift für Pflanzenernährung und Bodenkunde* **145**, 513–525.

Cockroft, B., Barley, K.P. and Greacen, E.L. (1969) The penetration of clays by fine roots and root tips. *Australian Journal of Soil Research* **7**, 333–348.

Crawford, M.C., Grace, P.R. and Oades, J.M. (2000) Allocation of carbon to shoots, roots, soil and rhizosphere respiration by barrel medic (*Medicago truncatula*) before and after defoliation. *Plant and Soil* **227**, 67–75.

Czarnes, S., Dexter, A.R. and Bartoli, F. (2000a) Wetting and drying cycles in the maize rhizosphere under controlled conditions. Mechanics of the root-adhering soil. *Plant and Soil* **221**, 253–271.

Czarnes, S., Hallett, P.D., Bengough, A.G. and Young, I.M. (2000b) Root- and microbial-derived mucilages affect soil structure and water transport. *European Journal of Soil Science* **51**, 435–443.

Darrah, P.R. (1993) The rhizosphere and plant nutrition: a quantitative approach. *Plant and Soil*, **155–156**, 3–22.

DeBano, L.F. (2000) Water repellency in soils: a historical overview. *Journal of Hydrology* **231–232**, 4–32.

de Neergaard, A. and Gorissen, A. (2004) Carbon allocation to roots, rhizodeposits and soil after pulse labelling: a comparison of white clover (*Trifolium repens* L.) and perennial ryegrass (*Lolium perenne* L.). *Biology and Fertility Soils* **39**, 228–234.

Dexter, A.R. (1987) Compression of soil around roots. *Plant and Soil* **97**, 401–406.

Dilkes, N.B., Jones, D.L. and Farrar, J.F. (2004) Temporal dynamics of carbon partitioning and rhizodeposition in wheat. *Plant Physiology* **134**, 706–715.

Dinkelaker, B., Römheld, V. and Marschner, H. (1989) Citric acid excretion and precipitation of calcium citrate in the rhizosphere of white lupin (*Lupinus albus* L.). *Plant, Cell and Environment* **12**, 285–292.

Dunham, R.J. and Nye, P.H. (1973) The influence of soil water content on the uptake of ions by roots. I. Soil water content gradients near a plane of onion roots. *Journal of Applied Ecology* **11**, 581–596.

Fan, T.W.-M., Lane, A.N., Pedler, J., Crowley, D.E. and Higashi, R.M. (1997) Comprehensive analysis of organic ligands in whole root exudates using nuclear magnetic resonance and gas chromatography-mass spectometry. *Analytical Biochemistry* **251**, 57–68.

Fan, T.W.-M., Lane, A.N., Shenker, M., Bartley, J.P., Crowley, D.E. and Higashi, R.M. (2001) Comprehensive chemical profiling of graminaceous plant root exudates using high-resolution NMR and MS. *Phytochemistry* **57**, 209–221.

Gahoonia, T.S. and Nielsen, N.E. (1992) The effects of root-induced pH changes on the depletion of inorganic and organic phosphorus in the rhizosphere. *Plant and Soil* **143**, 185–191.

Gardner, W.K., Barber, D.A. and Parberry, D.G. (1983) The acquisition of phosphorus by *Lupinus albus* L. III. The probable mechanism by which phosphorus movement in the soil/root interface is enhanced. *Plant and Soil* **70**, 107–124.

George, T.S., Gregory, P.J., Robinson, J.S. and Buresh, R.J. (2002a) Changes in phosphorus concentrations and pH in the rhizosphere of some agroforestry and crop species. *Plant and Soil* **246**, 65–73.

George, T.S., Gregory, P.J., Wood, M., Read, D.B. and Buresh, R.J. (2002b) Phosphatase activity and organic acids in the rhizosphere of potential agroforestry species and maize. *Soil Biology and Biochemistry* **34**, 1487–1494.

Gollany, H.T., Schumacher, T.E., Rue, R.R. and Liu, S.-Y. (1993) A carbon dioxide microelectrode for *in situ* pCo_2 measurement. *Microchemical Journal* **48**, 42–49.

Grayston, S.J., Vaughan, D. and Jones, D. (1996) Rhizosphere carbon flow in trees, in comparison with annual plants: the importance of root exudation and its impact on microbial activity and nutrient availability. *Applied Soil Ecology* **5**, 29–56.

Gregory, P.J. and Atwell, B.J. (1991) The fate of carbon in pulse-labelled crops of barley and wheat. *Plant and Soil* **136**, 205–213.

Grinsted, M.J., Hedley, M.J., White, R.E. and Nye, P.H. (1982) Plant-induced changes in the rhizosphere of rape (*Brassica napus* var. Emerald) seedlings. I. pH change and the increase in P concentration in the soil solution. *New Phytologist* **91**, 19–29.

Hacisalihoglu, G. and Kochian, L.V. (2003) How do some plants tolerate low levels of soil zinc? Mechanisms of zinc efficiency in crop plants. *New Phytologist* **159**, 341–350.

Hallett, P.D., Gordon, D.C. and Bengough, A.G. (2003) Plant influence on rhizosphere hydraulic properties: direct measurements using a miniaturized infiltrometer. *New Phytologist* **157**, 597–603.

Haynes, R.J. (1992) Relative ability of a range of crop species to use phosphate rock and monocalcium phosphate as P sources when grown in soil. *Journal of the Science of Food and Agriculture* **60**, 205–211.

Hedley, M.J., Kirk, G.J.D. and Santos, M.B. (1994) Phosphorus efficiency and the forms of soil phosphorus utilized by upland rice cultivars. *Plant and Soil* **158**, 53–62.

Hiatt, A.J. (1967) Relationship of cell sap pH to organic acid change during ion uptake. *Plant Physiology* **42**, 294–298.

Higuchi, T. (1982) Gaseous CO_2 transport through the aerenchyma and intercellular spaces in relation to the uptake of CO_2 by rice roots. *Soil Science and Plant Nutrition* **28**, 491–497.

Hiltner, L. (1904) Über neuere Erfahrungen und Probleme auf dem Gebiete der Bodenbakteriologie unter besonderer Berücksichtigung der Gründüngung und Brache. *Arbeiten der Deutschen Landwirtschaftlichen Gesellschaft* **98**, 59–78.

Hinsinger, P. (1998) How do plant roots acquire mineral nutrients? Chemical processes involved in the rhizosphere. *Advances in Agronomy* **64**, 225–265.

Hinsinger, P. (2001) Bioavailability of soil inorganic P in the rhizosphere as affected by root-induced chemical changes: a review. *Plant and Soil* **237**, 173–195.

Hinsinger, P. and Gilkes, R.J. (1996) Mobilization of phosphate from phosphate rock and alumina-sorbed phosphate by the roots of ryegrass and clover as related to rhizosphere pH. *European Journal of Soil Science* **47**, 533–544.

Hinsinger, P. and Jaillard, B. (1993) Root-induced release of interlayer potassium and vermiculitization of phlogopite as related to potassium depletion in the rhizosphere of ryegrass. *Journal of Soil Science* **44**, 525–534.

Hinsinger, P., Plassard, C., Tang, C. and Jaillard, B. (2003) Origins of root-mediated pH changes in the rhizosphere and their responses to environmental constraints: a review. *Plant and Soil* **248**, 43–59.

Hocking, P.J. and Jeffery, S. (2004) Cluster-root production and organic anion exudation in a group of old-world lupins and a new-world lupin. *Plant and Soil* **258**, 135–150.

Hoffland, E., van den Boogaard, R., Nelemans, J. and Findenegg, G. (1992) Biosynthesis and root exudation of citric and malic acids in phosphate-starved rape plants. *New Phytologist* **122**, 675–680.

Høgh-Jensen, H. and Schjoerring, J. K. (2001) Rhizodeposition of nitrogen by red clover, white clover and ryegrass leys. *Soil Biology and Biochemistry* **33**, 439–448.

Jaillard, B., Ruiz, L. and Arvieu, J.-C. (1996) pH mapping in transparent gel using color videodensitometry. *Plant and Soil* **183**, 85–95.

Jarvis, S.C. and Robson, A.D. (1983) The effects of nitrogen nutrition on plants on the development of acidity in Western Australian soils. I. Effects with subterranean clover grown under leaching conditions. *Australian Journal of Agricultural Research* **34**, 341–353.

Jensen, B. (1994) Rhizodeposition by field-grown winter barley exposed to $^{14}CO_2$ pulse-labelling. *Applied Soil Ecology* **1**, 65–74.

Johnson, J.F., Vance, C.P. and Allan, D.L. (1996) Phosphorus deficiency in *Lupinus albus*. Altered lateral root development and enhanced expression of phospho*enol*pyruvate carboxylase. *Plant Physiology* **112**, 31–41.

Jones, D.L. (1998) Organic acids in the rhizosphere – a critical review. *Plant and Soil* **205**, 25–44.

Jones, D.L. and Darrah, P.R. (1994) Amino-acid influx at the soil-root interface of *Zea mays* L. and its implications in the rhizosphere. *Plant and Soil* **163**, 1–12.

Jones, D.L., Darrah, P.R. and Kochian, L.V. (1996) Critical evaluation of organic acid mediated iron dissolution in the rhizosphere and its potential role in root iron uptake. *Plant and Soil* **180**, 57–66.

Jones, D.L., Hodge, A. and Kuzyakov, Y. (2004) Plant and mycorrhizal regulation of rhizodeposition. *New Phytologist* **163**, 459–480.

Keerthisinghe, G., Hocking, P.J., Ryan, P.R. and Delhaize, E. (1998) Effect of phosphorus supply on the formation and function of proteoid roots of white lupin (*Lupinus albus* L.). *Plant, Cell and Environment* **21**, 467–478.

Keith, H., Oades, J.M. and Martin, J.K. (1986) Input of carbon to soil from wheat plants. *Soil Biology and Biochemistry* **18**, 445–449.

Kirk, G.J.D. (1999) A model of phosphate solubilization by organic anion excretion from plant roots. *European Journal of Soil Science* **50**, 369–378.

Kirk, G.J.D. (2001) Plant-mediated processes to acquire nutrients: nitrogen uptake by rice plants. *Plant and Soil* **232**, 129–134.

Kirk, G.J.D. and Bajita, J.B. (1995) Root-induced iron oxidation, pH changes and zinc solubilization in the rhizosphere of lowland rice. *New Phytologist* **131**, 129–137.

Kirk, G.J.D. and Saleque, M.A. (1995) Solubilization of phosphate by rice plants growing in reduced soil: prediction of the amount solubilized and the resultant increase in uptake. *European Journal of Soil Science* **46**, 247–255.

Kirk, G.J.D., Santos, E.E. and Findenegg, G.R. (1999a) Phosphate solubilization by organic anion excretion from rice (*Oryza sativa* L.) growing in aerobic soil. *Plant and Soil* **211**, 11–18.

Kirk, G.J.D., Santos, E.E. and Santos, M.B. (1999b) Phosphate solubilization by organic anion excretion from rice growing in aerobic soil: rates of excretion and decomposition, effects on rhizosphere pH and effects on phosphate solubility and uptake. *New Phytologist* **142**, 185–200.

Kodama, H., Nelson, S., Yang, A.F. and Kohyama, N. (1994) Mineralogy of rhizospheric and non-rhizospheric soils in corn fields. *Clays and Clay Minerals* **42**, 755–763.

Kuchenbuch, R. and Jungk, A. (1984) Wirkung der Kaliumdüngung auf die Kaliumverfügbarkeit in der Rhizosphäre von Raps. *Zeitschrift für Pflanzenernährung und Bodenkunde* **147**, 435–448.

Kuzyakov, Y. and Cheng, W. (2001) Photosynthesis controls of rhizosphere respiration and organic matter decomposition. *Soil Biology and Biochemistry* **33**, 1915–1925.

Kuzyakov, Y., Ehrenberger, H. and Stahr, K. (2001) Carbon partitioning and below-ground translocation by *Lolium perenne*. *Soil Biology and Biochemistry* **33**, 61–74.

Lambers, H., van der Werf, A. and Konings, H. (1996) Respiratory patterns in roots in relation to their functioning. In: *Plant Roots: The Hidden Half* (eds Y. Waisel, A. Eshel and U. Kafkafi), 2nd edn, pp. 229–263. Marcel Dekker Inc., New York.

Lamont, B.B. (2003) Structure, ecology and physiology of root clusters – a review. *Plant and Soil* **248**, 1–19.

Li, J. and Copeland, L. (2000) Role of malonate in chickpeas. *Phytochemistry* **54**, 585–589.

Lorenz, S.E., Hamon, R.E. and McGrath, S.P. (1994) Differences between soil solutions obtained from rhizosphere and non-rhizosphere soils by water displacement and soil centrifugation. *European Journal of Soil Science* **45**, 431–438.

Lynch, J.M. and Whipps, J.M. (1990) Substrate flow in the rhizosphere. *Plant and Soil* **129**, 1–10.

Marschner, H. (1995) *Mineral Nutrition of Higher Plants*, 2nd edn. Academic Press, London.

Marschner, H. and Römheld, V. (1983) *In vivo* measurement of root-induced pH changes at the soil-root interface: effect of plant species and nitrogen source. *Zeitschrift für Pflanzenphysiologie* **111**, 241–251.

Marschner, H. and Römheld, V. (1994) Strategies of plants for acquisition of iron. *Plant and Soil* **165**, 261–274.

Marschner, H., Römheld, V., Horst, W.J. and Martin, P. (1986) Root-induced changes in the rhizosphere: importance for the mineral nutrition of plants. *Zeitschrift für Pflanzenernährung und Bodenkunde* **149**, 441–456.

Marschner, H., Treeby, M. and Römheld, V. (1989) Role of root-induced changes in the rhizosphere for iron acquisition in higher plants. *Zeitschrift für Pflanzenernährung und Bodenkunde* **152**, 197–204.

Martin, J.K. and Kemp, J.R. (1986) The measurement of C transfers within the rhizosphere of wheat grown in field plots. *Soil Biology and Biochemistry* **18**, 103–107.

Martin, J.K. and Merckx, R. (1992) The partitioning of photosynthetically fixed carbon within the rhizosphere of mature wheat. *Soil Biology and Biochemistry* **24**, 1147–1156.

Mayer, J., Buegger, F., Jensen, E.S., Schloter, M. and Hess, J. (2003) Estimating N rhizodeposition of grain legumes using a ^{15}N *in situ* stem labelling method. *Soil Biology and Biochemistry* **35**, 21–28.

Meharg, A.A. (1994) A critical review of labelling techniques used to quantify rhizosphere carbon-flow. *Plant and Soil* **166**, 55–62.

Mench, M., Morel, J.L. and Guckert, A. (1987) Metal binding properties of high molecular weight soluble exudates from maize (*Zea mays* L.) roots. *Biology and Fertility Soils* **3**, 165–169.

Mengel, K. and Scherer, H.W. (1981) Release of nonexchangeable (fixed) soil ammonium under field conditions during the growing season. *Soil Science* **131**, 226–232.

Mengel, K., Horn, D. and Tributh, H. (1990) Availability of interlayer ammonium as related to root vicinity and mineral type. *Soil Science* **149**, 131–137.

Neumann, G. and Martinoia, E. (2002) Cluster roots – and underground adaptation for survival in extreme environments. *Trends in Plant Science* **7**, 162–167.

Neumann, G., Massonneau, A., Martinoia, E. and Römheld, V. (1999) Physiological adaptations to phosphorus deficiency during proteoid root development in white lupin. *Planta* **208**, 373–382.

Nye, P.H. (1981) Changes of pH across the rhizosphere induced by roots. *Plant and Soil* **61**, 7–26.

Nye, P.H. (1986) Acid-base changes in the rhizosphere. *Advances in Plant Nutrition* **2**, 129–153.

Passioura, J.B. (1988) Water transport in and to roots. *Annual Review of Plant Physiology and Plant Molecular Biology* **39**, 245–265.

Pate, J.S. and Watt, M. (2002) Roots of *Banksia* spp. (Proteaceae) with special reference to functioning of their specialized proteoid root clusters. In: *Plant Roots: The Hidden Half* (eds Y. Waisel, A. Eshel and U. Kafkafi), 3rd edn, pp. 989–1006. Marcel Dekker Inc., New York.

Paterson, E. (2003) Importance of rhizodeposition in the coupling of plant and microbial productivity. *European Journal of Soil Science* **54**, 741–750.

Paterson, E. and Sim, A. (1999) Rhizodeposition and C-partitioning of *Lolium perenne* in axenic culture affected by nitrogen supply and defoliation. *Plant and Soil* **216**, 155–164.

Paynel, F., Murray, P.J. and Cliquet, J.B. (2001) Root exudates: a pathway for short-term N transfer from clover and reyegrass. *Plant and Soil* **229**, 235–243.

Peek, C.S., Robson, A.D. and Kuo, J. (2003) The formation, morphology and anatomy of cluster root of *Lupinus albus* L. as dependent on soil type and phosphorus supply. *Plant and Soil* **248**, 237–246.

Ponnamperuma, F.N. (1972) The chemistry of submerged soils. *Advances in Agronomy* **24**, 29–96.

Read, D.B., Bengough, A.G., Gregory, P.J., Crawford, J.W., Robinson, D., Scrimgeour, C.M., Young, I.M., Zhang, K. and Zhang, X. (2003) Plant roots release phospholipid surfactants that modify the physical and chemical properties of soil. *New Phytologist* **157**, 315–326.

Riley, D. and Barber, S.A. (1971) Effect of ammonium and nitrate fertilization on phosphorus uptake as related to root-induced pH changes at the root-soil interface. *Soil Science Society of America Proceedings* **35**, 301–306.

Roelofs, R.F.R., Rengel, Z., Cawthray, G.R., Dixon, K.W. and Lambers, H. (2001) Exudation of carboxylates in Australian Proteaceae: chemical composition. *Plant, Cell and Environment* **24**, 891–903.

Römheld, V. and Marschner, H. (1983) Mechanism of iron uptake by peanut plants. I. Fe^{III} reduction, chelate splitting, and release of phenolics. *Plant Physiology* **71**, 949–954.

Römheld, V. and Marschner, H. (1990) Genotypical differences among graminaceous species in release of phytosiderophores and uptake of iron phytosiderophores. *Plant and Soil* **123**, 147–153.

Rufyikiri, G., Dufey, J.E., Nootens, D. and Delvaux, B. (2001) Effect of aluminium on bananas (*Musa* spp.) cultivated in acid solutions. II. Water and nutrient uptake. *Fruits* **56**, 5–16.

Ryan, P.R., Delhaize, E. and Jones, D.L. (2001) Function and mechanism of organic anion exudation from plant roots. *Annual Review of Plant Physiology and Plant Molecular Biology* **52**, 527–560.

Saggar, S. and Hedley, C.B. (2001) Estimating seasonal and annual carbon inputs, and root decomposition rates in a temperate pasture following field ^{14}C pulse-labelling. *Plant and Soil* **236**, 91–103.

Saleque, M.A. and Kirk, G.J.D. (1995) Root-induced solubilization of phosphate in the rhizosphere of lowland rice. *New Phytologist* **129**, 325–336.

Scherer, H.W. and Ahrens, G. (1996) Depletion of non-exchangeable NH_4-N in the soil-root interface in relation to clay mineral composition and plant species. *European Journal of Agronomy* **5**, 1–7.

Schubert, S., Schubert, E. and Mengel, K. (1990) Effect of low pH of the root medium on proton release, growth, and nutrient uptake of field beans (*Vicia faba*). *Plant and Soil* **124**, 239–244.

Shen, J., Rengel, Z., Tang, C. and Zhang, F. (2003) Role of phosphorus nutrition in development of cluster roots and release of carboxylates in soil-grown *Lupinus albus*. *Plant and Soil* **248**, 199–206.

Shenker, M., Fan, T.W.-M. and Crowley, D.E. (2001) Phytosiderophores influence on cadmium mobilization and uptake by wheat and barley plants. *Journal of Environmental Quality* **30**, 2091–2098.

Shepherd, T. and Davies, H.V. (1993) Carbon loss from the roots of forage rape (*Brassica napus* L.) seedlings following pulse-labelling with $^{14}CO_2$. *Annals of Botany* **72**, 155–163.

Skene, K.R. (2000) Pattern formation in cluster roots: some developmental and evolutionary considerations. *Annals of Botany* **85**, 901–908.

Skene, K.R., Kierans, M., Sprent, J.I. and Raven, J.A. (1996) Structural aspects of cluster root development and their possible significance for nutrient acquisition in *Grevillea robusta* (Proteaceae). *Annals of Botany* **77**, 443–451.

Stirzaker, R.J. and Passioura, J.B. (1996) The water relations of the root-soil interface. *Plant, Cell and Environment* **19**, 201–208.

Swinnen, J. (1994) Evaluation of the use of a model rhizodeposition technique to separate root and microbial respiration in soil. *Plant and Soil* **165**, 89–101.

Swinnen, J., van Veen, J.A. and Merckx, R. (1994) ^{14}C pulse-labelling of field-grown spring wheat: an evaluation of its use in rhizosphere carbon budget estimations. *Soil Biology and Biochemistry* **26**, 161–170.

Swinnen, J., van Veen, J.A. and Merckx, R. (1995) Carbon fluxes in the rhizosphere of winter wheat and spring barley with conventional vs integrated farming. *Soil Biology and Biochemistry* **27**, 811–820.

Takagi, S., Nomoto, K. and Takemoto, T. (1984) Physiological aspect of mugineic acid, a possible phytosiderophore of graminaceous plants. *Journal of Plant Nutrition* **7**, 469–477.

Tang, C., McLay, C.D.A. and Barton, L. (1997) A comparison of proton excretion of twelve pasture legumes grown in nutrient solution. *Australian Journal of Experimental Agriculture* **37**, 563–570.

Thornton, B., Paterson, E., Midwood, A. J., Sim, A. and Pratt, S.M. (2004) Contribution of current carbon assimilation in supplying root exudates of *Lolium perenne* measured using steady-state ^{13}C labelling. *Physiologia Plantarum* **120**, 434–441.

Tinker, P.B. and Nye, P.H. (2000) *Solute Movement in the Rhizosphere*. Oxford University Press, Oxford, UK.

Trolldenier, G. (1988) Visualisation of oxidizing power of rice roots and of possible participation of bacteria in iron deposition. *Zeitschrift für Pflanzenernährung und Bodenkunde* **151**, 117–121.

Uhde-Stone, C., Gilbert, G., Johnson, J. M.-F., Litjens, R., Zinn, K.E., Temple, S.J., Vance, C.P. and Allan, D.L. (2003) Acclimation of white lupin to phosphorus deficiency involves enhanced expression of genes related to organic acid metabolism. *Plant and Soil* **248**, 99–116.

Uren, N.C. (2001) Types, amounts, and possible functions of compounds released into the rhizosphere by soil-grown plants. In: *The Rhizosphere: Biochemistry and Organic Substances at the Soil-Plant Interface* (eds R. Pinton, Z. Varanini and P. Nannipieri), pp. 19–39. Marcel Dekker, New York.

Vance, C.P., Uhde-Stone, C. and Allan, D.L. (2003) Phosphorus acquisition and use: critical adaptations by plants for securing a nonrenewable resource. *New Phytologist* **157**, 423–447.

Vetterlein, D. and Jahn, R. (2004) Gradients in soil solution composition between bulk soil and rhizosphere – *in situ* measurement with changing soil water content. *Plant and Soil* **258**, 307–317.

Warembourg, F.R. and Estelrich, H.D. (2001) Plant phenology and soil fertility effects on below-ground carbon allocation for an annual (*Bromus madritensis*) and a perennial (*Bromus erectus*) grass species. *Soil Biology and Biochemistry* **33**, 1291–1303.

Warembourg, F.R., Roumet, C. and Lafont, F. (2003) Differences in rhizosphere carbon-partitioning among plant species of different families. *Plant and Soil* **256**, 347–357.

Watt, M. and Evans, J.R. (1999a) Linking development and determinacy with organic acid efflux from proteoid roots of white lupin grown with low phosphorus and ambient or elevated atmospheric CO_2 concentration. *Plant Physiology* **120**, 705–716.

Watt, M. and Evans, J.R. (1999b) Proteoid roots. Physiology and development. *Plant Physiology* **121**, 317–323.

Whalley, W.R., Leeds-Harrison, P.B., Leech, P.K., Riseley, B. and Bird, N.R.A. (2004) The hydraulic properties of soil at root-soil interface. *Soil Science* **169**, 90–99.

Whipps, J.M. and Lynch, J.M. (1983) Substrate flow and utilization in the rhizosphere of cereals. *New Phytologist* **95**, 605–623.

Wild, A. (ed.) (1988) *Russell's Soil Conditions and Plant Growth*, 11th edn. Longman Scientific, Harlow, UK.

Wouterlood, M., Cawthray, G.R., Scanlon, T.T., Lambers, H. and Veneklaas, E.J. (2004) Carboxylate concentrations in the rhizosphere of lateral roots of chickpea (*Cicer arietinum*) increase during plant development, but are not correlated with phosphorus status of soil or plants. *New Phytologist* **162**, 745–753.

Young, I.M. (1998) Biophysical interactions at the root-soil interface: a review. *Journal of Agricultural Science, Cambridge* **130**, 1–7.

Zhang, F., Römheld, V. and Marschner, H. (1989) Effect of zinc deficiency in wheat on the release of zinc and iron mobilizing root exudates. *Zeitschrift für Pflanzenernährung und Bodenkunde* **152**, 205–210.

Zoysa, A.K.N., Loganathan, P. and Hedley, M.J. (1998) Effect of forms of nitrogen supply on mobilisation of phosphorus from a phosphate rock and acidification in the rhizosphere of tea. *Australian Journal of Soil Research* **36**, 373–387.

Chapter 8

Genetic Control of Root System Properties

Chapter 3 focused on describing the general characteristics of root systems in the field and differences between species. Within species, though, there are differences within natural populations, and between genotypes both in growth and in the response of root functions to the edaphic environment. Some of this variation has already been alluded to in relation to specific responses (e.g. lodging in section 4.1.2), but the purpose of this chapter is to detail some of the many genotypic differences that have been documented and the prospects for utilizing these in programmes of crop breeding.

The idea of 'improving' root systems of crops has been around for a long time but relative to shoots, progress has been slow and any improvements that there have been have, until now, been largely inadvertent rather than deliberate. Breeding programmes for shoot characters are common in many parts of the world and have delivered substantial benefits in terms of yield, but progress in selection for modified roots has been slower largely because: (i) the root characters required for a particular circumstance are rarely well defined; (ii) roots are concealed in soil so that characters cannot be readily measured; and (iii) soils vary spatially, which complicates interpretation of results when screening large numbers of plants and families. As with shoots, the essential features of a root system are determined by the genetic composition of the plant but there can be considerable interaction with the environment. Chapters 5 and 6 detail the many physical, chemical and biological factors that can influence root growth and contribute to the commonly large genotype × environment interaction found in many studies.

Several considerations are involved in designing efficient breeding and selection strategies aimed at genetic improvement of target traits including root morphology. These include issues such as: (i) the size and nature of the underlying gene action; (ii) whether there is sufficient additive genetic variance in the population/germplasm pool; (iii) the degree of narrow-sense heritability (that portion of the observed or phenotypic variance that is transmitted from parent to progeny) and whether it can be increased; and (iv) whether the genetic response sought can be achieved via indirect selection of secondary traits. Opportunities for use of linked molecular markers should also be considered where markers are robust, informative and account for a large proportion of the total genetic variance.

8.1 Genotypic differences in root systems

There is widespread evidence for genotypic diversity in the root characteristics of many crop species (O'Toole and Bland, 1987), and increasing interest in using this diversity to exploit soils more effectively. However, largely because of the difficulties associated with measuring root systems in any quantity in anything other than seedlings, the science of these studies is still at an early stage of development, with quantitative association of root trait and plant performance often absent. It is common for comparisons between genotypes of a species to be made for a particular characteristic (e.g. uptake of phosphorus) and their performance interpreted with respect to a measured or implied characteristic of the root system.

8.1.1 Size and architecture

The majority of studies on root genotypic variation have focused on establishing differences in the depth of rooting, size of the root system, or rate of increase of the size of the root system, because these are measures that are conceptually easy to relate to the exploitation of soil resources. Impetus for the selection of root traits that might improve resource use has largely come from the need to find superior genotypes for unfavourable environments, particularly drought-prone environments. O'Toole and Bland (1987) summarized the range of crops for which genotypic variation in root characters had been demonstrated and Table 8.1 provides more recent results of morphological differences that have been recorded.

Table 8.1 Some examples of reports of genotypic differences in growth and properties of root systems

Crop	Root system parameters	Genotypes	Reference
Barley	Dry weight, number, surface area	Maris Badger Proctor	Hackett (1968)
Barley	Length and mass distribution	Beecher Arabic abiad	Brown *et al.* (1987)
Chickpea	Mass, length and rates of extension	G 543, ILC 482, ILC 1929, ILC 3279, ILC 3397	Vincent and Gregory (1986)
Maize	Mass, length and distribution	LRS, LNS, P3905	Costa *et al.* (2002)
Maize	Length, numbers of laterals	La Posta, SA-3, TS-6, Pool-26	Bushamuka and Zobel (1998)
Rice (lowland)	Mass, length and distribution	IR20, NSG19, CT9993, IR62266, Mahsuri, FDML105, IR58821, IR52561	Azhiri-Sigari *et al.* (2000)
Rice (upland and lowland)	Mass, length, number of nodal roots and distribution of root length	Dular, IR20, IR72, IR43, IR64, UFLRi-5, Vandana, IR65598-112-2, IRAT216, Azucena, Moroberekan	Kondo *et al.* (2003)
Soyabean	Length, numbers of laterals	Perry, PI 416937, Weber, Maple Green	Bushamuka and Zobel (1998)
Soyabean	Hydraulic conductance, surface area and anatomy	PI 416937, H2L16, N95-SH-259, PI 407859-2, PI 471938, Young	Rincon *et al.* (2003)
Wheat (bread and durum)	Length and distribution	Tahara, Yallaroi, Kamilaroi, Beliouni, Excalibur, RH-880011	Zubaidi *et al.* (1999)

In environments subjected to drought, roots have an obvious direct role in moderating the supply of water through the depth of rooting and the quantity of roots in a particular layer. There are also less obvious, indirect effects in changing the rate at which the supply becomes available; both direct and indirect roles for root systems have been demonstrated experimentally. Issues relating to the timing of water use will be described in section 8.1.2, but where crops are grown on deep soils and water is stored throughout the soil profile, the depth of rooting has a major influence in determining the potential supply of soil water that is available to the crop. Frequently crops fill their grain during periods when rain is infrequent and stored soil moisture is utilized to sustain growth (terminal drought conditions). Such conditions were the impetus for the pioneering work of Hurd (1968, 1974), who was among the first to show genotypic variation in the rooting patterns of wheat. Based on glasshouse and field measurements, he concluded that drought resistance in Canadian wheat cultivars was associated with rapid downward growth of the root axes and a high proportion of total root weight at 0.2–0.3 m depth and, when stressed, below 0.3 m. Early measurements showed that the variety Thatcher had such a root system and later work confirmed the role of rooting characteristics in the observed drought resistance of the cultivar Pitic 62 (Fig. 8.1). This cultivar had a rapid rate of downward root growth and a substantial root mass whether it was droughted or not. Although this knowledge alone did not lead to direct selection in a breeding programme, the knowledge gained made a contribution in the breeding of new varieties of durum wheat such as Wascana and Wakooma that performed well under terminal drought conditions (Hurd *et al.*, 1972).

Deeper rooting and subsequent extraction of water can be an important contributor to drought tolerance because drought is effectively avoided. Sponchiado *et al.* (1989) demonstrated the importance of this effect in the relative drought tolerance of different cultivars of common bean (*Phaseolus vulgaris* L.). Their results showed that when grown on a slightly calcareous mollisol in irrigated conditions, all lines rooted to 0.7–0.8 m, but under drought conditions, the two drought-tolerant lines (BAT 85 and BAT 477) rooted to at least 1.2 m, whereas the two drought-sensitive lines (BAT 1224 and A70) reached only 0.8 m (Fig. 8.2). This difference in rooting depth between lines subjected to drought was associated with lower soil water contents at the end of the growing season (implying greater water extraction) and an almost threefold difference in grain yields. Complementary research using grafting techniques demonstrated that the drought tolerance was conferred by genes expressed in the root system and not by the shoot genotype indirectly affecting root characteristics (White and Castillo, 1989).

Drought is also a major abiotic constraint to yield improvement in upland and lowland rice, and there has been a prolonged effort to identify useful root traits that might improve the capture of water and yield stability. Chang *et al.* (1972) found that upland and lowland rice genotypes from Africa and Asia possessed similar variation in traits such that differences were not of type but of degree. There was a distribution of similar properties within the upland and lowland genotypes studied although, overall, upland types had greater length of root axes leading to deeper rooting, larger root diameter, and less branching. Chang and Vergara (1975) suggested that thick nodal axes might be an important root trait conferring drought resistance in upland rice and concluded that drought resistance would be assisted by combinations of two or more of the following root characteristics: (i) a high

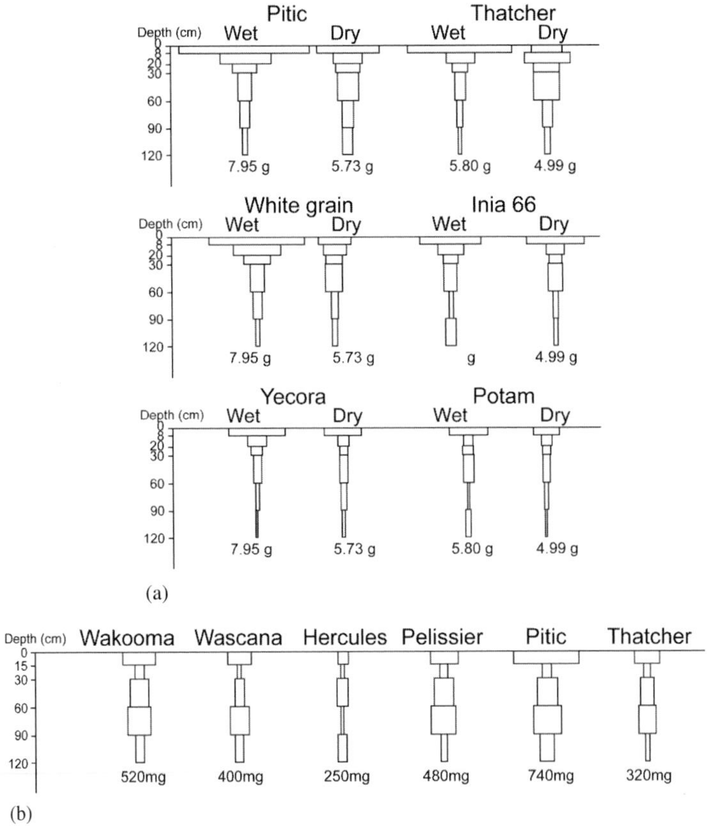

Fig. 8.1 The distribution of root dry weight with soil depth for various Canadian wheat cultivars measured close to maturity in (a) boxes in a glasshouse that were either kept near field capacity (wet) or allowed to dry after initial watering (dry); or (b) cores obtained from field plots. The total dry weight of roots recovered to the depth of sampling (1.2 m) is shown for each cultivar, and the width of the bars represents the amount of root weight at each level. (Reproduced with permission from Hurd, *Agricultural Meteorology*; Elsevier, 1974.)

proportion of thick roots; (ii) several very thick and long roots; (iii) a dense root system; (iv) uniform branching of fine roots from the axes. Yoshida and Hasegawa (1982) identified differences between genotypes in depth of rooting with a tendency for upland rice to be more deeply rooted than lowland rice. For example, lowland cultivar IR 20 rooted to 0.5 m while upland cultivar OS 4 rooted to 0.8 m under comparable conditions (Table 8.2). They also measured differences between cultivars in the vertical distribution of root length, root: shoot ratio, and specific root length. Yoshida and Hasegawa (1982) and Mambani and Lal (1983) regarded the development of a large and deep root system as an important means of ensuring water capture and hence yield stability of upland rice.

Thickness of roots and the ability of roots to penetrate compact layers have repeatedly been suggested as root traits linked to the ability of rice to avoid drought. Lafitte *et al.* (2001) found that a wide range of root characters including root thickness, xylem vessel diameter and patterns of root distribution differed between different types of rice. *Indica* rice types (typically lowland Asian cultivars) had thin, superficial roots with narrow vessels and a low root:shoot ratio, whereas *japonica* types (typically upland Asian cultivars)

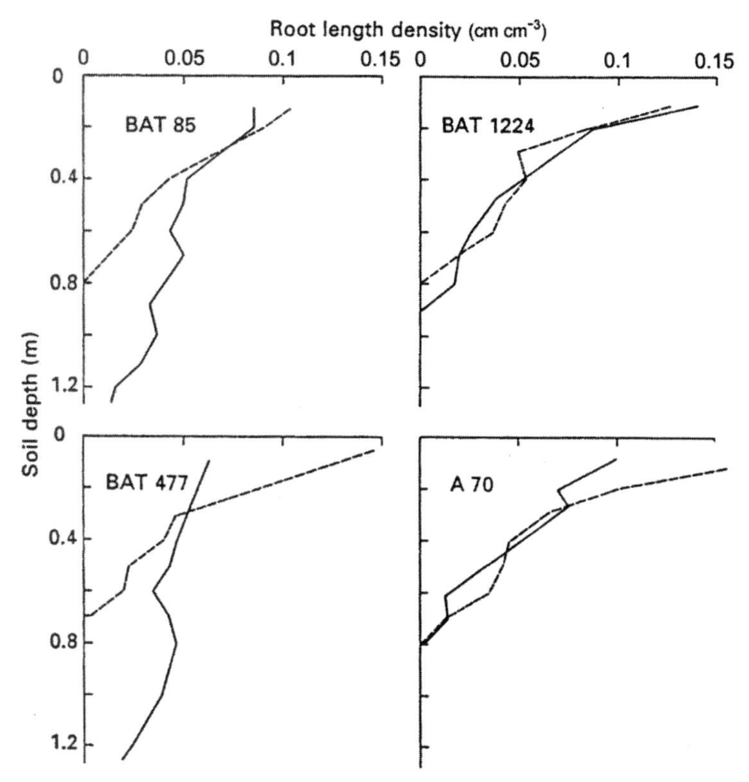

Fig. 8.2 Distribution of root length at 56 days after planting for four lines of common bean (*Phaseolus vulgaris* L.) grown under droughted (solid line) and irrigated (dashed line) conditions on a slightly calcareous mollisol. (Reproduced with permission from Sponchiado *et al.*, *Experimental Agriculture*; Cambridge University Press, 1989.)

Table 8.2 The distribution of root length (cm root cm^{-3} soil) with depth for seven genotypes of lowland and upland rice grown in a dryland field

	Lowland genotypes			Upland genotypes			
Depth (m)	IR20	IR2035-117-3	IR442-2-58	OS4	Moro-berekan	Salumpikit	20 A
0–0.1	14.4	22.7	16.8	12.6	11.8	16.2	19.8
0.1–0.2	2.8	5.8	7.1	1.4	2.3	5.5	2.6
0.2–0.3	0.9	0.8	1.2	0.8	0.9	1.9	0.9
0.3–0.4	0.4	0.1	0.3	0.9	0.8	1.4	0.8
0.4–0.5	0.1	0.1	0.1	0.8	0.6	0.8	0.9
0.5–0.6			0.1	0.5	0.8	0.6	0.9
0.6–0.7			0.1	0.5	0.4	0.3	0.6
0.7–0.8			0.1	0.5	0.2	0.1	0.4

From Yoshida and Hasegawa, 1982.

had thick roots with wider vessels, a greater proportion of root weight distributed below 15 cm and a larger root:shoot ratio, and *aus* types (upland, early cultivars from eastern India and Bangladesh) had intermediate root thickness, and a root:shoot ratio similar to *japonica* types. As determined earlier by Chang *et al.* (1972), while there were significant differences between types for all root traits, there was also significant variation within groups and the groups overlapped in all traits. Genotypic variation in the ability of rice roots to penetrate compact soil layers was assessed by Babu *et al.* (2001) using a system of wax-petrolatum layers in a potting mixture (developed by Yu *et al.*, 1995), who found that root penetration index (the ratio of penetrating axes to total root axis number) ranged from 0.08 to 0.45. *Japonica* (upland rice) accessions had higher root penetration indices than *indica* (lowland rice) types although one *indica* line had an index of 0.38.

In the sandy soils of the West African Sahel, both drought and nutrient shortages (especially phosphate) occur so that both the depth of rooting and the total length of the root system might be expected to affect resource acquisition. Brück *et al.* (2003) investigated differences in rooting parameters in eight varieties of pearl millet, selected to represent a broad range of times to maturity and plant height, grown on sandy alfisols in fields in Niger with different amounts of P fertilizers applied. There was substantial genotypic variation for root dry mass, total root length and length per unit soil volume, but not for depth of rooting, partitioning of roots between topsoil and subsoil, or for specific root length. In early experiments, the traditional landrace, Sadore local, had significantly greater total root length (3.06 km m^{-2}) than two earlier-maturing varieties (mean 2.12 km m^{-2}) at anthesis, although there were no significant differences by maturity when total length had declined slightly in two of the three varieties (mean 2.30 km m^{-2}). In subsequent studies with a wider range of varieties, root length was, as expected, considerably greater in the topsoil than the subsoil, but the varietal ranking was generally consistent irrespective of what depth root length was observed, and there were no measurable differences in the depth of rooting; Sadore local had the greatest root length and GB 8735 the smallest at all depths (Fig. 8.3). The consistency of ranking with depth meant that

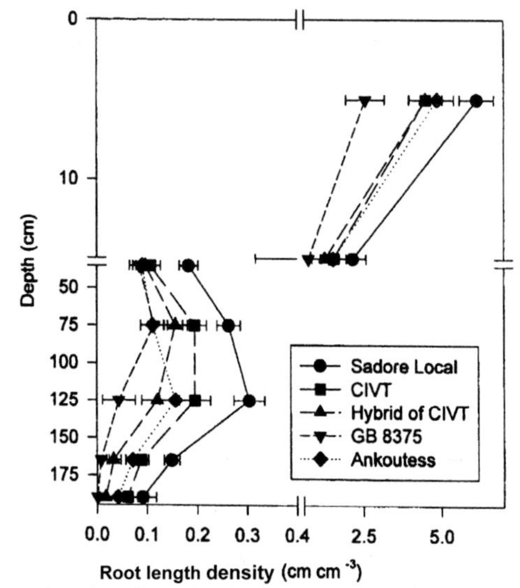

Fig. 8.3 Distribution of root length at flowering for five genotypes of pearl millet grown on an alfisol near Niamey, Niger. The results are the average for several P fertilizer treatments. (Reproduced with permission from Bruck *et al.*, *Plant and Soil*; Springer Science and Business Media, 2003.)

the partitioning of root length between topsoil and subsoil was almost constant for the varieties with 40–44% of root length in the upper 0.2 m in one season and 42–46% in the second season. Positive linear correlations were obtained between shoot dry matter at maturity and the total length of the root system but the extent and nature of these relations varied from season to season.

Genotype often interacts with environment, affecting the relative importance of genetic and environmental factors on root morphology and crop performance. For example, Kondo *et al.* (2003) investigated such interactions in a comparison of lowland and upland rice genotypes under field conditions using varieties identified in previous hydroponic and field studies as having contrasting maximum root length (Kondo *et al.*, 2000). There was substantial genotypic variation in root morphology between the 11 genotypes at 3 upland sites in the Philippines. Genotype accounted for the largest component of variation for the number of nodal roots, specific root weight and root:shoot ratio, whereas the environmental effect was relatively more important for total root dry weight and especially for the deep root length ratio (the ratio of root length below 0.3 m to total root dry weight). Among upland varieties, *japonica* types were characterized by a large total root dry weight (90–110 g m^{-2}), associated with high root:shoot ratio (0.18–0.23) and a high specific root weight (26–27 g m^{-1}), whereas *aus* and *indica* types (cv. Dular and Vandana, respectively) had a medium root:shoot ratio (0.14–0.17) and a large deep root length ratio (12.5–13.8 m g^{-1}). The results for the *aus* and *indica* types suggest the formation of a deep root system with only a limited allocation of biomass to the roots compared with *japonica* types. There was significant site × variety interaction, especially for root dry weight, with the implication that it is important to characterize site conditions in order to optimize the deployment of adaptive genotypes (Kondo *et al.*, 2003).

Soil properties also proved to be important in the expression of drought tolerance by common bean referred to earlier in this section (Sponchiado *et al.*, 1989). When the four lines were grown on an acidic oxisol, rooting depth was restricted to 0.4 m under irrigated conditions and to 0.7–0.8 m under droughted conditions, with no difference between drought-tolerant and drought-susceptible lines. There was also no significant difference in seed yields between the four lines grown on the oxisol. Deeper rooting, then, may confer an ability to avoid drought but only on soils where other properties do not restrict the expression of this character.

Relative to water uptake, less work has been undertaken on genotypic differences in root growth and their association with nutrient uptake. Lynch and van Beem (1993) grew four genotypes of common bean representing distinct shoot growth habits (erect determinate, erect indeterminate, prostrate indeterminate, and climbing) in containers of an oxisol and measured a range of root parameters up to 14 days after planting. Table 8.3 shows that there were significant differences between the genotypes after 14 days in root length and mass, number of roots arising from the base of the hypocotyls (basal roots), and in root growth and root elongation rates. The P-efficient genotype Tostado (which grows well in highly acidic, infertile soils in Rwanda) had the most vigorous seedling root system, which was highly branched and with numerous basal roots, whereas the landrace Porrillo Sintetico (which grows well on fertile soils in South America) had a smaller, less branched root system. Such observations have led to the development of quantitative models to investigate the effects of root architecture on

Table 8.3 Root growth parameters for four genotypes of common bean grown in containers of an oxisol for 14 days

	Genotype			
	Tostado	Porrillo Sintetico	Carioca	HAB 229
Total root dry weight (g plant^{-1})	0.38 (0.03)	0.23 (0.05)	0.27 (0.04)	0.28 (0.01)
Total root length (m plant^{-1})	65.9 (23.9)	23.9 (2.6)	35.1 (5.0)	49.6 (15.6)
Number of basal roots	252 (14)	171 (22)	271 (39)	216 (44)
Relative total root growth rate (d^{-1})	0.20 (0.01)	0.20 (0.04)	0.18 (0.01)	0.25 (0.01)
Relative total root elongation rate (d^{-1})	0.48 (0.03)	0.38 (0.03)	0.40 (0.03)	0.42 (0.02)

Values are the mean of four replicates with the standard error shown in brackets. From Lynch and van Beem, 1993.

the capture of nutrients such as phosphate (e.g. Lynch, 1995; Ho *et al.*, 2004) detailed in section 2.5. Substantial genetic variability exists for root gravitropic traits (determining the relative distribution of roots in different soil layers), which may influence the acquisition of resources such as P that are largely confined to shallow soil layers. Root gravitropic responses are also affected by soil P availability (Bonser *et al.*, 1996). Liao *et al.* (2004a) crossed a deep-rooted and a shallow-rooted genotype to obtain 86 recombinant inbred lines and found that lines with the highest P acquisition efficiency demonstrated more gravitropic plasticity or shallower root systems. At least two factors are believed to contribute to the greater P efficiency of shallower root systems compared with deeper root systems: (i) spatial coincidence of root and resource; and (ii) lower intra-plant inter-root competition. The latter is an important consideration because at typical planting densities of crops, inter-root competition has been found to be more important in determining the efficiency of P uptake than root competition between plants (Rubio *et al.*, 2001).

It will be clear from the preceding account that many of the studies that have investigated genotypic differences in size attributes of root systems have often deliberately chosen, or inadvertently included, materials that also differ in phenotype (for example, differences in time to maturity or, perhaps more importantly for studies with seedlings, differences in seed size – see section 3.2). Separating the phenotypic and genotypic interactions remains a major difficulty in specifying useful root traits for use in plant breeding programmes.

8.1.2 Functional properties

In addition to studies demonstrating genotypic differences in size and architecture of root systems, there have also been studies of differences in root functioning (i.e. water and nutrient uptake, and anchorage). As with the previous section, much of this work initially focused on drought, but nutrient use, especially of phosphorus, has also been examined.

Drought and water uptake

In the Mediterranean environment of northern Syria, crops are largely dependent on the use of growing season rainfall (very little water is stored from season to season) and several studies showed that pre-anthesis growth and water use were the keys to higher barley yields (Cooper *et al.*, 1987; Wahbi and Gregory, 1989). Studies of the root growth, water use and yield of the local landrace (Arabic abiad) at sites with typically <350 mm annual rainfall, showed that this genotype consistently had greater root lengths per unit soil volume at depths below 15 cm than other genotypes (Fig. 8.4) and that this was associated with greater water uptake in the 3–4 weeks before anthesis. In this environment, faster rates of growth before anthesis were reflected in higher crop yields; senescence of the crop before the grains could fill ('haying off') was not observed. Complementary studies with seedlings grown in nutrient solution in a glasshouse (Wahbi and Gregory, 1995) showed that two Syrian landraces produced more lateral roots with a faster rate of extension than other genotypes. The landraces Arabic abiad (2.25 lateral roots $°Cd^{-1}$and an extension rate of 22 mm $°Cd^{-1}$) and Arabic aswad (2.78 roots $°Cd^{-1}$and 27.8 mm $°Cd^{-1}$) were substantially more branched than genotypes such as Proctor (0.31 roots $°Cd^{-1}$ and 3.7 mm $°Cd^{-1}$) and Gerbel 'A' (0.72 roots $°Cd^{-1}$ and 7 mm $°Cd^{-1}$). Such rapid rates of proliferation may be advantageous in dry areas where water is available only early in the season. Landraces of barley and *Hordeum spontaneum* have proved to be very useful donors of traits in breeding for stressful conditions around the Mediterranean (Ceccarelli *et al.*, 1991; Ceccarelli and Grando, 1991) as they often out-yield other genotypes at dry sites. Grando and Ceccharelli (1995) found in very young seedlings grown on moist paper that landraces had similar numbers of seminal root axes to modern cultivars but that root length was significantly longer. *H. spontaneum* typically had only three seminal axes compared with the five to seven in landraces and modern cultivars of barley, and the axes were thin and of intermediate length. They suggested that the characteristics of the seminal roots could be one of the many traits contributing to the adaptation of landraces to difficult conditions. It is noteworthy, though, that *H. spontaneum* had considerably smaller seeds (mean 28.5 mg; range 19.8–36.5 mg) than modern cultivars and landraces (mean 50.5 mg; range 36.8–64.4 mg).

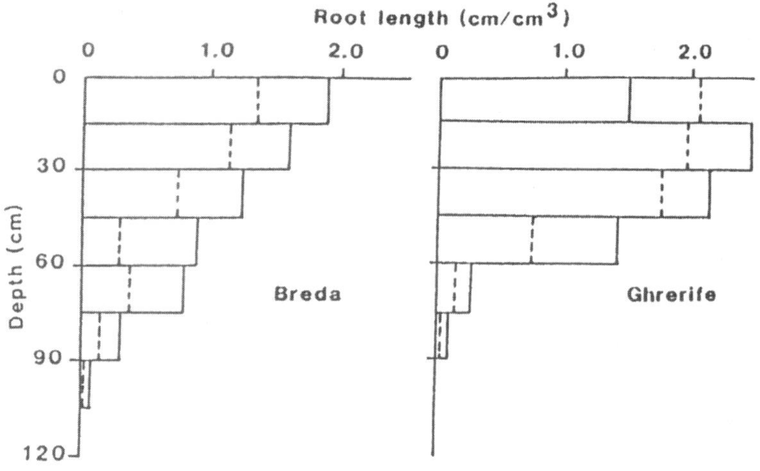

Fig. 8.4 Comparison of root distributions of the barley genotypes Arabic abiad (solid line) and Beecher (dashed line) grown at two sites in northern Syria with contrasting soils and rainfall. (Reproduced with permission from Gregory, *Drought Resistance in Cereals*; CAB International, 1989.)

In contrast to these Mediterranean conditions, where crops rely largely on water stored in the soil to fill their grain (e.g. in the north-eastern Australian wheatbelt), plants that are conservative in their early water use may out-yield more profligate users (Passioura, 1977). Stimulation of early growth (and water use) in these conditions, for example by application of N fertilizer, may result in the premature death of the crop before grain-filling can occur so that the crop 'hays off' (van Herwaarden *et al.*, 1998). Passioura (1972) proposed that a plant with a large hydraulic resistance in the roots might be advantageous by restricting early water use and conserving it for the grain-filling period. It was suggested that increased hydraulic resistance in the seminal axes would act as a throttle when conditions were dry, forcing the plants to use subsoil water slowly, but in wet seasons the nodal root system would proliferate in the topsoil and water uptake would occur unchecked. Axial resistance to water flow (see section 4.2.2) can be increased by either reducing the number of root axes or the diameter of xylem vessels. Richards and Passioura (1981a) found that only a few environmental factors (grain size, drought, and drought in the mother plant when the seed was produced), and no feasible agronomic practices, could greatly change the hydraulic resistance of axes, so that genotypic selection under controlled conditions should be possible. Screening of over 1000 wheat accessions demonstrated that increased resistance to water flow was unlikely to be achieved by reducing the number of seminal axes or increasing the proportion of multiple metaxylem vessels because of the absence of sufficient important genetic variation in these characters (Richards and Passioura, 1981b). However, screening for small diameter of the seminal xylem vessels was a more attractive character because: (i) vessel diameters substantially <60 μm were found in landraces; (ii) vessel diameter had a high heritability in most segregating populations (i.e. the portion of the observed phenotypic variation that was genetic in origin was high); and (iii) a significant response to selection was achieved in each of six different segregating populations.

Richards and Passioura (1989) undertook a backcross breeding programme using two commercial Australian varieties (Cook and Kite) which were crossed with the Turkish landrace (AUS 16871) to reduce the xylem vessel diameter of derivatives from about 65 to 55 μm. In field trials conducted over 5 years, selections with narrow vessels yielded between 3% and 11% more than the unselected controls, depending on genetic background; the yield advantage was greatest in the driest environments (with corresponding low yields) and disappeared in the wetter environments (Fig. 8.5). Although there were no measurements of water use in these experiments, most of the yield advantage of the narrow-vessel selections arose because of their greater harvest index, a parameter that Passioura (1977) demonstrated was associated with a greater proportion of water used after anthesis in this environment. This project is currently the only one for which a root character for a cereal crop has been specifically bred. Despite these promising results demonstrating the validity of the principle, the project was discontinued because disease became an issue in one of the varieties (Cook) and other traits were perceived as more important, and of higher heritability, to the improvement of yield.

Waterlogging

Genotypic differences within cereal species exist for tolerance to waterlogging, although environments differ in the timing, duration and intensity of waterlogging, leading to sub-

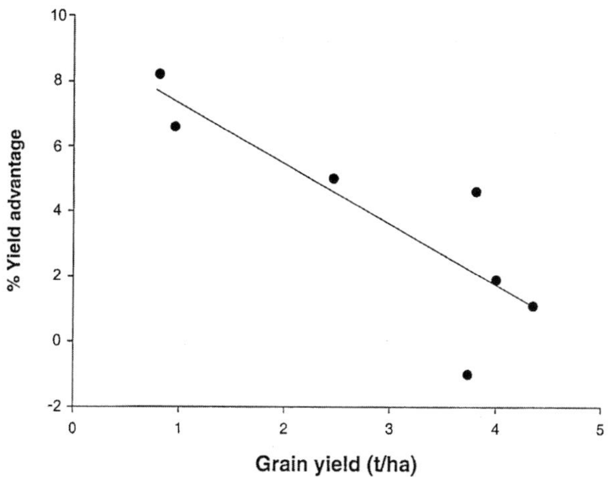

Fig. 8.5 The yield advantage of wheat lines selected for narrow xylem vessels. The values are from seven field environments and are averaged over two genetic backgrounds (cultivars Cook and Kite). (I am grateful to Dr R. Richards for this previously unpublished figure; to be published in Richards, 2005.)

stantially different potential impacts on crop growth. This means that proper environmental characterization is necessary to achieve gains from germplasm improvement (Setter and Waters, 2003). A wide range of physiological mechanisms exists for tolerance to waterlogging including adaptive traits related to phenology, morphology and anatomy (including the development of aerenchyma in roots), nutrition, metabolism, and post-anoxic damage and recovery (Setter and Waters, 2003). Despite the well-known changes to root anatomy induced by waterlogging via the formation of aerenchyma and the induction of barriers to radial oxygen loss (see section 5.4.3), there have been relatively few studies of differences between genotypes. Huang *et al.* (1994) demonstrated a positive relation between amount of growth following 21 days of hypoxia and the amount of aerenchymatous nodal roots for six wheat varieties grown in nutrient solution, while Setter *et al.* (1999) demonstrated a similar positive relationship between grain yield and the amount of aerenchyma for spring wheat cultivars grown under conditions of intermittent waterlogging in a field study in Western Australia (Fig. 8.6). However, the relationship between yield and aerenchyma development was not found in long-season wheat cultivars or in barley; if radial losses of O_2 exceed the capacity of the aerenchyma to diffuse O_2 to growing tissues, then aerenchyma development *per se* will be of little value (Setter and Waters, 2003).

Nutrient uptake

There has been a long-standing interest in varietal differences in the uptake of nutrients (especially of N and P) and of their utilization within the plant (e.g. Nielsen and Barber, 1978), but despite an appreciation of the major processes determining uptake (see section 4.3.3), progress has been slow in translating this into information that can be used in breeding programmes. For rice production systems, Kirk *et al.* (1998) summarized the root factors contributing to P uptake efficiency as: (i) root geometry – differences in root length and its distribution in soils, root hair length and density, root diameter, etc.; (ii)

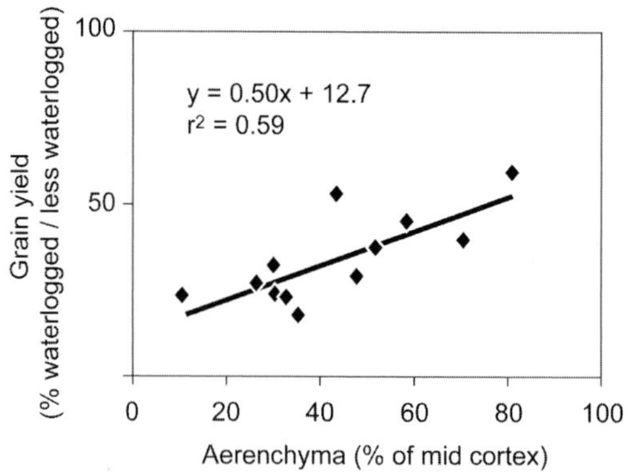

Fig. 8.6 Relationship between grain yield (expressed as a percentage of waterlogged plants experiencing waterlogging in the upper 30 cm of soil ≥160 cm d^{-1} to less waterlogged plants experiencing about 40 cm d^{-1}) of different spring wheat cultivars and the formation of aerenchyma (percentage of mid cortex of nodal axes) grown under field conditions at Esperance, Western Australia. (Reproduced with permission from Setter and Waters, *Plant and Soil*; Springer Science and Business Media, 2003.)

mycorrhizal effects – differences in the extent or rate of infection, or species of mycorrhzal fungus; and (iii) solubilization effects – differences in P solubility close to the root surface arising from changed soil chemical conditions. This is a complex set of properties and it is perhaps not surprising that because of the lack of appropriate screening techniques (among other things), Kirk *et al.* (1998) conclude that 'the evidence for germplasm differences in P uptake and use efficiencies in rice, and their genetic control, is not well advanced'.

There is evidence for genotypic diversity in both internal (quantity of biomass produced per unit of P taken up) and external (quantity of P taken up per unit of root) use efficiencies of P by rice, but these factors are highly inter-related because any additional P provided by externally efficient roots may also stimulate root growth. For example, model simulations of rice by Wissuwa (2003) showed that small changes (22%) in root diameter or internal efficiency had big effects (threefold) on P uptake. The same result could be achieved by a 33% increase in root external efficiency but only 10% of the threefold increase in P uptake was directly attributable to the direct effect of increased external root efficiency, with 90% due to enhanced root growth as a consequence of higher P uptake per unit of root. Wissuwa (2003) concluded that large genotypic differences in P uptake from P-deficient soils can result from small differences in tolerance mechanisms and that these small changes will be especially difficult to detect for external efficiency because they are likely to be overshadowed by the effects on root growth. Moreover, reliably detecting genotypic differences in nutrient use efficiency is made more difficult in the field because there are frequently large genotype × environment interactions that result in very variable results between growing seasons. For example, Fukai *et al.* (1999) grew rainfed lowland rice in multiple locations in Thailand and Laos and found that the genotype × environment interaction variance for grain yields was often greater than the genotype variance. Both Fukai *et al.* (1999) and Inthapanya *et al.* (2000) found significant genotypic variation in N and P use efficiency (grain yield per unit nutrient uptake) in rainfed lowland rice, and this was also related to the genotypic variation in grain yield. The contribution of root properties to these observed differences is unknown.

A root character that has been demonstrated to have a marked effect on the uptake of P is that of root hairs (see sections 2.3.3 and 4.3.4). As described in section 4.3.4, the primary effect of root hairs is to extend into the soil the surface of the root and thereby to shorten the pathway for P diffusion. Gahoonia and Nielsen (1997) demonstrated that differences in the length and density (number mm^{-1} root) of root hairs could affect P uptake by barley cultivars. In soil cv. Salka had root hairs that were 1.02 ± 0.22 mm long while those of cv. Zita were only 0.54 ± 0.14 mm long, and Salka depleted twice as much bicarbonate-extractable P from the soil as Zita (Fig. 8.7a). For Salka, the concentration of bicarbonate-extractable P was a uniform 0.2 mmole kg^{-1} soil from the root surface to 1 mm distant (about the length of the root hairs), while the uniform depletion profile of Zita was shorter (0.4 mm) reflecting the shorter root hairs. Both cultivars also depleted hydroxide-extractable P to 1 mm from the root surface, but the difference between cultivars was not significant (Fig. 8.7b). Subsequent studies with a hairless root mutant of barley (*bald root barley, brb*) and wild type have confirmed the importance of root hairs in a low-P soil, and the fact that both phenotypes grew and reproduced normally with a high nutrient supply may allow identification of candidate genes for root hair formation in cereals (Gahoonia and Nielsen, 2003). Field and hydroponic studies with barley genotypes with either long (about 1 mm) or short (about 0.5 mm) root hairs produced similar rankings for root hair length in both environments, and root hair length was significantly correlated with P uptake in the field (Gahoonia and Nielsen, 2004). Genotypes with long root hairs produced similar grain yields (about 6 t ha^{-1}) irrespective of the amount of P fertilizer applied (0, 10 and 20 kg ha^{-1}), while genotypes with short root hairs had lower yields at zero P addition but responded to applications of P fertilizer. It was concluded that long root hairs are likely to benefit yields on low-P soils and may also enhance the efficiency with which P fertilizer is used.

Genotypic differences in N uptake have been demonstrated in many species including wheat, rice and maize but the reasons for such differences are not fully understood. Both plant growth rate and soil N availability play a role in addition to any inherent differences in internal N use efficiency (Devienne-Barret *et al.*, 2000). Liao *et al.* (2004b) compared growth and N uptake of a vigorously growing breeding line with four commercial cultivars in field and column experiments and found that uptake efficiency at tillering was 16–22% greater in the breeding line than three of the four cultivars. They concluded that the inclusion of traits leading to vigorous early shoot and root growth into breeding programmes for wheat could result in improved N uptake efficiency and reduced N leaching from the sandy soils on which they worked.

Anchorage

In contrast to the water and nutrient uptake functions of root systems, genotypic variation in anchorage ability has only rarely been assessed. Genotypic differences for lodging resistance have been demonstrated (see section 4.1.2) but this is a complex characteristic involving both stem and root properties. Root pulling resistance, the vertical force required to pull a plant from the soil, was suggested by O'Toole and Soemartono (1981) as a means of evaluating the relative size and depth of rice root systems. They demonstrated that rice varieties with high root pulling resistance tended to be more drought-tolerant than those with

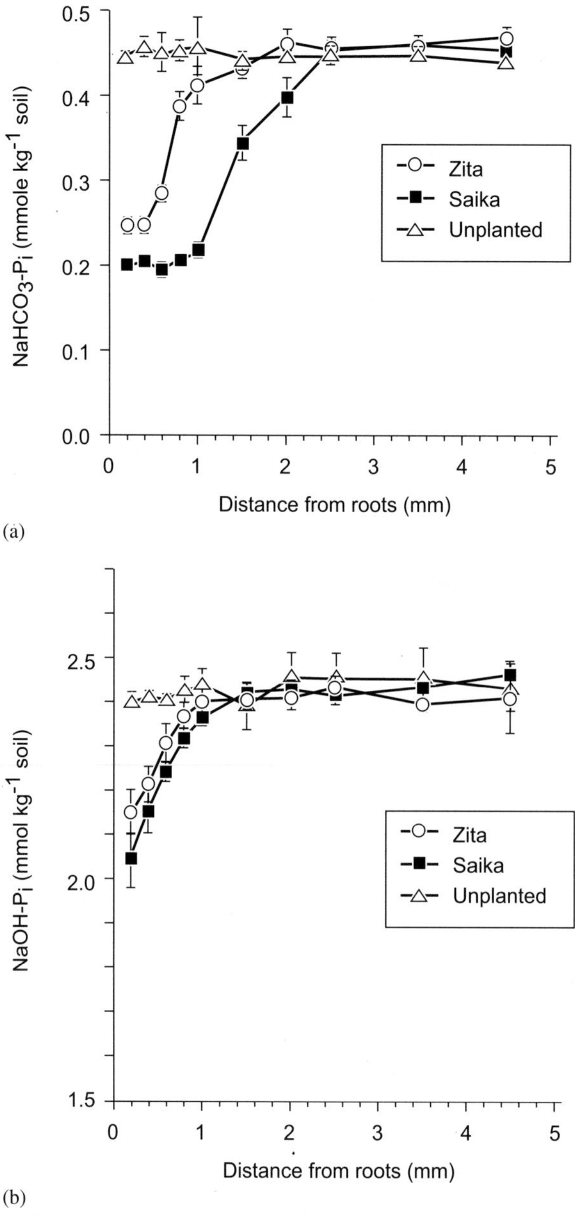

Fig. 8.7 Depletion profiles of soil phosphorus from soil by two barley cultivars differing in the length of their root hairs: (a) bicarbonate-extractable P (inorganic P); and (b) sodium hydroxide-extractable P (organic P). (Reproduced with permission from Gahoonia and Nielsen, *Euphytica*; Springer Science and Business Media, 1997.)

low root pulling resistance. Ekanayake *et al.* (1986) investigated the association between root pulling resistance and characteristics of pulled (i.e. removed from the soil) and unpulled (i.e. left in the soil after detachment of the remainder of the plant) components of the root systems of five lowland rice genotypes with different tolerance to drought grown on an irrigated clay soil. Substantial genotypic variation in root pulling resistance was measured,

Table 8.4 Root pulling resistance at two time points during the wet season of 1981 for five genotypes of lowland rice differing in susceptibility to drought

		Root pulling resistance (kg)	
Genotype	Response to drought	20 days after transplanting	31 days after transplanting
IR20	Susceptible	2.77	13.85
IR36	Moderately susceptible	3.32	12.18
IR442-2-58	Moderately tolerant	4.80	17.35
IR52	Tolerant	5.91	18.83
MGL2	Tolerant	4.06	18.09

From Ekanayake *et al.*, 1986.

with the drought-susceptible genotypes having lower root pulling resistance (Table 8.4). Genotypes with high root pulling resistance were characterized by larger, thicker roots and longer root systems. Pulled roots of genotypes with high root pulling resistance were thicker and heavier than those with low root pulling resistance, and unpulled roots had more root length per unit soil volume. Overall, the results suggested that high root pulling resistance was strongly related to the ability to rapidly develop roots in deeper soil layers; this characteristic was, in turn, associated with the drought tolerance of the genotypes.

8.2 Genetics of root systems

8.2.1 Genetic control of root development and growth

One of the major advances in crop breeding during the last century was the introduction of dwarfing genes into cereals leading to genotypes with shorter straw that could respond to applications of fertilizers without lodging. While the benefits of this advance have been widely felt, there have been some environments (predominantly droughted environments) in which semi-dwarf cereal varieties have not prospered and there has been a long-running debate about the possible effects of the dwarfing genes on root systems. For example, Cholick *et al.* (1977) investigated rooting patterns of two tall and three semi-dwarf (based on Norin 10, Norin 16 and Seu Seun 27 dwarfing sources) winter wheat cultivars to assess whether there were deleterious effects of dwarfing genes on root growth under dryland conditions at Akron, Colorado, USA. They found no significant differences between the cultivars in rooting depth, nor in the distribution of roots as assessed by ^{32}P translocation from shoots to roots. Similar findings for the size of semi-dwarf and tall wheat genotypes have also been reported by other workers when root systems were measured in field experiments in the UK (Lupton *et al.*, 1974), in glasshouse and field experiments in Australia (Richards, 1992), and in controlled-environment conditions in Canada (McCaig and Morgan, 1993). In a Mediterranean environment, Siddique *et al.* (1990) compared old (tall) and modern (semi-dwarf) varieties and near-isogenic lines for *Rht* dwarfing genes (KCD0 and KCD1). At anthesis, the older varieties (Purple Straw and Gamenya) had significantly greater root mass than the modern types (Kulin, KCD0 and KCD1), although there were no differences in rooting depth, water extraction or water use. Shoot dry weight was similar between varieties at anthesis (KCD1 and Kulin were lower) so that root dry weight as a proportion of total plant weight was slightly higher (not statistically significant) in older (0.39) than modern (0.35) varieties. However, the study was confounded by differences in phenology; time to anthesis ranged from 132 days in Purple Straw to 106 days in KCD types.

Nevertheless, the doubts about the performance of semi-dwarf cultivars persist in dry, rainfed environments. Lupton *et al.* (1974) stated that poor performance was partly associated with poor coleoptile development, which prevents satisfactory emergence of deep-sown crops, but might also indicate limited development of the root system. Richards (1992) reported no differences in water use between near-isogenic *Rht* lines grown in 1 m long tubes in a glasshouse, although there was some evidence that the shortest plants extracted slightly less water when grown in the field where depth of extraction extended to 1.9 m. The rate of water extraction reflected growth, with the taller plants that grew fastest extracting water from each depth earlier than dwarf plants. This theme of altered plant performance has been picked up again more recently by Rebetzke and Richards (1999) and Ellis *et al.* (2004). The reduced height associated with the *Rht-B1b* (formerly *Rht1*) and *Rht-D1b* (formerly *Rht2*) genes arises through their insensitivity to gibberellin (GA) growth hormones. The insensitivity of the stem and leaf cells to endogenous GA decreases cell elongation in juvenile leaf and stem tissues and ultimately reduces plant height, but in the short term it also results in a shorter coleoptile and a reduction in leaf area, namely reduced early vigour. Whether reduced early vigour of the shoot is also evident in the growth of roots is still an open question, but results of Manske and Vlek (2002) suggest that further study may be warranted. In a field study with near-isogenic lines in two cultivar backgrounds of bread wheat carrying different combinations of *Rht-B1b*, *Rht-D1b* and *Rht-B1c* genes, they found that the *Rht-B1b* and *Rht-B1c* genes were associated with reduced root length at tillering in both cultivars, while the *Rht-D1b* genes had a similar effect in only the Mexican cultivar Nainari 60 (Table 8.5). The double dwarfs had root lengths that were similar to the tall controls in both cultivars at tillering. Between tillering and anthesis, the root systems of genotypes with *Rht-B1b* and *Rht-B1c* genes grew more so that at anthesis there were no differences in the length.

Many high-yielding wheat cultivars contain a segment from the short arm of rye chromosome 1R (1RS) because it confers several useful characteristics such as resistance to pests and diseases. Rye is believed to have a well-developed root system, at least comparable with other temperate cereals, grows better than other temperate cereals on soils of low fertility, and is generally more tolerant of abiotic stresses such as drought, heat and cold

Table 8.5 Effects of dwarfing genes on the root length in the 0–0.2 m soil layer (expressed as cm root cm^{-3} soil) of two bread wheat cultivars at tillering

	cv. Nainari	cv. Maringa	Mean (dwarfing genes)
Tall control	3.7c	3.1b	3.4b
Rht-B1b	2.9b	2.0a	2.5a
Rht-D1b	2.7ab	3.0b	2.9ab
Rht-B1b and *Rht-D1b*	3.3bc	2.6ab	3.0ab
Rht-B1c	2.7ab	2.1a	2.4a
Mean	3.1b	2.6a	

The data are averaged over two growing seasons and the letters represent differences in significance at $p = 0.05$. From Manske and Vlek, 2002.

than cereals such as bread wheat (Nalborczyk and Sowa, 2001). The agronomic advantages of 1RS translocation lines have been inconsistent across wheat genotypes and environmental conditions, although there are some indications that addition of 1RS can affect the root system and improve tolerance to environmental stresses in the field. Ehdaie *et al.* (2003) investigated the effects of 1RS translocations in spring wheat (cv. Pavon) recommended for cultivation under irrigated conditions because of its susceptibility to drought. In pot experiments in a glasshouse, root dry matter at maturity was increased significantly in the 1RS.1AL and 1RS.1DL lines in well-watered conditions and in the 1RS.1AL line in droughted conditions (Table 8.6). The relative increases in root dry matter in 1RS translocations were consistent in both treatments and roots were more branched and thinner. Although there was no significant effect of 1RS translocation on grain yield or water use efficiency, there were significant positive correlations between root mass and grain yield under both well-watered and droughted conditions. Unfortunately roots were not measured in the accompanying field experiments but 1RS lines had higher grain yield and were more tolerant of environmental stresses. In contrast, translocations from chromosome 2 (2RS) had no beneficial effects on any of the characters measured in either glasshouse or the field.

Understanding of lateral root development at cellular and genetic levels is being rapidly advanced in the model plant *Arabidopsis*, in which wild-type and mutant phenotypes have been employed to characterize, for example, the specific effects of phytohormones such as IAA and ABA. Casimiro *et al.* (2003) list 28 genes that are involved in the regulation of lateral root development of *Arabidopsis* but highlight the issue that, to date, there have been comparatively few detailed morphological studies that demonstrate the cellular, rather than molecular, basis for a mutant's lateral root defect. Moreover, although several studies have shown the importance of IAA for both lateral root initiation and emergence, little is known about how auxin regulates the spacing of lateral roots. Root development in cereals (especially in rice and maize) is also being studied via a growing number of mutants that affect various developmental stages (Table 8.7). Figure 8.8 shows the effects of three genes in maize on the development of seminal and short-borne root axes, and on lateral roots (Hochholdinger *et al.*, 2004a). The mutant *rtcs* lacks all root axes but the tap root (the first seminal axis), and the *rt1* gene reduced shoot-borne root formation (nodal axes). Other mutants suggest that there are at least two pathways of lateral root formation for roots present in the embryo and those formed post-embryonically. The mutant *lrt1* results in an absence of lateral roots from the seminal axes and those at the coleoptile node; shoot-borne roots with normal branching appear from the second

Table 8.6 The effect of including different moieties of the short arm of rye chromosome 1 (1RS) in the spring wheat cultivar Pavon on root dry weight (g per pot) at maturity

	Pavon	1RS.1AL	1RS.1BL	1RS.1DL
Well-watered	2.5	3.4*	3.0	3.4*
Droughted	2.9	3.6*	3.0	3.3

The plants were grown in pots in a glasshouse. Results significant at $p = 0.05$ are shown with an asterisk. From Ehdaie *et al.*, 2003.

Table 8.7 Mutants of maize and rice that affect aspects of root development

Gene	Species	Phenotype
Shoot-borne roots		
Rtcs	Maize	Complete absence of shoot-borne roots
rt1	Maize	Fewer shoot-borne roots
crl1	Rice	Fewer shoot-borne roots
crl2	Rice	Fewer shoot-borne roots and reduced seminal root length
Lateral roots		
lrt1	Maize	No lateral roots on seminal roots or shoot-borne roots at first (coleoptile) node
slr1	Maize	No lateral root elongation on seminal root axes
slr2	Maize	No lateral root elongation on seminal root axes
rm109	Rice	Lateral root initiation blocked
Tap root (seminal axis 1)		
rm1	Rice	Reduced root elongation
rm2	Rice	Reduced root elongation
rrl1	Rice	Reduced root elongation
rrl2	Rice	Reduced root elongation
srt5	Rice	Reduced elongation of all root types
srt6	Rice	Reduced root elongation
Root hairs		
rth1	Maize	Reduced root hair elongation
rth2	Maize	Reduced root hair elongation
rth3	Maize	Reduced root hair elongation
rh2	Rice	Missing root hairs

Adapted from Hochholdinger *et al.*, 2004a.

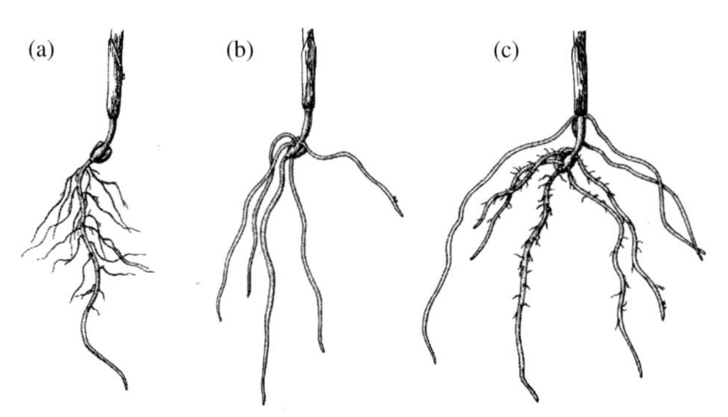

Fig. 8.8 Effects of recessive root mutants on root appearance: (a) the *rtcs* mutant in which only the first seminal root is present and other seminal and nodal roots are absent; (b) the *lrt1* mutant in which lateral roots on seminal roots are missing and nodal roots at the coleoptile node are affected while lateral root formation on higher nodes is normal; and (c) the *slr1* and *slr2* mutants in which lateral root elongation of seminal roots is reduced but that on shoot-borne nodal axes is normal. (Original drawings by Miwa Kojima of Iowa State University. Reproduced with permission from Hochholdinger *et al.*, *Trends in Plant Science*; Elsevier, 2004a.)

node onwards. The *slr1* and *slr2* mutants have reduced lateral root elongation only on the seminal axes, and form normal lateral roots on the post-embryonic shoot-borne root system (nodal axes; Fig. 8.8). When these mutations are considered at the level of the whole root system, then development can be described in embryonic, early post-embryonic and late post-embryonic terms (Table 8.8). For example, the mutants *rtcs* and *rum1* are both affected in embryogenesis and fail to initiate seminal roots other than the tap root. Both mutants also show a post-embryonic phenotype but while *rum1* is affected in early post-embryonic development and forms no laterals on the tap root, the *rtcs* mutant shows a complementary phenotype with laterals on the tap root but all later post-embryonic roots (including shoot-borne roots) are missing (Hochholdinger *et al.*, 2004b). Practical applications for these mutants are some years away, but identification of genes that are specific to root formation in cereals via comparison of the rice and *Arabidopsis* genome databases is already possible (Hochholdinger *et al.*, 2004a).

8.2.2 Genetic control of root properties

A wide range of genes has been identified as being involved in aspects of the development and functioning of plant roots, especially in relation to the exclusion of toxic ions and the uptake of nutrients (Aeschbacher *et al.*, 1994; Gallais and Hirel, 2004). Much of this research is at an early stage of development and the remainder of this section focuses on two examples of how the emerging knowledge of genetic control of functions might be exploited. The first example focuses on the use of a gene already present in some crop plants that appears to regulate the production of organic acids from roots and thereby allow the roots to cope with high concentrations of Al^{3+} in soil solution. The second example details the potential offered by genetic manipulation to change the ability of plant roots to access P from organic sources in soil; it also describes some of the difficulties in transferring this technology from laboratory media to soils.

Table 8.8 Summary of root types that are affected in the monogenic root mutants of maize that have been identified

	Embryonic		Early post-embryonic			Late post-embryonic			
			Lateral roots		Nodal roots node 1	Nodal roots node 2+	Above-ground nodal (brace) roots	Lateral roots	
Locus	Tap root	Seminal roots	Tap root	Seminals				Nodal	Brace
rtcs		dark grey		dark grey	dark grey	dark grey	dark grey	dark grey	dark grey
rt1					dark grey	dark grey	dark grey	dark grey	dark grey
lrt1			dark grey	dark grey	dark grey				
rum1		dark grey	dark grey	dark grey					
slr1,2			light grey	light grey					
rth1,2,3	light grey	light grey	light grey	light grey	light grey	light grey		light grey	

Dark grey indicates root types that are completely missing in the affected mutants, with light grey indicating root types that are formed but are affected by the mutation. All the affected roots of the *rth* mutants show impaired root hair formation but are otherwise normal. From Hochholdinger *et al.*, 2004b.

Aluminium toxicity

As described in section 5.5.2, many plants have evolved mechanisms to enable them to grow on soils in which toxic concentrations of Al^{3+} can limit plant growth. Such soils are prevalent in tropical regions where heavy rainfall over long periods has washed many of the more soluble bases (calcium and magnesium) from the soils. Differences between plant species and between genotypes within a species in tolerance to aluminium have been recognized for a long time, and attempts have been made to screen tolerant cultivars (see Foy *et al.*, 1992, for soyabean). A practical difficulty with this approach has been to develop a screening technique that can cope with the large numbers of plants involved and to achieve consistency of results; genotypic rankings for Al tolerance often vary among soil types. Villagarcia *et al.* (2001) describe the issues involved and show that a hydroponic-based screen of soyabean seedlings produced an inflated range of genotypic response, and different tolerance rankings, compared with sand culture.

Discovering the genetic basis for the observed genotypic variation in response to Al has been difficult but is yielding practical benefits. Garvin and Carver (2003) reviewed the progress made with breeding Al-tolerant wheat and highlighted a major paradox, i.e. that when designed crosses are studied, Al tolerance is commonly inherited in a simple fashion, suggesting that tolerance, or a large part of it, derives from a single dominant gene (e.g. Al-tolerant lines derived from Brazilian materials such as BH1146, Riede and Anderson, 1996). However, when Al tolerance is evaluated in a range of germplasm, a broad range of response is found from highly tolerant to highly sensitive, suggesting that multiple genes are involved in the response. Garvin and Carver (2003) suggest that this paradox is a major unanswered question in Al tolerance genetics, but conclude that the broad and continuous distribution of response 'is likely to be due to the presence of multiple genes that have evolved independently, multiple allele series for these genes, interactions between loci, or combinations of all these factors'.

A mechanism that plants use to overcome aluminium toxicity is the release of organic acids from roots into the rhizosphere to chelate aluminium (Ma *et al.*, 2001; see also section 5.5.2), and there have been several studies reporting a positive relationship between Al tolerance and organic acid efflux (e.g. Delhaize *et al.*, 1993b, malate efflux from wheat; Jorge and Arruda, 1997, citrate and malate efflux from maize). Silva *et al.* (2001) evaluated this mechanism using up to eight genotypes of soyabean grown in hydroponics and concluded that differential tolerance to Al was associated both with sustaining high rates of citrate release into the external solution, and high concentrations of citrate in the root tip over time. In this study malate concentrations in roots and root tips were higher than citrate concentrations but citrate efflux was much greater than that of malate.

Delhaize *et al.* (1993a, 1993b) investigated the mechanism of Al tolerance using near-isogenic lines of wheat differing in Al tolerance at a single locus designated as *Alt1*. This locus co-segregates with an Al-activated malate efflux from root apices, a process which is mediated by Al-activated anion channels in the plasma membrane. The *Alt1* gene was believed to encode a malate-permeable anion channel or a protein that could modulate the activity of such a channel. The exuded malate formed a stable complex with Al which explained the observed tolerance to Al. Subsequent work with the same near-isogenic lines has led to the identification of a gene, *ALMT1* (aluminium-activated malate transporter), which encodes a membrane protein which is constitutively expressed in the root apices of the Al-tolerant line at higher levels than in the Al-sensitive line (Sasaki *et al.*, 2004). The gene co-segregated with Al tolerance in F_2 and F_3 populations of wheat derived from crosses of the near-isogenic lines when grown in solution culture (Fig. 8.9). When *ALMT1* was introduced into tobacco cells, Al-activated malate efflux was observed, and the transgenic cells accumulated less Al than the control cells and had a greater capacity for regrowth after Al treatment. In contrast, when the gene was introduced into rice roots, Al activated the efflux of malate from roots but tolerance of the transgenic plants to Al was not increased. This result may have occurred because rice is among the most tolerant cereals to Al so that the amount of malate released may have been insufficient to increase the Al tolerance above its already high endogenous level.

Barley is a much more Al-sensitive species than rice, and transgenic barley expressing the *ALMT1* gene has been shown to increase Al tolerance both in hydroponics and soils (Delhaize *et al.*, 2004). When seedlings of the cultivar Golden Promise, which is normally very susceptible to Al toxicity, were grown in an acid subsoil (pH 3.9 in $CaCl_2$), root elongation in barley lines with the *ALMT1* gene was about 30 mm over a 4-day period compared with only 10–15 mm in the wild-type and two near-isogenic lines differing in Al tolerance. Al-activated malate efflux was evident in the transgenic barley and disruption of the root apex was substantially reduced in hydroponic conditions.

Phosphate uptake

Plants have developed a wide range of mechanisms to enhance their ability to acquire P from soils including the symbiotic association with mycorrhiza, and the production of root hairs, exudates and enzymes (Vance *et al.*, 2003). Root exudates appear to assist mainly in the solubilization of inorganic and organic P, while phosphatase enzymes hydrolyse organic forms of P. In many soils, organic P is a substantial component of total soil P and some plant species (e.g. *Tithonia diversifolia* and *Crotalaria grahamiana*, used as green manures in parts of East Africa) appear to have a greater ability than others to utilize this P source. George *et al.* (2002a, 2002b) found that *T. diversifolia* and *C.*

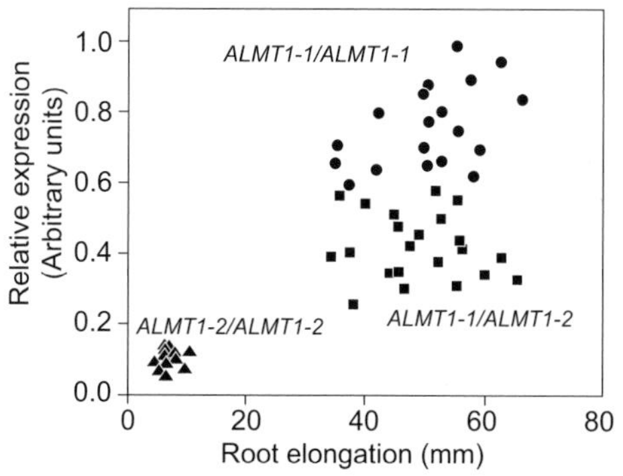

Fig. 8.9 Relationship between the relative expression of ALMT1 in root apices of wheat lines (normalized against the values obtained for the 28S rRNA transcript) and root elongation in a solution containing 10 μM Al. The ALMT1 expression co-segregates with Al tolerance. (Reproduced with permission from Sasaki *et al.*, *The Plant Journal*; Blackwell Publishing Ltd, 2004.)

grahmiana had enhanced levels of phosphatase activity in their rhizospheres compared with maize, and that there was depletion of organic P that was subsequently identified as orthophosphate monoesters. The predominant form of monoester P found in most soils is phytate (derivatives of inositol hexaphosphate). Paradoxically, plants have a limited capacity to utilize phytate as a P source due either to poor substrate availability or to their limited capacity to exude functional extracellular phytase. Richardson *et al.* (2001) demonstrated that transgenic *Arabidopsis* plants which expressed a phytase gene (*phyA*) from the soil fungus *Aspergillus niger* secreted extracellular phytase and were able to use P supplied as phytate in agar. When combined with the carrot extension (*ex*) gene, there was a 20-fold increase in root phytase activity in *ex::phyA* plants resulting in growth and P content being similar to that of control plants supplied with inorganic phosphate. In this initial study, all tissues of the plant produced phytase, but Mudge *et al.* (2003) found a similar response in *Arabidopsis* when the gene was promoted by a *Pht1;2-ex::phytase* construct that secreted phytase predominantly from root epidermal cells when grown on a low-P medium. This transgene enabled *Arabidopsis* to grow on a medium containing phytate as the sole P source.

George *et al.* (2004) showed that incorporation of the *phyA* gene into five lines of subterranean clover increased exudation of extracellular phytase by an average of 77-fold, and uptake of P from phytate by 1.3 to 3.6-fold compared with controls when plants were grown in agar. However, only one of them showed increased shoot biomass or P uptake when grown in soil. Subsequent work (George *et al.*, 2005a) showed that phytase activity in solution was lost within 10 minutes when phytase was added to three soils, because of rapid sorption of phytase onto the solid phase (Fig. 8.10). There was less sorption to soil collected from the rhizosphere of transgenic subterranean clover plants expressing *phyA*, suggesting that the rhizosphere may maintain phytase activity in solution, possibly by altering pH. Rapid sorption of phytase by soils may, then, limit the ability of roots to acquire P from soil phytate. Studies in which the *phyA* gene was expressed in tobacco showed that when two soils were amended with either phytate or phosphate combined with lime, then transgenic plants contained up to 52% more P than controls (George *et al.*, 2005b). Therefore, in addition to extracellular phytase, there is a requirement for the phytate to be in an available form. Interestingly, wild type subterranean clover compensated for the lack of *phyA* by producing a longer root system, although the size of this effect changed with time. Initially plants with the *phyA* gene had shorter root systems, but once the phytate was depleted, root growth caught up with that of the wild types. Thus, the ability to obtain phytate appears to confer only short-term advantage.

Together these studies indicate that the inclusion of a gene to modify root function may work in simple growing media such as agar, but that the biology of root function in soils is a complex matter. Limitations to the performance of plants transformed to secrete extracellular phytase include: (i) immobilization of phytase activity by sorption on soil components; (ii) poor availability of substrate; (iii) effects of rhizosphere microbes which may compete for released orthophosphate or release their own phytase; and (iv) suppression of compensatory effects of other P efficiency mechanisms such as the greater root lengths found in early growth of wild types. For full advantage to be accrued from introducing a capacity of plants to produce extracellular phytase, it will also be desirable to find means of improving the availability of native soil phytate.

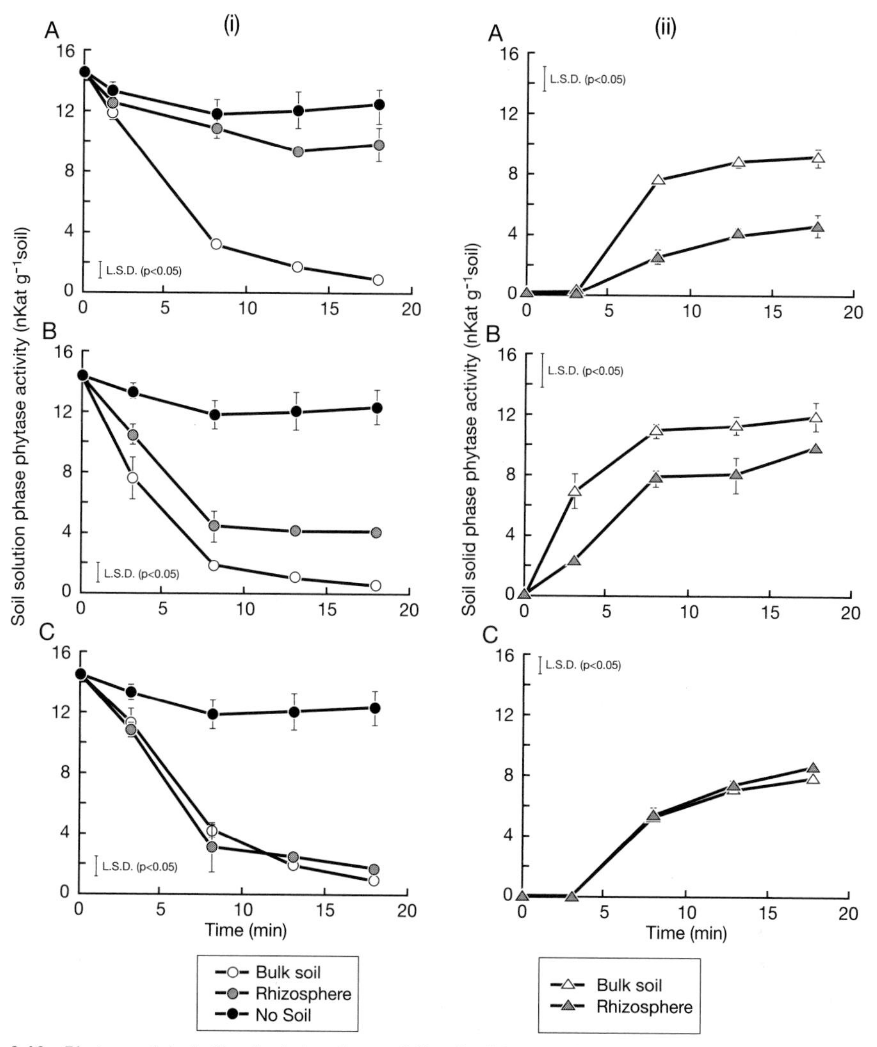

Fig. 8.10 Phytase activity in (i) soil solution phase and (ii) soil solid phase after addition of phytase to bulk and rhizosphere samples of three soil types; (A) spodosol, (B) alfisol and (C) oxisol and in solution in the absence of soil. Data points are the mean of four replicates with bars representing two standard errors. Differences between soil sources (within soil types) were established using ANOVA. LSD ($p<0.05$) is presented as a bar for each soil type and phase within soil. (Reproduced with permission from George *et al.*, *Soil Biology and Biochemistry*; Elsevier, 2005a.)

8.3 Breeding better root systems

While there are many observations of genotypic differences in root systems and their functioning, incorporating this information into routine breeding and crop improvement programmes is fraught with practical difficulties. O'Toole and Bland (1987) drew on other work to outline the necessary steps in a hypothetical programme seeking to improve root systems via genetic selection. Their suggested steps were:

(1) *Definition of the problem* – define the perceived problem and the opportunity for improvement of the production system, then define the factor(s) most limiting root system function in the system. It is important that a problematic environmental condition and a specific yield-determining growth stage should coincide.
(2) *Evaluation of root system parameters* – identify those aspects of the root system (e.g. rooting depth, length distribution of system, rate of downward growth) that are potential solutions to the problem and develop a conceptual model. Mechanistic crop simulations can further aid selection of parameters, give advance warning of the potential importance of genotype × environment interactions, and serve as a caution against single-factor solutions.
(3) *Determination of existence, level and nature of genetic variation* – examine germplasm, including exotic germplasm and landraces, for evidence of variation. Seedling screening may be appropriate for some characters but others (e.g. rooting depth) may require time-consuming methods which may not be possible in field environments. It may be necessary to develop an effective selection method and to assess its potential for use in the selection process.
(4) *Hybridization and selection* – root traits can be handled in the same way as other quantitatively inherited traits, although there are practical difficulties with selecting roots in the F_2 and F_3 generations for anything other than traits evident in seedlings. For complex traits such as root distribution with depth, only screening of parent lines and advanced lines may be possible.
(5) *Field evaluation of resultant genotypes for validation* – in addition to yield, the evaluation should ideally include assessment of the processes and root parameters included in the initial conceptual framework.

Screening for complex root characteristics in large numbers of breeding lines is problematic although simpler characteristics, such as Al, P or B concentrations, are no more difficult than those in shoots. Soil-based screening is difficult because: (i) the roots cannot easily be seen unless they are grown against a sloping glass wall (e.g. Hurd, 1968); (ii) different soils may produce different rankings of genotypes (e.g. screening for Al tolerance, Villagarcia *et al.*, 2001); and (iii) large quantities of uniform media are required if anything other than seedlings are grown. For these reasons many workers have preferred to develop non-soil systems such as pouches (McMichael *et al.*, 1985), hydroponics (Vincent and Gregory, 1986), sand systems (Villagarcia *et al.*, 2001), and aeroponics (Armento-Soto *et al.*, 1983) for initial ranking and selections, followed by observation of a smaller number of genotypes in soil. There are few studies where the ranking of genotypes at different stages of growth or in different media have been compared, but this is a growing topic of research.

The low heritability of root traits is also an issue that concerns plant breeders but this is an issue that involves both sufficient genotypic diversity and appropriate gene action for the trait to be selected, and an appropriate technique for assessing or measuring the trait in a large number of individuals. There is no reason to suppose that heritability of root traits is less than that of comparable shoot traits and the issue can usually be resolved by devising appropriate selection methods that maximize genetic variability and also enable rapid selection, thereby substantially increasing both heritability and selection response. This approach was adopted by Richards and Passioura (1981b), who demonstrated high heritability for xylem vessel diameter. However, the choice of growing medium and measurement technique may pose difficulties that are not as amenable to solutions that are as easy as with shoots. Ehdaie *et al.* (2001) found that root biomass of two bread wheat crosses had a relatively high heritability under both well-watered and dry conditions, but as will be shown below, many traits are only expressed in particular environments, so that high heritability in a particular environment is not a good indicator of utility.

While root characters are determined genetically, their full expression is dependent on the environment, and many studies have indicated substantial genotype × environment interactions. Breeders often select for a character (e.g. drought resistance) in the agroclimatic zone of interest but soils vary within regions so that what is thought to be a response to drought may also include responses to other environmental factors such as subsoil salinity, soil pH, reduced P availability, root disease, etc., unless the soil is also characterized. The full exploitation of root traits depends, then, on appropriate targeting for soil properties in much the same way that appropriate phenology is currently linked with agroclimatic analysis.

8.3.1 Use of markers and QTL

The development of DNA-based molecular markers has opened up opportunities for identifying the genetic factors (quantitative trait loci, QTL) underpinning various root traits. This technology makes it possible to investigate the inheritance of root traits that are either single gene or polygenic in nature, and to select for chromosomal regions associated with the characters of interest. Again, this science is at an early stage of development for root traits, but significant progress has been made in studies with rice and, to a lesser extent, with maize (see Tuberosa *et al.*, 2002). This section will focus on rice.

The tolerance of rice to P deficiency was investigated by Wissuwa *et al.* (1998) using 98 backcross inbred lines derived from a cross of the traditional Indian *indica* variety Kasalath that is tolerant of P deficiency with the modern Japanese *japonica* variety Nipponbare which does not yield well under low P conditions. The lines were grown on P-deficient soil under upland conditions and dry weight and P concentration were measured. Three QTL explained 45% of the variation in dry weight and four QTL explained 55% of the variation in P uptake, and for both traits the quantitative trait locus linked to marker C443 on chromosome 12 had a major effect. A complementary study with a wide range of rice cultivars in which tiller number under P deficiency was assessed confirmed the identity of the major quantitative trait locus on chromosome 12 (C443) for tolerance of P deficiency. Interestingly, one of the minor QTL associated with tolerance to P deficiency coincided with a locus found in another study to account for 10% of the variation in total root number. Subsequent transference of the major quantitative trait locus linked to marker C443 on chromosome 12 into Nipponbare by three backcrosses resulted in an improved line that increased P uptake by 170% and grain yield by 230% compared with Nipponbare when grown in the field on a low P andosol (Wissuwa and Ae, 2001). Using this marker-based approach, then, it was possible to combine the high P uptake of the traditional variety (Kasalath) with the high harvest index of the modern variety to produce a plant that, although only 58% of the shoot weight and 68% of the root weight of Kasalath, yielded 21% more grain even under low P conditions.

Much work with rice has focused on root traits associated with drought avoidance in both upland and lowland types. Early studies by Champoux *et al.* (1995) set out to identify QTL associated with five aspects of rice root morphology (maximum rooting depth, nodal root diameter, root:shoot ratio, dry weight of roots on tiller 1, and deep root dry weight of tiller 1), and to determine whether or not these QTL were located in the same chromosomal regions as QTL associated with drought avoidance. They produced 203 recombinant inbred lines from a cross between the lowland *indica* cultivar Co39 (susceptible to drought, with shallow, thin roots) with the upland *japonica* cultivar Moroberekan (drought-tolerant with deep, thick roots) and assessed root traits in glasshouse experiments and leaf rolling (a measure of drought tolerance/avoidance) in field experiments in which drought was applied at the seedling, early vegetative and late vegetative growth stages. All QTL which exerted a positive effect on root characters were derived from Moroberekan. Root diameter was the trait with the greatest number of identified QTL, and maximum rooting depth the least. Figure 8.11 shows that many of the putative QTL for the five root traits overlap, suggesting that these regions either contain genes that interact to exert a general effect on root morphology or a gene that has multiple effects on root characteristics. Twelve of the 14 chromosomal regions containing putative QTL associated with drought avoidance in the field also contained QTL associated with root traits, suggesting that selecting for Moroberekan alleles at three or four of the loci with the largest effects on root morphology and crossing them into an *indica* cultivar might improve drought tolerance in lowland rice production. Subsequent studies with an *indica* (Co39) × *japonica* (IAC165) cross has confirmed the importance of genomic regions carrying QTL for several root traits on chromosomes 1, 4, 9, 11 and 12 (Courtois *et al.*, 2003), and there are now several studies with similar QTL linked to the five root traits measured in the other populations studied (Table 8.9) (see also Babu *et al.*, 2003). These studies have indicated, then, that there are many genomic regions of small genetic effect associated with root morphology. Combining these many alleles into one genetic background through conventional breeding is likely to be very slow.

Although there appears to be some commonality in the QTL regions identified (Table 8.9), water stress is not usually an all or nothing phenomenon, but builds slowly with time so that QTL for adaptive responses will change with time. This means that QTL for both drought and root traits are likely to vary depending on the environment. For example, Hemamalini *et al.* (2000) found no common loci for root morphological traits in a population derived from an IR64 × Azucena cross when grown in either well-watered conditions or conditions that allowed stress to develop slowly. This absence of common QTL for root traits suggests the existence of parallel genetic pathways operating at different moisture regimes. Similarly, in a comparison of growth of a population from an IR552 × Azucena cross under flooded and water-deficit conditions, Zheng *et al.* (2003) showed that of 18 QTL detected for seminal root length, nodal root number, and lateral root length and lateral root number on the seminal root at a specified depth, only one was detected under both water conditions. Five limited-water-induced QTL for four root traits were identified, and comparison with other studies using Azucena as a parent indicated several identical QTL for root elongation. The importance of QTL × environment interaction is also illustrated in the work of Price *et al.* (2002a, 2002b) with rice. They screened 140 recombinant inbred lines and the parental varieties Bala and Azucena for root growth in glass-sided chambers in a glasshouse under conditions of early and late water deficit. There were major differences

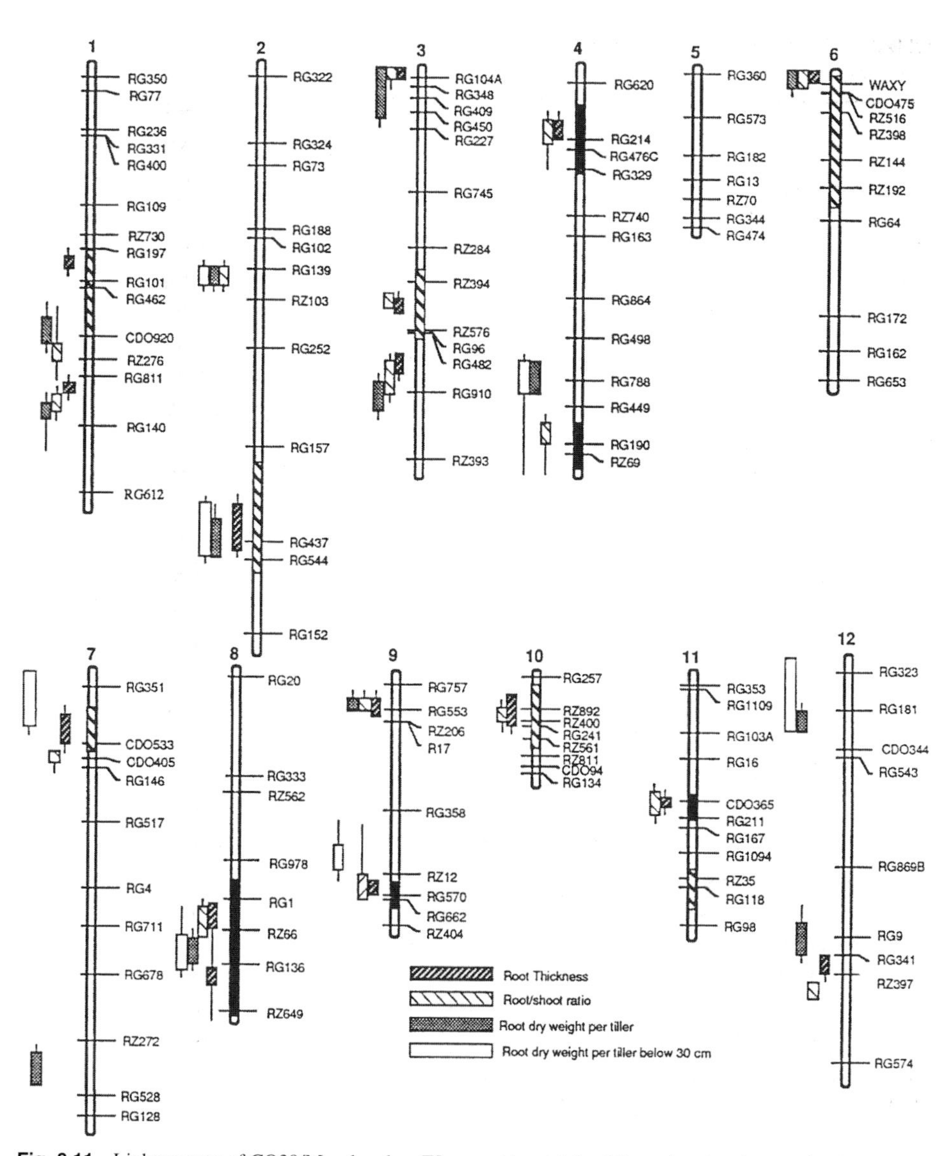

Fig. 8.11 Linkage map of CO39/Moroberekan F9 recombinant inbred lines showing the putative location of QTL associated with root characters for each of the 12 chromosomes of rice as measured in glasshouse experiments. Boxes represent 1-LOD support intervals surrounding each putative QTL and lines show the associated 2-LOD intervals. Darkened areas within chromosomal bars represent putative QTL associated with drought avoidance in three field experiments, and hatched areas represent regions associated with drought avoidance in two of the three field experiments. (Reproduced with permission from Champoux *et al.*, *Theoretical and Applied Genetics*; Springer Science and Business Media, 1995.)

in root:shoot partitioning and root mass between treatments with effective water extraction only at root lengths >0.4 cm cm^{-3}. They identified 6 QTL for root weight below 90 cm and maximum root length, 11 for root:shoot ratio, 12 for number of roots deeper than 1.0 m, and 14 for root diameter, occupying a total of 24 chromosome regions. However, some were revealed only in individual experiments and/or for individual traits, while others were

Table 8.9 Comparison of the location of the main QTLs for various constitutive root traits under unstressed conditions determined for several mapped populations of rice crosses

			Cross						
Root trait	Chromosome	Region	IAC165 × Co39	Co39 × Moro.	IR64 × Azucena	Bala × Azucena (F_2)	Bala × Azucena (RILs)	IR58821 × IR52561	CT9993 × IR62266
Diameter	1	RZ801-RZ14	THK	THK	THK	–	THK	THK	THK
	4	RZ69-RG499	THK	–	–	–	THK	–	–
	4	RZ23-RM225	THK	–	–	–	THK	THK	THK
	7	RZ337C-RM234	THK	–	–	–	THK	–	–
Deep root	1	RZ690B-RZ730	MRL DRT	–	MRL DRW DRS DRT	–	DRW	MRL	DRW DRT
	2	RG171-RG157	DRT	DRT	DRT MRL	MRL	–	DRT	DRW DRT MRL
	2	RM250-RZ123	DRS	DRT	–	–	–	DRW	MRL
	4	RM261-RG449	DRW DRS DRT	DRT	–	–	MRL DRW	DRT MRL	–
	7	RZ488-RG477	DRW	–	–	–	–	DRT	–
	8	RZ987-RZ143	DRW	–	DRT	–	–	–	–
	9	R570-RG667	MRL DRS	DRT	DRW MRL DRT DRS	–	MRL DRW	DRW	–
	12	RM181-RG95	MRL	DRT	–	–	–	–	–

The root traits assessed were root diameter (referred to as thickness in most studies, THK), maximum root length (MRL), deep root weight (DRW), deep root weight per tiller (DRT), and deep root weight:shoot dry weight (DRS). Adapted from Courtois *et al.* (2003), where all sources of data are cited.

common to different traits or experiments. It is also important to note that the number of QTL apparently identified can depend on the threshold value specified.

All of the studies referred to found a large number of QTL for root morphology and that there were usually many QTL for each root trait (see also Liao *et al.*, 2004a, for QTL associated with root traits and P uptake efficiency of common bean). This demonstrates that the traits sought were polygenic. The marked QTL × environment interaction shows that the genetic control of root growth is complex and, conversely, that environmental conditions may have a large effect on the morphology of root systems. To be useful in breeding programmes, stability of QTL expression across soil environments is essential and this remains a challenge.

References

Aeschbacher, R.A., Schiefelbein, J.W. and Benfey, P.N. (1994) The genetic and molecular basis of root development. *Annual Review of Plant Physiology and Plant Molecular Biology* **45**, 25–45.

Armento-Soto, J., Chang, T.T., Loresto, G.C. and O'Toole, J.C. (1983) Genetic analysis of root characters in rice. *SABRAO Journal* **15**, 103–116.

Azhiri-Sigari, T., Yamauchi, A., Kamoshita, A. and Wade, L.J. (2000) Genotypic variation in response of rainfed lowland rice to drought and rewatering. II. Root growth. *Plant Production Science* **3**, 180–188.

Babu, R.C., Shashidhar, H.E., Lilley, J.M., Thanh, N.D., Ray, J.D., Sadasivam, S., Sarkarung, S., O'Toole, J.C. and Nguyen, H.T. (2001) Variation in root penetration ability, osmotic adjustment and dehydration tolerance among accessions of rice adapted to rainfed lowland and upland ecosystems. *Plant Breeding* **120**, 233–238.

Babu, R.C., Nguyen, B.D., Chamarerk, V., Shanmugasundaram, P., Chezian, P., Jeyaprakash, P., Ganesh, S.K., Palchamy, A., Sadasivam, S., Sarkarung, S., Wade, L.J. and Nguyen, H.T. (2003) Genetic analysis of drought resistance in rice by molecular markers: association between secondary traits and field performance. *Crop Science* **43**, 1457–1469.

Bonser, A.M., Lynch, J.P. and Snapp, S. (1996) Effect of phosphorus deficiency on growth angle of basal roots in *Phaseolus vulgaris*. *New Phytologist* **132**, 281–288.

Brown, S.C., Keatinge, J.D.H., Gregory, P.J. and Cooper, P.J.M. (1987) Effects of fertilizer, variety and location on barley production under rainfed conditions in northern Syria. 1. Root and shoot growth. *Field Crops Research* **16**, 53–66.

Brück, H., Sattelmacher, B. and Payne, W.A. (2003) Varietal differences in shoot and rooting parameters of pearl millet on sandy soils in Niger. *Plant and Soil* **251**, 175–185.

Bushamuka, V.N. and Zobel, R.W. (1998) Differential genotypic and root type penetration of compacted soil layers. *Crop Science* **38**, 776–781.

Casimiro, I., Beeckman, T., Graham, N., Bhalerao, R., Zhang, H., Casero, P., Sandberg, G. and Bennett, M.J. (2003) Dissecting *Arabidopsis* lateral root development. *Trends in Plant Science* **8**, 165–171.

Ceccarelli, S. and Grando, S. (1991) Environment of selection and type of germplasm in barley breeding for low-yielding conditions. *Euphytica* **57**, 207–219.

Ceccarelli, S., Acevedo, E. and Grando, S. (1991) Breeding for yield stability in unpredictable environments: single traits, interaction between traits, and architecture of genotypes. *Euphytica* **56**, 169–185.

Champoux, M.C., Wang, G., Sarkarung, S., Mackill, D.J., O'Toole, J.C., Huang, N. and McCouch, S.R. (1995) Locating genes associated with root morphology and drought avoidance in rice via linkage to molecular markers. *Theoretical and Applied Genetics* **90**, 969–981.

Chang, T.T. and Vergara, B.S. (1975) Varietal diversity and morpho-agronomic characteristics of upland rice. In: *Major Research in Upland Rice,* pp. 72–90. IRRI, Los Banos, The Philippines.

Chang, T.T., Loresto, G.C. and Tagumpay, O. (1972) Agronomic and growth characteristics of upland and lowland rice varieties. In: *Rice Breeding*, pp. 645–661. IRRI, Los Banos, The Philippines.

Cholick, F.A., Welsh, J.R. and Cole, C.V. (1977) Rooting patterns of semi-dwarf and tall winter wheat cultivars under dryland field conditions. *Crop Science* **17**, 637–639.

Cooper, P.J.M., Gregory, P.J., Tully, D. and Harris, H.C. (1987) Improving water use efficiency of annual crops in the rainfed farming systems of West Asia and North Africa. *Experimental Agriculture* **23**, 113–158.

Costa, C., Dwyer, L.M., Zhou, X., Dutilleul, P., Hamel, C., Reid, L.M. and Smith, D.L. (2002) Root morphology of contrasting maize genotypes. *Agronomy Journal* **94**, 96–101.

Courtois, B., Shen, L., Petalcorin, W., Carandang, S., Mauleon, R. and Li, Z. (2003) Locating QTLs controlling constitutive root traits in the rice population IAC 165 × Co39. *Euphytica* **134**, 335–345.

Delhaize, E., Craig, S., Beaton, C.D., Bennet, R.J., Jagadish, V.C. and Randall, P.J. (1993a) Aluminium tolerance in wheat (*Triticum aestivum* L.). I. Uptake and distribution of aluminium in root apices. *Plant Physiology* **103**, 685–693.

Delhaize, E., Ryan, P.R. and Randall, P.J. (1993b) Aluminium tolerance in wheat (*Triticum aestivum* L.). II. Aluminium-stimulated excretion of malic acid from root apices. *Plant Physiology* **103**, 695–702.

Delhaize, E., Ryan, P.R., Hebb, D.M., Yamamoto, Y., Sasaki, T. and Matsumoto, H. (2004) Engineering high-level aluminium tolerance in barley with the *ALMT1* gene. *Proceedings of the National Academy of Science* **101**, 15249–15254.

Devienne-Barret, F., Justes, E., Machet, J.M. and Mary, B. (2000) Integrated control of nitrate uptake by crop growth rate and soil nitrate availability under field conditions. *Annals of Botany* **86**, 995–1005.

Ehdaie, B., Barnhart, D. and Waines, J.G. (2001) Inheritance of root and shoot biomass in a bread wheat cross. *Journal of Genetics and Breeding* **55**, 1–10.

Ehdaie, B., Whitkus, R.W. and Waines, J.G. (2003) Root biomass, water-use efficiency, and performance of wheat-rye translocations of chromosomes 1 and 2 in spring bread wheat 'Pavon'. *Crop Science* **43**, 710–717.

Ekanayake, I.J., Garrity, D.P. and O'Toole, J.C. (1986) Influence of deep root density on root pulling resistance in rice. *Crop Science* **26**, 1181–1186.

Ellis, M.H., Rebetzke, G.J., Chandler, P., Bonnett, D., Spielmeyer, W. and Richards, R.A. (2004) The effect of different height reducing genes on the early growth of wheat. *Functional Plant Biology* **31**, 583–589.

Foy, C.D., Duke, J.A. and Devine, T.E. (1992) Tolerance of soybean germplasm to an acid Tatum subsoil. *Journal of Plant Nutrition* **15**, 527–547.

Fukai, S., Inthapanya, P., Blamey, F.P.C. and Khunthasuvon, S. (1999) Genotypic variation in rice grown in low fertile soils and drought-prone, rainfed lowland environments. *Field Crops Research* **64**, 121–130.

Gallais, A. and Hirel, B. (2004) An approach to the genetics of nitrogen use efficiency in maize. *Journal of Experimental Botany* **55**, 295–306.

Gahoonia, T.S. and Nielsen, N.E. (1997) Variation in root hairs of barley cultivars doubled soil phosphorus uptake. *Euphytica* **98**, 177–182.

Gahoonia, T.S. and Nielsen, N.E. (2003) Phosphorus (P) uptake and growth of a root hairless barley mutant (bald root barley, brb) and wild type in low- and high-P soils. *Plant, Cell and Environment* **26**, 1759–1766.

Gahoonia, T.S. and Nielsen, N.E. (2004) Barley genotypes with long root hairs sustain high grain yields in low-P field. *Plant and Soil* **262**, 55–62.

Garvin, D.F. and Carver, B.F. (2003) Role of the genotype in tolerance to acidity and aluminium toxicity. In: *Handbook of Soil Acidity* (ed. Z. Rengel), pp. 387–406. Marcel Dekker Inc., New York.

George, T.S., Gregory, P.J., Robinson, J.S., Buresh, R.J. and Jama, B. (2002a) Utilisation of soil organic P by agroforestry and crop species in the field, western Kenya. *Plant and Soil* **246**, 53–63.

George, T.S., Gregory, P.J., Wood, M., Read, D.B. and Buresh, R.J. (2002b) Phosphatase activity and organic acids in the rhizosphere of potential agroforestry species and maize. *Soil Biology and Biochemistry* **34**, 1487–1494.

George, T.S., Richardson, A.E., Hadobas, P.A. and Simpson, R.J. (2004) Characterization of transgenic *Trifolium subterraneum* L. which expresses phyA and releases extracellular phytase: growth and P nutrition in laboratory media and soil. *Plant, Cell and Environment* **27**, 1351–1361.

George, T.S., Richardson, A.E. and Simpson, R.J. (2005a) Behaviour of plant-derived extracellular phytase upon addition to soil. *Soil Biology and Biochemistry* **37**, 977–988.

George, T.S., Simpson, R.J., Hadobas, P.A. and Richardson, A.E. (2005b) Expression of a fungal phytase gene in *Nicotiana tabacum* improves phosphorus nutrition of plants grown in amended soils. *Plant Biotechnology Journal* **3**, 129–140.

Grando, S. and Ceccarelli, S. (1995) Seminal root morphology and coleoptile length in wild (*Hordeum vulgare* ssp. *spontaneum*) and cultivated (*Hordeum vulgare* ssp. *vulgare*) barley. *Euphytica* **86**, 73–80.

Gregory, P.J. (1989) The role of root characteristics in moderating the effects of drought. In: *Drought Resistance in Cereals* (ed. F.W.G. Baker), pp. 141–150. CABI, Wallingford.

Hackett, C. (1968) A study of the root system of barley. I. Effects of nutrition on two varieties. *New Phytologist* **67**, 287–299.

Hemamalini, G.S., Shashidhar, H.E. and Hittalmani, S. (2000) Molecular marker assisted tagging of morphological and physiological traits under two contrasting moisture regimes at peak vegetative stage in rice (*Oryza sativa* L.). *Euphytica* **112**, 69–78.

Ho, M.D., McCannon, B.C. and Lynch, J.P. (2004) Optimization modeling of plant root architecture for water and phosphorus acquisition. *Journal of Theoretical Biology* **226**, 331–340.

Hochholdinger, F., Park, W.J., Sauer, M. and Woll, K. (2004a) From weeds to crops: genetic analysis of root development in cereals. *Trends in Plant Science* **9**, 42–48.

Hochholdinger, F., Woll, K., Sauer, M. and Dembinsky, D. (2004b) Genetic dissection of root formation in maize (*Zea mays*) reveals root-type specific developmental programmes. *Annals of Botany* **93**, 359–368.

Huang, B.-R., Johnson, J.W., NeSmith, D.S. and Bridges, D.C. (1994) Root and shoot growth of wheat genotypes in response to hypoxia and subsequent resumption of aeration. *Crop Science* **34**, 1538–1544.

Hurd, E.A. (1968) Growth of roots of seven varieties of spring wheat at high and low moisture levels. *Agronomy Journal* **60**, 201–205.

Hurd, E.A. (1974) Phenotype and drought tolerance in wheat. *Agricultural Meteorology* **14**, 39–55.

Hurd, E.A., Townley-Smith, T.F., Patterson, L.A. and Owen, C.H. (1972) Techniques used in producing Wascana wheat. *Canadian Journal of Plant Science* **52**, 689–691.

Inthapanya, P., Sipaseuth, Sihavong, P., Sihathep, V., Chanphengsay, M., Fukai, S. and Basnayake, J. (2000) Genotype differences in nutrient uptake and utilisation for grain yield production of rainfed lowland rice under fertilised and non-fertilised conditions. *Field Crops Research* **65**, 57–68.

Jorge, R.A. and Arruda, P. (1997) Aluminium-induced organic acids exudation by roots of an aluminium-tolerant maize. *Phytochemistry* **45**, 675–681.

Kirk, G.J.D., George, T., Courtois, B. and Senadhira, D. (1998) Opportunities to improve phosphorus efficiency and soil fertility in rainfed lowland and upland rice ecosystems. *Field Crops Research* **56**, 73–92.

Kondo, M., Aguilar, A., Abe, J. and Morita, S. (2000) Anatomy of nodal roots in tropical upland and lowland rice varieties. *Plant Production Science* **3**, 437–445.

Kondo, M., Pablico, P.P., Aragones, D.V., Agbisit, R., Abe, J., Morita, S. and Courtois, B. (2003) Genotypic and environmental variations in root morphology in rice genotypes under upland field conditions. *Plant and Soil* **255**, 189–200.

Lafitte, H.R., Champoux, M.C., McLaren, G. and O'Toole, J.C. (2001) Rice root morphological traits are related to isozyme group and adaptation. *Field Crops Research* **71**, 57–70.

Liao, H., Yan, X., Rubio, G., Beebe, S.E., Blair, M.W. and Lynch, J.P. (2004a) Genetic mapping of basal root gravitropism and phosphorus acquisition efficiency in common bean. *Functional Plant Biology* **31**, 959–970.

Liao, M., Fillery, I.R.P. and Palta, J.A. (2004b) Early vigorous growth is a major factor influencing nitrogen uptake in wheat. *Functional Plant Biology* **31**, 121–129.

Lupton, F.G.H., Oliver, R.H., Ellis, F.B., Barnes, B.T., Howse, K.R., Welbank, P.J. and Taylor, P.J. (1974) Root and shoot growth of semi-dwarf and taller winter wheats. *Annals of Applied Biology* **77**, 129–144.

Lynch, J.P. (1995) Root architecture and plant productivity. *Plant Physiology* **109**, 7–13.

Lynch, J.P. and van Beem, J.J. (1993) Growth and architecture of seedling roots of common bean genotypes. *Crop Science* **33**, 1253–1257.

Ma, J.F., Ryan, P.R. and Delhaize, E. (2001) Aluminium tolerance in plants and the complexing role of organic acids. *Trends in Plant Science* **6**, 273–278.

McCaig, T.N. and Morgan, J.A. (1993) Root and shoot dry matter partitioning in near-isogenic wheat lines differing in height. *Canadian Journal of Plant Science* **73**, 679–689.

McMichael, B.L., Burke, J.J., Berlin, J.D., Hatfield, J.L. and Quisenberry, J.E. (1985) Root vascular bundle arrangement among cotton strains and cultivars. *Environmental and Experimental Botany* **25**, 23–30.

Mambani, B. and Lal, R. (1983) Response of upland rice varieties to drought stress. III. Estimating root system configuration from soil moisture data. *Plant and Soil* **73**, 95–104.

Manske, G.G.B. and Vlek, P.L.G. (2002) Root architecture – wheat as a model plant. In: *Plant Roots: The Hidden Half* (eds Y. Waisel, A. Eshel and U. Kafkafi), 3rd edn, pp. 249–259. Marcel Dekker Inc., New York.

Mudge, S.R., Smith, F.W. and Richardson, A.E. (2003) Root-specific and phosphate-regulated expression of phytase under the control of a phosphate transport promoter enables *Arabidopsis* to grow on phytate as a sole P source. *Plant Science* **165**, 871–878.

Nalborczyk, E. and Sowa, A. (2001) Physiology of rye. In: *Rye: Production, Chemistry and Technology* (ed. W. Bushuk), 2nd edn, pp. 53–68. American Association of Cereal Chemists Inc., St Paul, Minnesota, USA.

Nielsen, N.E. and Barber, S.A. (1978) Differences among genotypes of corn in the kinetics of P uptake. *Agronomy Journal* **70**, 695–698.

O'Toole, J.C. and Bland, W.L. (1987) Genotypic variation in crop plant root systems. *Advances in Agronomy* **41**, 91–145.

O'Toole, J.C. and Soemartono (1981) Evaluation of a simple technique for characterizing rice root systems in relation to drought resistance. *Euphytica* **30**, 283–290.

Passioura, J.B. (1972) The effect of root geometry on the yield of wheat growing on stored water. *Australian Journal of Agricultural Research* **23**, 745–752.

Passioura, J.B. (1977) Grain yield, harvest index and water use of wheat. *Journal of the Australian Institute of Agricultural Science* **43**, 117–120.

Price, A.H., Steele, K.A., Gorham, J., Bridges, J.M., Moore, B.J., Evans, J.L., Richardson, P. and Jones, R.G.W. (2002a) Upland rice grown in soil-filled chambers and exposed to contrasting water-deficit regimes. I. Root distribution, water use and plant water status. *Field Crops Research* **76**, 11–24.

Price, A.H., Steele, K.A., Moore, B.J. and Lones, R.G.W. (2002b) Upland rice grown in soil-filled chambers and exposed to contrasting water-deficit regimes. II. Mapping quantitative trait loci for root morphology and distribution. *Field Crops Research* **76**, 25–43.

Rebetzke, G.J. and Richards, R.A. (1999) Genetic improvement of early vigour in wheat. *Australian Journal of Agricultural Research* **50**, 291–301.

Richards, R.A. (1992) The effect of dwarfing genes in spring wheat in dry environments. II. Growth, water use and water-use efficiency. *Australian Journal of Agricultural Research* **43**, 529–539.

Richards, R.A. (2005) Physiological traits used in the development of new cultivars for water-scarce environments. *Agricultural Water Management*. In press.

Richards, R.A. and Passioura, J.B. (1981a) Seminal root morphology and water use of wheat. I. Environmental effects. *Crop Science* **21**, 249–252.

Richards, R.A. and Passioura, J.B. (1981b) Seminal root morphology and water use of wheat. II. Genetic variation. *Crop Science* **21**, 253–255.

Richards, R.A. and Passioura, J.B. (1989) A breeding program to reduce the diameter of the major xylem vessel in the seminal roots of wheat and its effect on grain yield in rain-fed environments. *Australian Journal of Agricultural Research* **40**, 943–950.

Richardson, A.E., Hadobas, P.A. and Hayes, J.E. (2001) Extracellular secretion of Aspergillus phytase from Arabidopsis roots enables plants to obtain phosphorus from phytate. *The Plant Journal* **25**, 641–649.

Riede, C.R. and Anderson, J.A. (1996) Linkage of RFLP markers to an aluminium tolerance gene in wheat. *Crop Science* **36**, 905–909.

Rincon, C.A., Raper, C.D. and Patterson, R.P. (2003) Genotypic differences in root anatomy affecting water movement through roots of soybean. *International Journal of Plant Science* **164**, 543–551.

Rubio, G., Walk, T., Ge, Z., Yan, X., Liao, H. and Lynch, J.P. (2001) Root gravitropism and below-ground competition among neighbouring plants: a modelling approach. *Annals of Botany* **88**, 929–940.

Sasaki, T., Yamamoto, Y., Ezaki, B., Katsuhara, M., Ahn, S.J., Ryan, P.R., Delhaize, E. and Matsumoto, H. (2004) A wheat gene encoding an aluminium-activated malate. *The Plant Journal* **37**, 645–653.

Setter, T.L. and Waters, I. (2003) Review of prospects for germplasm improvement for waterlogging tolerance in wheat, barley and oats. *Plant and Soil* **253**, 1–34.

Setter, T.L., Burgess, P., Waters, I. and Kuo, J. (1999) Genetic diversity of barley and wheat for waterlogging tolerance in Western Australia. In: *9th Barley Technical Symposium*, pp. 2.17.1–2.17.7. Melbourne, Australia, September 1999.

Siddique, K.H.M., Belford, R.K. and Tennant, D. (1990) Root:shoot ratios of old and modern, tall and semi-dwarf wheats in a mediterranean environment. *Plant and Soil* **121**, 89–98.

Silva, I.R., Smyth, T.J., Raper, C.D., Carter, T.E. and Rufty, T.W. (2001) Differential aluminium tolerance in soybean: an evaluation of the role of organic acids. *Physiologia Plantarum* **112**, 200–210.

Sponchiado, B.N., White, J.W., Castillo, J.A. and Jones, P.G. (1989) Root growth of four common bean cultivars in relation to drought tolerance in environments with contrasting soil types. *Experimental Agriculture* **25**, 249–257.

Tuberosa, R., Sanguineti, M.C., Landi, P., Giuliani, M.M., Salvi, S. and Conti, S. (2002) Identification of QTLs for root characteristics in maize grown in hydroponics and analysis of their overlap with QTLs for grain yield in the field at two water regimes. *Plant Molecular Biology* **48**, 697–712.

van Herwaarden, A.F., Farquhar, G.D., Angus, J.F., Richards, R.A. and Howe, G.N. (1998) 'Haying-off', the negative grain yield response of dryland wheat to nitrogen fertiliser. I. Biomass, grain yield, and water use. *Australian Journal of Agricultural Research* **49**, 1067–1081.

Vance, C.P., Uhde-Stone, C. and Allan, D.L. (2003) Phosphorus acquisition and use: critical adaptations by plants for securing a nonrenewable resource. *New Phytologist* **157**, 423–447.

Villagarcia, M.R., Carter, T.E.J., Rufty, T.W., Niewoehner, A.S., Jennette, M.W. and Arrellano, C. (2001) Genotypic rankings for aluminium tolerance of soybean roots grown in hydroponics and sand culture. *Crop Science* **41**, 1499–1507.

Vincent, C. and Gregory, P.J. (1986) Differences in the growth and development of chickpea seedling roots (*Cicer arietinum*). *Experimental Agriculture* **22**, 233–242.

Wahbi, A. and Gregory, P.J. (1989) Genotypic differences in root and shoot growth of barley (*Hordeum vulgare*). II. Field studies of growth and water use of crops grown in northern Syria. *Experimental Agriculture* **25**, 389–399.

Wahbi, A. and Gregory, P.J. (1995) Growth and development of young roots of barley (*Hordeum vulgare* L.) genotypes. *Annals of Botany* **75**, 533–539.

White, J.W. and Castillo, J.A. (1989) Relative effect of root and shoot genotypes on yield of common bean under drought stress. *Crop Science* **29**, 360–362.

Wissuwa, M. (2003) How do plants achieve tolerance to phosphorus deficiency? Small causes with big effects. *Plant Physiology* **133**, 1947–1958.

Wissuwa, M. and Ae, N. (2001) Genotypic variation for tolerance to phosphorus deficiency in rice and the potential for its exploitation in rice improvement. *Plant Breeding* **120**, 43–48.

Wissuwa, M., Yano, M. and Ae, N. (1998) Mapping of QTLs for phosphorus-deficiency tolerance in rice (*Oryza sativa* L.). *Theoretical and Applied Genetics* **97**, 777–783.

Yoshida, S. and Hasegawa, S. (1982) The rice root system: its development and function. In: *Drought Resistance in Crops with Emphasis on Rice*, pp. 97–114. IRRI, Los Banos, The Philippines.

Yu, L.-X., Ray, J.D., O'Toole, J.C. and Nguyen, H.T. (1995) Use of wax-petrolatum layers for screening rice root penetration. *Crop Science* **35**, 684–687.

Zheng, B.S., Yang, L., Zhang, W.P., Mao, C.Z., Wu, Y.R., Yi, K.K., Liu, F.Y. and Wu, P. (2003) Mapping QTLs and candidate genes for rice root traits under different water-supply conditions and comparative analysis across three populations. *Theoretical and Applied Genetics* **107**, 1505–1515.

Zubaidi, A., McDonald, G.K. and Hollamby, G.J. (1999) Shoot growth, root growth and grain yield of bread and durum wheat in South Australia. *Australian Journal of Experimental Agriculture* **39,** 709–720.

Chapter 9

Root Systems as Management Tools

Major advances have been made in understanding how plant canopies intercept light and convert its energy to dry matter and yield, leading to better crop management and also better management of plants grown in mixed communities. Increasingly, the management of crops is concerned not only with production and its associated economic returns, but also with reducing the environmental impacts of production practices so that sustainability can be ensured. For example, the ability to specify optimal canopy sizes at particular stages of crop development provides a basis for targeting fertilizer and pesticide inputs and reducing wasteful, and potentially harmful, applications (see Sylvester-Bradley *et al.* [1997] for N fertilizer and Dimmock and Gooding [2002] for fungicides). Below ground, the interactions between root systems and resources are more difficult to specify, but interest in managing root systems to effect efficient capture of nutrients is mounting as pressures increase for more environmentally sustainable production systems. Hoad *et al.* (2001) demonstrated the difficulties of defining a single optimal root system for cereals because of the influences of soil type, cultural practices and weather. Nevertheless, some progress has been made in managing root systems to achieve particular ends, and this chapter provides a few case studies and examples of both past successes and future possibilities.

9.1 Optimal root systems and competition for resources

The issue of the size of root system necessary to take up resources has been around in the crop production literature for a long time. However, as shown in section 1.2.1 it is impossible to give a single answer because it depends on the size, architecture and activity of the roots as well as the behaviour of the particular resource under consideration in the soil. In general, a large, more intensely branched root system can extract the plant's requirements from a soil more efficiently than a smaller root system, but the optimal size for a particular resource varies so that there can appear to be an element of redundancy or overprovision in many systems if a mobile resource is used as the basis for comparison. For example, van Noordwijk (1983) calculated that a root length of 0.1–1.0 cm cm^{-3} throughout the upper 0.2 m soil layer would be sufficient to supply the N requirements of most crop plants, whereas a length of 1–10 cm cm^{-3} would be required for P. For water, the required root length was similar to that of nitrate when roots were in intimate contact with the soil but rose to 1–5 cm cm^{-3} if there was an appreciable soil/root contact resistance. Similarly, the 'required' root length varied depending on the potential rate of evaporation and the degree of water stress experienced by the plant, so that a range of root lengths from 0.5 to 10 cm cm^{-3} were optimal under different conditions. These values cover the range of root lengths commonly measured for a range of crops in the cultivated layer of many soils (section 3.3.1).

The area over which a plant takes up resources such as water and nutrients, or otherwise alters its environment, is considered as its zone of influence. Characterizing this zone is important because its size and shape determine the total resources available to an individual, and the overlapping of zones determines the probability of competition between neighbouring plants (Casper *et al.*, 2003). Bray (1954) was among the first to appreciate that the zones of influence, and hence competition, for neighbouring plants would depend on the mobility of the resource under consideration; zones of influence for mobile resources such as nitrate being much greater than those for immobile nutrients such as phosphate. In most ecosystems, resources are not uniformly distributed and patchiness is common. For example, Jackson and Caldwell (1993) measured the concentration of extractable soil phosphorus in a steppe dominated by sagebrush (*Artemisia tridentata*) and found that it varied by 40% over the 0.5 × 0.5 m site, and that there were distinct patches <10 cm in diameter of high concentration (39–31 mg kg^{-1}) surrounded by more extensive areas in which concentration was <26 mg kg^{-1}. Plants can respond to this heterogeneity of resource

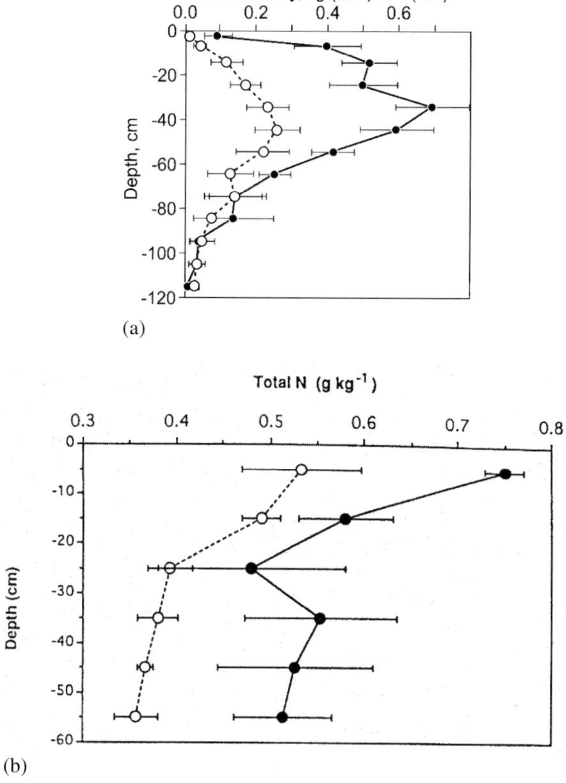

Fig. 9.1 The distributions with depth in the soil of (a) root mass, and (b) total soil N outside the canopy edge (○) and beneath canopies of clumps of trees (●) in a savanna dominated by the palm tree *Borassus aethiopum*. (Reproduced with permission from Modelet *et al.*, *Plant and Soil*; Springer Science and Business Media, 1996.)

distribution in two ways (Robinson, 1996). First, by stimulating nutrient uptake per unit of root (a quick and reversible response that is nutrient-specific and related to the extent of non-uniformity in nutrient availability); and second, by stimulating root growth or proliferation in the nutrient-rich zone (a slower and irreversible response that is generally less nutrient-specific). Robinson (1996) analysed several experiments including trees, shrubs and crops and found that the roots of plants locally supplied with nutrients increased their inflow (uptake per unit length of root) more as the proportion of roots with access to the nutrient decreased. This meant that the physiological response of stimulating inflow was a continuous, inverse function of the degree of localization, which provided a compensatory flux of nutrients to the plant. The degree of compensation was nutrient-specific, but for N, P and K, a threefold stimulation of inflow was typical. Root proliferation, however, appeared a less precise response with little if any nutrient specificity and an apparent overproduction of roots in the case of mobile nutrients. However, using a simple model of inter-specific competition, Robinson (2001) was able to show that for the same N capture, nitrate inflow must increase proportionately more than root length, so that a 10% increase in root length equated to a 20% to 20-fold increase in N inflow, depending on the initial conditions. If C is not the limiting factor, then, root proliferation is the better plant response to maximize a competitor's nitrate capture.

This strategy of root proliferation as a means of increasing competitive advantage is seen in the soil exploration patterns (root foraging strategies) of many species. For example, Mordelet *et al.* (1996) measured the distribution of roots and of N in the patchy savanna of Côte d'Ivoire dominated by the palm tree *Borassus aethiopum* (Mart.). Root mass and total N concentration were significantly greater under clumps of trees (and termite mounds) than outside the same clumps and mounds (Fig. 9.1). Palm trees extended their roots as far as 20 m towards the nutrient-rich patches where they proliferated. This foraging strategy of root proliferation under tree clumps or termite mounds results in both a large area explored and efficient resource exploitation, because high root lengths only occur in nutrient-rich patches.

9.2 Intercropping and agroforestry

A major aim of intercropping and agroforestry practices is to increase the overall productivity of land and/or its sustainability by making best use of the environmental resources (light, water and nutrients) used by plants for growth. In theory, there are several means of achieving these ends including: (i) minimizing the amount of below-ground competition by separating the root systems either in space (spatial complementarity) or time (temporal complementarity); (ii) one component making available the other resources that would otherwise be unavailable; and (iii) utilizing resources that would otherwise be lost from the system. In practice, below-ground interactions are complex and difficult to measure, so that progress in designing improved and sustainable systems of production has been slow, although some general principles are beginning to emerge.

A key issue in crop/crop, tree/crop or tree/pasture systems is the amount of competition or, conversely, complementarity that exists in the distribution of roots and in the activities of the root systems (van Noordwijk *et al.*, 1996; Willey, 1996). Temporal complementarity is the best documented mechanism for increasing yields in intercropping

(Willey, 1996). For example, in the sorghum/pigeonpea system, the two crops have maturity periods of typically 100 and 180–200 days, so that the major demands for resources such as light, water and nutrients differ in time. The result is that the intercrops make better use of resources over time than the two sole crops. This is easily demonstrated for light (a resource that must be instantaneously intercepted if it is to be utilized) where appreciably greater interception of radiation has been measured in many temporal systems (Natarajan and Willey, 1980). The advantage is more difficult to assess below ground because of the time-consuming and destructive measurements involved. Katayama *et al.* (1996a) measured the root growth of four crops (sorghum, pearl millet, groundnut and cowpea) grown either as sole crops or intercropped with pigeonpea. The total root length and the distribution of root length in the profile for the two cereals were unaffected by the cropping system, whereas those for the legumes were significantly reduced by intercropping. The largest reduction in length was for pigeonpea in all crop combinations, especially before harvest of the companion crops; root growth of pigeonpea improved after harvest of the companion crops to reach similar values to the sole crops at maturity. Katayama *et al.* (1996a) concluded that the temporal separation of the root systems in these pigeonpea-based intercropping systems would be advantageous in sharing limited resources such as soil N. This was demonstrated by measurements of the N balance, which showed no significant difference in total N content of sole or intercrops (except groundnut which was higher in the sole crop), but that intercropped pigeonpea acquired less N from both soil and fertilizer than sole pigeonpea crops (Table 9.1) (Katayama *et al.*, 1996b). Correspondingly, the intercropped pigeonpea derived a greater proportion of its total N uptake from biological N fixation.

Spatial complementarity of resource use has been found in canopies. For example, Reddy and Willey (1981) showed that an intercrop of pearl millet (canopy height 1.5 m) and groundnut (0.3–0.4 m) improved the efficiency of light conversion to dry matter by >20%, resulting in a yield increase of similar size. A similar 'two-tier' effect for root systems in which a shallow-rooted species is combined with a deep-rooted species might provide similar benefits in an intercrop if the deep-rooted species did not also explore the upper profile. In practice, this combination has proved impossible to find (Willey, 1996). Nevertheless, root lengths of intercropped systems are often greater than those of sole crops (for pearl millet/groundnut see Gregory and Reddy, 1982) resulting in an overall enhancement of resource use. In legume/non-legume systems, this may arise from a specific complementarity of resource use in which the legume meets some of its N requirement by fixation, allowing the non-legume to use more soil N and resulting in a greater combined N content of the crops (Reddy and Willey, 1981). Furthermore, there may be transfer of fixed N from the legume to the non-legume, also enhancing the overall growth of the intercrop.

Cannell *et al.* (1996) suggested that agroforestry systems may be more productive than sole crop systems if the trees are able to access resources that are under-utilized by crops. Where trees and crops are grown together (i.e. simultaneously), it has been suggested that exploiting the different rooting depths of trees and crops might increase resource capture without introducing severe below-ground competition. Ideally, to minimize competition between the tree and the crop, the tree should have a deep root system with little root proliferation near the top of the soil profile, thereby allowing the crop to utilize resources at

Table 9.1 A comparison of total N content, and the contribution of different sources of N utilized by sole and intercropped crops of pigeonpea, sorghum, pearl millet, groundnut and cowpea

Crop	Total N content (kg ha^{-1})	N from fertilizer (%)	N from soil (%)	N fixed (%)
Pigeonpea				
Sole	122.8	3.9	65.7	30.4
Intercropped with sorghum	89.0	1.6	39.2	59.2
Intercropped with pearl millet	97.4	1.6	25.3	73.1
Intercropped with groundnut	100.2	1.5	23.1	75.4
Intercropped with cowpea	113.1	1.9	41.9	56.2
SE	18.0 (NS)	0.3*	9.6*	9.6*
Sorghum				
Sole	85.3	8.2	91.8	
Intercropped with pigeonpea	84.8	10.3	89.7	
SE	8.9 (NS)	2.8 (NS)	2.9 (NS)	
Pearl millet				
Sole	57.6	11.0	89.0	
Intercropped with pigeonpea	52.9	10.7	89.3	
SE	16.5 (NS)	3.8 (NS)	3.8 (NS)	
Groundnut				
Sole	156.0	4.3	49.4	46.3
Intercropped with pigeonpea	109.1	3.6	26.9	69.5
SE	4.2*	0.6 (NS)	13.7 (NS)	13.4 (NS)
Cowpea				
Sole	93.6	5.4	37.3	57.3
Intercropped with pigeonpea	87.3	5.3	25.6	69.1
SE	9.7 (NS)	1.9 (NS)	9.4 (NS)	7.7 (NS)

SE, standard error of the mean; NS, not significant. *$p<0.05$. From Katayama *et al.*, 1996b.

the top of the profile while the tree accesses resources in deeper layers (Schroth, 1995). However, several studies have shown that achieving such a desirable outcome is difficult to realize, not least because in many parts of the world soils are shallow and tree root systems occupy much the same soil as crop roots. For example, Smith *et al.* (1999) grew *Grevillea robusta* trees that were 4–6 years old and maize on a sandy clay loam (alfisol) overlying hard gneiss at depths from 0.2 to 2 m at Machakos, Kenya and found that root populations in mixed plots were dominated by *G. robusta* at all times, all depths and all distances from trees and maize (Fig. 9.2). Mean root biomass at anthesis of maize was 53 g m^{-2} for sole maize, 410 g m^{-2} for trees only, and 447 g m^{-2} for the tree plus crop system of which only 23 g m^{-2} was maize roots. The lack of spatial separation of the root systems meant that there was competition between them for resources and the impact of the trees on the recharge of the profile during the rainy season was marked. Recharge of lower soil layers was much reduced in the mixed plots relative to the maize monocrop, with the high density of tree roots in the upper profile capturing more of the water that fell as rain. Smith *et al.* (1999) concluded that spatial separation of tree and crop root systems may be unattainable so that

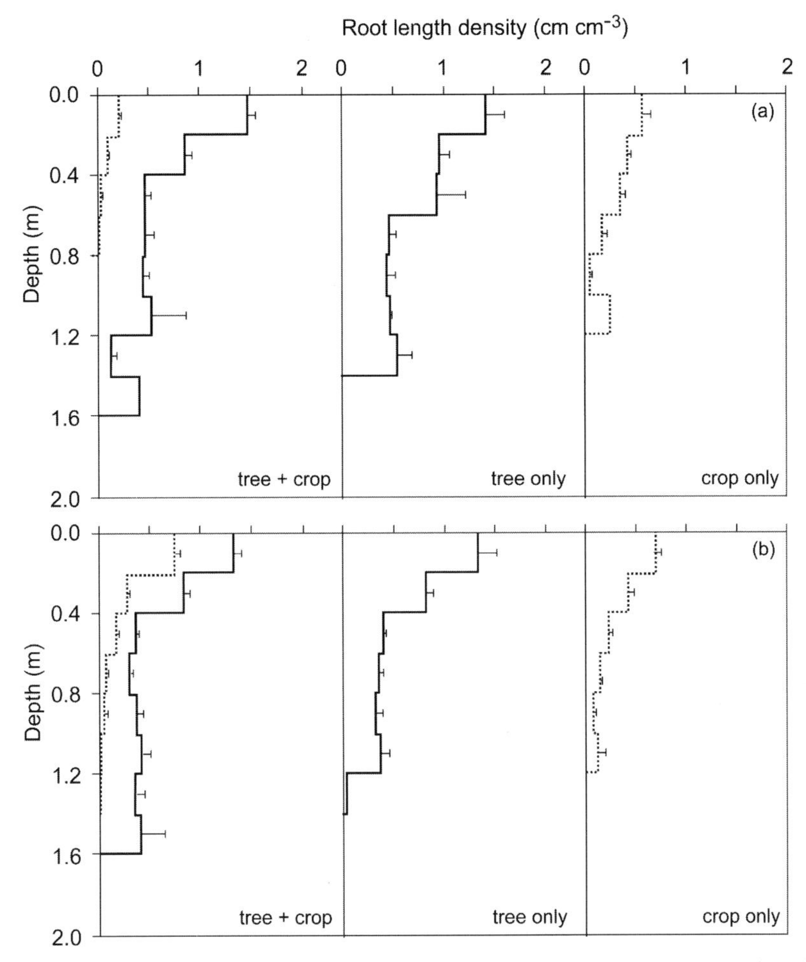

Fig. 9.2 Profiles of root length for *Grevillea robusta* (solid lines) and maize (dotted lines) at the time of maize anthesis during two growing seasons (a – the short rains; and b – the subsequent long rains) at Machakos, Kenya. Differences in the maximum depths of tree roots between treatments were a consequence of variations between plots in the depth of the underlying bedrock. Error bars are +1 standard error of the mean. (Reproduced with permission from Smith *et al*., *Plant and Soil*; Springer Science and Business Media, 1999.)

complementarity of water use could not be expected just because maximum rooting depths are greater for trees. In mixtures, complementarity would only result if there were an alternative source of water (e.g. groundwater) only accessible to the trees.

Although there appears to be very little experimental evidence to support the spatial separation of tree and crop roots, tree roots do not always dominate crop roots. In the same area of semi-arid Kenya as Smith *et al.*'s study, Odhiambo *et al.* (2001) found that while *Grevillea robusta* and *Gliricidia sepium* roots dominated populations grown with maize at the start of the growing season, tree roots died and maize root grew during the cropping season, so that amounts of tree and crop roots were similar at the end of the season. Maximum root lengths for both trees and maize were in the top 0.2 m of the soil profile and maize yields were reduced by about 50% with *Gliricidia* and by about 40% with *Grevillea* compared with maize plots without trees (however, yields were very variable and differences were significant at $p = 0.10$). There was, then, no spatial separation of the rooting zones of the trees and crops but there was a temporal separation of root activity between species. Plots with trees had less soil water than cropped plots, especially at the end of the dry season, and there was a difference between the two tree species with plots of *Gliricidia* drier than those of *Grevillea*.

Interactions of trees and crops are likely to differ depending on the tree species because trees are managed in different ways. Livesley *et al.* (2000, 2002, 2004) investigated the interactions between 3-year-old *Grevillea robusta* (managed as a tree row with lower branches removed annually to reduce shading) and *Senna spectabilis* (managed as a hedgerow that was pruned to a 20 cm high stump before each cropping season) trees with maize in the sub-humid highlands of western Kenya on a deep, very fine kaolinitic oxisol. Tree roots were sampled to 3.0 m at the start and end of the cropping season with maize roots sampled to 1.8 m depth at three developmental stages. As in Odhiambo *et al.*'s study, the length of fine tree roots decreased during the cropping season by about 35% for *Grevillea* and by 65% for *Senna*; the length of *Senna* decreased at all depths during the cropping season while that of *Grevillea* only decreased in the crop rooting zone (Fig. 9.3). The decrease in root length occurred because of a combination of competition with maize roots, manual weeding with hoes, seasonal senescence and, in the case of *Senna*, pruning regime. As in the previously mentioned studies, the trees depleted soil water in the dry season and intercepted rain in the wet season so that the amount of water stored in the cropping season was greater under sole maize than under the agroforestry systems. Soil water content at the end of the cropping season was similar to that at the start of the season in the *Grevillea*–maize system, but about 50 mm greater in the *Senna*–maize system and 80 mm greater in the sole maize system (Livesley *et al.*, 2004). The distribution of soil water in the profile also differed between the tree species. Soil water content beneath the *Grevillea*–maize system increased with increasing distance from the tree row (probably because of preferential uptake beneath the canopy and interception of rain by the canopy), while in the *Senna*–maize system it was high near the tree row and decreased between 0.75 and 3.0 m from the hedgerow. There were differences, too, between the systems in the dynamics and uptake of soil inorganic nitrogen. During the cropping season, nitrate plus ammonium decreased by 94 kg N ha^{-1} in the *Senna*–maize system and by 33 kg N ha^{-1} in the *Grevillea*–maize system, with N uptake by the trees of 39 and 23 kg N ha^{-1}, respectively, and N uptake by the maize of 65 and 85 kg N ha^{-1}, respectively. Differences in the availability and dynamics of inorganic N were primarily determined by the amount of N removed through the pruning management of the canopies (Livesley *et al.*, 2002).

The scope for achieving spatial complementarity between tree and crop root systems appears, then, to be very limited, but there are ways of reducing below-ground competition in simultaneous tree–crop systems by, for example, crown and root pruning (Schroth, 1999; Ong *et al.*, 2002). Crown pruning, pollarding and removal of selected branches may all improve timber quality (depending on the species) and reduce competition with crops. Physical root barriers have rarely proved effective in reducing below-ground competition for more than a season or so (see Ong *et al.*, 2002, for a summary) but root pruning can be effective in alley cropped systems if labour is available.

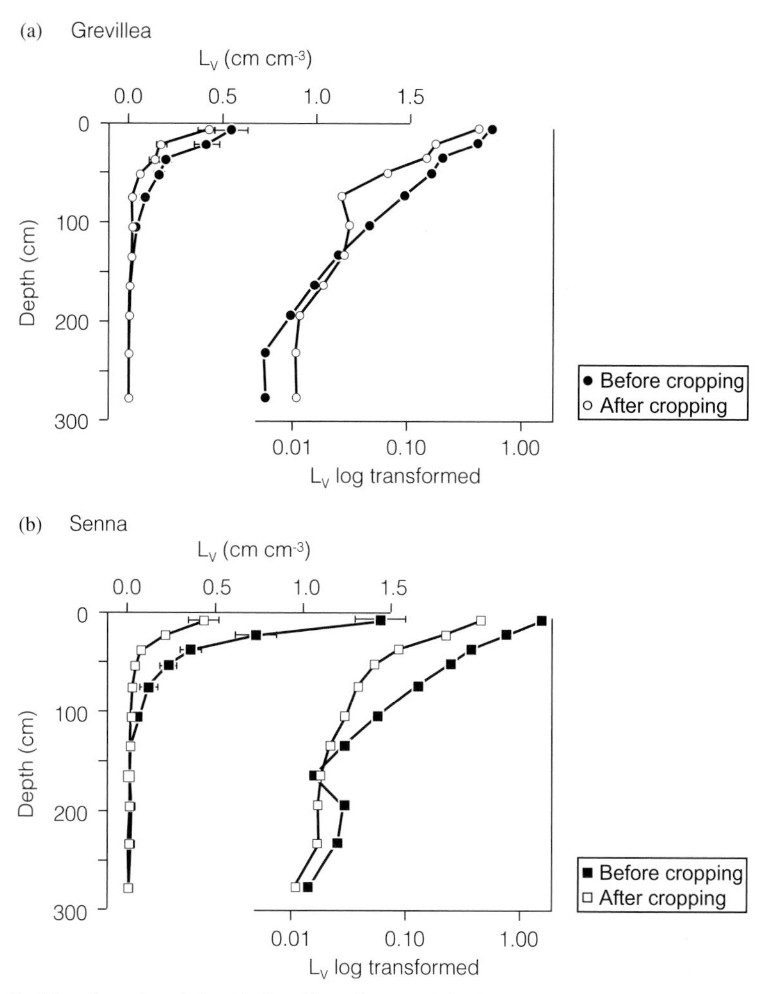

Fig. 9.3 Profiles of root length for (a) *Grevillea robusta* and (b) *Senna spectabilis* measured before and after the maize growing season in western Kenya. (Reproduced with permission from Livesley *et al.*, *Plant and Soil*; Springer Science and Business Media, 2000.)

In many cropping systems, the supply of nutrients such as N mineralized from organic matter is rarely completely synchronized with the nutrient demand of the crop. If supply exceeds demand then temporary storage is required in the rooting zone. In seasonally dry regions, there is frequently a burst of N mineralization coincident with the first rains (Birch, 1960; Cui and Caldwell, 1997) and because rainfall frequently exceeds evaporation during the early part of the season, then mineralized N may be leached to deeper soil layers. Generally the rooting depth of annual crops is shallower than that of perennial trees, and it is not uncommon for nutrients to be leached beyond the crop roots (Fig. 9.4). In such circumstances, deeper-rooted components of the mixed system can then act as a 'safety net' intercepting N moving deeper into the soil (van Noordwijk *et al.*, 1996). The importance of this loss of mineralized N from cropping systems was assessed by Shepherd *et al.* (2000) in a study of 14 smallholder farms in the sub-humid highlands of Kenya where there is little

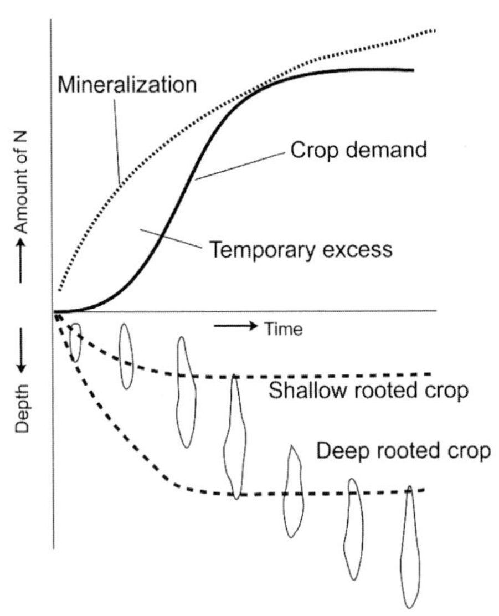

Fig. 9.4 A diagrammatic representation of the synchrony hypothesis. The time course of nitrogen mineralization and crop demand frequently do not match, leaving a temporary stock of mineral nitrogen that can be leached to deeper layers out of the reach of shallow-rooted crops. (Reproduced with permission from van Noordwijk *et al.*, *Tree-Crop Interactions*; CAB International, Wallingford, UK, 1996.)

use of fertilizers. Measurement of nitrate-N in the soils beneath different types of vegetation showed that subsoil (0.5–4 m depth) concentrations were low (<0.2 mg kg^{-1}) beneath hedgerows and woodlots, intermediate beneath weed fallows, banana and markhamia trees (*Markhamia lutea*), but high beneath both poor- and good-looking maize (1–3 mg kg^{-1}) (Fig 9.5). Subsoil nitrate was only 2 kg N ha^{-1} beneath woodlots and hedgerows but 27 (good-) to 37 (poor-looking) kg N ha^{-1} beneath maize.

Several studies have demonstrated that deep-rooted plants can effectively capture nitrate leached below the root zone of crops. For example, Jama *et al.* (1998) demonstrated that the fast-growing trees sesbania (*Sesbania sesban* L.) and calliandra (*Calliandra calothyrsus*) reduced soil nitrate in the upper 2 m of a kaolinitic oxisol in a humid region of Kenya by 150 and 200 kg N ha^{-1}, respectively, 11 months after establishment; this recovered nitrogen can be used in subsequent crops by applications of leaf litter and other tree residues (biomass transfer). Similarly, tree fallows using coppiced tree legumes sequentially with crops in Zambia have been found to retrieve subsoil nitrogen with benefits to the subsequent crops (Chintu *et al.*, 2004). Chintu *et al.* (2004) compared four nitrogen-fixing trees (*Gliricidia sepium*, *Leucaena leucocephala*, *Acacia angustissima* and *Sesbania sesban*) with natural fallows and continuous maize and found that there was more inorganic N in the topsoil of coppiced fallows than in unfertilized maize plots and that there was less subsoil nitrate. Grain yield of continuous maize (eighth successive year) was 0.90 t ha^{-1} with yields following 2–3-year-long coppiced tree fallows of about 3.3 t ha^{-1}.

The safety net aspect of tree root systems has also been evident in hedgerow and alley cropping systems. Rowe *et al.* (1999) placed ^{15}N at various depths in the soil between mixed hedgerows of *Gliricidia sepium* (shallow-rooted) and *Peltophorum dasyrrhachis*

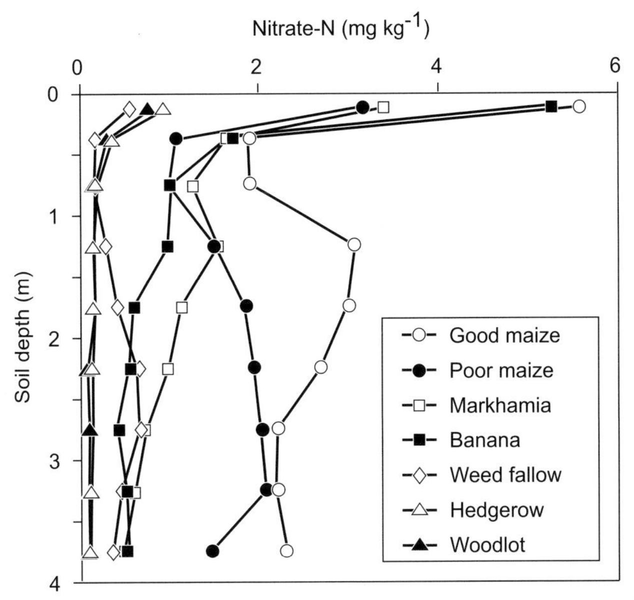

Fig. 9.5 Nitrate concentrations in soils at the time of maize harvesting under seven vegetation types on smallholder farms in western Kenya. (Reproduced with permission from Shepherd *et al.*, *Biology and Fertility of Soils*; Springer Science and Business Media, 2000.)

(deep-rooted) grown on an ultisol in North Lampung, Indonesia. *Gliricidia* roots shared much the same soil volume as the roots of the groundnut crop but there was no evidence of strong competition with the crop for topsoil N, in part because of the large proportion (0.44–0.58) of N obtained from biological fixation. In contrast *Peltophorum* took up a large proportion (0.42) of its N from soil beneath the main crop rooting zone and acted in the role of a safety net. When maize was cropped between the hedgerows, the vertical root distributions were closely related to the recovery of ^{15}N, with more recovered by maize and *Gliricidia* from placement at 0.05 m than from placements at 0.45 or 0.65 m (Rowe *et al.*, 2001). *Peltophorum* recovered similar amounts of ^{15}N from all three depths. Similar positive effects of tree roots in reducing N leaching have been determined in the temperate alley cropping system of pecan (*Carya illinoensis*) trees with cotton in north-western Florida (Allen *et al.*, 2004).

Models of root growth and resource capture by components of agroforestry systems demonstrate the potential for intercepting nutrients by trees and thereby improving the nutrient use efficiency at system level by closing the nutrient cycle. For example, the WaNuLCAS model allows the spatial and temporal dynamics of water, nutrients and light to be investigated in a range of systems (van Noordwijk and Lusiana, 1999). It can be used, too, to estimate the root length of tree required in the subsoil to function as an efficient safety net. Simulations using the model in Lampung, Indonesia showed that assuming that there is a demand by the tree for the N which might be taken up, the effectiveness of the root system in a deep soil layer (0.5–1.0 m) in filtering out any leached N increases above that of the crop when tree root length exceeds 0.001 cm cm^{-3}, and reaches a maximum at 1 cm cm^{-3} (Fig. 9.6a). Tree biomass benefited from the deeper roots as root length increased to 0.03 cm cm^{-3}, a root length apparently sufficient to ensure the survival of the tree in the dry season (Fig. 9.6b) (van Noordwijk and Cadisch, 2002).

9.3 Crop rotations

In crop production systems, it is usual to rotate crops on the same piece of land so that weeds and diseases do not persist, and nutritional differences between crops are exploited. A range of functions has been suggested for roots to maximize the performance of crops in different situations and especially on soils that are normally inhospitable to crop roots.

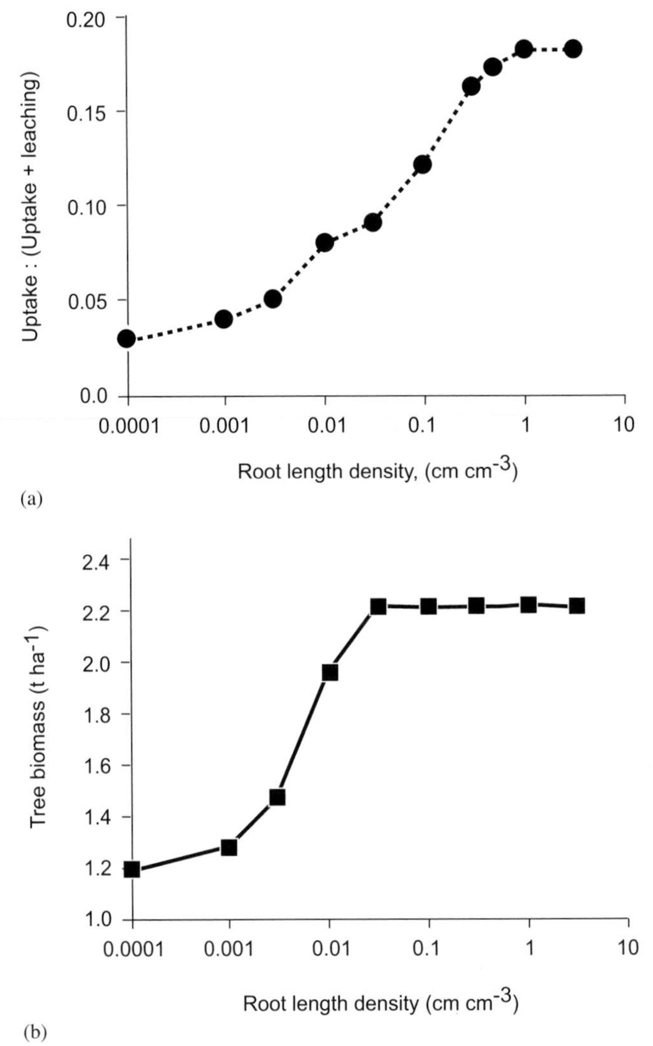

Fig. 9.6 Model predictions of the relations between (a) the ratio of N uptake to uptake plus leaching, and (b) tree biomass, with root length at 0.5 m depth. The simulation shows that the capture of N and tree biomass increased as root length density increased to about 0.1 cm cm^{-3}. (Reproduced with permission from van Noordwijk and Cadisch, *Plant and Soil*; Springer Science and Business Media, 2002.)

This section considers some of the management interventions available via the rooting properties of particular plants.

9.3.1 Biological drilling

In some parts of the world, soils may be deep but their subsoils dense resulting in a physical impediment to the downward growth of root systems of annual crops. Such soils are often associated with long periods of weathering leading to clay subsoils; in Australia they are called duplex soils and pose a range of problems for farmers (Tennant *et al.*, 1992). Mechanical disruption of the subsoil by deep ripping can reduce the physical constraints imposed by such soils but the process is often expensive and the benefits may be short-lived. An option that has been proposed is to grow plants with roots that can penetrate the subsoil to create stable pores which, following decay of the roots, leave open biopores that can be exploited by the roots of subsequent crops (Plate 9.1). This biological drilling, through incorporation of appropriate species in the rotation, might be economically advantageous.

Cresswell and Kirkegaard (1995) summarized results from several studies in eastern Australia and found no evidence for stable biopore formation in annual crop rotations, concluding 'that tap rooted annual crops such as lupins or canola are unlikely to be able to improve B-horizon porosity in dense, duplex soil'. Similarly Gregory (1998) found only sparse rooting for a range of annual oilseed and legume crops in the subsoil of a duplex soil in Western Australia; only 1.3% of the total root length was present below 0.4 m in one season and 3.8% the next. Cresswell and Kirkegaard (1995) speculated, however, that deep-rooted perennial species such as lucerne might be more successful in performing this function. The ability of lucerne roots to create stable macropores for use by subsequent crops was noted by Rasse and Smucker (1998) who grew crops on a fine-loamy alfisol in lysimeters; while the subsoil was a clay loam, it was not especially hostile for roots. Maize grown after a 1-year crop of alfalfa had patterns of root distribution with depth similar to those of the previous lucerne crop, whereas successive maize crops had root systems with different patterns of distribution with depth from one year to the next. The influence of the lucerne on the distribution of roots was partially via the macropores induced by the roots. The proportion of roots recolonizing macropores induced by roots of the previous crop averaged only 18% for maize following maize, and 22% for lucerne following maize, but 41% for maize following lucerne.

McCallum *et al.* (2004) demonstrated the formation of stable biopores by perennial pastures on a duplex soil with a dense subsoil (starting at 0.1–0.15 m depth) in New South Wales, Australia. After a period of 10 years of phalaris grass and 4 years of lucerne, both of the perennial pasture species had considerably more pores >2 mm diameter at the top of the B horizon than the annual control (canola, wheat, canola in the three previous seasons) (Table 9.2), although the phalaris result was not significantly different from the control at $p = 0.05$. Moreover, the time taken to pond water on the top of the B horizon and the rate of water infiltration into the B horizon were significantly greater following the perennial species. Other data showed that the total number of biopores at the top of the B horizon (170–180 m^{-2}) after lucerne remained unchanged for 2 years of wheat/canola cropping, although the mean diameter of the channels formed by lucerne tap roots declined steadily from 8.0 to 5.7 to 3.7 mm over a 3-year cropping cycle. These results, then, clearly indicate

Table 9.2 Effect of annual crops, phalaris grass and lucerne on soil properties at the top of the B horizon (100–120 mm) on a duplex soil near Temora, Australia

Parameter	Annual crop	Phalaris	Lucerne	LSD (p = 0.05)
Bulk density (mg m^{-3})	1.66	1.58	1.64	0.05
Pores >2 mm m^{-2}	68	190	228	145
Number of lucerne tap root channels m^{-2}	3	0	53	14
Mean diameter of lucerne channels (mm)	–	–	8	–
Time to ponding (min)	6	19	61	10
Steady-state infiltration (mm h^{-1})	2.3	6.8	9.3	0.6

From McCallum *et al.*, 2004.

the ability of perennial species such as lucerne to create stable macropores that could be used by subsequent annual crops. Unfortunately the root distributions of the crops following lucerne were not measured in these studies so it is impossible to state unequivocally that the macropores created by the lucerne were actually occupied. Ward *et al.* (2002) working on a duplex soil near Katanning in Western Australia found no major difference in the distribution of root dry weight with depth in canola crops following either subterranean clover or lucerne although root dry weight between 0.45 and 0.65 m was greater following clover. These contrasting studies show that the value of biopores in terms of subsoil water use and yield benefits to subsequent crops has yet to be demonstrated. In the study by McCallum *et al.* (2004), there was no winter waterlogging or water stress during grain-filling, so the additional subsoil pores resulted in increased subsoil water use in only 2 out of 12 annual crops studied, and increased grain yield of canola in only one of those seasons. Thus the benefits of increasing subsoil porosity and subsoil rooting by biological drilling may be very dependent on seasonal weather conditions.

9.3.2 Utilization of subsoil water

The downward growth of roots into the soil profile is a major means of securing an uninterrupted supply of water should rainfall cease during crop growth and, indeed, most crops rely on water stored in the soil for at least part of their growing season. Attempts to increase the depth of rooting through breeding and varietal selection have been detailed in Chapter 8, but agronomic means have also been attempted with moderate success. Deeper roots will only be beneficial if the depth of wetting is greater than the usual depth of rooting so that in northern Syria, for example, where wetting is generally limited to the upper 0.8–1.0 m, deeper roots in cereals such as wheat and barley would offer no advantage (Cooper *et al.*, 1987). In the wheatbelt of Western Australia, Asseng *et al.* (2002; cited by Turner, 2004) showed by simulation analysis that deeper roots gave the greatest benefit on sandy soils in high-rainfall zones where water and nitrogen would otherwise pass below the root zone. On clay soils, though, there was little or no advantage. Despite the often suggested benefits of deeper rooting or a different distribution of root length in the soil profile, there is little information about the additional resources that would then be acquired and the value to the crop. Typically, the maximum value of wheat grain produced is 20–25 kg ha^{-1} mm^{-1} but it is not known whether such efficiency is achieved by water extracted by deep roots. King *et al.* (2003) attempted to evaluate the production and economic benefits of deeper rooting and/or a greater proportion of roots deeper in the profile using a quantitative model. Based on the growth and performance of winter wheat crops in the UK, the model demonstrated that a larger investment in fine roots at depth, and less proliferation in surface layers, would improve yields by accessing extra resources. Their economic analysis indicated that the value of additional water captured was about twice that of the additional N captured, but the model assumed an even distribution of resources through the profile; a condition likely to be met for water but not for N.

Some agronomic practices have little effect on rooting depth or the use of subsoil water. For example, Gregory and Eastham (1996) on a duplex soil in Western Australia measured an average rate of downward extension of 5.2 mm d^{-1} for narrow-leafed lupin and 8.7 mm d^{-1} for wheat in three seasons with a maximum rooting depth of 0.8 m for both crops. These values were unaffected by differences in sowing date of 3–6 weeks. In such environments, the principal effect of agronomic management on water use has been to increase the proportion of water transpired and to reduce other pathways of loss such as evaporation from the soil surface rather than to increase the total evaporation (Eastham and Gregory, 2000; Asseng *et al.*, 2001).

In several parts of the world, replacement of the native vegetation with relatively shallow-rooted annual crops and pastures has led to significant changes in the water balance with greater drainage resulting in higher water-tables and greater runoff. This problem is especially acute in southern Australia where salt stored in the profile is carried to the surface with the rising water-table resulting in secondary salinization of the land. Potential solutions to this problem are to re-introduce woody perennials into the landscape to capture and use the water that would otherwise drain (a safety net – see previous section) or to introduce perennial pasture plants into the production system. Lucerne, with its ability to penetrate hard subsoils (see previous section), is known to extract water from deeper in the soil profile and appears to use at least 50 mm more water than annual pastures at several sites in Australia (Angus *et al.*, 2001). Over a 3-year period of pasture, Ward *et al.* (2002) showed that lucerne depleted soil water content during summer and autumn by about 60 mm more than the annual pasture dominated by subterranean clover (Fig. 9.7). This water was drawn mainly from depths >0.6 m (water uptake by lucerne extended to 1.7 m) and although the difference was reduced during winter, sometimes to zero, the difference was re-established during summer. In the subsequent cropping phase, while soil previously under lucerne was much drier than soil under clover when wheat was sown, this difference disappeared by late winter as the profile re-wet. During the spring drying phase, wheat after lucerne extracted more water from the subsoil than wheat after clover especially at depths >0.7 m. When canola was grown in the next season, there was no difference to 1.2 m in soil water content between crops grown after lucerne or clover, suggesting that the root channels created by the lucerne had closed (Fig. 9.7). Averaged over the 5-year rotation, annual water drainage below 1.9 m was 45 mm for the clover-wheat-canola system and 17 mm for the lucerne-wheat-canola system – a 60% decrease in potential groundwater recharge. The dry soil buffer zone created by the lucerne persisted for the wheat crop but drainage occurred from this soil before the buffer zone was fully re-wetted, suggesting that bypass flow may occur through the larger pores. When lucerne is fully established (i.e. in the second season of growth and onwards), its use of water can be as great, if not greater at particular times, as the native perennial woody vegetation. On a deep sand, lucerne dried

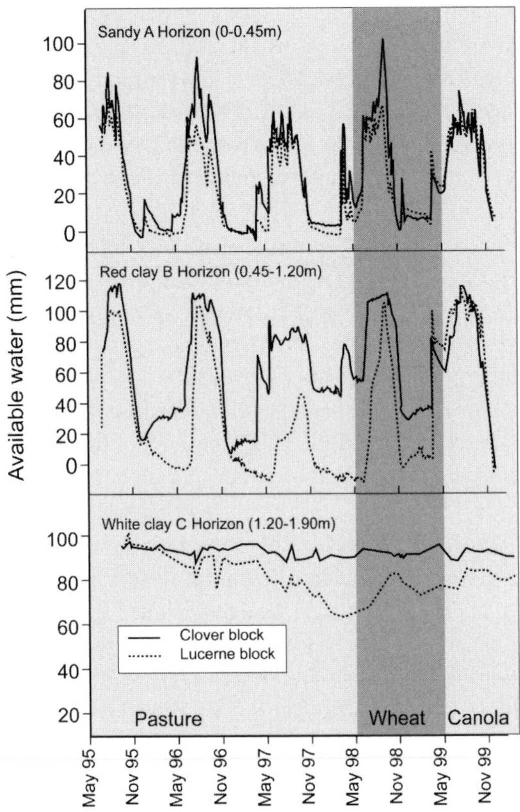

Fig. 9.7 Seasonal changes in the available water stored in the horizons of a soil on which either subterranean clover (solid line) or lucerne (dotted line) were grown during the pasture phase of the rotation. (Reproduced with permission from Ward *et al.*, *Agricultural Water Management*; Elsevier, 2002.)

the soil to the same extent as *Banksia* woodland vegetation and drainage from both vegetation types to soil layers deeper than 4.5 m and groundwater was recovered in a subsequent dry growing season (Ward *et al.*, 2003).

9.3.3 Allelopathy

A wide range of chemicals is released from roots with diverse effects on microbial populations, plant pathogens and, sometimes, on roots of other species. The phenomenon of allelopathy refers to the chemical interactions that occur among plants mediated by the release of chemicals into the soil (Bertin *et al.*, 2003). The word is applied to a wide range of interactions with the chemicals involved leached from leaves and other tissues, released from decomposing plant materials and exuded from roots.

Some plants are well known for reducing the growth of neighbouring competitors, and chemicals released from roots have been identified in several species. To date, only a few crops (mostly cereals) have been screened for allelopathic potential and, in particular, for their ability to suppress weeds. Among crops, sorghum has long been known to suppress the growth of some weeds and, depending on conditions, succeeding crops such as maize. Water extracts of mature sorghum plants applied in field experiments were found to control from 18% to 50% of weeds and to increase maize yields by 11–44% compared with a no-spray control (Cheema *et al.*, 2004), although the effects on crop growth and yields can be variable. For example, Roth *et al.* (2000) found that sorghum residues tilled into soil delayed development of the subsequent wheat crop although grain yields were unaffected – probably because the allelopathic compounds degraded in the soil. Sorghum roots exude several biologically active compounds one of which – sorgoleone (2-hydroxy-5-methoxy-3-[(8'Z,11'Z)-8',11',14'-pentadecatriene]-*p*-benzoquinone) – is a potent bioherbicide that is inhibitory to broadleaf and grass weeds at concentrations as low as 10 μM in hydroponic assays (Nimbal *et al.*, 1996). The compound appears to be synthesized in the root hairs and then deposited between the cell wall and plasmalemma of root hairs from whence it is exuded (Plate. 9.2) (Weston and Duke, 2003). Its mode of action is as a biological inhibitor of oxygen evolution, inhibiting photosystem II electron transport, similar to that of diuron-type herbicides (Czarnota *et al.*, 2001).

Several varieties of rice (*Oryza sativa*) have also been found to inhibit the growth of other plants. Dilday *et al.* (1991; cited by Olofsdotter, 2001) found that rice was highly effective in the suppression of aquatic weeds including ducksalad (*Heteranthera limosa*) in Arkansas, and subsequent screening has found that about 3.5% of 12 000 accessions were effective against ducksalad. Similarly, in the Philippines, rice cultivars were evaluated for the suppression of barnyardgrass (*Echinochloa crus-galli*) in field and laboratory tests. There was considerable variation between genotypes, with rice height the most important factor influencing weed growth, but allelopathic potential as measured in laboratory screening followed closely (Olofsdotter *et al.*, 1999). Experiments conducted with the same genotypes in Korea confirmed these results, suggesting that the allelopathic effect was influenced more by genetics than environment. Together the results also show that weed suppression in the field is likely to be most effective when both competition and allelopathy are optimized (Olofsdotter, 2001). Rice root exudates and decomposing rice straw produce a number of secondary metabolites including phenolics, aromatic acids, benzene derivatives, long-chain hydrocarbons and fatty acids, and several sterols (Rimando and Duke, 2003) although few approach phytotoxic levels under normal growing conditions. Rimando and Duke (2003) consider that the most phytotoxic compounds identified to date in rice are the momilactones (tricarbocyclic diterpenes) extracted from rice husks. Kato-Noguchi (2004) found that momilactone B was released from rice roots into solutions from seedlings from which the husks were removed and suggested that this chemical may play an important role in rice allelopathy.

Among trees, the presence of black walnut (*Juglans nigra* L.) in landscapes has long been known to be detrimental to a wide range of vegetable and crop species including tomato, potato, pea, beans, maize, wheat and barley. The toxicity is associated with the presence of juglone (5-hydroxy-1,4-napthoquinone). In living tissues, juglone is present in a reduced, non-toxic form, but when released from a tissue it is oxidized and becomes toxic (Rietveld, 1983). The highest concentrations of juglone are produced and associated with actively growing roots and because it is sparingly soluble in water, it does not diffuse far away from roots and can persist beneath a tree canopy for some while. The build up of juglone in the soil means that it can be toxic to other plant roots for at least a year following the removal of the tree.

Beside their role in agriculture, allelochemicals released from roots may also play a role in the establishment of some invasive species. For example, the ability of diffuse knapweed (*Centaurea diffusa*) and spotted knapweed (*Centaurea maculosa*) to invade grasslands in the western United States compared to grasslands in Eurasia where they are native, appears to be due to the adverse effects of their root exudates on native North American species compared to Eurasian species (Callaway and Aschehoug, 2000). In experiments using activated carbon to adsorb or deactivate the root exudates, Callaway and Aschehoug (2000) demonstrated that the advantage of diffuse knapweed over others was, at least in part, due to chemical mediation. Similarly, Ridenour and Callaway (2001) found that root exudates from spotted knapweed inhibited growth of a native bunchgrass (*Festuca idahoensis*). Although the chemicals responsible for the inhibition were not characterized in this study, sesquiterpene lactones have been isolated from foliar parts of *Centaurea* spp. (Weston and Duke, 2003).

9.3.4 Biofumigation by brassicas

Several field studies have demonstrated that amendment of soils with green manures containing brassicas such as cabbage, kale and mustard reduces the severity of diseases such as Aphanomyces root rot of peas (Chan and Close, 1987; Muehlchen, 1990). This effect has been linked to the production of sulphur-containing volatiles such as isothiocyanates (ITCs) during the decomposition of the *Brassica* shoots in soils leading to suppression of the pathogen (Lewis and Papavizas, 1970). Field measurements in southern Australia following crops of Indian mustard (*B. juncea*) and canola (*B. napus*) found that shoot biomass of wheat at anthesis was increased by 29% (range 12–46% depending on site, season and crop) compared with a preceding wheat crop; root length in the upper 10 cm was also increased by about 40% (in five of the six cases studied) following the *Brassica* crops (Kirkegaard *et al.*, 1994). The reasons for the improved early growth following the *Brassica* crops were uncertain, but it was not related to amounts of soil mineral N, nor to the incidence of the diseases take-all and *Rhizoctonia solani*. It was possible, however, that other inhibitory organisms were present because treatment with methyl bromide increased wheat growth and the ranking of the crop effect was maintained with Indian mustard > canola > wheat.

In accompanying laboratory studies, Angus *et al.* (1994) demonstrated that the presence of root tissues of canola and Indian mustard inhibited growth of pure cultures of the take-all fungus, and identified volatile ITCs released from the roots. This biocidal action was referred to as biofumigation. Studies of *in vitro* toxicity with six ITCs on the mycelial growth of five cereal root pathogens found that aromatic ITCs were more toxic than aliphatic ITCs when dissolved in agar, and that the toxicity of aliphatic ITCs decreased with increasing length of the side chain (Table 9.3) (Sarwar *et al.*, 1998). The fungi differed in sensitivity to the ITCs with *Gaeumannomyces graminis* var. *tritici* (take-all) the most sensitive, *Rhizoctonia solani* and *Fusarium graminearum* intermediate, and *Bipolaris sorokiniana* and *Pythium irregulare* the least sensitive. Suppression of some fungi by the two main products of glucosinolate hydrolysis in *Brassica* tissues (propenyl ITC and 2-phenylethyl ITC) was better than that achieved by the synthetic fumigant methyl-ITC, suggesting an important role for these compounds in pathogen suppression. However, although subsequent field studies found significant reductions in the quantities of take-all fungus under

Table 9.3 Concentrations (μmol l^{-1} agar) of isothiocyanates (ITCs) dissolved in agar causing 50% growth suppression (SD_{50D}) and death (LD_D) of five cereal fungal pathogens

	2-Propenyl ITC		3-Butenyl ITC		4-Pentenyl ITC		Benzyl ITC		2-Phenylethyl ITC	
Fungus	SD_{50D}	LD_D	SD_{50D}	LD_D	SD_{50D}	LD_D	SD_{50D}	LD_D	SD_{50D}	LD_D
Gaemannomyces	3.7	20	5.3	30	4.4	30	1.7	5	3.6	10
Rhizoctonia	7.1	30	7.6	40	8.4	40	1.8	10	3.3	10
Fusarium	4.3	40	7.6	60	10.8	70	3.8	10	3.9	20
Bipolaris	8.3	40	11.6	60	19.4	80	7.2	30	6.6	30
Pythium	14.0	40	20.6	60	41.0	90	56.2	80	34.2	60
LSD $p = 0.05$	0.5		0.2		1.0		0.8		0.8	

From Sarwar *et al.*, 1998.

Indian mustard and canola crops coincident with the period of root decay around maturity, this apparent biofumigation did not result in any differences in the expression of the disease or in the yields of subsequent wheat crops (Kirkegaard *et al.*, 2000).

ITCs released from *Brassica* roots also affect a wide range of organisms other than fungi. For example, Rumberger and Marschner (2004) studied seasonal patterns of the release of 2-phenylethyl ITC from roots of four cultivars of canola and found significant effects on the bacterial community composition of the rhizosphere. Some of the changes in bacterial community composition in the rhizosphere were correlated with the ITC concentration, indicating that ITC can be a selective factor for microbes in the rhizosphere. Similarly, *in vitro* studies have demonstrated that 2-phenylethyl glucosinolate (precursor of ITC) from *Brassica* roots has a biocidal effect on the root-lesion nematode *Pratylenchus neglectus* (Potter *et al.*, 1998). Pot experiments showed a significant negative correlation between root concentrations of 2-phenylethyl glucosinolate above a threshold of 12 μmol g^{-1} root tissue and plant susceptibility to *P. neglectus*, suggesting that selection for cultivars with high levels of 2-phenylethyl glucosinolate could reduce the susceptibility of the plants during the growing season and increase the nematicidal impact of the degrading root tissues (Potter *et al.*, 1999).

However, field-based evidence for a biofumigation effect of brassicas on crop yields because of root-released allelochemicals remains scant. Thus while Smith *et al.* (2004) found many positive benefits for *Brassica*, and other crops, in the crop rotation with wheat, no significant additional effects related to biofumigation could be identified. Similarly while *Brassica* and legume crops reduced levels of crown rot (*Fusarium pseudograminearum*) in subsequent wheat crops by 3.4–41.3% and increased yields by 0.24–0.89 t ha^{-1}, there was no evidence that biofumigation associated with ITCs released by the brassicas reduced crown rot inoculum in the soil (Kirkegaard *et al.*, 2004). In this study, *Brassica* crops were generally more effective than chickpea in reducing crown rot infection in a susceptible durum wheat cultivar (by 6–18.4%) and increasing grain yield (by 0.27–0.58 t ha^{-1} for some *Brassica* cultivars). The reasons for the lower levels of crown rot infection following the *Brassica* crops include: (i) more rapid breakdown of cereal residues under brassicas thereby diminishing the quantity of host material available; (ii) promotion of changes in the soil microbial community conducive to inoculum decline (e.g. increased populations of antagonists such as *Trichoderma*); or (iii) higher soil N status after a legume crop inducing higher rates of infection. This study illustrates the multiple factors operating when crops are grown and the difficulties of specifying unequivocally the magnitude of particular rhizosphere effects.

So, while the direct ITC-related disease suppression related to biofumigation as originally proposed has been difficult to substantiate in dryland crops, probably due to the limited ITC release from intact root systems of senescing dryland canola (other studies have indicated that substantial tissue maceration is required to release the ITCs in any quantity), the low concentrations in the rhizosphere may still cause changes in the soil environment (Rumberger and Marschner, 2004) which significantly influence following cereal crops.

9.4 Phytoremediation

Phytoremediation is an emerging technology that uses plants for the remediation of contaminated land. Phytoremediation of some organic compounds is possible (see Salt *et al.*, 1998), but most attention has been focused on inorganic metals and metalloids where the use of phytostabilization to reduce the flow of contaminants in the environment and phytoextraction to remove the pollutant from the environment are used (McGrath and Zhao, 2003). Efficient phytoextraction is determined by two factors: the biomass produced and the element concentration in the biomass relative to that in the soil (the bioconcentration factor). Roots cannot be harvested easily, so it is the size of the shoot system and the ability of a plant to take up and transport metals to the shoots that determine the phytoextraction potential (McGrath *et al.*, 2002). Most plants have a bioconcentration factor for metals and metalloids of <1 so their use as phytoextractors is not feasible because the number of growing cycles that would be required to reduce the concentration of a metal in the topsoil by 50% would be >100. Assuming a shoot biomass production of 10 t ha^{-1} per crop, the bioconcentration factor would need to be 20 to achieve a halving of soil metals in less than 10 crops (McGrath and Zhao, 2003). Two strategies may be used to achieve phytoextraction: (i) the use of natural hyperaccumulators with good biomass production and a high bioconcentration factor; and (ii) the use of various chemicals that enhance uptake by species that are not normally accumulating species.

Metal hyperaccumulation is rare, with about 400 plant species so identified representing <0.2% of all flowering plants (Brooks, 1998). The mechanisms leading to metal hyperaccumulation are still emerging but there is evidence that Zn and Cd hyperaccumulation in *Thlaspi caerulescens* involves enhanced root uptake of the metals; for Zn this is possibly via transporters that are minimally down-regulated even when intercellular levels of Zn are high (Pence *et al.*, 2000; Zhao *et al.*, 2002). Enhanced root to shoot transport is another key component of hyperaccumulation and is achieved by vacuolar sequestration in root cells or by enhanced xylem loading (Lasat *et al.*, 1998). Following transport to the shoot, internal detoxification occurs by compartmentation and complexation in leaf vacuoles with enhanced tonoplast transport of metals into vacuoles appearing to play an important role.

Most metals have low solubility in soils, and although the addition of chelators such as ethylene-diaminetetraacetic acid (EDTA) can increase solubility and hence availability to plants, it is not generally practical to use such an approach because of the increased risk of leaching to groundwater (Wenzel *et al.*, 2003). To date there have been few attempts to manipulate the rhizosphere to increase metal bioavailability, but root exudates collected from hydroponically grown *T. caerulescens* did not increase the availability of soil metals

(Zhao *et al.*, 2001). This remains a topic to be researched. One aspect of root behaviour that might be exploited more, though, is the ability of hyperaccumulators such as *T. caerulescens* to proliferate roots in Zn-rich patches of soil (Whiting *et al.*, 2000). This response may partially explain the ability of this plant to acquire fairly large amounts of Zn even from soils with low Zn, and the trait may be useful even though the mechanism for sensing Zn in soil has yet to be established.

References

Allen, S.C., Jose, S., Nair, P.K.R., Brecke, B.J., Nkedi-Kizza, P. and Ramsey, C.L. (2004) Safety-net of tree roots: evidence from a pecan (*Carya illinoensis* K. Koch)-cotton (*Gossypium hirsutum* L.) alley cropping system in the southern United States. *Forest Ecology and Management* **192**, 395–407.

Angus, J.F., Gardner, P.A., Kirkegaard, J.A. and Desmarchelier, J.M. (1994) Biofumigation: isothiocyanates released from *Brassica* roots inhibit growth of the take-all fungus. *Plant and Soil* **162**, 107–112.

Angus, J.F., Gault, R.R., Peoples, M.B., Stapper, M. and van Herwaarden, A.F. (2001) Soil water extraction by dryland crops, annual pastures, and lucerne in south-eastern Australia. *Australian Journal of Agricultural Research* **52**, 183–192.

Asseng, S., Dunin, F.X., Fillery, I.R.P., Tennant, D. and Keating, B.A. (2001) Potential deep drainage under wheat crops in a Mediterranean climate. II. Management opportunities to control drainage. *Australian Journal of Agricultural Research* **52**, 57–66.

Bertin, C., Yang, X. and Weston, L.A. (2003) The role of root exudates and allelochemicals in the rhizosphere. *Plant and Soil* **256**, 67–83.

Birch, H.F. (1960) Nitrification in soils after different periods of dryness. *Plant and Soil* **12**, 81–96.

Bray, R.H. (1954) A nutrient mobility concept of soil-plant relationships. *Soil Science* **78**, 9–22.

Brooks, R.R. (1998) *Plants that Hyperaccumulate Heavy Metals*. CAB International, Wallingford, UK.

Callaway, R.M. and Aschehoug, E.T. (2000) Invasive plants versus their new and old neighbours: a mechanism for exotic invasion. *Science* **290**, 521–523.

Cannell, M.G.R., van Noordwijk, M. and Ong, C.K. (1996) The central agroforestry hypothesis: the tree must acquire resources that the crop would not otherwise acquire. *Agroforestry Systems* **34**, 27–31.

Casper, B.B., Schenk, H.J. and Jackson, R.B. (2003) Defining a plant's belowground zone of influence. *Ecology* **84**, 2313–2321.

Chan, M.K.Y. and Close, R.C. (1987) *Aphanomyces* root rot of peas 3. Control by the use of cruciferous amendments. *New Zealand Journal of Agricultural Research* **30**, 225–233.

Cheema, Z.A., Khaliq, A. and Saeed, S. (2004) Weed control in maize (*Zea mays* L.) through sorghum allelopathy. *Journal of Sustainable Agriculture* **23**, 73–86.

Chintu, R., Mafongoya, P.L., Chirwa, T.S., Mwale, M. and Matibini, J. (2004) Subsoil nitrogen dynamics as affected by planted coppicing tree legume fallows in eastern Zambia. *Experimental Agriculture* **40**, 327–340.

Cooper, P.J.M., Gregory, P.J., Keatinge, J.D.H. and Brown, S.C. (1987) Effects of fertilizer, variety and location on barley production under rainfed conditions in northern Syria. 2. Soil water dynamics and crop water use. *Field Crops Research* **16**, 67–84.

Cresswell, H.P. and Kirkegaard, J.A. (1995) Subsoil amelioration by plant roots – the process and the evidence. *Australian Journal of Soil Research* **33**, 221–239.

Cui, M. and Caldwell, M.M. (1997) A large ephemeral release of nitrogen upon wetting of dry soil and corresponding root responses in the field. *Plant and Soil* **191**, 291–299.

Czarnota, M.A., Paul, R.N., Dayan, F.E., Nimbal, C.I. and Weston, L.A. (2001) Mode of action, localization of production, chemical nature, and activity of sorgoleone: a potent PSII inhibitor in *Sorghum* spp. root exudates. *Weed Technology* **15**, 813–825.

Dimmock, J.P.R.E. and Gooding, M.J. (2002) The effects of fungicides on rate and duration of grain filling in winter wheat in relation to maintenance of flag leaf green area. *Journal of Agricultural Science* **138**, 1–16.

Eastham, J. and Gregory, P.J. (2000) The influence of crop management on the water balance of lupin and wheat crops on a layered soil in a Mediterranean climate. *Plant and Soil* **221**, 239–251.

Gregory, P.J. (1998) Alternative crops for duplex soils: growth and water use of some cereal, legume, and oilseed crops, and pastures. *Australian Journal of Agricultural Research* **49**, 2–32.

Gregory, P.J. and Eastham, J. (1996) Growth of shoots and roots, and interception of radiation by wheat and lupin crops on a shallow, duplex soil in response to time of sowing. *Australian Journal of Agricultural Research* **47**, 427–447.

Gregory, P.J. and Reddy, M.S. (1982) Root growth in an intercrop of pearl millet/groundnut. *Field Crops Research* **5**, 241–252.

Hoad, S.P., Russell, G., Lucas, M.E. and Bingham, I.J. (2001) The management of wheat, barley, and oat root systems. *Advances in Agronomy* **74**, 193–246.

Jackson, R.B. and Caldwell, M.M. (1993) Geostatistical patterns of soil heterogeneity around individual perennial plants. *Journal of Ecology* **81**, 683–692.

Jama, B., Buresh, R.J., Ndufa, J.K. and Shepherd, K.D. (1998) Vertical distribution of roots and soil nitrate: tree species and phosphorus effects. *Soil Science Society of America Journal* **62**, 280–286.

Katayama, K., Adu-Gyamfi, J., Devi, G., Rao, T.P. and Ito, O. (1996a) Root system development of component crops in intercropping. In: *Dynamics of Roots and Nitrogen in Cropping Systems of the Semi-Arid Tropics* (eds O. Ito, C. Johansen, J. Adu-Gyamfi, K. Katayama, J.V.D. Kumar Rao and T.J. Rego), pp. 199–209. Japan International Research Center for Agricultural Sciences, Tsukuba, Ibaraki, Japan.

Katayama, K., Adu-Gyamfi, J., Devi, G., Rao, T.P. and Ito, O. (1996b) Balance sheet of nitrogen from atmosphere, fertilizer, and soil in pigeonpea-based intercrops. In: *Dynamics of Roots and Nitrogen in Cropping Systems of the Semi-Arid Tropics* (eds O. Ito, C. Johansen, J. Adu-Gyamfi, K. Katayama, J.V.D. Kumar Rao and T.J. Rego), pp. 341–350. Japan International Research Center for Agricultural Sciences, Tsukuba, Ibaraki, Japan.

Kato-Noguchi, H. (2004) Allelopathic substance in rice root exudates: rediscovery of momilactone B as an allelochemical. *Journal of Plant Physiology* **161**, 271–276.

King, J., Gay, A., Sylvester-Bradley, R., Bingham, I., Foulkes, J., Gregory, P.J. and Robinson, D. (2003) Modelling cereal root systems for water and nitrogen capture: towards an economic optimum. *Annals of Botany* **91**, 383–390.

Kirkegaard, J.A., Gardner, P.A., Angus, J.F. and Koetz, E. (1994) Effect of *Brassica* break crops on the growth and yield of wheat. *Australian Journal of Agricultural Research* **45**, 529–545.

Kirkegaard, J.A., Sarwar, M., Wong, P.T.W., Mead, A., Howe, G. and Newell, M. (2000) Field studies on the biofumigation of take-all by *Brassica* break crops. *Australian Journal of Agricultural Research* **51**, 445–456.

Kirkegaard, J.A., Simpfendorfer, S., Holland, J., Bambach, R., Moore, K.J. and Rebetzke, G.J. (2004) Effect of previous crops on crown rot and yield of durum and bread wheat in northern NSW. *Australian Journal of Agricultural Research* **55**, 321–334.

Lasat, M.M., Baker, A.J.M. and Kochian, L.V. (1998) Altered Zn compartmentation in the root symplasm and stimulated Zn absorption into the leaf as mechanisms involved in Zn hyperaccumulation in *Thlaspi caerulescens*. *Plant Physiology* **118**, 875–883.

Lewis, J.A. and Papavizas, G.C. (1970) Evolution of volatile sulfur-containing compounds from decomposition of crucifers in soil. *Soil Biology and Biochemistry* **2**, 239–246.

Livesley, S.J., Gregory, P.J. and Buresh, R.J. (2000) Competition in tree row agroforestry systems. 1. Distribution and dynamics of fine root length and biomass. *Plant and Soil* **227**, 149–161.

Livesley, S.J., Gregory, P.J. and Buresh, R.J. (2002) Competition in tree row agroforestry systems. 2. Distribution, dynamics and uptake of soil inorganic nitrogen. *Plant and Soil* **247**, 177–187.

Livesley, S.J., Gregory, P.J. and Buresh, R.J. (2004) Competition in tree row agroforestry systems. 3. Soil water distribution and dynamics. *Plant and Soil* **264**, 129–139.

McCallum, M.H., Kirkegaard, J.A., Green, T.W., Cresswell, H.P., Davies, S.L., Angus, J.F. and Peoples, M.B. (2004) Improved subsoil macroporosity following perennial pastures. *Australian Journal of Experimental Agriculture* **44**, 299–307.

McGrath, S.P. and Zhao, F.J. (2003) Phytoextraction of metals and metalloids from contaminated soils. *Current Opinion in Biotechnology* **14**, 277–282.

McGrath, S.P., Zhao, F.J. and Lombi, E. (2002) Phytoremediation of metals, metalloids, and radionuclides. *Advances in Agronomy* **75**, 1–56.

Mordelet, P., Barot, S. and Abbadie, L. (1996) Root foraging strategies and soil patchiness in a humid savanna. *Plant and Soil* **182**, 171–176.

Muehlchen, A.M. (1990) Evaluation of crucifer green manures for controlling Aphanomyces root rot of peas. *Plant Disease* **74**, 651–654.

Natarajan, M. and Willey, R.W. (1980) Sorghum-pigeonpea intercropping and the effects of plant population density. II. Resource use. *Journal of Agricultural Science, Cambridge* **95**, 59–65.

Nimbal, C.I., Pedersen, J.F., Yerkes, L.A., Weston, L.A. and Weller, S.C. (1996) Phytotoxicity and distribution of sorgoleone in grain sorghum germplasm. *Journal of Agricultural and Food Chemistry* **44**, 1343–1347.

Odhiambo, H.O., Ong, C.K., Deans, J.D., Wilson, J., Khan, A.A.H. and Sprent, J.I. (2001) Roots, soil water and crop yield: tree crop interactions in a semi-arid agroforestry system in Kenya. *Plant and Soil* **235**, 221–233.

Olofsdotter, M. (2001) Rice – a step toward use of allelopathy. *Agronomy Journal* **93**, 3–8.

Olofsdotter, M., Navarez, D., Rebulanan, M. and Streibig, J.C. (1999) Weed-suppressing rice cultivars – does allelopathy play a role? *Weed Research* **39**, 441–454.

Ong, C.K., Wilson, J., Deans, J.D., Mulayta, J., Raussen, T. and Wajja-Musukwe, N. (2002) Tree-crop interactions: manipulation of water use and root function. *Agricultural Water Management* **53**, 171–186.

Pence, N.S., Larsen, P.B., Ebbs, S.D., Letham, D.L.D., Lasat, M.M., Garvin, D.F., Eide, D. and Kochian, L.V. (2000) The molecular physiology of heavy metal transport in the Zn/Cd hyperaccumulator *Thlaspi caerulescens*. *Proceedings of the National Academy of Science* **97**, 4956–4960.

Potter, M.J., Davies, K.A. and Rathgen, A.J. (1998) Suppressive impact of glucosinolates in *Brassica* vegetative tissues on root lesion nematodes (*Pratylenchus neglectus*). *Journal of Chemical Ecology* **24**, 67–80.

Potter, M.J., Vanstone, V.A., Davies, K.A., Kirkegaard, J.A. and Rathgen, A.J. (1999) Reduced susceptibility of *Brassica napus* to *Pratylenchus neglectus* in plants with elevated root levels of 2-phenylethyl glucosinolate. *Journal of Nematology* **31**, 291–298.

Rasse, D.P. and Smucker, A.J.M. (1998) Root recolonization of previous root channels in corn and alfalfa rotations. *Plant and Soil* **204**, 203–212.

Reddy, M.S. and Willey, R.W. (1981) Growth and resource use studies in an intercrop of pearl millet/groundnut. *Field Crops Research.* **4**, 13–24.

Ridenour, W.M. and Callaway, R.M. (2001) The relative importance of allelopathy in interference: the effects of an invasive weed on a native bunchgrass. *Oecologia* **126**, 444–450.

Rietveld, W.J. (1983) Allelopathic effects of juglone on germination and growth of several herbaceous and woody species. *Journal of Chemical Ecology* **9**, 295–308.

Rimando, A.M. and Duke, S.O. (2003) Studies on rice allelochemicals. In: *Rice: Origin, History, Technology and Production* (ed. C.W. Smith), pp. 221–244. John Wiley & Sons, New York.

Robinson, D. (1996) Variation, co-ordination and compensation in root systems in relation to soil variability. *Plant and Soil* **187**, 57–66.

Robinson, D. (2001) Root proliferation, nitrate inflow and their carbon costs during nitrogen capture by competing plants in patchy soil. *Plant and Soil* **232**, 41–50.

Roth, C.M., Shroyer, J.P. and Paulsen, G.M. (2000) Allelopathy of sorghum on wheat under several tillage systems. *Agronomy Journal* **92**, 855–860.

Rowe, E.C., Hairiah, K., Giller, K.E., van Noordwijk, M. and Cadisch, G. (1999) Testing the safety-net role of hedgerow tree roots by ^{15}N placement at different soil depths. *Agroforestry Systems* **43**, 81–93.

Rowe, E.C., van Noordwijk, M., Suprayogo, D., Hairiah, K., Giller, K.E. and Cadisch, G. (2001) Root distributions partially explain ^{15}N uptake patterns in *Gliricidia* and *Peltophorum* hedgerow intercropping systems. *Plant and Soil* **235**, 167–179.

Rumberger, A. and Marschner, P. (2004) 2-Phenylethylisothiocyanate concentration and bacterial community composition in the rhizosphere of field-grown canola. *Functional Plant Biology* **31**, 623–631.

Salt, D.E., Smith, R.D. and Raskin, I. (1998) Phytoremediation. *Annual Review of Plant Physiology and Plant Molecular Biology* **49**, 643–668

Sarwar, M., Kirkegaard, J.A., Wong, P.T.W. and Desmarchelier, J.M. (1998) Biofumigation potential of brassicas. III. In vitro toxicity of isothiocyanates to soil-borne fungal pathogens. *Plant and Soil* **201**, 103–112.

Schroth, G. (1995) Tree root characteristics as criteria for species selection and systems design in agroforestry. *Agroforestry Systems* **30**, 125–143.

Schroth, G. (1999) A review of below-ground interactions in agroforestry, focussing on mechansims and management options. *Agroforestry Systems* **43**, 5–34.

Shepherd, G., Buresh, R.J. and Gregory, P.J. (2000) Land use affects the distribution of soil inorganic nitrogen in smallholder production systems in Kenya. *Biology and Fertility of Soils* **31**, 348–355.

Smith, B.J., Kirkegaard, J.A. and Howe, G.N. (2004) Impacts of *Brassica* break-crops on soil biology and yield of following wheat crops. *Australian Journal of Agricultural Research* **55**, 1–11.

Smith, D.M., Jackson, N.A., Roberts, J.M. and Ong, C.K. (1999) Root distributions in a *Grevillea robusta*-maize agroforestry system in semi-arid Kenya. *Plant and Soil* **211**, 191–205.

Sylvester-Bradley, R., Scott, R.K., Stokes, D.T. and Clare, R.W. (1997) The significance of crop canopies for N nutrition. *Aspects of Applied Biology* **50**, 103–116.

Tennant, D., Scholz, G., Dixon, J. and Purdie, B. (1992) Physical and chemical characteristics of duplex soils and their distribution in the south-west of Western Australia. *Australian Journal of Experimental Agriculture* **32**, 827–843.

Turner, N.C. (2004) Agronomic options for improving rainfall-use efficiency of crops in dryland farming systems. *Journal of Experimental Botany* **55**, 2413–2425.

van Noordwijk, M. (1983) Functional interpretation of root densities in the field for nutrient and water uptake. In: *Root Ecology and its Application*, pp. 207–226. International Society for Root Research, International Symposium, Gumpenstein, 1982. Bundesanstalt Gumpenstein, Irdning, Austria.

van Noordwijk, M. and Cadisch, G. (2002) Access and excess problems in plant nutrition. *Plant and Soil* **247**, 25–40.

Van Noordwijk, M. and Lusiana, B. (1999) WaNuLCAS, a model of water, nutrient and light capture in agroforestry systems. *Agroforestry Systems* **43**, 217–242.

van Noordwijk, M., Lawson, G., Soumaré, A., Groot, J.J.R. and Hairiah, K. (1996) Root distribution of trees and crops: competition and/or complementarity. In: *Tree-Crop Interactions: A Physiological Approach* (eds C.K. Ong and P. Huxley), pp. 319–364. CAB International, Wallingford, UK.

Ward, P.R., Dunin, F.X. and Micin, S.F. (2002) Water use and root growth by annual and perennial pastures and subsequent crops in a phase rotation. *Agricultural Water Management* **53**, 83–97.

Ward, P.R., Fillery, I.R.P., Maharaj, E.A. and Dunin, F.X. (2003) Water budgets and nutrients in a native *Banksia* woodland and an adjacent *Medicago sativa* pasture. *Plant and Soil* **257**, 305–319.

Wenzel, W.W., Unterbrunner, R., Sommer, P. and Sacco, P. (2003) Chelate-assisted phytoextraction using canola (*Brassica napus* L.) in outdoors pot and lysimeter experiments. *Plant and Soil* **249**, 83–96.

Weston, L.A. and Duke, S.O. (2003) Weed and crop allelopathy. *Critical Reviews in Plant Sciences* **22**, 367–389.

Whiting, S.N., Leake, J.R. and McGrath, S.P. (2000) Positive responses to Zn and Cd by roots of the Zn and Cd hyperaccumulator *Thlaspi caerulescens*. *New Phytologist* **145**, 199–210.

Willey, R.W. (1996) Intercropping in cropping systems: major issues and research needs. In: *Dynamics of Roots and Nitrogen in Cropping Systems of the Semi-Arid Tropics* (eds O. Ito, C. Johansen, J. Adu-Gyamfi, K. Katayama, J.V.D. Kumar Rao and T.J. Rego), pp. 93–102. Japan International Research Center for Agricultural Sciences, Tsukuba, Ibaraki, Japan.

Zhao, F.J.,, Hamon, R.E. and McLaughlin, M.J. (2001) Root exudates of the hyperaccumulator *Thlaspi caerulescens* do not enhance metal mobilization. *New Phytologist* **151**, 613–620.

Zhao, F.J., Hamon, R.E., Lombi, E. and McLaughlin, M.J. (2002) Characteristics of cadmium uptake in two contrasting ecotypes of the hyperaccumulator *Thlaspi caerulescens*. *Journal of Experimental Botany* **53**, 535–543.

Index

Note: page numbers in *italics* refer to figures, those in **bold** refer to tables.